and Molecular Biology

A Comprehensive Review

Second Edition

Essential Biochemistry and Molecular Biology

A Comprehensive Review

Rudolf Werner, PhD
Professor
Department of Biochemistry and Molecular Biology
University of Miami School of Medicine
Miami, Florida

Prentice-Hall International Inc.

Notice: The author(s) and the publisher of this volume have taken care that the information and recommendations contained herein are accurate and compatible with the standards generally accepted at the time of publication. Nevertheless, it is difficult to ensure that all the information given is entirely accurate for all circumstances. The publisher disclaims any liability, loss, or damage incurred as a consequence, directly or indirectly, of the use and application of any of the contents of this volume.

93 94 95 96 97 / 10 9 8 7 6 5 4 3 2 1

Prentice Hall International (UK) Limited, *London*
Prentice Hall of Australia Pty. Limited, *Sydney*
Prentice Hall Canada, Inc., *Toronto*
Prentice Hall Hispanoamericana, S.A., *Mexico*
Prentice Hall of India Private Limited, *New Delhi*
Prentice Hall of Japan, Inc., *Tokyo*
Simon & Schuster Asia Pte. Ltd., *Singapore*
Editora Prentice Hall do Brasil Ltda., *Rio de Janeiro*
Prentice Hall, *Englewood Cliffs, New Jersey*

ISBN 0-8385-2246-7

9 780838 522462 90000

PRINTED IN THE UNITED STATES OF AMERICA

CONTENTS

PREFACE

Because of the continuing explosion of scientific knowledge, many of today's textbooks are becoming too voluminous and thus tend to lose their usefulness as study guides. Students, therefore, look for smaller review books that cover all course material in a concise form. This allows the student to review the material covered during a lecture in a more efficient way. Full-size textbooks are still required as information sources and for in-depth study of certain topics.

This book provides the student with a comprehensive yet concise description of all the biochemistry and molecular biology that is medically relevant. It allows the student to find important facts without having to read through more than 1000 pages of text. In order to achieve this comprehensive coverage, I have sought to condense the entire field of medical biochemistry into essentials and thereby offer students a comprehensive coverage of all topics of medical biochemistry. Despite its limited format, the book contains all the information a medical student must have to pass the National Board Examination Part I, including topics often not found in a single textbook.

Despite its emphasis on medical biochemistry, the book should appeal to all undergraduate students in the life sciences. Since the first edition of this book, there have been numerous scientific advances in the area of molecular biology. In this second edition, I have tried to include molecular biological approaches in most chapters, and have added an additional chapter (Chapter 7) to describe the latest developments in molecular biology as they apply to modern medicine.

The first seven chapters deal with the basic biochemical and molecular biological mechanisms; the last six chapters cover more medically oriented topics. Medical correlations are mentioned whenever appropriate. (Undergraduate students, especially those who are not premedical or predental students, may want to skip over these sections.) Each chapter is preceded by a set of objectives that help students to recognize the relevant points and, at the same time, to test their understanding of the subject. References to specific chapters of the major biochemistry textbooks as well as to review aticles and original research papers enable the student to consult the more extensive coverage when desired and thus deepen his or her understanding of a particular topic.

While writing this book I received valuable advice and criticism from my colleagues at the Department of Biochemistry and Molecular Biology at the University of Miami and from my students. I am especially indebted to David Anderson, Bonnie Blomberg, Keith Brew, Gerhard Dahl, Ian Dickerson, Frans Huijing, David Puett, and Fred Woessner, who critically read sections of the manuscript.

Rudolf Werner, PhD

CHAPTER

1 Amino Acids, Proteins, and Enzymes

OBJECTIVES

After studying this chapter, the student should be able to

1. Draw the structures of the 20 amino acids found in natural proteins and understand their properties.
2. Remember the approximate pK values of conjugate acid-base pairs found in amino acids.
3. Be able to calculate the pI of an amino acid from the pKs of its ionizable groups.
4. Describe the nature of the peptide bond.
5. Explain the types of intramolecular forces that determine protein conformation.
6. Define the terms "primary," "secondary," "tertiary," and "quaternary" structures of proteins and name examples for each.
7. Remember the general rules that govern the structure of globular proteins.
8. Describe some of the techniques used for the purification and characterization of proteins and understand their rationale.
9. Understand how a mixture of amino acids can be analyzed.
10. Explain some of the techniques used for the determination of amino acid sequences in proteins.
11. Name the site of cleavage in a polypeptide chain by cyanogen bromide and by trypsin.
12. Understand the basis of polyacrylamide gel electrophoresis in the presence of SDS and why it can be used to determine molecular weights of protein subunits.
13. Explain the terms "activation energy" and "transition state intermediate" and relate these to enzyme action.
14. Understand the basis of Michaelis-Menten enzyme kinetics and distinguish the different types of enzyme inhibition from Lineweaver-Burk plots.
15. State the evidence for the existence of enzyme-substrate complexes.
16. Understand the mechanism of allosteric regulation of enzyme activity.
17. Name examples of the labeling and identification of active sites in enzymes.
18. Give an example of an isoenzyme.
19. Define the terms "prochiral" and "chiral centers" and explain their significance in enzyme-substrate interactions.
20. Give examples of a multi-subunit enzyme and a multienzyme complex.

Proteins are polymers of amino acids. They constitute by far the largest class of polymers found in living cells and play a central role in the cell's structure and function. In this chapter we will discuss the structures and properties of proteins. Their biosynthesis will be covered extensively in the chapter on biochemical genetics (Chapter 6).

Examples of protein function are

1. Catalysis of chemical reactions by enzymes
2. Provide structure (hair, bone, connective tissue)
3. Transport ions and metabolites (permeases), gases (hemoglobin), electrons (cytochromes), and lipids (lipoproteins) across membranes and between organs and tissues
4. Regulate metabolism (cell membrane receptors, hormones)
5. Provide contractibility (actin and myosin in muscle)
6. Transmit nerve impulses (neurotransmitters, excitable assemblies)
7. Body defense mechanisms (antibodies)
8. Maintain functions (blood clotting factors)
9. Perception (visual pigments, chemoreceptors)
10. Storage (ferritin for iron storage and ovalbumin in egg white and casein in milk for amino acid storage)

A protein or functional entity consists of one or more polypeptide chains folded into a specific three-dimensional structure (conformation) that is determined by the sequence of the amino acids in the polypeptide chain(s). Before we can discuss protein structure, we must acquaint ourselves with the structures of the amino acids found in proteins.

1.1 STRUCTURES AND PROPERTIES OF AMINO ACIDS

There are 20 amino acids that can be polymerized into linear polypeptide chains, which make up native proteins. Nineteen of these amino acids are α amino acids of the general structure shown in Figure 1.1. (The letter "α" indicates that the amino group is attached to the first carbon atom after the carboxyl group.) The twentieth amino acid, proline, is an α imino acid.

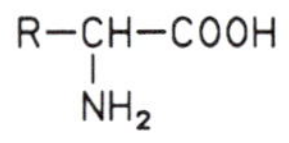

Figure 1.1.
General structure of an α amino acid.

Table 1.1 shows the structures of the natural amino acids. They are classified according to the nature of their side chains, which in Figure

Table 1.1 Amino Acids Used in Protein Synthesis

Amino Acid	Symbols	Structural Formula	pK$_1$	pK$_2$	pK$_3$	pI
Aliphatic side chains						
Glycine	Gly (G)	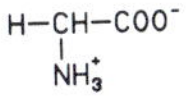	2.34	9.60		5.97
Alanine	Ala (A)	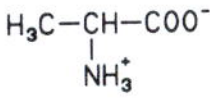	2.34	9.69		6.00
Valine	Val (V)	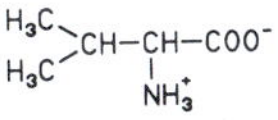	2.32	9.62		5.96
Leucine	Leu (L)	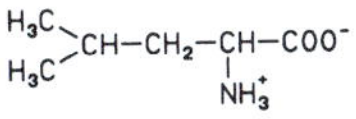	2.36	9.60		5.98
Isoleucine	Ile (I)	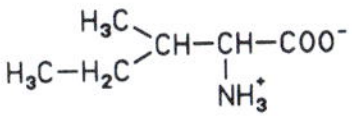	2.36	9.60		6.02
Proline	Pro (P)	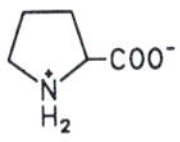	1.99	10.60		6.30
Aromatic side chains						
Phenylalanine	Phe (F)	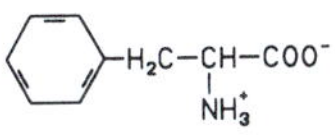	1.83	9.13		5.48
Tyrosine	Tyr (Y)	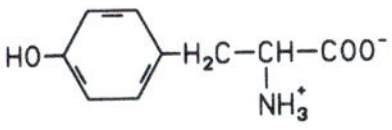	2.20	9.11		5.66
Tryptophan	Trp (W)	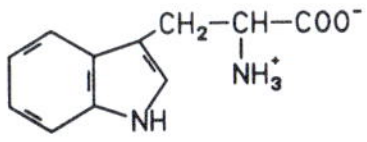	2.83	9.39		5.89
Side chains with other groups						
Serine	Ser (S)	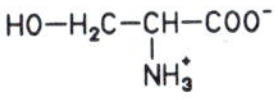	2.21	9.15		5.68
Threonine	Thr (T)	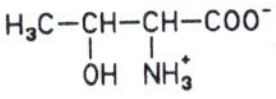	2.09	9.10		5.60
Side chains containing sulfur						
Cysteine	Cys (C)	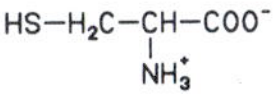	1.96	10.28	8.33 ↑ (SH)	5.07
Methionine	Met (M)	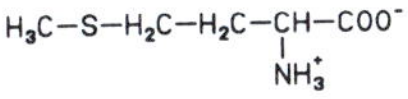	2.28	9.21		5.74
Acidic amino acids and their amides						
Aspartate	Asp (D)	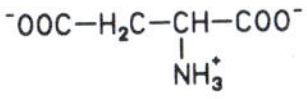	1.88	3.65 ↑ (β-*COOH*)	9.60 ↑ (α-*NH*$_2$)	2.77
Asparagine	Asn (N)	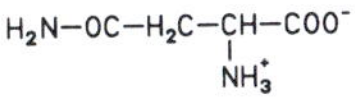	2.02	8.80		5.41
Glutamate	Glu (E)	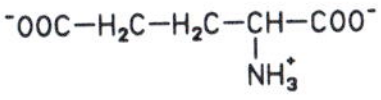	2.19	4.25 ↑ (γ-*COOH*)	9.67 ↑ (α-*NH*$_2$)	3.22
Glutamine	Gln (Q)	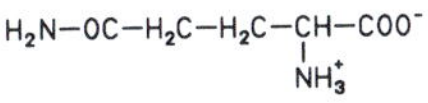	2.17	9.13		5.65
Basic amino acids						
Arginine	Arg (R)	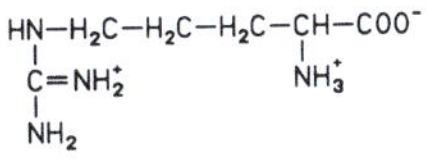	2.17	9.04 ↑ (α-*NH*$_2$)	12.48 ↑ (*guanidinium*)	10.76
Lysine	Lys (K)	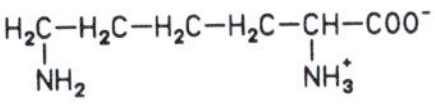	2.18	8.95 ↑ (α-*NH*$_2$)	10.53 ↑ (ε-*NH*$_2$)	9.74
Histidine	His (H)	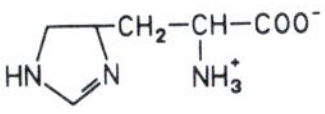	1.82	6.00 ↑ (*imidazole*)	9.17 ↑ (α-*NH*$_2$)	7.59

1.1. are denoted R. For reasons of brevity, their commonly used three-letter abbreviations will be used throughout this book. Amino acids with hydrophobic or nonpolar side chains are Phe, Trp, Pro, Met, Val, Leu, and Ile. All other amino acids have hydrophilic or polar side chains, with the exception of Gly and Ala, which have intermediate character.

1.1.1 Stereochemistry of Amino Acids

The α **carbon atom** of all amino acids with the exception of glycine is **asymmetric** because it is linked to four different groups. Therefore, two possible stereoisomers exist for each amino acid (except glycine) that are mirror images of each other (Figure 1.2). The two stereoisomers have very similar physicochemical properties but rotate the axis of polarized light in opposite directions. The two configurations are designated L and D. These letters refer to the absolute configuration of the isomers, not the direction of optical rotation.

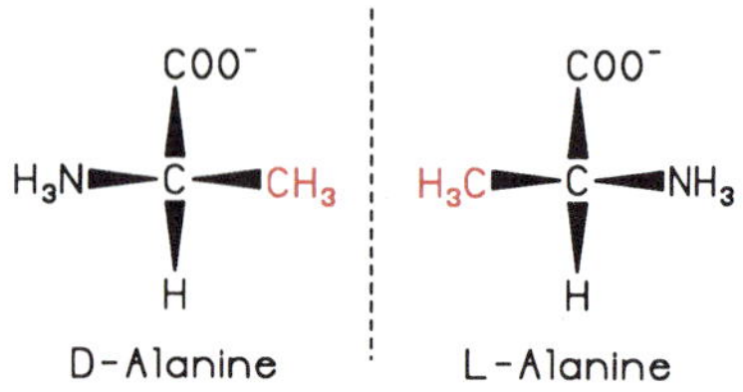

Figure 1.2.
Stereoisomers of an α amino acid.

All amino acids found in natural proteins have the L configuration. Two amino acids, Thr and Ile, possess two asymmetric centers, thus giving rise to four possible **diastereoisomers.** However, in each case only one form, the L form, is found in proteins.

Some D amino acids are found in substances other than proteins. Bacterial cell walls, for example, contain a network of polysaccharide chains that are cross-linked by short oligopeptides containing D-Ala and D-Glu. The antibiotic **penicillin** inhibits the biosynthesis of these cross-links, thus causing the formation of a defective cell wall in growing bacteria, which results in cell lysis.

1.1.2 Acid-Base Properties

Amino acids are **amphoteric** in nature, containing both acidic and basic groups. Under physiologic conditions (pH 7.4), amino acids exist as dipolar ions, called **zwitterions** (Figure 1.3).

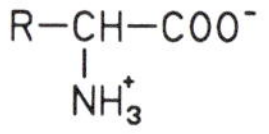

Figure 1.3.
Zwitterion form of an α amino acid.

Depending on their state of ionization, carboxyl and amino groups may be regarded as weak bases or weak acids, and their dissociation can be described by the single equation

$$HA \rightleftharpoons H^+ + A^-$$

HA denotes the proton donor or **conjugate acid** (—COOH or —NH_3^+) and A^- the proton acceptor or **conjugate base** (—COO^- or —NH_2).

The extent of this ionization reaction depends on its equilibrium constant or dissociation constant

$$K = \frac{[H^+][A^-]}{[HA]}$$

in which [ ] denotes the molar concentration.

By rearrangement, we obtain

$$[H^+] = \frac{K[HA]}{[A^-]}$$

Table 1.2. pK Values of Some Common Acids

Acid	pK
Phosphoric acid (pK_1)	2.0
Acetic acid	4.7
Carbonic acid (pK_1)	6.1
Phosphoric acid (pK_2)	6.8
Ammonium ion	9.3
Phosphoric acid (pK_3)	12.3

or

$$\log[H^+] = \log K + \log[HA] - \log[A^-]$$

Since pH is defined as $-\log[H^+]$ and pK as $-\log K$, we obtain the **Henderson-Hasselbalch equation**

$$pH = pK + \log \frac{[A^-]}{[HA]}$$

Remembering that $\log 1 = 0$, we can see that the pK value of an acid is the pH value at which it is 50% dissociated (i.e., $[A^-] = [HA]$; Table 1.2).

When a weak acid, HA, is titrated with a strong base, increasing amounts of HA will dissociate. Titration curves, which may be obtained experimentally or calculated from the Henderson-Hasselbalch equation, show a zone where the pH of the system resists change when small amounts of H^+ or OH^- ions are added. In other words, the conjugate acid-base pair acts as a **buffer** in this pH region. Buffering capacity is maximal (i.e., the system shows the smallest change in pH) when the concentration of the proton donor equals that of the proton acceptor (i.e., when pH = pK; Figure 1.4).

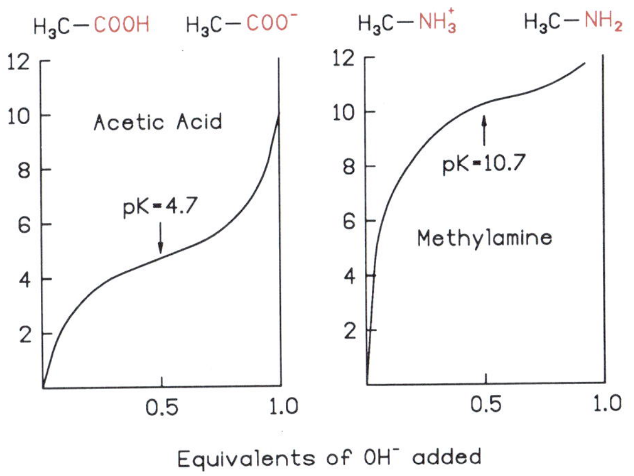

Figure 1.4. Titration of acetic acid and methylamine with NaOH.

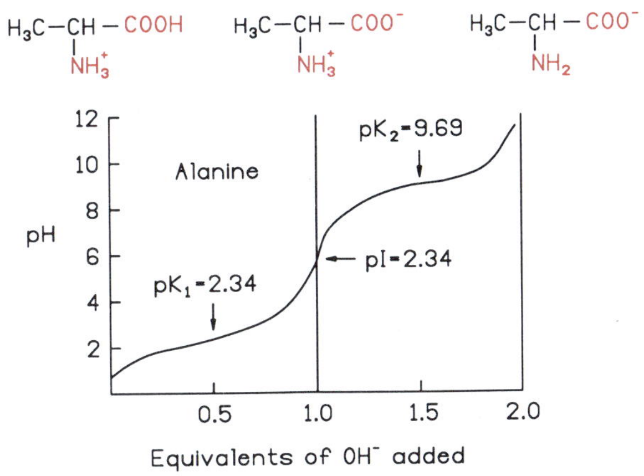

Figure 1.5. Titration curve of alanine.

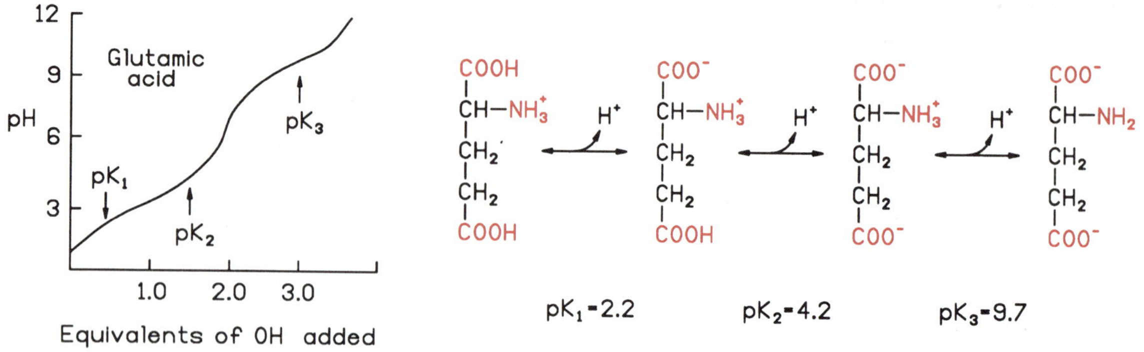

Figure 1.6.
Titration curve of glutamate.

Titration of the amino acid **alanine** (Figure 1.5) shows two regions of relatively slow pH change. They arise from the dissociation of the carboxyl group ($pK_1 = 2.34$) and that of the ammonium group ($pK_2 = 9.69$).

The **isoelectric point** of a molecule, **pI**, is the pH at which the molecule carries no net charge. For a simple monoamino monocarboxylic acid (such as alanine), the pI is simply the arithmetic mean of pK_1 and pK_2, i.e., $pI = \frac{1}{2}(pK_1 + pK_2)$.

For **glutamic acid,** a monoamino dicarboxylic acid, the isoelectric point is the arithmetic mean between the pK values of the two carboxyl groups. At this point, the two carboxyl groups are half dissociated and together carry a net charge of -1, which is just enough to neutralize the single positive charge of the ammonium group (Figure 1.6). The isoelectric point for **lysine** can be calculated in a similar fashion. It lies halfway between the pK values for the two amino groups (Figure 1.7). The calculation of the pI for **histidine** is analogous to that for lysine (Figure 1.8). Some pK values of ionizable groups found in amino acids are worth remembering (see Table 1.3).

1.1.3 Methods for Separating Amino Acids

The amino acid composition of a protein can be analyzed quantitatively, after complete hydrolysis of the protein in 6 *M* HCl at 110°C (see

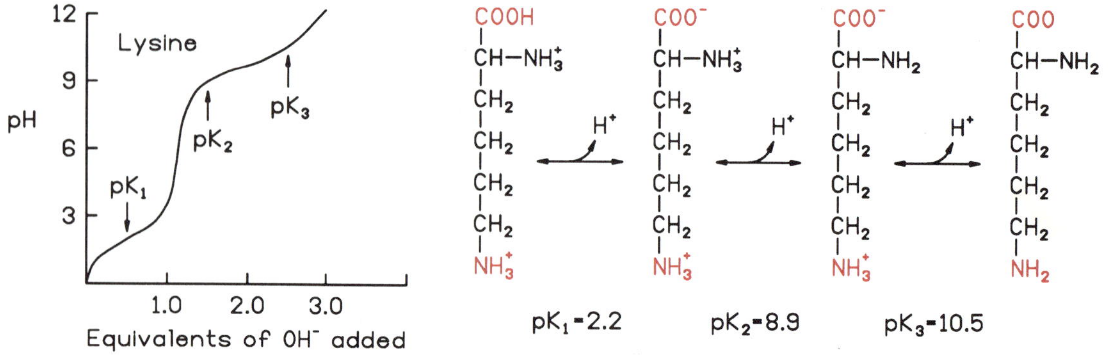

Figure 1.7.
Titration curve of lysine.

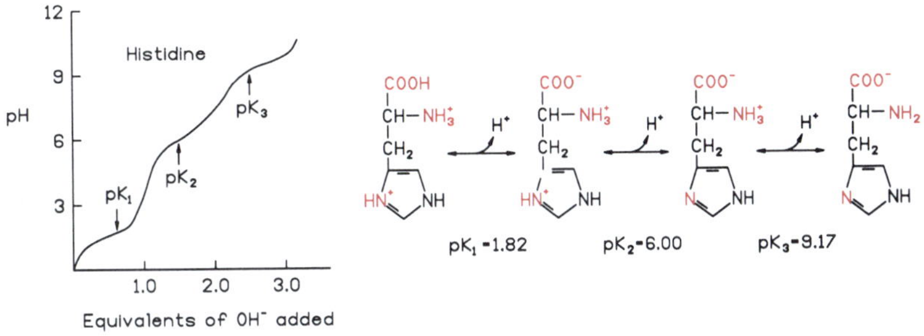

Figure 1.8.
Titration curve of histidine.

Section 1.4), by electrophoretic or chromatographic separation of the released amino acids.

Electrophoresis separates amino acids according to their electric charge. It is often used in conjunction with chromatography for two-dimensional separations.

In **partition chromatography,** amino acids are separated according to their different partition coefficients between a stationary aqueous phase and a mobile organic phase. The amino acid mixture is usually spotted on filter paper or a similar support and then "developed" by letting a solvent mixture pass through it. Two-dimensional chromatography yields the best separation (Figure 1.9). Amino acid spots on the developed chromatogram can be visualized and quantitated with the **ninhydrin reagent.**

Ion-exchange chromatography, like electrophoresis, is also based on charge differences between amino acids. This method was used in automatic amino acid analyzers. A mixture of amino acids in a buffer of pH 2.0 is applied to a column of sulfonated polystyrene resin (containing —SO_3^- residues). At this pH, all amino acids are positively charged and, therefore, are retained on the column. They are subsequently eluted with buffers of increasing pH and ionic strength starting

Table 1.3. pK Values of Ionizable Groups Found in Amino Acids

Group	Conjugate Acid	Conjugate Base	pK
α Carboxyl	R—CH(NH_2)—COOH	R—CH(NH_2)—COO^-	2.1 ± 0.5
Non-α carboxyl	R—COOH	R—COO^-	4.0 ± 0.3
Imidazolium (His)	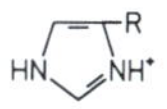	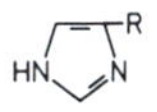	6.0
α Amino	R—CH(NH_3^+)—COO^-	R—CH(NH_2)—COO^-	9.8 ± 1.0
ϵ Amino	R—NH_3^+	R—NH_2	10.5
Guanidinium (Arg)	R—HN—C(=NH_2^+)—NH_2	R—HN—C(=NH)—NH_2	12.5
Sulfhydryl (Cys)	R—SH	R—S^-	8.3

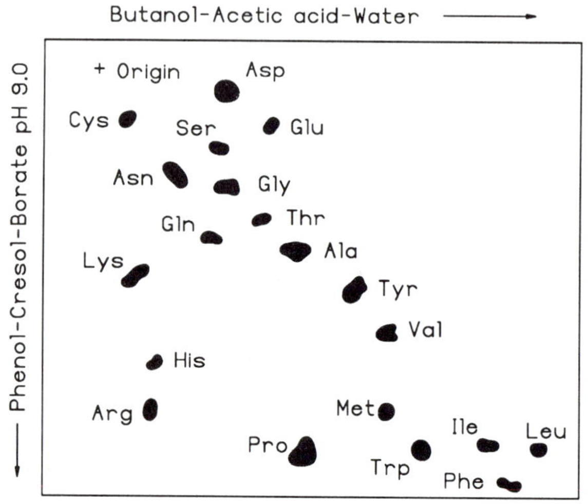

Figure 1.9.
Two-dimensional chromatogram of an amino acid mixture. Development in the first dimension occurs from left to right.

at pH 3.25. Acidic and neutral, aromatic, and finally basic amino acids emerge from the column in that order. (The separation of neutral amino acids is due to their different nonpolar affinities for the resin.)

The favored method for analyzing mixtures of amino acids in automatic amino acid analyzers is to derivatize amino acids with phenylisothiocyanate and separate the resulting phenylthiocarbamate derivatives by high-performance liquid chromatography (HPLC); see the Edman reaction discussed in Section 1.4.

1.2 STRUCTURE OF PROTEINS

1.2.1 The Peptide Bond

Polypeptides consist of linear sequences of amino acids that are joined together by peptide bonds. The peptide bond is formed between the carboxyl group of one amino acid and the α amino group of a second amino acid (Figure 1.10). The equilibrium of this reaction lies almost completely on the left side. Thus, synthesis of polypeptides requires an input of free energy (see Chapter 2).

$$H_3\overset{+}{N}-\underset{R_1}{CH}-COO^- + H_3\overset{+}{N}-\underset{R_2}{CH}-COO^- \rightleftharpoons H_3\overset{+}{N}-\underset{R_1}{CH}-CO-NH-\underset{R_2}{CH}-COO^- + H_2O$$

Figure 1.10.
Peptide bond formation.

The CO—NH linkage of the peptide bond has partial (40%) double-bond character. As a consequence, there is essentially no freedom of rotation around the bond itself. This makes the peptide group a planar group. Free rotation can occur only at the bonds of the C_α atom connecting two peptide groups. Of the possible configurations, *cis* and *trans*, only the *trans* peptides are found in proteins (Figure 1.11). An exception are the C_α atoms flanking a peptide bond that includes the imino group of proline, which can assume both the *trans* and *cis* configuration. However, the *trans* configuration still is the predominant form. These are specific peptidyl **prolyl *cis-trans* isomerases** that catalyze the interconversion of the two forms of the prolyl-peptide bond. Proline *cis-trans* isomerization is a recognized slow step in protein folding.

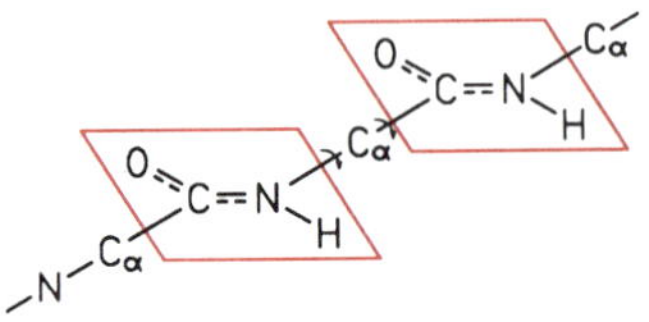

Figure 1.11.
Configuration of the peptide bond.

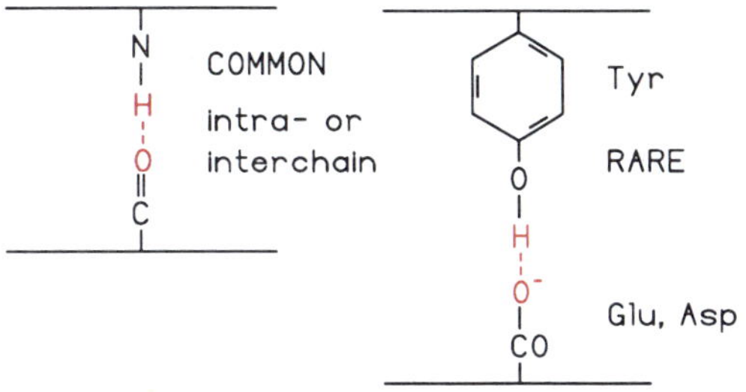

Figure 1.12.
Examples of hydrogen bond formation.

1.2.2 Protein Conformation

The three-dimensional folding pattern (conformation) of a protein is completely determined by the amino acid sequence of the polypeptide(s) and represents the most stable conformation (i.e., the lowest state of energy). This is demonstrated by the fact that many proteins that have been denatured (i.e., have lost their three-dimensional structure) renature and regain activity upon removal of the denaturing agent (see below).

1.2.3 Intermolecular Forces

The forces that contribute to the maintenance of the three-dimensional structure of a protein include van der Waals forces, hydrogen bonds, ionic bonds, hydrophobic interactions, and disulfide bonds.

Van der Waals interactions are nonspecific attractive forces that occur between atoms of molecules when brought into close contact (3–4 Å). Although rather weak in nature, they can amount to a significant force since there are so many of these interactions in a given protein.

Hydrogen bonds are somewhat stronger. They arise from the tendency of a bound hydrogen atom to share the electrons of a nearby oxygen atom (Figure 1.12).

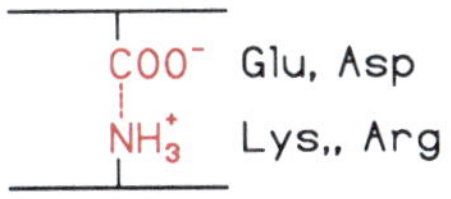

Figure 1.13.
Ionic interactions.

Ionic interactions occur between charged side chains of amino acids located at different positions in the polypeptide chains (Figure 1.13). The interactions probably play only a minor part in the maintenance of protein structure because most charged groups are at the surface of proteins where they interact with water.

Hydrophobic interactions are due to the tendency of nonpolar molecules to cluster in an aqueous environment (Figure 1.14). Segregation of lipophilic groups away from solvent water causes an increased disorder of water structure (increase in entropy). Globular proteins might thus be thought of as grease balls with lipophilic groups inside and hydrophilic residues outside interacting with water. Thermodynamically, the formation of hydrophobic bonds is the main driving force in protein folding. Ionic and hydrogen bond-forming groups can interact with water molecules just as well as with other groups on the protein.

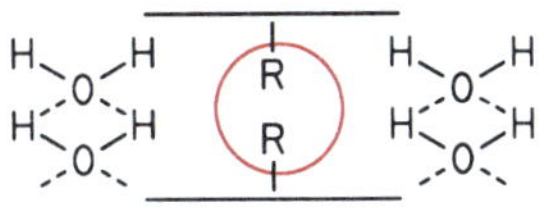

Figure 1.14.
Hydrophobic interactions. R = nonpolar aminoacyl side chain.

Disulfide bonds are covalent bonds occurring between two cysteinyl residues (Figure 1.15). They form spontaneously in the presence of air and can be cleaved by reducing reagents such as β-mercaptoethanol or dithiothreitol.

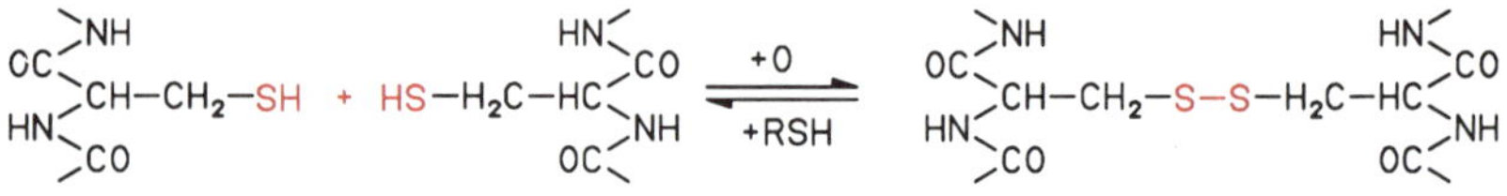

Figure 1.15.
Formation of disulfide bonds.

1.2.4 Primary and Secondary Structures

The conformation of a protein is described by four substructures of increasing complexity: the primary, secondary, tertiary, and quaternary structures.

The **primary structure** of a protein describes the sequence of amino acid residues in the polypeptide chain(s). Its determination will be discussed in Section 1.4.

The **secondary structure** arises from the folding of the polypeptide chain into three-dimensionally ordered structures, which are maintained by hydrogen bonds between the carbonyl groups (—CO—) of one peptide bond and the imido groups (—NH—) of another. There are four principal types of secondary structures: the helix structure, the β-pleated sheet, the β turn, and the triple helix of collagen.

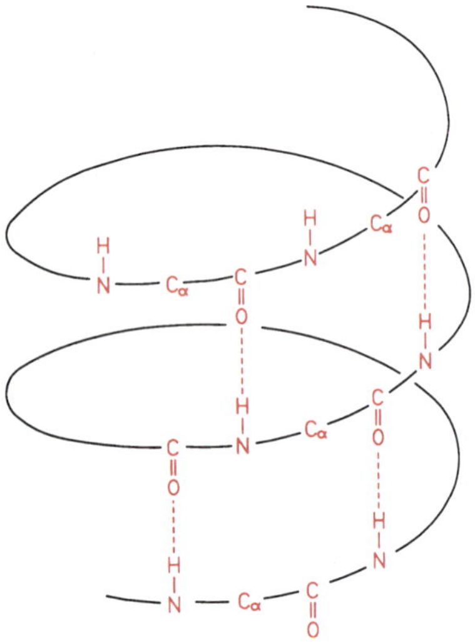

Figure 1.16.
α Helix structure.

The formation of hydrogen bonds within the same polypeptide chain (intrachain) can produce a helical structure. In the most common helix, the α **helix,** the carbonyl group of amino acid n forms a hydrogen bond with the imido group of amino acid $n + 4$, and amino acid $(n + 1)$ interacts with amino acid $(n + 1) + 4$, and so on (Figure 1.16). The resulting right-handed helix contains 3.6 amino acid residues per turn. The α helix is also called the 3.6_{13} helix, the principal number indicating the number of residues per turn, the subscript indicating the number of atoms in the ring that forms when the hydrogen bond is formed. Many globular and fibrous proteins contain α-helical segments in their polypeptide chains.

A few fibrous proteins consist almost exclusively of α helices. For example, **α-keratin,** the main component of hair, consists of right-handed α helices that are wound around each other to form multistrand ropes. Interstrand disulfide bonds further stabilize such structures.

Although the α helix is the most stable helical polypeptide conformation, other helical structures with different numbers of residues per turn exist [e.g., the π helix (4.4_{16}), the 3_{10} helix, and the 2_7 ribbon].

Some amino acids do not fit easily into an α-helical conformation because of steric hindrance, charge, or other factors. They tend to prevent the formation of an α helix. Glu, Met, and Ala are the strongest α helix formers, whereas Gly, Pro, and Asn are least likely to be found in α helices. Proline is an extreme case because of its ring imino group.

The α-helical content of a protein can be estimated by optical rotation measurement. Native α-helical proteins are significantly more dextrorotatory than randomly coiled (i.e., without organized structure) polypeptide chains.

β-Pleated sheet structures arise when hydrogen bonds between neighboring polypeptide chains, or between different regions of the same chain, are formed. The intermolecular (or intramolecular) hydrogen bonds are oriented perpendicular to the long axis of the polypeptide chains (Figure 1.17).

This type of secondary structure is found in relatively pure form in silk fibroin, a member of the β-**keratin** family. The β-keratins contain

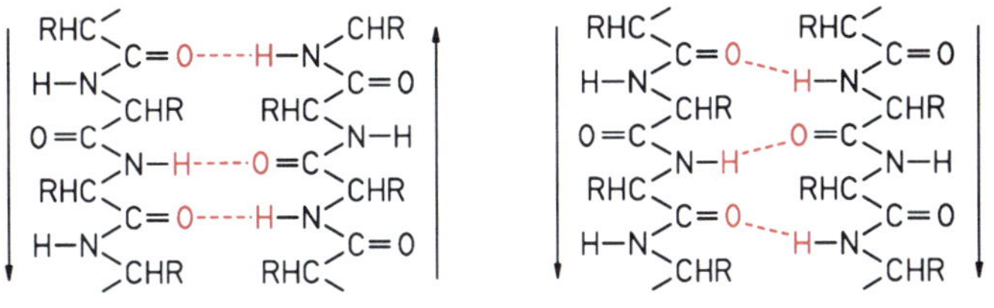

Figure 1.17.
β-Pleated sheet structure.

no cysteine residues but mostly amino acids with small side chains such as alanine, glycine, and serine. Adjacent polypeptide chains in β-keratin are **antiparallel**, that is, they run in opposite N-terminal to C-terminal directions. Other β-pleated sheet structures, often found in the interior regions of globular proteins, contain **parallel** polypeptide chains. These structures are intrinsically less stable than antiparallel structures because of the presence of nonlinear hydrogen bonds.

Another type of secondary structure is the **β bend.** When a polypeptide chain is folded into a globular protein, α-helical regions and β sheet structures are often connected by sharp bends in the polypeptide chain. Glycine, asparagine, serine, and proline are most frequently observed in such β bend conformations.

Collagen has a high content of proline and hydroxyproline, making α helic formation impossible. Instead, collagen forms a **supercoiled triple helix.** Its structure will be discussed in section 11.2.

1.2.5 Tertiary Structure

The **tertiary structure** of a protein refers to the way a polypeptide chain, already containing segments of secondary structure, is folded further to give the molecule its overall three-dimensional shape. Tertiary structure is maintained by noncovalent interactions of amino acid side chains, such as hydrogen bonds and van der Waals, ionic, and hydrophobic interactions, and by covalent disulfide bridges.

The **conformation** (secondary and tertiary structure) of a protein can be determined by x-ray diffraction. In this method, a beam of x-rays is directed at the crystal of a protein. The scattering (diffraction) of the x-ray beam by the atoms of the protein crystal produces a series of diffraction maxima whose locations and intensities give information about the three-dimensional arrangement of protein molecules in the crystal and of atoms in the protein. In the case of globular proteins, isomorphous incorporation of electron-dense heavy metal ions into the crystal lattice is used to help determine the phases of the scattered x-ray beams.

In addition to x-ray diffraction, **nuclear magnetic resonance (NMR)** is increasingly used to determine the structures of smaller polypeptides. It is based on the ability of atoms with paramagnetic nuclei, such as ^{1}H or ^{31}P, to align with a strong external magnetic field. Different orientations of these nuclear magnets, called spins, in the magnetic field have different energies. Microwave radiation can flip a nucleus from one orientation to the other: this is called nuclear magnetic resonance. At a given magnetic field strength, the wavelength at which this flip occurs corresponds to the energy difference between the two spin states. This energy difference is also influenced by the surrounding atoms. For example, a proton in a —CH_3 group spins at a different energy level than a proton in a —CHOH group. This shift of the resonance energy is called **chemical shift** and allows the identification of individual peaks in an NMR spectrum. More sophisticated applications of this technique, such as **two-dimensional NMR,** are used for determining protein structures.

1.2.6 Quaternary Structure

Some proteins consist of several polypeptide chains, joined by noncovalent interactions or disulfide bonds. This quaternary structure of a multi-subunit protein arises from specific interactions of exposed groups on the folded polypeptide chains. The polypeptide subunits may be identical or different. Hemoglobin, for example, contains four subunits, two α and two β chains, each having a molecular weight of about 16,000 (see Chapter 9). The polypeptide chains of different subunits are not intertwined. The holoprotein is formed by association of prefolded subunits through specific interaction sites.

The number of subunits in a protein can be determined by comparing its total molecular weight with the molecular weights of the individual subunits (SDS gels, equilibrium sedimentation in 8 *M* urea) or from the number of N-terminal amino acids per unit weight of protein. N-terminal amino acid analysis may also yield information on whether or not the subunits are identical.

1.2.7 Globular Proteins

As the name implies, globular proteins have a more or less spherical shape. However, they contain many of the secondary structural features found in fibrous proteins discussed previously. Local regions of α helices and β sheets are interspersed with β bends or unstructured polypeptide chains connecting them. Groups of α helices or β sheets often form higher order tertiary structures that are more compact. Such segments of the protein are called **domains** and often represent a functional unit of the protein.

Little water is found inside the globular protein structure, although some proteins contain water molecules immobilized within their structures. Polar amino acids are found predominantly on the surface, whereas nonpolar amino acids are buried in the hydrophobic center of the molecule.

1.2.8 Membrane Proteins

In contrast to globular proteins, proteins that pass through the plasma membrane do not have to be folded in such a way that nonpolar amino acids are buried inside the protein molecule. The membrane bilayer is highly hydrophobic and thus can accommodate uncharged nonpolar amino acids. The transmembrane domains of a protein are generally α-helical in nature, consisting of predominantly nonpolar amino acids. Such regions in a protein sequence can be identified with relatively high accuracy from the hydropathy profile (see below).

1.2.9 Posttranslational Modification

Many proteins are synthesized in a form that is different from that found in the mature protein. Chymotrypsin, for example, a protein-digesting enzyme, is synthesized in the pancreas as an inactive zymogen, chymotrypsinogen, that is subsequently cleaved in the small intestine by other proteases to form the active enzyme. Other proteins are modified through the covalent attachment of carbohydrate chains, lipids, phosphate, hydroxyl, carboxyl, acetyl, or other groups. A large proportion of intracellular proteins have an acyl (e.g., formyl or acetyl) group at their amino terminus. These posttranslational modifications will be discussed in later chapters.

1.2.10 Glycoproteins

Proteins with carbohydrate side chains attached to some of their amino acids are called glycoproteins. One distinguishes between N-linked and O-linked glycoproteins depending on whether the carbohydrate side chains are attached to the amido nitrogen of asparagine or to the hydroxyl oxygen of serine and threonine, respectively. (In collagen, carbohydrates are found linked to hydroxylysine.) As a rule, N-linked glycoproteins are rather complex, carrying long, branched carbohydrate side chains. They are found in many cell membrane proteins and in secreted proteins such as antibodies. In contrast, O-linked glycoproteins contain relatively short carbohydrate side chains. The biosynthesis and function of glycoproteins will be discussed in more detail in Chapter 2.

1.2.11 Denaturation of Proteins

The loss of the highly ordered three-dimensional structure of a protein is called denaturation. The resulting nonordered structure is often called a **random coil.** Denaturation is usually accompanied by loss of biological activity, decreased solubility, and increased susceptibility to enzymatic degradation (proteolysis). (If caused by extremes of pH or temperature, denaturation of proteins may be irreversible because some covalent rearrangements have occurred.)

Denaturation can be achieved with high concentrations of urea (8 *M*), guanidinium salts (4–6 *M*), or detergents (sodium dodecyl sulfate, or SDS). If the denaturing agent is removed slowly, for instance by dialysis, the protein often renatures and regains biological activity. Certain proteins, however, cannot regain their native conformation because they were originally formed from a larger precursor molecule of which only part is still present in the mature form of the protein.

1.2.12 Computer Analysis of Primary Structures and Secondary Structure Prediction

Although some progress toward understanding the mechanism of protein folding has been made, it is still not possible to predict accurately the conformation of a protein from its amino acid sequence. The fact that protein folding occurs within a relatively short time span (seconds) suggests that folding occurs in a partially ordered pathway. If folding occurred by random trials of all possible configurations around peptide bonds, it would take far too long for proteins to fold.

The folding process is probably initiated from nucleation points, such as α-helical segments or a β sheet structures that form first, and then proceeds to include additional segments of secondary structure that interact with each other. Analysis of proteins with known three-dimensional structures indicated that different amino acids have different probabilities of occurring in α-helical, β sheet, or β turn structures; Table 1.4 lists these probabilities. Any amino acid sequence that has several contiguous residues with high P_α, P_β, or P_{turn} values is likely to form that particular secondary structure in the protein.

Although there exist several computer programs for structure prediction of proteins from the amino acid sequence, almost all of these predictions are still very imprecise. The most reliable predictions are those of membrane-spanning domains of membrane proteins. As mentioned before, transmembrane domains are α helices. Kyte and Doolittle came up with a simple scheme for predicting such domains. Each amino acid is assigned a **hydrophobicity index, *h*.** This index is then plotted as an average over seven contiguous amino acids $[H_7(i)]$. If $H_7(i) \geq 1.6$ over a region of 20 residues (long enough to span a phospholipid membrane), a transmembrane domain is predicted. Figure 1.18 shows

Table 1.4. Probabilities that Amino Acids Will Occur in Protein Secondary Structures

Amino Acid	α Helix P_α	β Sheet P_β	β Turn P_{turn}
α Helix formers			
Ala	1.29	0.90	0.78
Cys	1.11	0.74	0.80
Leu	1.30	1.02	0.59
Met	1.47	0.97	0.39
Glu	1.44	0.75	1.00
Gln	1.27	0.80	0.97
His	1.22	1.08	0.69
Lys	1.23	0.77	0.96
β Sheet formers			
Val	0.91	1.49	0.47
Ile	0.97	1.45	0.51
Phe	1.07	1.32	0.58
Tyr	0.72	1.25	1.05
Trp	0.99	1.14	0.75
Thr	0.82	1.21	1.03
β Turn formers			
Gly	0.56	0.92	1.64
Ser	0.82	0.95	1.33
Asp	1.04	0.72	1.41
Asn	0.90	0.76	1.23
Pro	0.52	0.64	1.91
No preference			
Arg	0.96	0.99	0.88

Source: Levitt M. *Biochemistry.* 1978; 17:4277–4285.

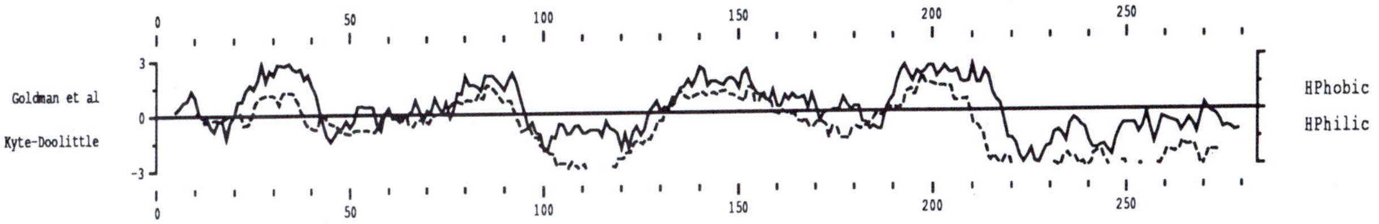

Figure 1.18.
Hydropathy plot of the rat liver cell-cell channel protein, connexin32. The solid line shows the Kyte and Doolittle plot. (Kyte J, Doolittle RF. A simple method for displaying the hydropathic character of a protein. *J Mol Biol.* 1982; 157:105–132.) The dotted line represents the Goldman-Engelman-Steiz plot that identifies nonpolar transmembrane helices (*Annu Rev Biophys Chem.* 1986; 15:321–353.) Both curves indicate hydrophobic regions when the line is in the upper half of the frame.

the computer printout of such a **hydropathy plot** for connexin32, the liver cell-cell channel protein. This protein has four transmembrane segments that are easily recognized on the plot.

1.3 PURIFICATION AND CHARACTERIZATION OF PROTEINS

1.3.1 Purification of Proteins

Methods for the separation and purification of individual proteins from a mixture of proteins are based upon differences in such properties as size, acid-base character, solubility, and function (e.g., specific affinity for other molecules).

Since the α carboxyl and α amino groups of internal amino acids are involved in peptide bond formation and, therefore, are no longer present in ionizable form, the acid-base properties of a protein are solely determined by its terminal amino acids and by those amino acids that have ionizable side chains (Glu, Asp, Lys, Arg, His, Cys, and Tyr).

These side chains determine the overall electrical charge a protein carries at a particular pH value. Proteins of different charge move at different rates when subjected to an electric field. This property is used in **polyacrylamide gel electrophoresis (PAGE).** The rate, v, at which a protein moves in PAGE depends on its net charge, c, its frictional coefficient, f (which is determined by its molecular size and the pore size of the gel, $f = k \times Mr$), and the magnitude of the electric field, E.

$$v = \frac{Ec}{f}$$

The semisolid polyacrylamide acts to prevent diffusion of the proteins and to retard their migration in a manner related to their size. After electrophoresis, bands corresponding to different proteins can be visualized on the gel by staining with Coomassie blue, a commonly used protein stain.

For every protein there is a particular pH at which it has no net charge and, therefore, does not move in an electric field. This **isoelectric point** depends to some extent on the ionic composition of the solvent, since proteins usually have ions bound to them. (The isoionic point defines the pH at which the number of positive charges in the protein molecule itself equals the number of negative charges.) The isoelectric point of a protein can be utilized to separate it from other proteins by **isoelectric focusing.** In this method, a mixture of proteins is subjected to electrophoresis in a polyacrylamide gel that contains a pH gradient.

Under the influence of an electric field, individual proteins migrate as long as they possess a net charge. When a protein reaches the region of the gel where the pH corresponds to its isoelectric point, it loses it charge and stops migrating, thus forming a sharp band.

In combination with SDS-polyacrylamide gel electrophoresis (see below), which separates proteins according to their molecular weight, isoelectric focusing is used for the analysis of complex protein mixtures by **two-dimensional gel electrophoresis.** In the first dimension, proteins are separated according to their isoelectric points; in the second dimension, in SDS, proteins are separated according to their size. The method allows the separation and identification of several thousand protein species on a single polyacrylamide gel (Figure 1.19).

Chromatographic separation of proteins on **ion exchange** columns, such as diethylaminoethyl (DEAE) cellulose or carboxymethyl (CM) cellulose, is also based upon differences in their acid-base properties. DEAE-cellulose is an anion-exchange resin. Proteins bind to this support at pH values where they carry a negative charge (i.e., above their isoelectric point). CM-cellulose, being a cation-exchange resin, will bind proteins at pH values below their isoelectric points. In both cases proteins can be eluted from the column by raising the ionic strength (salt gradient) or by altering the pH.

Proteins and peptides can be purified quickly by **high-performance liquid chromatography V(HPLC).** In this technique, the chromatography is done under very high pressure (5,000–10,000 psi). This reduces the time during which the proteins may diffuse and thereby cause a decrease in resolution. Ion-exchange columns can be used as well as hydrophobic columns, in which proteins are separated according to their hydrophobicity.

One of the most powerful methods for the purification of proteins is **affinity chromatography,** which is based on the specific binding of a protein to a ligand (usually a substrate, substrate analog, or inhibitor of the protein) that is covalently linked to a carrier such as cellulose or dextran. After binding, the protein can be eluted with high concentrations of free substrate or by conditions that disrupt the specific interaction. The method is highly specific.

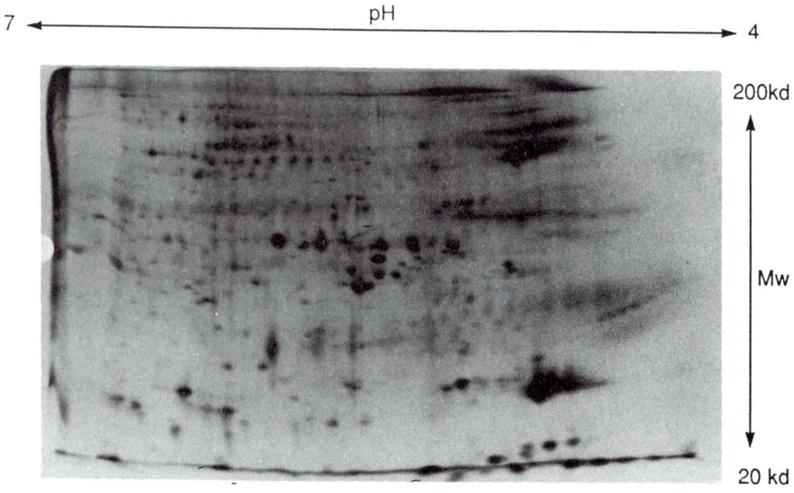

Figure 1.19.
Two-dimensional gel electrophoresis of ^{35}S-labeled fast axonally transported proteins in the frog *Rana pipiens* optic nerve. First dimension (left to right): isoelectric focusing; second dimension: SDS polyacrylamide gel electrophoresis. (Data from G.S. Perng and D. Wilson, University of Miami.)

Proteins can also be separated according to their differences in **solubility.** The four most important factors that influence the solubility of proteins are

1. *pH.* A protein is least soluble at its isoelectric point.
2. *Ionic strength.* Low concentrations of neutral salts increase the solubility of proteins (salting in), while high concentrations of salts may precipitate proteins (salting out). Divalent and trivalent ions are most effective in salting out. Ammonium sulfate is most frequently used for the differential precipitation of proteins.
3. *Dielectric constant of the solvent.* The addition of water-miscible organic solvents, such as ethanol or acetone, also decreases the solubility of most proteins and can be used for precipitation. The decrease in the dielectric constant of the solvent causes a reduction in the ionization of the protein and, thereby, the removal of the repulsive forces between similar charges on protein molecules that keep proteins in solution.
4. *Temperature.* Raising the temperature usually increases the solubility of proteins up to about 40°C. At higher temperatures, most proteins are denatured and precipitate. Heat treatment can be used to enrich a mixture selectively for those proteins that are more heat resistant.

Differences in their molecular size are utilized for the separation of proteins by **gel filtration.** In this method, a mixture of proteins is applied to the top of a column filled with small **Sephadex** beads and slowly filtered through this column. Each Sephadex bead consists of cross-linked dextran molecules (a polysaccharide made from glucose). The degree of cross-linking determines the pores of the Sephadex beads and thus the size of the protein molecules that can enter the dextran meshwork. Larger protein molecules are **excluded** from entering the Sephadex beads; as a result they have a shorter path through the column and elute ahead of the smaller molecules (Figure 1.20).

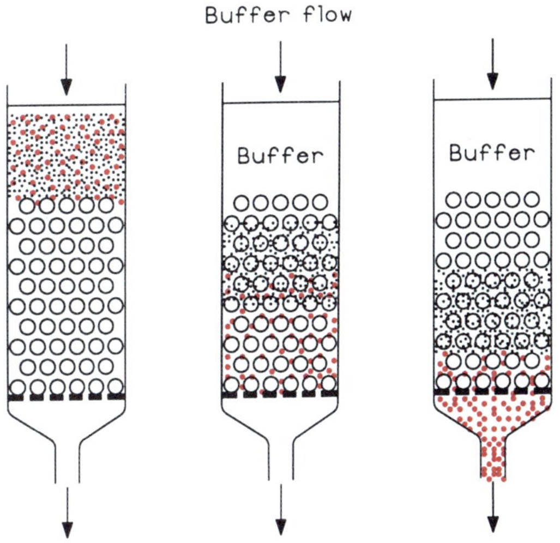

Figure 1.20.
Gel filtration. A mixture of large and small molecules applied to the top of a column of gel particles separates as the mixture passes through the column. The small molecules can enter the gel particles and thus are retained longer on their path through the column than the large molecules, which are excluded from the interior of the gel particles.

The elution behavior of a molecule is usuallv described bv its **distribution coefficient (K_d).** K_d is defined by

$$K_d = \frac{V_e - V_0}{V_t - V_0}$$

where V_0 is the **void volume,** that is, the volume of the solvent in the column that is outside the gel particles. It can be determined by measuring the volume of buffer that must pass through the column before a substance that cannot enter the gel particles appears in the eluate. V_t is the **total volume** of the column, including both intragel and extragel space, and V_e is the **elution volume** of a particular substance that can enter, partially or completely, the gel particles.

1.3.2 Molecular Weight Determination

Precise determinations of molecular weights of proteins are obtained by measuring **sedimentation coefficients** in the analytical ultracentrifuge. The sedimentation coefficient s is defined as

$$s = \frac{dx/dt}{\omega^2 x}$$

where

dx/dt = rate of sedimentation ($cm \cdot s^{-1}$)
x = distance of the protein boundary from the center of rotation (cm)
t = time (s)
ω = angular velocity (in radians$\cdot s^{-1}$)

Typical $s_{20,w}$ values (measured at 20°C, in water) of proteins lie between 1×10^{-13} and 200×10^{-13} s. A sedimentation coefficient of 10^{-13} is called a **Svedberg unit.** The relationship between ***s* value** and molecular weight is expressed in the **Svedberg equation**

$$M = \frac{sRT}{D(1 - \bar{v}\rho)}$$

where

R = gas constant
T = absolute temperature
D = diffusion coefficient
$\bar{v}$ = partial specific volume of protein in solution
ρ = density of the solvent

The often time-consuming and tedious determination of the diffusion coefficient can be avoided by **sedimentation to equilibrium.** After a sufficient period of centrifugation, an equilibrium is reached when the rate of sedimentation of the protein is exactly balanced by its rate

Table 1.5. Classification of Enzymes

Type of Enzyme	Type of Reaction
Oxidoreductases	$S_{ox} + R_{red} \rightleftharpoons S_{red}R_{ox}$
Transferases	A—B + C ⇌ A—C + B
Hydrolases	A—B + H_2O ⇌ A—OH + B—H
Lyases	A═B + Ck—D ⇌ C—A—B—D
Isomerases	A ⇌ B
Ligases	X + Y + ATP ⇌ X—Y + ADP + P_i

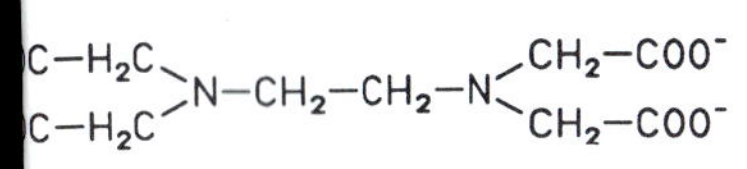

ure 1.25. ylenediamine tetraacetate (EDTA).

grees of affinity. When firmly bound to the enzyme, either covalently or noncovalently, a cofactor is referred to as a **prosthetic group.** The enzyme-cofactor complex is called a **holoenzyme.** Removal of the cofactor leaves an **apoenzyme** or core enzyme. Active alcohol dehydrogenase, for example, contains Zn^{2+} ions. Ethylenediamine tetraacetate (EDTA) **chelates** (i.e., forms tight complexes with) divalent cations, such as Zn^{2+} or Mg^{2+} (Figure 1.25). Thus, in the presence of EDTA, alcohol dehydrogenase has no catalytic activity.

Many coenzymes should be regarded as cosubstrates [e.g., nicotinamide-adenine dinucleotide (NAD), coenzyme A] because they are freely diffusible and are required in stoichiometric amounts (i.e., for each substrate molecule to be converted to product, one cosubstrate is required). Coenzymes usually act as transfer agents, transferring groups or ions from one substrate to another. (Individual coenzymes will be discussed in Chapter 2.)

1.6 ENZYME KINETICS

Enzymes do not affect the extent to which a chemical reaction can proceed. The reaction equilibrium is determined solely by thermodynamic parameters. Enzymes increase the rate constants of both the forward and reverse reactions equally (i.e., $K' = k_1/k_2 =$ constant).

Chemical reactions proceed through **transition states** that have a higher free energy (for a definition of free energy, see Chapter 2) than both substrate and reaction product:

$$A \rightarrow A^* \rightarrow B$$

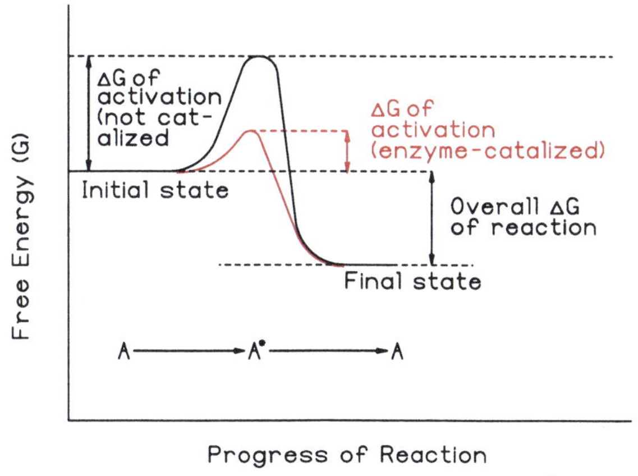

ure 1.26. e-energy changes during a emical reaction; activation en-gy.

of diffusion in the opposite direction. The molecular weight of the protein is then determined by the formula

$$M = \frac{2RT \ln(c_2)/c_1)}{\omega^2(1 - \bar{v}\rho)(x_2^2 - x_2^2)}$$

where c_1 and c_2 are the concentrations of the protein at radius x_1 and x_2 from the center of rotation.

In order to determine the size of a protein quickly, gel filtration is often used. This is not an absolute method, however, since the elution pattern of a protein also depends on its shape. Therefore, this method can only be used to compare the elution pattern of an unknown protein to that of a similarly shaped protein of known size and thereby estimate the size of the unknown protein. In practice, a calibration curve is constructed in which the K_d values of a series of compounds of similar molecular shape and density are plotted against the logarithms of their molecular weights. Such calibration curves usually give linear lines. The molecular weight of a substance of similar molecular shape and density can then be estimated by marking its K_d value on this calibration curve and reading its corresponding molecular weight.

A more reliable, and widely used, method for determining approximate molecular weight values of small amounts of protein is **SDS-gel electrophoresis.** In the presence of the detergent sodium dodecyl sulfate, proteins denature, and the lipophilic SDS chains form micelles around the polypeptide chain, resulting in the formation of rod-shaped complexes (Figure 1.21). Since each SDS chain carries a negative charge, the net charge per unit length of polypeptide chain is constant; as a consequence, the rate of migration of the SDS-protein complex in an electric field is determined only by the length of the polypeptide. A linear inverse relationship exists between the logarithm of the molecular weight of the polypeptide chain and its electrophoretic mobility. Of course, standards of known molecular weights are required by this method. The presence of prosthetic groups, such as carbohydrate residues, can result in aberrant molecular weight values.

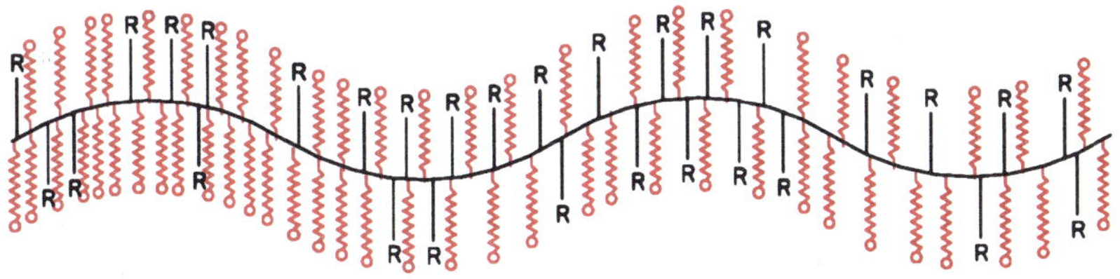

Figure 1.21. Polypeptide-SDS complx. R = aminoacyl side chain.

The most exact molecular weight determination of a protein is the determination of its amino acid sequence. This will be discussed in the next section.

1.4 AMINO ACID SEQUENCE DETERMINATION

The overall amino acid composition of a protein is determined after complete acid hydrolysis (6 *N* HCl, 110°C, 24–96 h). During this pro-

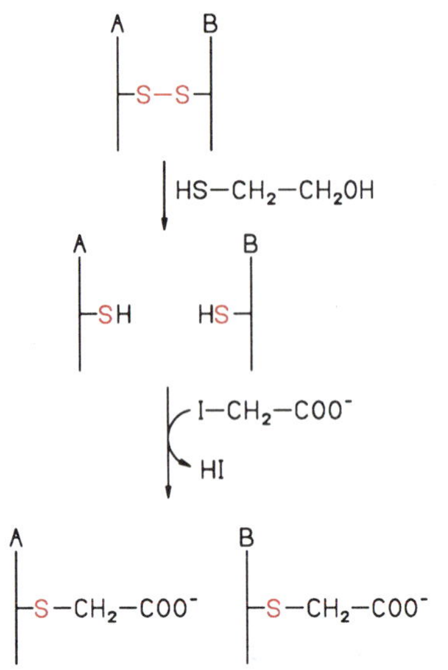

Figure 1.22.
Cleavage of disulfide by mercaptoethanol, followed by carboxymethylation with iodoacetate.

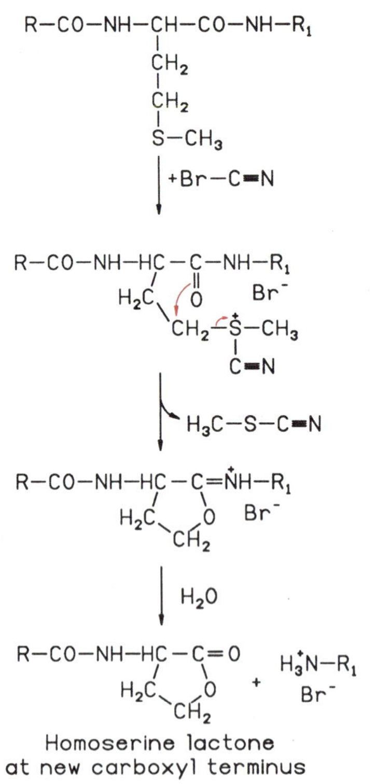

Figure 1.23.
Cleavage of polypeptides at methionyl residues with cyanogen bromide.

cedure, Trp is destroyed and must be estimated by other methods, for example, by alkaline hydrolysis or by ultraviolet (UV) absorption spectroscopy. (Only Trp and Tyr contribute significantly to the UV absorption of proteins in the 280 nm region.) In addition, Gln and Asn are hydrolyzed to Glu and Asp, respectively. Thus, only the sum Glu + Gln and the sum Asp + Asn are revealed.

If the protein has more than one subunit, it is necessary to separate the polypeptide chains with high concentrations of **urea, guanidinium hydrochloride,** or **detergents.** These reagents destroy noncovalent interactions, dissolve aggregations, and unfold (denature) the polypeptide chains. The most widely used detergent is SDS. Polypeptide chains that are covalently linked together by disulfide bonds are first cleaved by reduction with mercaptoethanol and subsequently carboxymethylated with iodoacetate (Figure 1.22) in order to prevent the reformation of the disulfide bonds in the presence of oxygen (air).

To determine the amino acid sequence of a long polypeptide chain, it is necessary first to cleave it into smaller fragments using proteases, such as trypsin, and/or cyanogen bromide (see below). The smaller peptide fragments are then separated by gel filtration, electrophoresis, or ion-exchange chromatography and sequenced individually.

Trypsin hydrolyzes polypeptides on the carboxyl side of lysine and arginine. **Cyanogen bromide** cleaves polypeptide chains specifically on the carboxyl side of methionine residues, yielding peptides with C-terminal homoserine lactone (Figure 1.23).

N-terminal sequences of proteins and peptides can be determined by the **Edman degradation** procedure, which is used in automatic peptide sequenators (Figure 1.24). **Phenyl isothiocyanate (PITC)** or **Edman's reagent** is reacted with the polypeptide. The PITC derivative of the peptide is then treated with strong anhydrous acid (trifluoroacetic or heptafluorobutyric acid) that releases the anilinothiazolinone derivative of the N-terminal amino acid and a polypeptide chain one residue shorter than the original. The anilinothiazolinone derivative is then extracted with organic solvents and converted to the phenylthiohydantoin derivative by treatment with dilute aqueous acid at 80°C. The phenylthiohydantoin (PTH) is then identified directly by high-performance liquid chromatography. A second Edman reaction is performed to identify the next residue from the N-terminus. Repeated automated application of the reaction may allow the sequence determination of as many as 30–60 amino acid residues. Manual application can only go to about 15 residues before heterogeneity in the PTH-amino acid generated, which results from incomplete chemical yields during the degradation cycle and nonspecific internal cleavages, obscures the sequence.

C-terminal amino acids are determined with **carboxypeptidase,** an enzyme that hydrolyzes polypeptides sequentially from the carboxyl terminus. This method, however, is only useful for determining the first one to three amino acids because of lack of synchrony in the degradation of individual chains.

The vast majority of protein sequences that are currently being elucidated are obtained by translation of their cDNA or genomic sequences, which can be determined more easily by DNA sequencing. Direct protein sequencing is now limited to obtaining partial sequence data that can then be used to isolate and identify the desired cDNA or

Figure 1.24.
Edman degradation of polypeptides from the amino terminus with phenyl isothiocyanate.

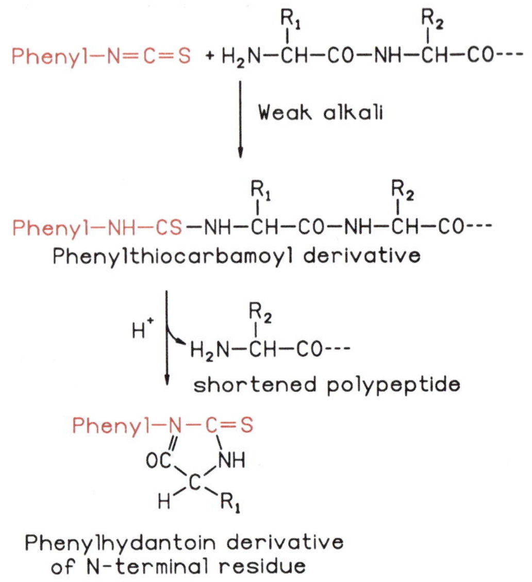

genomic clones. In addition, it is used to identify sites of
tional modification. For smaller proteins it is occasionally
elucidate the sequence directly. This is done by generating
lapping fragments that, after sequencing, can be used to a
overall sequence.

Intrachain disulfide linkages are located by digestin
polypeptide chains without disulfide cleavage and isolatin
fide-linked peptides. After purification, the disulfide bridges
and the individual peptides resulting from the cleavage are
to determine which cysteines in the polypeptide chains
together.

Interchain disulfide linkages can be located by dige
native protein followed by disulfide cleavage. Since disulfide
a tendency to open and reform at alkaline pH, thus gene
interstrand links, initial hydrolysis and separation of disulfide
peptides must be performed at low pH values (<6.5).

1.5 ENZYME CLASSIFICATION

Enzymes are proteins that catalyze biochemical reactions. Mo
catalyzed reactions proceed in aqueous solution, at physi
peratures and pH values. In the absence of an enzyme, th
actions often proceed at undetectable or insignificant rates
shows the six basic types of reaction that are catalyzed by e
which are used as the basis for enzyme classification.

Most enzymes are highly specific for their **substrates** (
that is, they recognize only one type of molecule and can acc
only limited structural variations in the substrate. They are
ways stereospecific and act upon, or form, only one stereoi
Most enzymes require additional molecules, called **cofact**
enzymes, for catalytic activity. A cofactor may be a metal io
molecular-weight substance that binds to the enzyme with

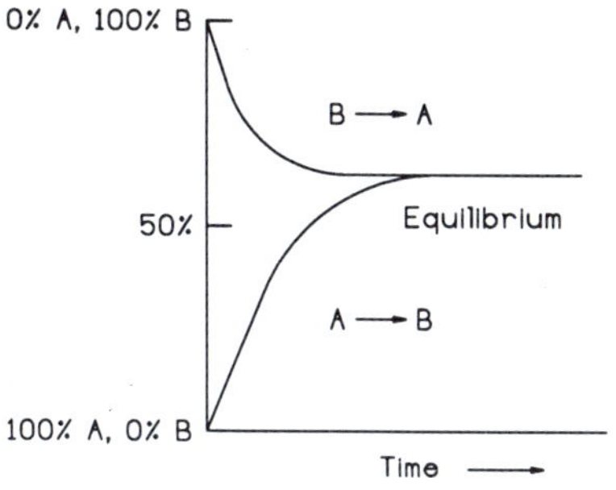

Figure 1.27.
Time course of a reversible reaction.

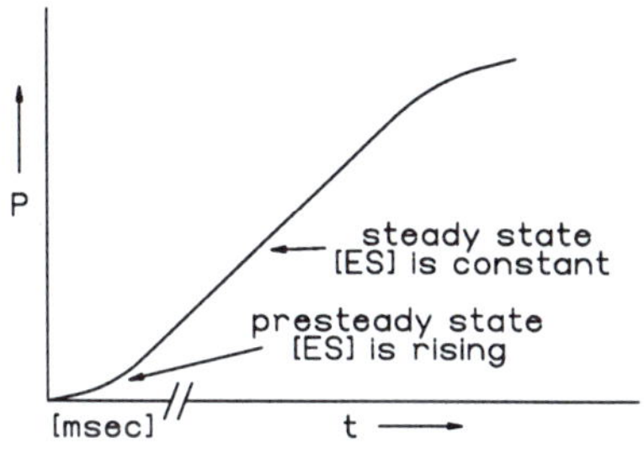

Figure 1.28.
Time course during the early phase of a biochemical reaction.

The attainment of the transition state thus requires the input of **activation energy.** Enzymes function by allowing the formation of enzyme-bound transition states that require lower activation energy and therefore are formed more readily (Figure 1.26).

In reversible biochemical reactions, the rate of the forward reaction decreases while the rate of reverse reaction increases as the reaction equilibrium is approached. For example, in the reaction A $\rightleftharpoons$ B, with an equilibrium constant $K' = 3.0$. The forward and reverse reactions become identical at equilibrium (Figure 1.27). This is why the measurement of **steady state reaction rates,** taken before any appreciable amount of substrate has been converted, is more meaningful for determining reaction rate parameters than rates measured at times when the reaction has progressed significantly. The steady state rate is reached very quickly, usually within milliseconds (Figure 1.28).

When reaction conditions (temperature, pH, ionic strength) are kept constant, the steady state velocity of an enzyme-catalyzed reaction depends on two factors: (1) the enzyme concentration [E] and (2) the substrate concentration [S]. At constant [E], a **maximum velocity,** V_{max}, is reached as [S] is increased. This saturation effect was interpreted by **Michaelis and Menten** as a finite number of substrate-binding sites on the enzyme. After the formation of an **enzyme-substrate complex** (ES), the transition state intermediate, this complex breaks down in a second step to yield free enzyme E and product P.

$$E + S \underset{k_2}{\overset{k_1}{\rightleftharpoons}} ES \underset{k_4}{\overset{k_3}{\rightleftharpoons}} E + P \quad (1.1)$$

The second step of the reaction, the breakdown of ES, is essentially one-directional during the initial stages of the reaction since the back reaction, P + E → ES, can be ignored because of the low concentration of P. The initial rate, v, of product formation thus is determined directly by the concentration of ES.

$$v = k_3[ES] \quad (1.2)$$

We can express [ES] as a function of (1) the rate of its formation, $k_1[E][S]$, and (2) the rate of its breakdown, $k_2[ES] + k_3[ES]$.

If the rate of formation of ES from S and E is constant and if the breakdown of ES to P is essentially "irreversible," the concentration of the enzyme-substrate complex [ES] must also remain constant. Both conditions are met at the beginning of the reaction because at that time [S] is very large and its rate of change is negligible while [P] is very small, so that the back reaction, P + E → ES, is negligible. This situation is called the **steady state.** (Before ES is built up, the rate of product formation is very low. This pre–steady state period usually lasts only a few milliseconds and can be ignored when measuring the initial reaction rates.) At constant [ES], therefore,

$$k_1[E][S] = (k_2 + k_3)[ES] \quad (1.3)$$

or

$$\frac{k_2 + k_3}{k_1} = \frac{[E][S]}{[ES]} = K_m \quad (1.4)$$

Since the concentration of free enzyme, [E], equals its total concentration, $[E_{tot}]$, minus the amount bound in ES,

$$[E] = [E_{tot}] - [ES] \tag{1.5}$$

and

$$[ES] = \frac{[E_{tot}) - [ES][S]}{K_m} \tag{1.6}$$

or

$$[ES] = [E_{tot}]\frac{[S]}{[S] + K_m} \tag{1.7}$$

Substituting equation (1.7) into equation (1.2), we obtain the Michaelis-Menten equation:

$$v = k_3[E_{tot}]\frac{[S]}{[S] + K_m} \tag{1.8}$$

or

$$v = V_{max}\frac{[S]}{[S] + K_m} \quad \text{with} \quad V_{max} = k_3[E_{tot}] \tag{1.9}$$

At high substrate concentrations, essentially all enzyme is contained in ES, so that $[E_{tot}]$ equals [ES] and $v = V_{max}$. When $v = \frac{1}{2}V_{max}$, $K_m =$ [S]. In other words, K_m is the substrate concentration at which the reaction rate is half-maximal (Figure 1.29).

A high K_m value generally indicates that the enzyme has a low affinity for the substrate since a high substrate concentration is required to saturate the enzyme. The K_m value is a characteristic property of an enzyme with regard to a particular substrate. It is independent of the substrate or enzyme concentration and can be determined with only partially purified preparations. A K_m value is not a dissociation constant but is composed of three kinetic constants [see equation (1.4)]. Only if the value of k_3 is much smaller than k_2 will K_m approach the dissociation constant of the ES complex, $K_d = k_2/k_1$.

Figure 1.29.
Relationship between reaction velocity, *v*, and substrate concentration, [S], at constant enzyme concentration.

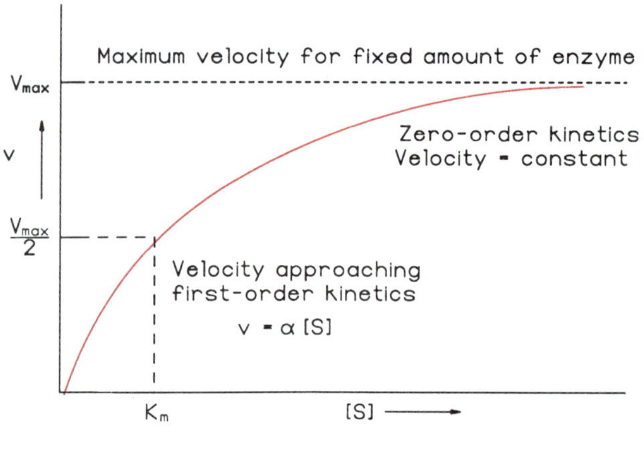

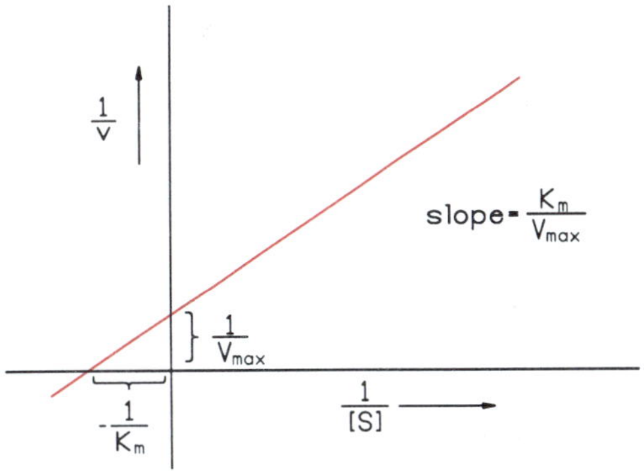

Figure 1.30.
Lineweaver-Burk plot.

In contrast to K_m, V_{max} values are dependent on the enzyme concentration and enzyme purity. When expressed as the number of substrate molecules reacted per single enzyme molecule in one second, V_{max} becomes the **turnover number,** a characteristic property of the enzyme (see below).

For determination of K_m and V_{max} values, a double-reciprocal plot of $1/v$ against $1/[S]$ **(Lineweaver-Burk plot)** is used that produces a straight line (Figure 1.30):

$$\frac{1}{v} = \frac{K_m}{V_{max}} \cdot \frac{1}{[S]} + \frac{1}{V_{max}} \tag{1.10}$$

Enzyme activity is measured under conditions in which the rate varies in direct proportion to the enzyme concentration. This occurs at high substrate concentrations, where the velocity approaches V_{max} and zero order kinetics are observed (see Figure 1.29). A suitable substrate concentration for enzyme assays is about 10 times the K_m value. (Some enzymes have to be assayed at lower substrate concentrations because of substrate inhibition.)

An **international unit** of enzyme activity is the amount of enzyme that converts one micromole of substrate per minute under specific assay conditions (pH, temperature, ionic strength). Often it is easier to measure the rate of product formation than that of disappearance of substrate.

Specific activity is the number of enzyme units per milligram of protein. It is a measure of enzyme purity; its value increases during enzyme purification. The **turnover number** is the number of substrate molecules that are converted to products by one enzyme molecule within one second.

Each enzyme has a characteristic **pH optimum** for maximum activity. Often the relationship between pH and activity of an enzyme shows a bell-shaped curve. The optimal pH for an enzyme depends on the **pK values** of the amino acids at its active site, which are involved in substrate binding and catalysis, and on the stability of the protein conformation at different pH values.

Temperature affects enzyme activity by raising or lowering the activation energy. Enzyme activity usually increases with rising temperature (about twofold for every 10°C rise in temperature) until an optimal temperature is reached. At higher temperatures, enzyme activity decreases because of denaturation of the protein.

1.7 ENZYME INHIBITION

Enzyme inhibitors reduce the activity of an enzyme. Studies of enzyme inhibition often yield information on the mechanism of action of an enzyme and allow certain conclusions about its active site.

Enzyme inhibition may be reversible or irreversible. Only the **reversible** type of inhibition, in which the inhibitor interacts with the enzyme or the enzyme-substrate complex in a reversible manner, can be described by the Michaelis-Menten theory. **Irreversible** enzyme inhibition is usually caused by covalent enzyme modification. Examples of irreversible inhibition will be discussed in Section 1.11.

There are three main types of reversible enzyme inhibition:

1. Competitive
2. Noncompetitive
3. Uncompetitive

Competitive inhibitors are compounds that are structurally related to the substrate. These **substrate analogs** compete with the normal substrate for the active site of the enzyme. The extent of competitive inhibition is dependent on the ratio of inhibitor to substrate rather than the absolute concentration of the inhibitor.

The classic example of competitive inhibition is the inhibition of **succinate dehydrogenase** by malonate. The structures of succinate and malonate differ by only one methylene group:

$$^{-}OOC\text{—}CH_2\text{—}CH_2\text{—}COO^{-} \qquad\qquad ^{-}OOC\text{—}CH_2\text{—}COO^{-}$$

Succinate — Malonate

Malonate binds to the enzyme, but since it cannot be acted upon by the enzyme, it blocks the active site.

A clinically important example of competitive enzyme inhibition is the treatment of **ethylene glycol poisoning** with ethanol. (Ethylene glycol is the main ingredient of automobile antifreeze.) In the body, ethylene glycol is oxidized to oxalic acid, a rather toxic compound. The first step in the oxidation reaction is catalyzed by alcohol dehydrogenase, an enzyme that normally oxidizes ethanol to acetaldehyde. Ingested alcohol (ethanol) will compete with this reaction, thus preventing the oxidation of ethylene glycol and allowing its excretion in the kidney. The treatment of **methanol poisoning** with ethanol is based on the same principle.

According to the Michaelis-Menten theory, the first step in competitive enzyme inhibition is the formation of an enzyme-inhibitor complex (EI).

$$E + I \rightleftharpoons EI \qquad (1.11)$$

The dissociation constant of this complex, K_i, is defined as

$$K_i = \frac{[E][I]}{[E]} \qquad (1.12)$$

(Note that K_i is a dissociation equilibrium constant and as such is not necessarily comparable to K_m.)

The Michaelis-Menten equation for a competitively inhibited enzymatic reaction is

$$v = \frac{V_{max}[S]}{K_m\{1 + ([I]/K_i)\} + [S]} \tag{1.13}$$

and its Lineweaver-Burk modification is

$$\frac{1}{v} = \frac{K_m\{1 + ([I]/K_i)\}}{V_{max}[S]} + \frac{1}{V_{max}} \tag{1.14}$$

In the Lineweaver-Burk plot, different concentrations of inhibitor produce a series of straight lines that intercept the ordinate at the same point (Figure 1.31). In other words, V_{max} is not altered by the presence of the inhibitor. The inhibitory effect can be overcome by raising the substrate concentration. On the other hand, a competitive inhibitor raises the apparent K_m value for the substrate. K_i can be determined from the slope or the intercept with the abscissa. The lower the K_i value of an inhibitor, the greater its degree of inhibition.

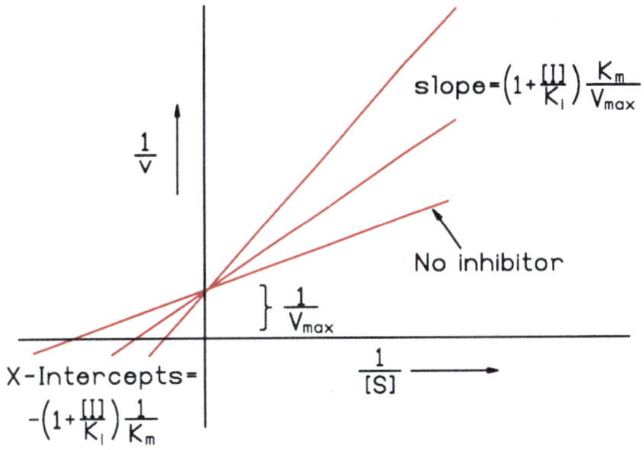

Figure 1.31. Competitive enzyme inhibition (Lineweaver-Burk plot).

In **noncompetitive inhibition** the inhibitor binds to the enzyme at a site different from the substrate binding site. The inhibitor can thus react with either the free enzyme or the enzyme-substrate complex; that is,

$$E + I \rightleftharpoons EI \tag{1.15}$$

and

$$ES + I \rightleftharpoons ESI \tag{1.16}$$

The dissociation constants, K_i, of these complexes are not necessarily equal. For identical K_i values, the Lineweaver-Burk equation is

$$\frac{1}{v} = \frac{K_m\{1 + ([I]/K_i)\}}{V_{max}[S]} + \frac{1}{V_{max}}\left(1 + \frac{[I]}{K_i}\right) \tag{1.17}$$

As can be seen from Figure 1.32, K_m remains constant in noncom-

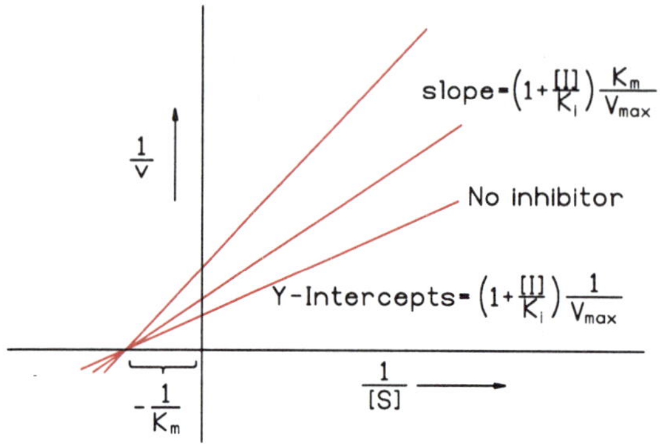

Figure 1.32. Noncompetitive enzyme inhibition (Lineweaver-Burk plot).

petitive inhibition, whereas V_{max} decreases with increasing inhibitor concentrations.

Uncompetitive inhibition is easily recognized by the parallel lines in the Lineweaver-Burk plot (Figure 1.33). The inhibitor reacts only with the enzyme-substrate complex. The relevant equations are

$$ES + I \rightleftharpoons ESI \qquad (1.18)$$

and

$$\frac{1}{v} = \frac{K_m}{V_{max}[S]} + \frac{1}{V_{max}}\left(1 + \frac{[I]}{K_i}\right) \qquad (1.19)$$

Both K_m and V_{max} values change with different inhibitor concentrations. This type of inhibition is frequently found in two-substrate reactions (i.e., reactions in which two different substrates, reacting with one another, are bound to the enzyme at the same time). It occurs rarely in single-substrate reactions. In multisubstrate reactions, an inhibitor may be competitive with one substrate and noncompetitive or uncompetitive with the other.

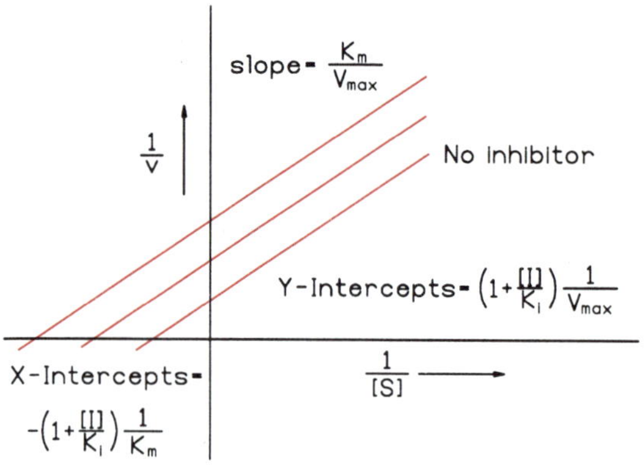

Figure 1.33. Uncompetitive enzyme inhibition (Lineweaver-Burk plot).

1.8 THE ENZYME-SUBSTRATE COMPLEX

Evidence for the transient formation of an enzyme-substrate complex comes from a variety of observations. For example, labile enzymes are often found to be stabilized by the addition of substrate. The light absorption maxima of other enzymes changes or shifts as they combine with their substrates. **Tryptophan synthetase** is such an enzyme. It catalyzes the formation of tryptophan from serine and indole. Upon binding to serine, the enzyme shows a large increase in the fluorescence of its pyridoxal phosphate prosthetic group (see Chapter 2). Addition of the second substrate, indole, reduces the fluorescence to levels below those of the free enzyme. (Fluorescence is the emission of light, usually at longer wavelengths, by molecules that have become excited by the absorption of shorter wavelength light.)

In the case of **aldolase,** an enzyme of glycolysis that reversibly cleaves fructose 1,6-diphosphate into two three-carbon sugar phosphates, the enzyme-substrate complex can be trapped as a stable derivative. When the enzyme is incubated with its substrate dihydroxyacetone phosphate in the presence of the reducing chemical sodium borohydride, the unstable Schiff base intermediate, formed by the interaction of the keto group of the substrate with the ε amino group of a lysine residue of the enzyme, is reduced to a stable secondary amine (Figure 1.34).

Finally, enzyme-substrate complexes have been demonstrated by electron microscopy (e.g., RNA polymerase on DNA) and by x-ray diffraction. For the latter, the enzyme is usually complexed with a competitive inhibitor or a poor substrate that forms a more stable complex or with substrate at low temperature.

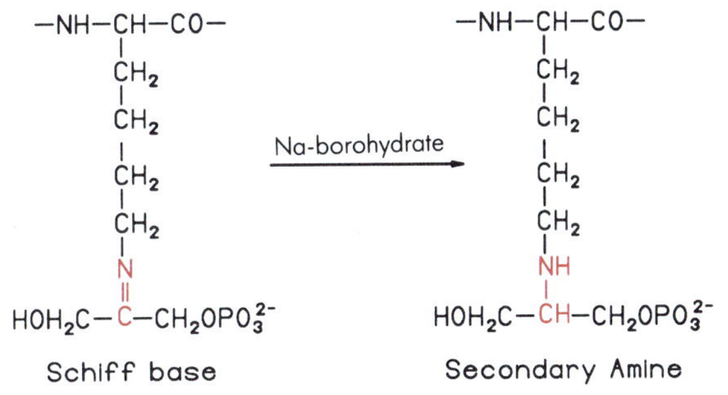

Figure 1.34.
Hydrogenation of a Schiff base intermediate between a lysyl residue of aldolase and dihydroacetone phosphate.

1.9 THE ACTIVE SITE

The active site of an enzyme can be probed by covalent modification of the amino acids at that site and by studying the effects of this modification upon substrate binding and catalysis. This is most easily accomplished when covalent enzyme-substrate intermediates are formed. An example of this approach is the formation of acetylated chymotrypsin during the hydrolysis of *p*-nitrophenyl acetate. The acetyl group in the modified enzyme is attached to a serine residue. Another example is the labeling of active sites of serine proteases with the inhibitor diisopropyl fluorophosphate (DIPF). Both examples will be discussed in detail later in this chapter.

Covalent attachment of labels to active sites of enzymes using general chemical reagents usually is less informative. For example, inactivation of cysteine-containing enzymes with *p*-hydroxymercuribenzoate may yield an inactive enzyme, but often it is not clear whether the active site itself is blocked or whether mercuration of sulfhydryl groups elsewhere in the molecule caused a general conformational change. The difficulty may be overcome by first treating the enzyme with the reagent in the presence of substrate or a competitive inhibitor. After removal of excess reagent and substrate (or inhibitor), readdition of radioactively labeled reagent will modify the exposed amino acids in the active site, thus allowing their identification. Further examples of active site characterization will be discussed later.

Different enzymes interact with their respective substrate molecules in different ways. We have already seen examples in which the substrate is covalently bound to the active site of the enzyme. Amino acids involved in the formation of covalent intermediates include serine, cysteine, histidine, and lysine. Alternatively, the substrate may be bound noncovalently to amino acids in the active site (e.g., by hydrogen bonds). In either case, the susceptible bond of the substrate molecule is activated by the proximity of other functional groups acting as proton donors or acceptors. The conformation of the active site may also induce strain or distortion in the susceptible bond of the substrate molecule, thus weakening it.

Emil Fisher compared the interaction between enzyme and substrate to a **lock and key relationship.** He proposed that the active site of an enzyme provides a rigid, tight-fitting, three-dimensional template for the substrate and will not accept molecules that differ in size or charge. This would explain the high degree of specificity of enzymes for their substrates.

An alternative model was later proposed by **Koshland.** He suggested that the enzyme molecule might undergo considerable conformational changes upon binding to its substrate. The substrate-induced changes in the active site of the enzyme bring the catalytic groups into position for catalytic action. This **induced-fit theory** is supported by x-ray diffraction studies of carboxypeptidase. The binding of glycyl-phenylalanine to this enzyme causes the Tyr-248 residue to move 12 Å to a position where it is in direct contact with the bound substrate. Many enzymes, however, show only small changes of their active sites upon substrate binding, so for them the lock and key analogy appears reasonable.

1.10 REGULATION OF ENZYME ACTIVITY

For the proper functioning of the cell's machinery, its biochemical reactions have to be coordinated. This is accomplished through the control of enzyme activity.

There are at least five different mechanisms by which the activity of an enzyme can be regulated:

1. *Product inhibition.* In this rather simple type of inhibition, the reaction product inhibits its own formation. It is most often observed in reversible reactions.
2. *Variation of the amount of enzyme.* This is accomplished by *de novo* biosynthesis of enzymes and may be considered a

long-term regulation. Its mechanism will be discussed in Chapter 6.

3. *Covalent modification of enzymes.* The activity of an enzyme can be changed drastically through the covalent attachment of specific groups, such as phosphate, adenylate, or acetate. For example, glycogen phosphorylase is activated by **phosphorylation.** In contrast, glycogen synthase is inactivated by phosphorylation. Often, but not always, the phosphorylation of an enzyme is triggered by specific hormones reacting with cell membrane receptors and is mediated intracellularly through **cyclic AMP** (for specific examples, see later chapters), which activates **protein kinases. Protein phosphatases** remove the phosphate group from the enzyme, thus reversing the action of the kinases.
4. *Protein-protein interaction.* In this type of regulation, the specificity of an enzyme is altered by the noncovalent binding of a regulatory protein to the enzyme (e.g., galactosyl transferase–lactalbumin complex; see Chapter 2).
5. *Allosteric regulation.* This is probably the most widespread cellular control mechanism. An **allosteric enzyme** is an enzyme whose activity can be modified through the noncovalent binding of a low-molecular-weight **effector** or **modulator.** As a rule, this modulator molecule binds to the enzyme at a site that is different from the catalytic site. The modulator may have a positive or negative effect upon the enzyme's activity. Some allosteric enzymes interact with one or more positive modulators and one or more negative modulators simultaneously, each modulator binding to its own specific site on the enzyme (see the example of aspartate transcarbamoylase). A special case of allosteric regulation is **feedback inhibition.** In this common form of enzyme inhibition, the end product of a metabolic pathway acts as an inhibitory effector on the first enzyme of the reaction sequence, thus indirectly slowing its own production (Figure 1.35)

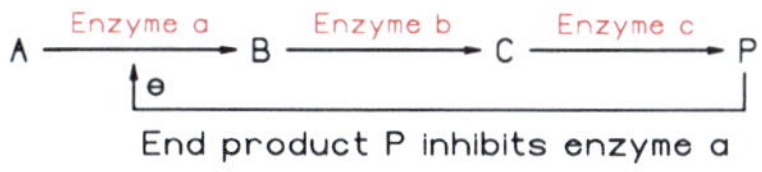

Figure 1.35.
Feedback inhibition.

All known allosteric enzymes are multi-subunit enzymes. Characteristically, they often exhibit sigmoidal Michaelis-Menten kinetics, indicating cooperativity between different substrate-binding sites of the enzyme. In other words, binding of one substrate molecule affects the binding properties of the other sites. In contrast, in nonregulated enzymes, the substrate-binding sites function independently. Such enzymes show normal hyperbolic Michaelis-Menten kinetics (see Figure 1.29).

There are three classes of allosteric enzymes:

1. **Homotropic enzymes,** in which the substrate itself is the modulator (positive or negative)
2. **Heterotropic enzymes,** in which the modulator is different from the substrate; for example, the end product of the pathway
3. **Homotropic-heterotropic** or **mixed enzymes,** in which the

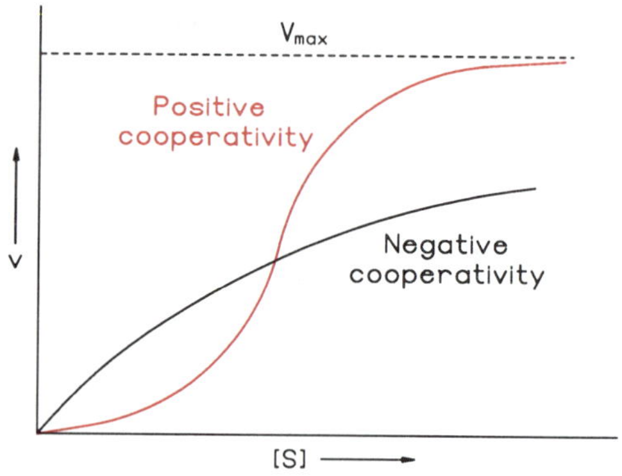

Figure 1.36.
Homotropic enzyme kinetics.

substrate is one of two or more modulators that affect the enzyme

In the case of **positive cooperativity,** or positive homotropic regulation, the binding of one substrate molecule to the enzyme facilitates the binding of the next substrate molecule by increasing the affinities of the other, still vacant, binding sites. This results in a sigmoidal Michaelis-Menten curve (Figure 1.36).

Most **heterotropic allosteric enzymes** respond to the presence of a modulator by changing their apparent K_m (Figure 1.37); variations in V_{max} are observed less frequently.

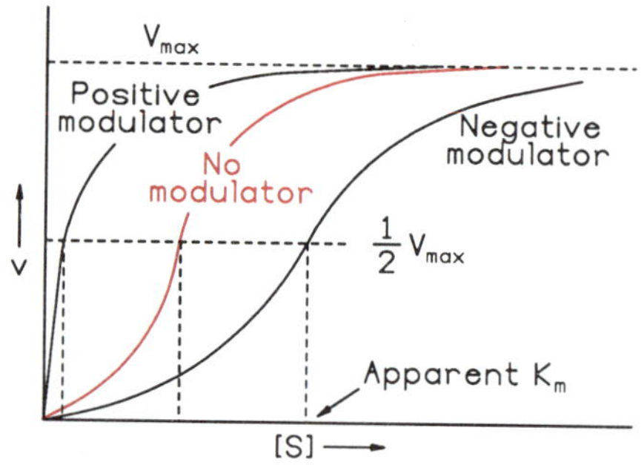

Figure 1.37.
Heterotropic enzyme kinetics illustrated on aspartate transcarbamoylase.

There are several models for the mechanism of interaction of the subunits in allosteric proteins. All have in common that each catalytic subunit of the enzyme can assume two conformational states, one active and the other inactive. The **symmetric model** proposed by **Monod** postulates that the binding of a substrate molecule to one subunit increases the tendency of all subunits to convert to the other conformational state in an all-or-none event. The **sequential model** of **Koshland,** on the other hand, does not require the simultaneous conformational change of all subunits from an inactive to an active state (or vice versa). Other more complex models have been devised to explain the behavior of enzymes that fit neither the symmetric nor the sequential model.

1.11 INDIVIDUAL ENZYMES

Different aspects of enzyme structure, function, and regulation are best discussed by using individual enzymes as examples. The biochemical reactions catalyzed by some of the enzymes discussed here will be described in detail in later chapters. Students without any knowledge of intermediary metabolism should postpone reading this section of Chapter 1 until they have completed studying at least Chapter 2.

The following four enzymes have been selected as examples:

Enzyme	Molecular Weight	Number of Subunits
Ribonuclease	1.35×10^4	1
Chymotrypsin	2.35×10^4	1 (3 peptides connected by S—S bonds)
Lactate dehydrogenase	1.4×10^5	4
Aspartate transcarbamoylase	3.1×10^5	12

1.11.1 Ribonuclease

Ribonuclease (RNase) from the pancreas is an endonuclease that hydrolyzes ribonucleic acid (RNA). It is a relatively simple enzyme containing a single polypeptide chain.

Substrate Analogs for the Assay of Enzyme Activity

RNase hydrolyzes phosphodiester bonds in RNA that are located at the 3′ side of pyrimidine nucleotides (Figure 1.38). RNA itself is not a good substrate on which to study enzyme kinetics, because the heterogeneity of the bases that are attached to the ribose makes the chemical analysis of intermediates difficult. Better substrates for enzyme studies are the 2′,3′ cyclic pyrimidine nucleotides, which are intermediates in the hydrolysis of RNA (see Figure 1.38). The course of the reaction can be easily followed by measuring the change in optical absorbance that occurs or by titrating the H^+ ions released in the process.

Figure 1.38.
Mechanism of hydrolysis of RNA.

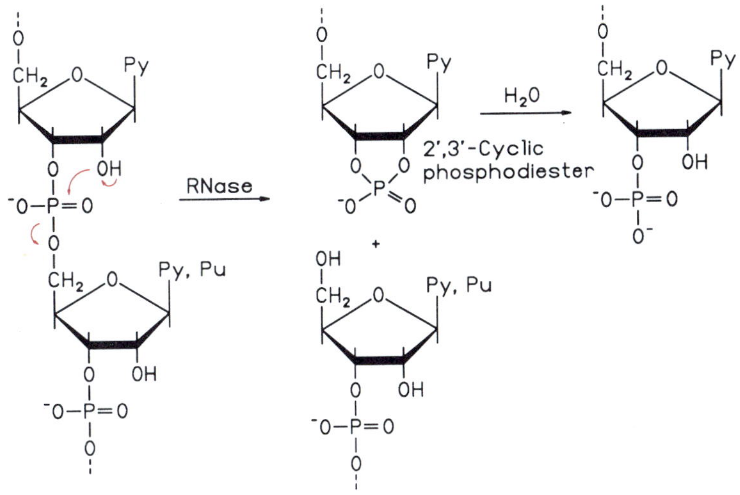

Table 1.6. List of Amino Acid Residues that React with Bromoacetate and Iodoacetate

Amino Acid Residue	Site of Carboxymethylation
Histidinyl	Carbons 1 and 3 of the ring
Lysyl	ε Amino group
Cysteinyl	Sulfhydryl group
Methionyl	S, to form sulfonium ion
Glutamyl	γ Carboxyl group
Terminal α amino group	α Amino group

Chemical Modification

Chemical modification of exposed amino acids can be useful in identifying the active site of the enzyme. Bromoacetate or iodoacetate are alkylating reagents (contributing carboxymethyl groups) that react selectively with a few amino acid residues in proteins (Table 1.6).

Various R groups are carboxymethylated in different reaction environments (e.g., pH, concentration of reagents). The microenvironment of residues in the protein frequently also provides selectivity.

When RNase is incubated at pH 5.5 with bromoacetate, enzymatic activity is lost, and the enzyme is alkylated in one of two major positions, His-12 or His-119 (Figure 1.39). Apparently, the two His residues are so close to one another that alkylation of one residue sterically hinders alkylation of the second one. The loss of enzymatic activity suggests that the two His residues may be close to the active site.

Active Site Studies with Enzyme Inhibitors

If the two reactive histidine residues are located at the active site of the enzyme, then compounds that resemble the substrate but are not chemically transformed (i.e., competitive inhibitors) should protect the active site from inactivation from bromoacetate. This expectation was found to be true for competitive inhibitors of RNase with low K_i values. In other words, the formation of stable enzyme-inhibitor complexes prevents alkylation. We can conclude, therefore, that the two His residues, although separated by over 100 amino acids in the primary protein structure, are instrumental in the catalytic mechanism and thus are probably part of the active site.

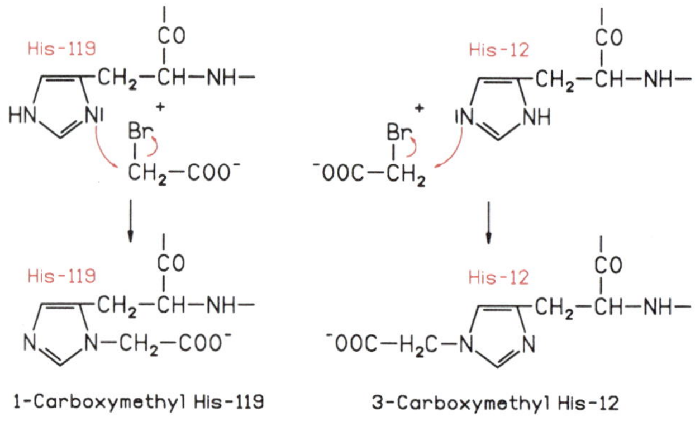

Figure 1.39.
Carboxymethylation of RNase with bromoacetate.

Enzyme Conformation

It was stated previously that the conformation of a protein is determined by its amino acid sequence. This can be demonstrated with RNase by disrupting its four disulfide bonds through reduction with mercaptoethanol and by disrupting its hydrogen bonds by high concentrations of urea (8*M*) or guanidine hydrochloride. Gradual removal of the denaturing agents, e.g., by dialysis, allows the protein to renature to its thermodynamically most stable conformation. Catalytic activity often returns with renaturation.

Once folded into its catalytic conformation, the polypeptide chain may even be broken without loss of activity. For example, the bacterial protease subtilisin cleaves RNase A specifically between residues 20 and 21. The product, called RNase S′, retains enzymatic activity because the two peptide chains are still held together by noncovalent bonds. The two chains can be separated by gel filtration in the presence of acetic acid. When mixed together again, the chains reassociate and form an active enzyme. This experiment demonstrates that enzyme conformation is not entirely dependent on an intact primary structure. Once formed, different parts of an enzyme can recognize and combine with one another because of their complementary three-dimensional shapes.

$$\underset{\text{(active)}}{\text{RNase A}} \xrightarrow{\textit{Subtilisin}} \underset{\text{(active)}}{\text{RNase S}'} \rightleftharpoons \underset{\text{(inactive)}}{\text{S protein} + \text{S peptide}}$$

The three-dimensional structure of RNase derived from X-ray diffraction studies confirms the positions of certain amino acid residues ascertained from chemical studies in solution. In addition to finding His-12 and His-119 residues at the active site, two lysine residues, Lys-7 and Lys-41, have also been found to be very close to each other. (They will cross-link with bifunctional reagents.)

The conformation of a protein in solution appears to be similar to that present in a crystal. This can be concluded from the fact that the catalytic activity of enzyme crystals is almost the same as that observed in solution, after appropriate corrections are made for the restricted diffusion of substrate and product in and out of the crystal lattice.

1.11.2 Chymotrypsin

Chymotrypsin is a **serine protease** that hydrolyzes polypeptides at aromatic amino acid residues. It is first synthesized as an inactive enzyme precursor, a zymogen, called chymotrypsinogen.

Enzyme Activation

Chymotrypsinogen is synthesized and secreted by the pancreas. It is a single polypeptide chain containing five disulfide bridges. Activation to chymotrypsin is catalyzed by trypsin, which cleaves a single peptide bond between amino acids 15 and 16. The resulting π **chymotrypsin** autocatalytically cleaves three additional peptide bonds, yielding three polypeptide chains of chymotrypsin (A, B, and C) that are held together by disulfide bonds. Only the first tryptic cut at site T (Figure 1.40) is

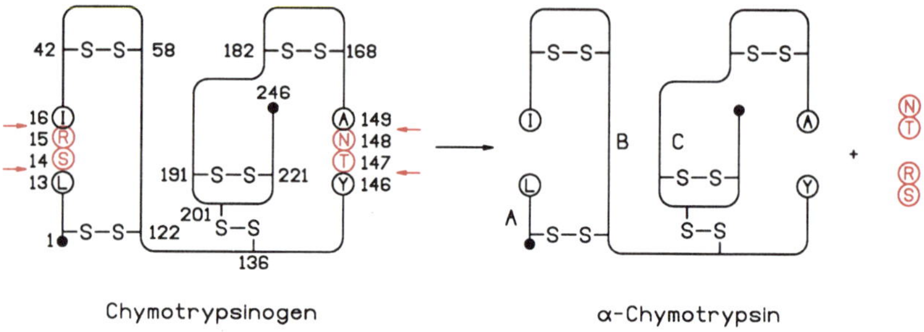

Figure 1.40.
Activation of chymotrypsin by trypsin. (Redrawn from Hartley BS, et al. *Structure and Activity of Enzymes.* New York, NY: Academic Press.)

essential for enzymatic activity, since the Ile-16 residue is required as a new amino terminus for the conformation of the active site.

Covalent Enzyme-Substrate Intermediate

Chymotrypsin also hydrolyzes esters of aromatic amino acids. A useful substrate for studying the enzyme's mechanism of action is *p*-nitrophenyl acetate. Although a very poor substrate for chymotrypsin, the colorless substance can be used for an extremely sensitive assay since the reaction product, *p*-nitrophenol, is intensely yellow in color. The time course of hydrolysis of *p*-nitrophenyl acetate is biphasic, with an initial burst before reaching steady state (Figure 1.41).

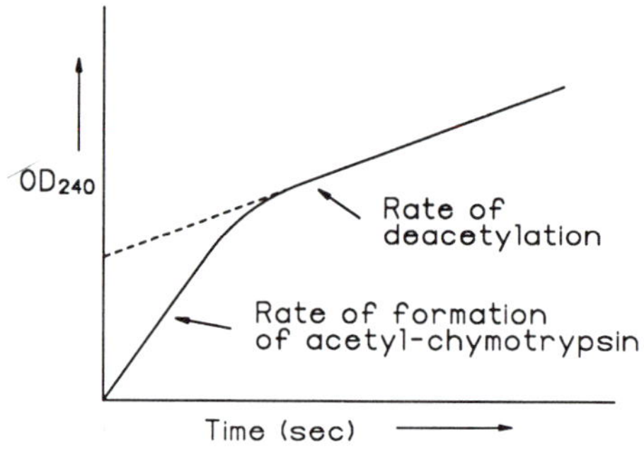

Figure 1.41.
Time course of hydrolysis of *p*-nitrophenyl acetate by chymotrypsin. (After Spencer T, Sturtevant JM. *J Am Chem Soc.* 1959; 81:1874.)

This initial burst of *p*-nitrophenol production is interpreted as the result of a fast reaction that is followed by a relatively slow dissociation of the enzyme-product complex, i.e., $k_3 \gg k_5$ in the following reaction scheme:

$$\mathrm{E} + \mathrm{CH_3{-}CO{-}X} \underset{k_2}{\overset{k_1}{\rightleftharpoons}} \underset{\text{(ES)}}{\mathrm{CH_3{-}CO{-}E{:}X}}$$

$$\underset{k_4}{\overset{k_3}{\rightleftharpoons}} \underset{\text{(ES*)}}{\mathrm{CH_3{-}CO{-}E}} + \mathrm{XH} \underset{k_6}{\overset{k_5}{\rightleftharpoons}} \mathrm{E} + \mathrm{CH_3{-}COOH}$$

where X = *p*-nitrophenyl residue.

ES*, the acetylated enzyme, probably represents a large proportion of total E. The acetylated enzyme has been isolated. It has the acetyl group covalently attached to a single serine residue (Ser-195). This serine thus seems to play a critical role in the catalysis occurring at the active site of the enzyme.

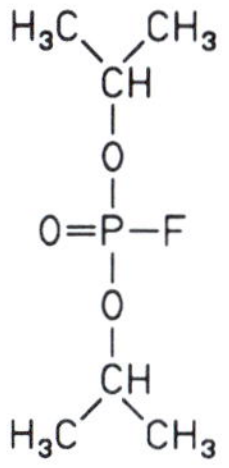

Figure 1.42.
Diisopropyl fluorophosphate.

Irreversible Inhibitors

Diisopropyl fluorophosphate (Figure 1.42) irreversibly inhibits chymotrypsin by covalently modifying the Ser-195 residue. The other 27 serine residues in the chymotrypsin molecule do not react with DIPF. In the presence of substrate or inhibitor, the Ser-195 residue is protected from reacting with DIPF. This suggests that Ser-195 is especially reactive and may function in the active site of the enzyme.

Affinity Labeling

Substrate analogs with reactive substituents might be expected to bind at the catalytic site and react with neighboring amino acids of the enzyme. This is called **affinity labeling** and is demonstrated by the chloromethylketone analog of phenylalanine, which binds to the active site because of its resemblance to a phenylalanine ester. However, the —CO—CH_2Cl group is not hydrolyzed. Instead, it reacts with the imidazole ring of the neighboring His-57 residue, thus implicating His-57 in the active site (Figure 1.43).

Mechanism of Action

X-ray diffraction analysis of the crystal structure of chymotrypsin suggests the possibility of a **charge relay system** between the amino acids Asp-102, His-57, and Ser-195 (Figure 1.44). This system is thought to be responsible for making the Ser-195 residue especially reactive through the formation of an alkoxide ion.

Homologous Enzymes and Evolution

The conformation of several proteases from widely different organisms is surprisingly similar. For example, **chymotrypsin, trypsin, thrombin, bacterial α-lytic protease,** and ***Streptomyces* trypsin** all exhibit almost superimposable three-dimensional structures with only minor changes in surface loops. Although their primary sequence homology is low (about 40%), all these enzymes contain a uniquely reactive serine residue in the sequence *-Asp-Ser-Gly-* and a nearby His residue in the sequence *-Ala-Ala-His-Cys-*. They are thus capable of forming the Asp···His···Ser charge relay system. Throughout evolution these **serine**

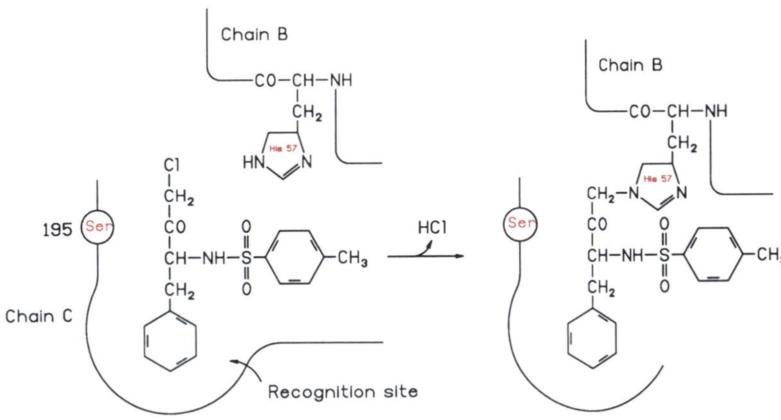

Figure 1.43.
Affinity-labeling of the active site of chymotrypsin by a chloromethylketone analog of phenylalanine.

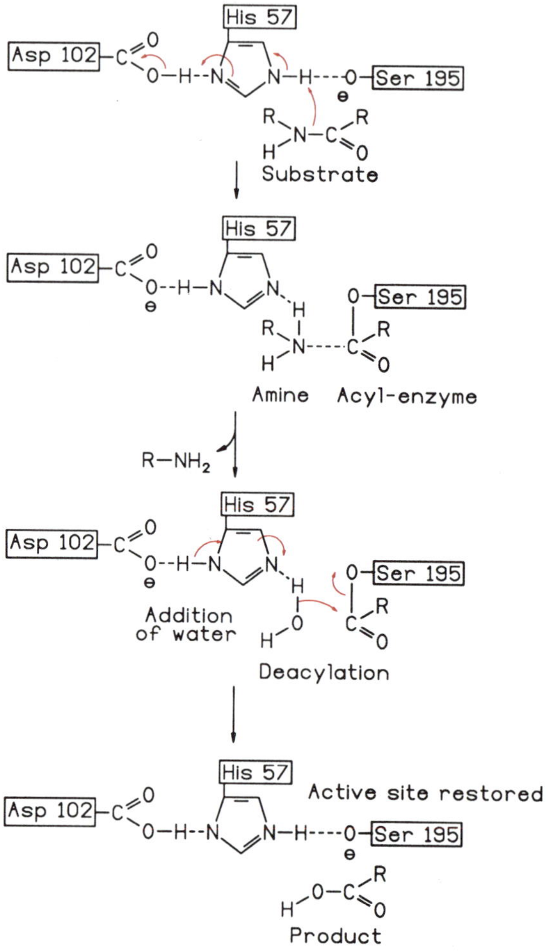

Figure 1.44.
The charge relay system at the active site of chymotrypsin.

proteases, as they have been named, seem to have preserved their basic conformation and their mechanism of action while exhibiting very different substrate specificities.

1.11.3 Lactate Dehydrogenase

Lactate dehydrogenase catalyzes the reversible dehydrogenation of lactate by NAD^+ (see Chapter 2).

$$\text{L-Lactate} + NAD^+ \rightleftharpoons \text{Pyruvate} + NADH + H^+$$

Isozymes and Subunit Structure

The enzyme is a tetramer of four subunits, each of molecular weight 35,000. There are two genes coding for two distinct subunits, designated M and H. Differential expression of these genes in different tissues produces five combinations of M and H subunits in the tetrameric enzyme. The resulting **isoenzymes,** M_4, M_3H, M_2H_2, MH_3, and H_4, can be separated electrophoretically because M and H chains have many amino acid differences and consequently differ considerably in charge.

All five isoenzymes are found in skeletal muscle in approximately equal amounts. In contrast, the predominant enzyme in heart muscle

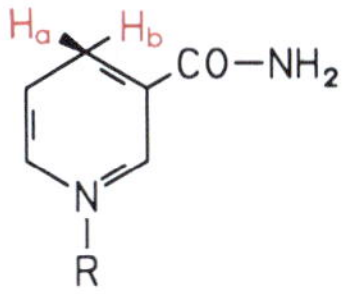

Figure 1.45.
Prochiral center in NADH.

is H_4, and liver cells contain mostly M_4. The isoenzymes exhibit different kinetic parameters depending on their content of H and M subunits. the M_4 isoenzyme, for example, has a low K_m value and a high V_{max} for pyruvate, while the H_4 enzyme has a high K_m and low V_{max}. The significance of this difference is obscure.

Stereospecificity at Prochiral Centers

Carbon atoms with two identical substituents are called **prochiral centers** (asymmetric carbon atoms are called chiral centers). The carbon 4 position in the pyrimidine ring of NADH is a prochiral center (Figure 1.45). Lactate dehydrogenase, which uses NADH as a cosubstrate, is capable of distinguishing between the two hydrogen atoms attached to this carbon atom even though the molecule has bilateral symmetry. The enzyme will always transfer the H_a hydrogen to its substrate.

All known lactate dehydrogenases possess this preference for the H_a hydrogen of NADH. This suggests a common ancestral protein from which these enzymes are derived, with a catalytic mechanism that has been conserved throughout evolution. Malate, isocitrate, and D-glycerate dehydrogenases, which use the same cosubstrate, show the same specificity for the H_a hydrogen, whereas glucose-6-phosphate, glutamate, 6-phosphogluconate, and 3-phosphoglyceraldehyde dehydrogenases select the H_b hydrogen.

1.11.4 Aspartate Transcarbamoylase

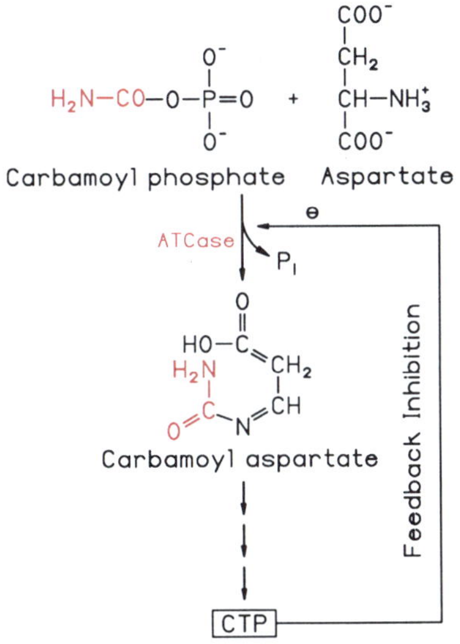

Figure 1.46.
The reaction catalyzed by aspartate transcarbamoylase.

Aspartate transcarbamoylase (ATCase) is one of the first enzymes in the biosynthetic pathway of pyrimidine nucleotides (see Chapter 5). It is inhibited by cytidine triphosphate (CTP), one of the end products of the pathway, and is activated by adenosine triphosphate (ATP), the end product of the biosynthetic pathway of purine nucleotides (Figure 1.46).

Subunit Structure and Function

ATCase (of *Escherichia coli*) contains six **catalytic subunits** (protein C) and six **regulatory subunits** (protein **R**), for a total of 12 subunits and a total molecular weight of 310 kd. In the presence of Hg^{2+} ions, the active enzyme dissociates into two C_3 aggregates (each consisting of three polypeptide chains of 34 kd) and three R_2 aggregates (each containing two polypeptide chains of 17 kd). In the absence of Hg^{2+} ions, the catalytic and regulatory subunits recombine to form the native enzyme (Figure 1.47). The C_3 aggregate is enzymatically active but no

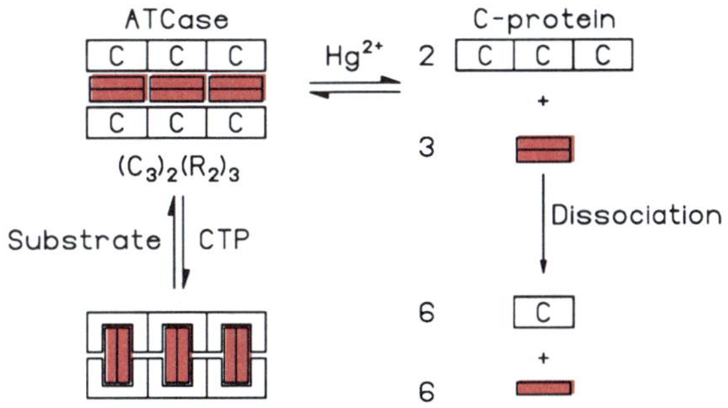

Figure 1.47.
Subunit structure of aspartate transcarbamoylase.

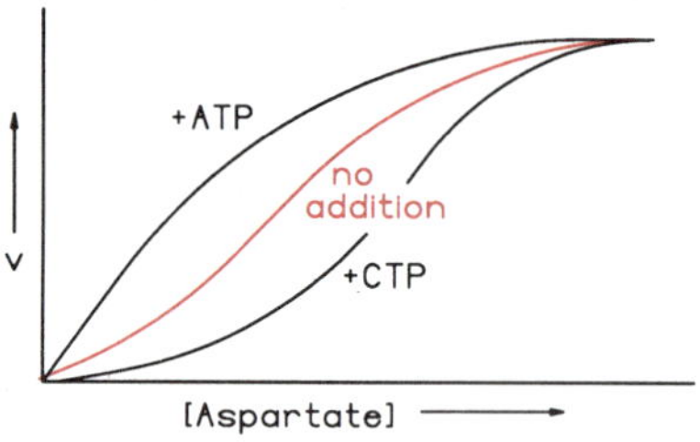

Figure 1.48.
Effects of ATP and CTP on the kinetics of aspartate transcarbamoylase.

longer responsive to regulation by CTP or ATP. (R_2 binds CTP or ATP). It shows normal Michaelis-Menten kinetics. In contrast, the complete enzyme $[(C_3)_2\text{-}(R_2)_3]$ exhibits sigmoidal kinetics.

Allosteric Regulation

The two regulatory effectors CTP and ATP compete for the same binding sites on the regulatory subunits of the enzyme. In the presence of the inhibitor CTP, the sigmoidicity of the velocity-substrate concentration curve is enhanced. The activator ATP, on the other hand, abolishes the sigmoidicity. In the absence of any of these heterotropic modulators, the subunits still exhibit some homotropic cooperativity as indicated by the sigmoidal kinetics. Note that only K_m is affected by ATP or CTP, whereas V_{max} remains constant (Figure 1.48).

The binding of CTP to the regulatory subunits causes a measurable conformational change in the enzyme. The active enzyme carries exposed SH groups, whereas the inhibited enzyme has its SH groups masked and is more compact, as indicated by its greater rate of sedimentation.

For a more extensive discussion of amino acids, proteins, and enzymes, consult the following textbooks:

Brendon C, Tooze G. (1991) *Introduction to Protein Structure.* Garland Publishing, New York.

Mathews CK, van Holde KE. (1990) *Biochemistry.* Benjamin/Cummings, Redwood City: Chapters 5, 6, 7, 10, 11

Stryer L. (1988) *Biochemistry,* 3rd ed. Freeman, New York: Chapters 2, 3, 9, 10

Further References

Anfinsen CB. Principles that govern the folding of protein chains. *Science.* 1973; 181:223–230.

Blake CCF, Johnson LN. Protein structure. *Trends Biochem Sci.* 1984; 9:147–151.

Chou PY, Fasman GD. Empirical predictions of protein structure. *Annu Rev Biochem.* 1978; 47:251–276.

Cohen P. The role of protein phosphorylation in neural and hormonal control of cellular activity. *Nature.* 1982; 296:613–620.

Doolittle R. Proteins. *Sci Am.* 1985; 253:88–96.

Fersht AR, Leatherbarrow RJ, Wells TNC. Binding energy and catalysis: a lesson from protein engineering of tyrosyl-tRNA synthetase. *Trends Biochem Sci.* 1986; 11:321–325.

Harrison SC, Durbin R. Is there a single pathway for the folding of a polypeptide chain? *Proc Natl Acad Sci USA.* 1985; 82:4028–4030.

Karplus M, McCannon JA. The dynamics of proteins. *Sci Am.* 1986; 254:42–51.

Kyte J, Doolittle RF. A simple method for displaying the hydropathic character of a protein. *J Mol Biol.* 1982;57:105–132.

Levitt M. Protein conformation dynamics and folding by computer simulation. *Annu Rev Biophys Bioeng.* 1982; 11:251–271.

Monod J, Changeux JP, Jacob F. Allosteric proteins and cellular control systems. *J Mol Biol.* 1963; 6:306–329.

Richards FM. The protein folding problem. *Sci Am.* 1991; 264:54–63.

Rose GD, Geselowitz AR, Lesser GJ, Lee RH, Zehfus MH. Hydrophobicity of amino acid residues in globular proteins. *Science.* 1985; 229:834–838.

Rothman JE. Polypeptide chain binding proteins: catalysts of protein folding and related processes in cells. *Cell.* 1990; 59:591–601.

Wallace BA, Cascio M, Mielke DL. Evaluation of methods for the prediction of membrane protein secondary structures. *Proc Natl Acad Sci USA.* 1986; 83:9423–9427.

Wetzel R. Harnessing disulfide bonds using protein engineering. *Trends Biochem Sci.* 1988; 12:478–482.

CHAPTER

2 Carbohydrate Metabolism and Energy Supply

OBJECTIVES

After studying this chapter, the student should be able to

1. Understand the thermodynamic principles that govern biochemical reactions and predict the direction of a reaction from the change in the free energy of its participants.
2. Explain the relationship between catabolic and anabolic reactions.
3. Understand the terms "high-energy compound" and "high-energy bond."
4. Understand the roles of ATP, NAD^+, and FAD in catabolism and those of ATP and NADPH in anabolism.
5. List the coenzymes that act as carriers of activated groups or molecules in metabolic reactions.
6. Draw both the linear and cyclic structures of glucose.
7. Define the term aldose, ketose, pyranose, furanose, D and L sugar, epimer, anomer, mutarotation, deoxysugar, amino sugar, -onic acid, -uronic acid, hemiacetal, acetal, glycoside, reducing end, nonreducing end.
8. Name all intermediates, end products, and enzymes of the aerobic and anaerobic glycolysis of glucose.
9. Recognize the high-energy intermediates in glycolysis and explain the term "substrate level oxidative phosphorylation."
10. Explain the Pasteur effect and its relation to the control of glycolysis.
11. State the cellular location of the enzymes involved in glycolysis.
12. Understand the link between glyceraldehyde-3-phosphate dehydrogenase and lactate dehydrogenase during anaerobic glycolysis.
13. Explain how electrons are shuttled from the cytoplasm into the mitochondrion.
14. Understand the physiologic basis for the different affinities for glucose of glucokinase and hexokinase.
15. Draw all the structures of the citric acid cycle intermediates.
16. Name the enzymes involved in the citric acid cycle and state their cellular location.
17. Identify the enzymatic steps in the citric acid cycle that are subject to regulation.
18. Give examples of amphibolic and anaplerotic reactions.
19. Explain why in thiamine deficiency the levels of pyruvate and lactate in blood are elevated.
20. Trace the fate of the individual carbon atoms when glucose is converted into acetyl CoA and CO_2.
21. Understand why most of the fat in humans cannot be converted into carbohydrate.
22. Understand the types of reaction that are catalyzed by kinases, mutases, isomerases, aldolases, and dehydrogenases.

23. Describe the components and functions of the complexes of the electron-transport chain.
24. State the locations of the electron-transport enzymes within the mitochondrion.
25. Describe Mitchell's chemiosmotic theory of oxidative phosphorylation.
26. Name some "uncouplers" of oxidative phosphorylation and explain their actions.
27. Identify the steps that contribute to the generation of 36 mol of ATP during the complete oxidation of 1 mol of glucose.
28. Understand the participation of inorganic ions in the conversion of glucose into CO_2 and water.
29. Distinguish the reactions catalyzed by dehydrogenases, oxidases, and oxygenases.
30. Understand the metabolic significance of the two arms of the pentose phosphate pathway.
31. Name the principal tissues where both arms of the pentose phosphate pathway function and those where only the nonoxidative arm functions.
32. Understand the purpose of gluconeogenesis.
33. Describe the reactions and enzymes involved in gluconeogenesis, and name those that are different from glycolysis.
34. Understand the interplay between the citric acid cycle and gluconeogenesis, and describe the events that may lead to ketosis.
35. Explain the mechanisms of contrary regulation of glycolysis and gluconeogenesis.
36. Remember the pathways of carbohydrate interconversions.
37. Name the high-energy intermediates involved in the acetylation, sulfation, phosphorylation, and nucleotidylation of sugars.
38. Name the amino acids and monosaccharides found in protein-carbohydrate linkages in humans.
39. Describe the basic structure of glycogen and the mechanism of its biosynthesis and degradation.
40. Explain the mechanisms that regulate glycogen synthesis and degradation.

This chapter deals with the molecular mechanisms that allow a cell to derive energy from the degradation of carbohydrates, an important component of food, and to use this energy for the synthesis of cellular constituents necessary for the maintenance of life. The metabolic pathways for the degradation of lipids and proteins, the other two major sources of energy, will be covered in the subsequent two chapters. Their points of convergence with the carbohydrate metabolism, however, will be indicated here. In addition, the structures and mechanisms of synthesis of complex carbohydrates are discussed.

2.1 INTRODUCTION TO METABOLISM

The cell extracts energy from the components of food by oxidizing them via a complicated network of intermediates to the end products CO_2 and water (Figure 2.1). This process, called **catabolism**, generally occurs in three stages:

1. Degradation of macromolecules (starch, protein, and fat) to monomers (monosaccharides, amino acids, and fatty acids, respectively).

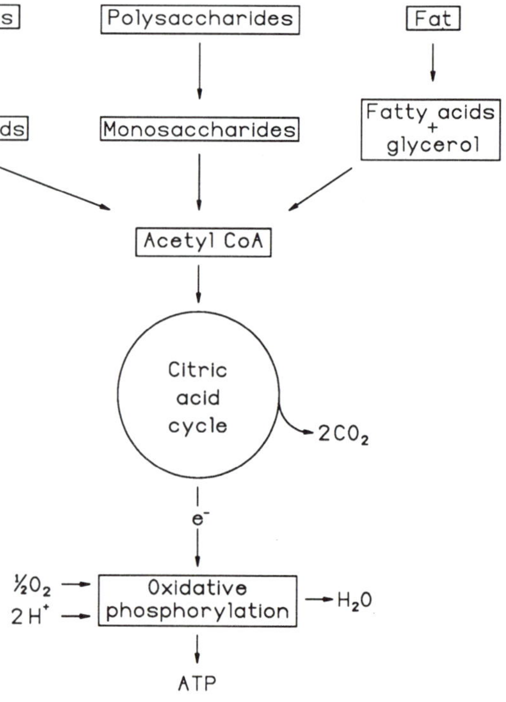

Figure 2.1.
The generation of metabolic energy in the form of ATP.

2. Degradation of these monomers to the common metabolic intermediate acetyl coenzyme A (CoA).
3. Oxidation of acetyl CoA in the citric acid cycle to CO_2 and H_2O, a process that generates energy in the form of the high-energy compound adenosine triphosphate (ATP).

2.1.1 Thermodynamic Principles

In order to understand the meaning of the term **high-energy compound**, it is necessary to review briefly the thermodynamic principles that underlie metabolic reactions. A chemical reaction will occur spontaneously only if the **free energy change** ΔG of the system is negative. The **Gibbs law** states that

$$\Delta G = \Delta H - T\Delta S \qquad (2.1)$$

The **enthalpy change**, ΔH, is the change of the "inner energy" of the system, ΔE (which is negative if the system generates heat and positive if the system consumes heat), minus the energy the system has lost in the form of work performed, as expressed by the product of pressure times volume change, $P\Delta V$.

$$\Delta H = \Delta E - P\Delta V \qquad (2.2)$$

T is the absolute temperature in Kelvin, and ΔS is the **entropy change** of the system. (Entropy is a function of the system's state of order; it increases with increasing disorder.)

The ΔG of a biochemical reaction, for example, the oxidation of glucose to CO_2 and H_2O, is independent of the pathway followed; it is determined solely by the difference between the free energies of the reactants and those of the reaction products.

The free energy changes associated with different reactions may be compared by measuring ΔG under "standard conditions" (1 *M*, 25°C, 1 atm). This **standard free energy change**, ΔG^{0}, is defined as the free energy generated in a reaction where all participants are present at molar concentrations. Biochemists use the term $\Delta G^{0\prime}$, which is the standard free energy change observed at pH 7.0.

ΔG varies with the concentration (C) of the reactants according to the equation

$$\Delta G = \Delta G^{0\prime} + RT \ln C \qquad (2.3)$$

[Note that if all reactants are present at molar concentrations, the second term in equation (2.3) disappears, so that $\Delta G = \Delta G^{0\prime}$.]

For the chemical reaction

$$A + B \rightleftharpoons C + D$$

equation (2.3) becomes

$$\Delta G = \Delta G^{0\prime} + RT \ln \frac{[C][D]}{[A][B]} = \Delta G^{0\prime} + RT \ln K' \qquad (2.4)$$

where K' is the equilibrium constant of the reaction, and R is the gas constant. (The square brackets designate concentrations.)

At equilibrium, ΔG is zero, so that

$$\Delta G^{0\prime} = -RT \ln K' = -2.303\, RT \log K \qquad (2.5)$$

Since at first approximation

$$K' \sim \frac{10^{\Delta G^{0\prime}}}{10}$$

it is easy to estimate roughly the equilibrium constant of a reaction from its $\Delta G^{0\prime}$, or vice versa. For example, the $\Delta G^{0\prime}$ for the hydrolysis of ATP to adenosine diphosphate (ADP) and inorganic phosphate (see below) is -7.3 kcal. That means that at equilibrium there will be about a million molecules of ADP per molecule of ATP.

Reactions with a negative ΔG are called **exergonic**, and those with a positive ΔG are called **endergonic**. Endergonic reactions do not occur spontaneously; in order to proceed, they have to be coupled to an exergonic reaction so that the overall ΔG of the system becomes negative. Reactions with a slightly positive $\Delta G^{0\prime}$ may still occur spontaneously if the concentrations are changed, to make ΔG negative [see equation (2.4)].

2.1.2 High-Energy Compounds

The hydrolysis of ATP is accompanied by a very large negative change in free energy.

$$\text{ATP} + H_2O \rightarrow \text{ADP} + P_i + H^+ \qquad \Delta G^{0\prime} = -7.3 \text{ kcal}$$

This large $\Delta G^{0\prime}$ is not only due to the energy that is stored in the P—0 bond between the phosphate groups of ATP but also to the fact that the free phosphate ion is resonance-stabilized, as compared with a phosphate group in ATP, and thus is at a lower energy state. In addition, the many negative charges in ATP tend to repel each other and thus make the molecule less stable.

Because of its high $\Delta G^{0\prime}$ value for hydrolysis and high potential for transferring its phosphate groups to other molecules, ATP is often referred to as a high-energy compound. When using this term we should remember, however, that a carbon-carbon bond represents considerably more free energy ($\Delta G^{0\prime} \sim 100$ kcal/mol), but that this energy is not readily available because of the high activation energy required for breaking this bond. In contrast, ATP, although relatively stable in aqueous solution, is readily hydrolyzed enzymatically. Thus, in reactions that are coupled to the hydrolysis of ATP, the reaction equilibrium is shifted to the right (product side).

The standard free energy changes observed during the hydrolysis of different metabolic intermediates are listed in Table 2.1 There is a trend toward expressing energy in kilojoules (kJ) rather than in kilocalories (kcal). Since this unit has not yet been generally accepted, the kilo-

Table 2.1 Free Energy Changes Observed during the Hydrolysis of Biologically Important Molecules

Compound	ΔG^0 (kcal/mol)
Phosphoenolpyruvate	−14.8
Cyclic AMP	−12.0
Acetyl phosphate	−10.3
Creatine phosphate	−10.3
S-adenosyl methionine	−10.0
Acetyl CoA	−7.5
ATP (hydrolysis to ADP + P_i)	−7.3
UDP-glucose	−7.3
Glucose 1-phosphate	−5.0

Note: AMP = adenosine monophosphate, UDP-glucose = uridine diphosphate glucose.

calorie will be used as the unit for energy in the examples presented in this book.

$$1 \text{ kcal} = 4.184 \text{ kJ} \ (1 \text{ J} = 1 \text{ kg} \cdot \text{m}^2 \cdot \text{s}^{-2})$$

2.1.3 Electron Carriers and Reducing Agents

The synthesis of cellular constituents is called anabolism. Since many of these compounds usually are more complex and more reduced than their precursors, this process requires an input of energy, which it obtains by being coupled to the hydrolysis of ATP. It also requires the participation of a reducing agent such as reduced nicotinamide adenine dinucleotide phosphate (NADPH) (Figure 2.2).

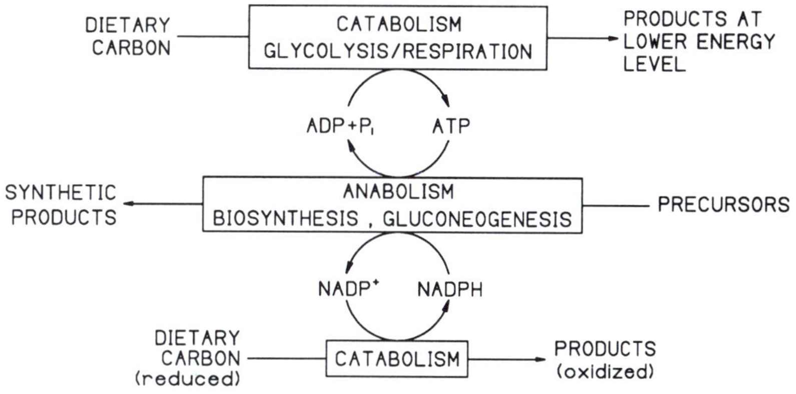

Figure 2.2. Interplay between anabolic and catabolic processes. (Modified from Whelan WJ. Biochemistry lecture notes. Miami FL: University of Miami; 1982, mimeographed.)

NADPH and **reduced nicotinamide adenine dinucleotide (NADH)** are formed from **NADP$^+$** and **NAD$^+$** (see Figure 2.6), respectively, during the oxidation of various substrates. (For an explanation of the nature of oxidation-reduction reactions, see Section 2.5.) Each molecule of NAD$^+$ (or NADP$^+$) accepts a hydrogen ion (proton) and two electrons. NAD$^+$ is the major cellular electron acceptor (Figure 2.3). The resulting NADH is used almost exclusively for the production

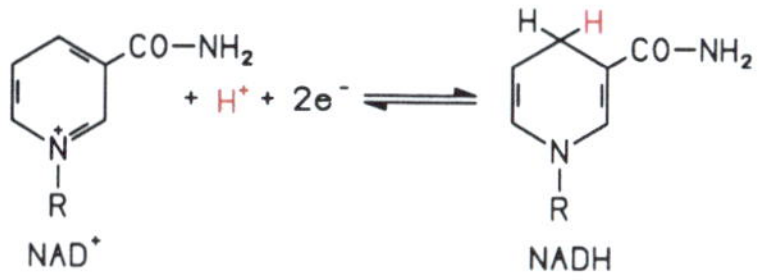

Figure 2.3. Electron transfer between pyridine nucleotides.

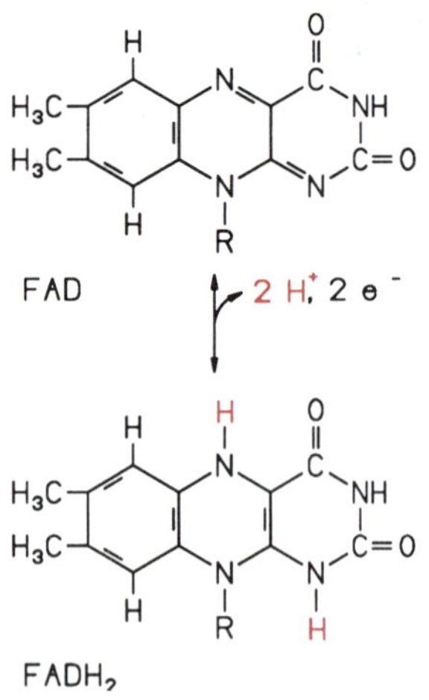

Figure 2.4.
Electron transfer between flavin nucleotides.

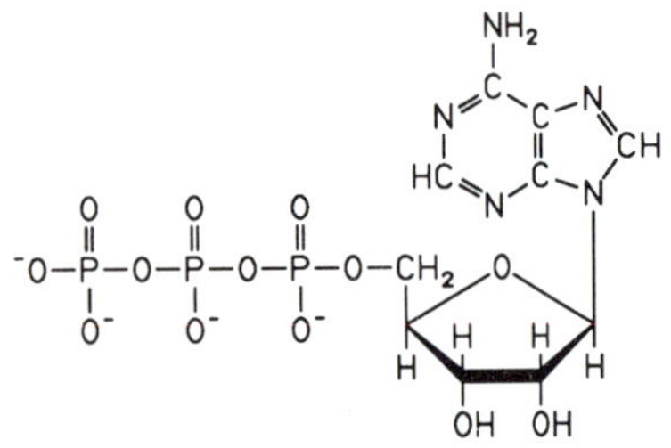

Figure 2.5.
Adenosine 5′-triphosphate (ATP).

of ATP in a process called **oxidative phosphorylation**. In contrast, NADPH is the primary electron donor for the biosynthesis of cellular constituents.

Reduced pyridine nucleotides, NADH or NADPH, absorb light at 340 nm while the oxidized pyridine nucleotides do not. This property provides the basis for a very convenient photometric assay. Any NAD^+- or $NADP^+$-coupled reaction can be monitored directly by measuring the change in absorption at 340 nm.

A third major electron carrier is flavin adenine dinucleotide (FAD). It accepts two electrons as well as two hydrogens (Figure 2.4).

2.1.4 Coenzymes

In addition to electrons and hydrogen ions, many other groups, or building units, are activated by being attached to a variety of carrier molecules. This activation requires the expenditure of energy, derived directly or indirectly from the hydrolysis of ATP. The carrier then delivers, in an exergonic reaction, the building unit to the molecule that is being synthesized. The overall ΔG of this reaction usually is so negative that the equilibrium lies far on the side of the synthesized products. The carriers act as **coenzymes,** or **prosthetic groups** on enzymes, in a large number of biochemical reactions. Many coenzymes are derived from **water-soluble vitamins**. Table 2.2 lists important coenzymes, some of which will be discussed only in later chapters.

Adenosine triphosphate (Figure 2.5) is the most important carrier of phosphate groups. Hydrolysis of both of its phosphoanhydride bonds yields a large amount of free energy. ATP can be regenerated from two molecules of ADP by the enzyme adenylate kinase (myokinase).

Nicotinamide adenine dinucleotide (**NAD^+**) and its phosphorylated form (**$NADP^+$**) are the major cellular electron acceptors (Figure 2.6; see also Figure 2.3).

Table 2.2. Important Coenzymes

Carrier Molecule	Vitamin	Group Transferred
ATP		Phosphate
NADH, NADPH	Nicotinate (niacin)	Electrons
$FADH_2$	Riboflavin (vitamin B_2)	Electrons
Ubiquinone		Electrons
Coenzyme A	Pantothenate	Acetyl, fatty acyl
Lipoamide		Acetyl, succinyl
Thiamine-PP	Vitamin B_1	Aldehyde
Biotin	Biotin	CO_2
Tetrahydrofolate	Folate	C_1 units
S-adenosyl methionine		Methyl
Pyridoxal phosphate	Vitamin B_6	Amino groups
UDP-sugar		Sugars
CDP-diacylglycerol		Phosphatidic acid
Cobalamin	Vitamin B_{12}	Methyl, molecular rearrangements
Carnitine		Long-chain fatty acyl
Dolichol phosphate		Oligosaccharide

Note: Some of these coenzymes will be discussed in later chapters.
CDP = cytosine diphosphate.

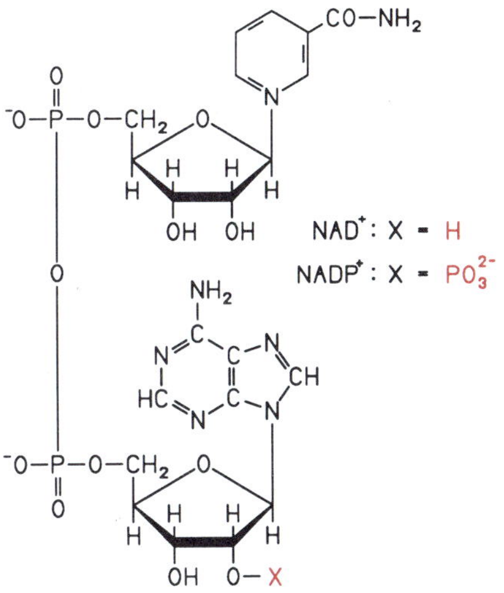

Figure 2.6. Nicotinamide adenine dinucleotide (NAD^+) and its phosphorylated form ($NADP^+$).

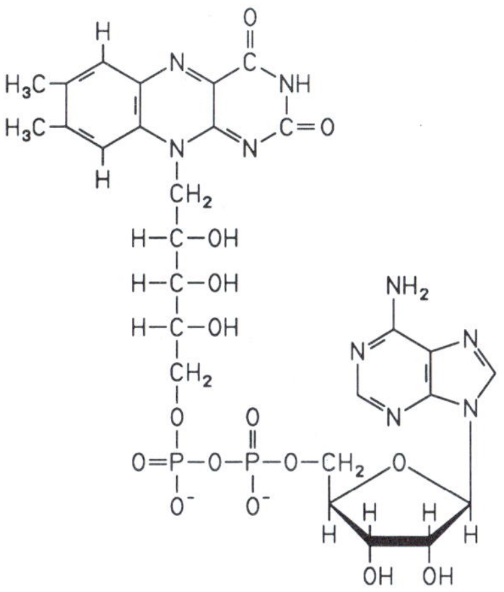

Figure 2.7. Flavin adenine dinucleotide (FAD).

Flavin adenine dinucleotide (FAD) (Figure 2.7) functions as an electron acceptor in many dehydrogenases. In some of these enzymes, for example, succinate dehydrogenase, the flavin nucleotide is covalently linked to the protein through one of its methyl groups.

Coenzyme A (CoA) (Figure 2.8) is the most important carrier of acyl groups. It forms thioesters with carboxylic acids and participates in the degradation and biosynthesis of fatty acids. Acetyl CoA is the final common intermediate in the catabolism of all major food components.

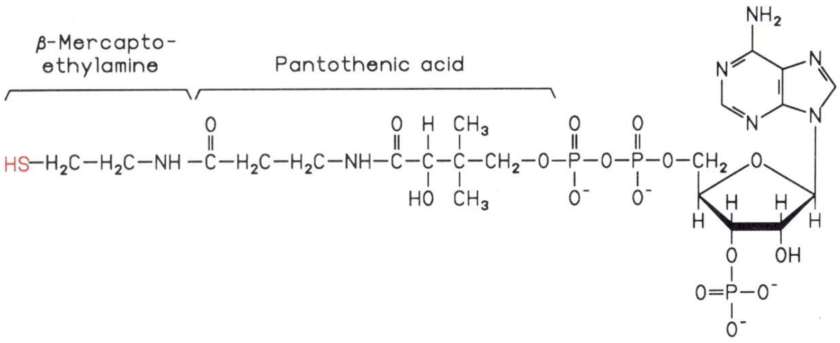

Figure 2.8. Coenzyme A.

Lipoamide (Figure 2.9) consists of lipoic acid linked through a peptide bond to the ϵ amino group of a lysine residue in dihydrolipoyl transacetylase, an enzyme that is part of the pyruvate dehydrogenase complex (see Chapter 1). It can carry an acyl group on its proximal SH group and transfer it to CoA.

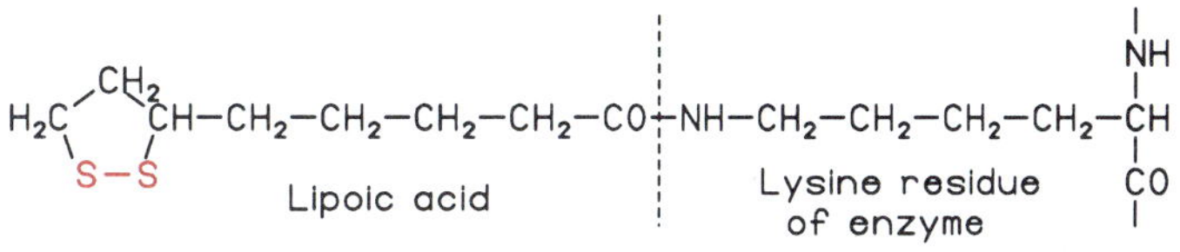

Figure 2.9. Lipoamide.

Thiamine pyrophosphate (TPP) (Figure 2.10) is a carrier of aldehydes and also functions as a carrier of pyruvate during its decarboxylation in the pyruvate dehydrogenase complex. (Additional functions of TPP will be discussed in Section 2.8.)

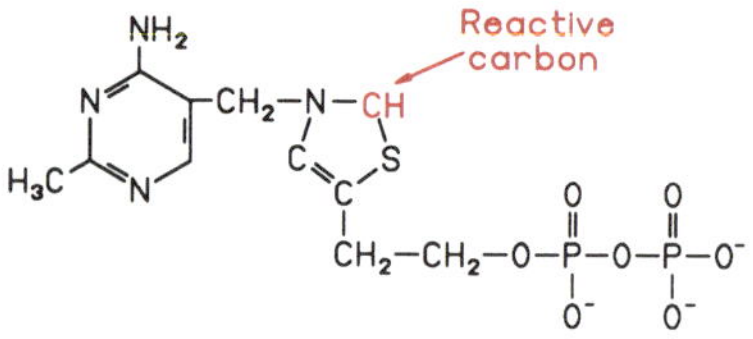

Figure 2.10.
Thiamine pyrophosphate.

Biotin (Figure 2.11), the carrier of CO_2, functions in many carboxylation reactions. It usually is covalently bound to the enzyme by an amide linkage to a lysine side chain.

2.1.5 Regulation of Metabolism

Cellular metabolism, catabolic as well as anabolic, must be responsive to changes in the cell's requirements for energy or biosynthetic intermediates and thus be subject to regulation. This is achieved by controlling the level of key enzymes through de novo protein synthesis and degradation, by regulating the catalytic activities of some enzymes through allosteric interactions with effectors or through covalent enzyme modification, by controlling substrate concentrations, and by using compartmentation and distinct pathways for biosynthesis and degradation.

ATP, ADP, and AMP are the substrates or products of many reactions. Thus, these reactions may be controlled coordinately by the amount of ATP available in the cell, which is often expressed as the **energy charge** of the cell, a strictly mathematical concept that is defined as

$$\text{Energy Charge} = \frac{[\text{ATP}] + \frac{1}{2}[\text{ADP}]}{[\text{ATP}] + [\text{ADP}] + [\text{AMP}]}$$

When the energy charge is high, ATP-generating reactions (catabolic) are inhibited whereas ATP-utilizing pathways (anabolic) are favored. The normal energy charge of most cells lies between 0.80 and 0.95. (Bacteria cannot survive a drop in energy charge to below 0.5.)

Finally, the overall metabolic activity of a complex organism such as a human is regulated by cooperative interactions between different organs and tissues of the body. This is accomplished by the endocrine system, which consists of endocrine glands (e.g., pancreas, thyroid) that respond to metabolic stimuli by secreting hormones into the bloodstream. These hormones, in turn, trigger specific biochemical reactions in the target tissues. Through such mechanisms, glucose, for example, may be synthesized in the liver, released into the bloodstream, and utilized by the brain. Another example is the production of lactate in muscle and its uptake and metabolism by the liver and other organs (see the Cori cycle, Section 2.9).

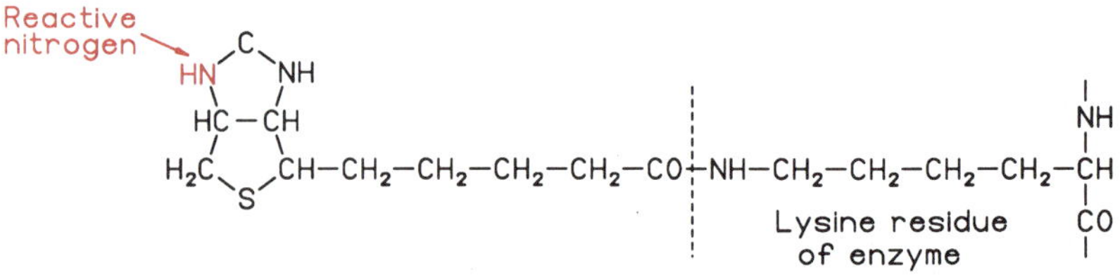

Figure 2.11.
Biotin.

2.2 STRUCTURE AND NOMENCLATURE OF CARBOHYDRATES

Most carbohydrates have the empirical formula $(CH_2O)_n$. (Their name originated from the mistaken belief that carbohydrates were adducts of carbon and water.) Monomeric carbohydrates, also called sugars or **monosaccharides**, can polymerize into **oligosaccharides** and **polysaccharides**. (Compare these with the function of amino acids as monometric units of proteins.)

Chemically, carbohydrates are polyhydroxyaldehydes or polyhydroxyketones. The simplest carbohydrate is D-**glyceraldehyde**, which is called an **aldose** because it contains an aldehyde group. **Dihydroxyacetone**, on the other hand, is called a **ketose** because the molecule carries a keto group. Both compounds are called **trioses** because they are composed of three carbon atoms (Figure 2.12). Tetroses, pentoses, hexoses, and heptoses are higher sugars containing 4, 5, 6, and 7 carbon atoms, respectively.

D-Glyceraldehyde Dihydroxyacetone

Figure 2.12. Structures of the simplest aldose and ketose.

Of the naturally occurring sugars, **glucose**, **mannose**, and **galactose** are aldohexoses, **fructose** is a ketohexose, and **ribose**, found in nucleotides, is an aldopentose.

Sugars (monosaccharides) contain one or more asymmetric carbon atoms and thus are optically active. The assignment of the D or L configuration to a sugar is determined only by the configuration of the asymmetric carbon atom farthest removed from the aldehyde or keto group. (Note that this assignment is not related at all to the optical rotational properties of the sugar.) In the linear **Fischer projection** of a sugar, this carbon atom must be rotated so that its bonds with the neighboring carbon atoms project into the plane of the paper, whereas the bonds to the hydrogen and hydroxyl groups reach out of it. By convention, the sugar has the D configuration if, after this arrangement, the hydroxyl group is located to the right of the carbon atom.

The predominant form of sugars in solution is not the linear form but a ring structure that arises from the interaction of the aldehyde or keto group with a hydroxyl group in the same molecule (Figure 2.13). Depending on the size of the ring, the structure is called a **pyranose** (six-member) or a **furanose** (five-member). Ring formation creates an additional asymmetric (anomeric) center derived from the most oxidized carbon atom. Glucose thus forms two **anomers**, the α and β form. In solution, these two forms slowly interconvert because of their equilibrium with the open form (0.02%). This process is called **mu-**

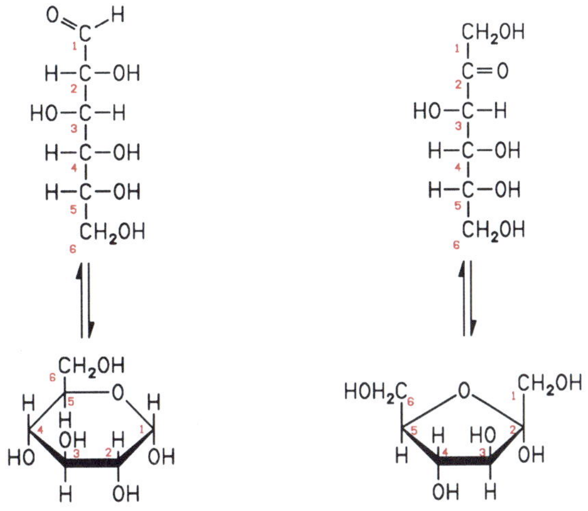

α-D-Glucose α-D-Fructose

Figure 2.13. Open and ring structures of an aldohexose (glucose) and a ketohexose (fructose).

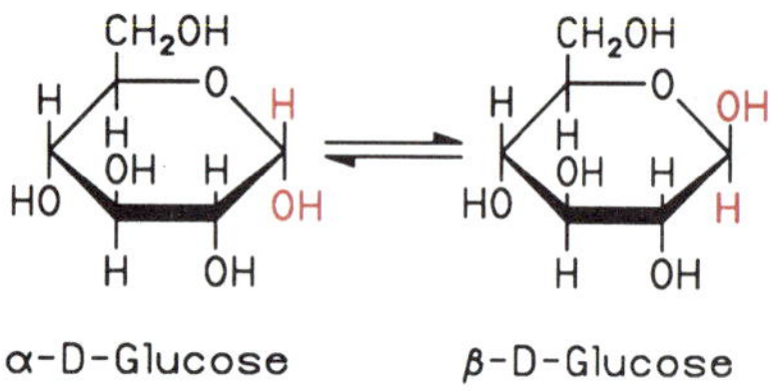

Figure 2.14.
Mutarotation of glucose.

tarotation; it can be accelerated by the action of the enzyme **mutarotase**. At equilibrium, a glucose solution contains 38% α form and 63% β form (Figure 2.14).

The pyranose form of glucose represents an intramolecular **hemiacetal**, which is an adduct between an alcohol and an aldehyde. Reaction with a second alcohol produces a full **acetal** (Figure 2.15). In an analogous reaction, aldopyranoses, such as glucose, can react with alcohols to form acetals that are called **glycosides**. In the case of glucose, the compound is called a glucoside (Figure 2.16).

Ketofuranoses, such as fructose, are **hemiketals**; they are also capable of forming glycosides. Instead of reacting with alcohols, sugars can also form glycosidic bonds with the hydroxyl groups of other sugars. The milk sugar **lactose**, for example, is a **disaccharide** formed from glucose and galactose. This **glucose 1,4-β-D-galactoside** is called a **reducing sugar** because its glucosyl residue still has a free anomeric carbon atom that can be oxidized to the carboxylic acid. Such an oxidation is not possible in **sucrose**, a disaccharide made from glucose and fructose, because both anomeric carbon atoms are engaged in the formation of the glycosidic bond (Figure 2.17). **Maltose** is a disaccharide made from two glucose residues linked by a 1,4-α-glucosidic bond.

Oxidation of monosaccharides at carbon 1 produces **-onic acids**. Glucose, for example, becomes gluconic acid (see Figure 2.41). Oxi-

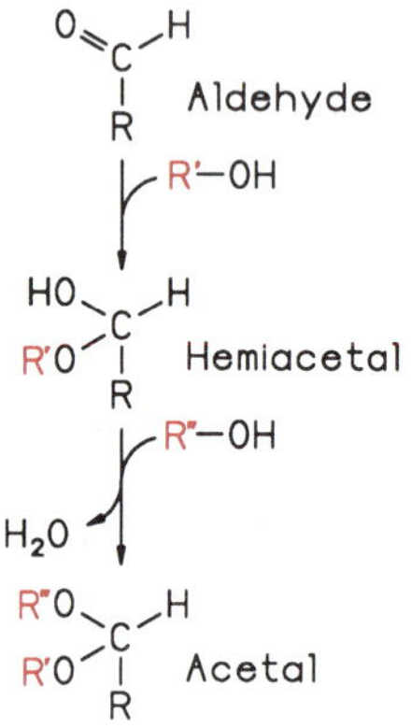

Figure 2.15.
Hemiacetal and acetal formation.

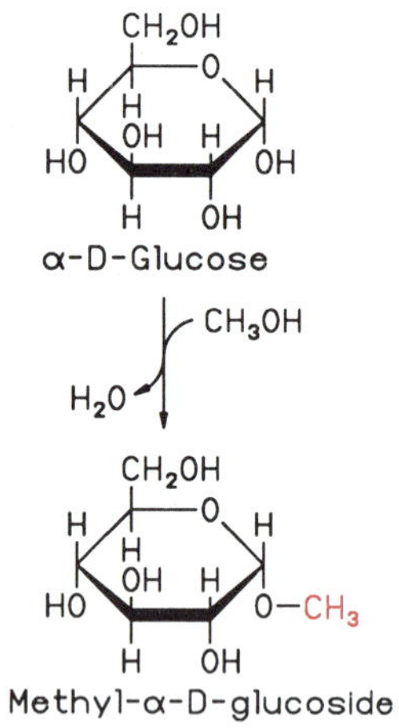

Figure 2.16.
Formation of a glycoside.

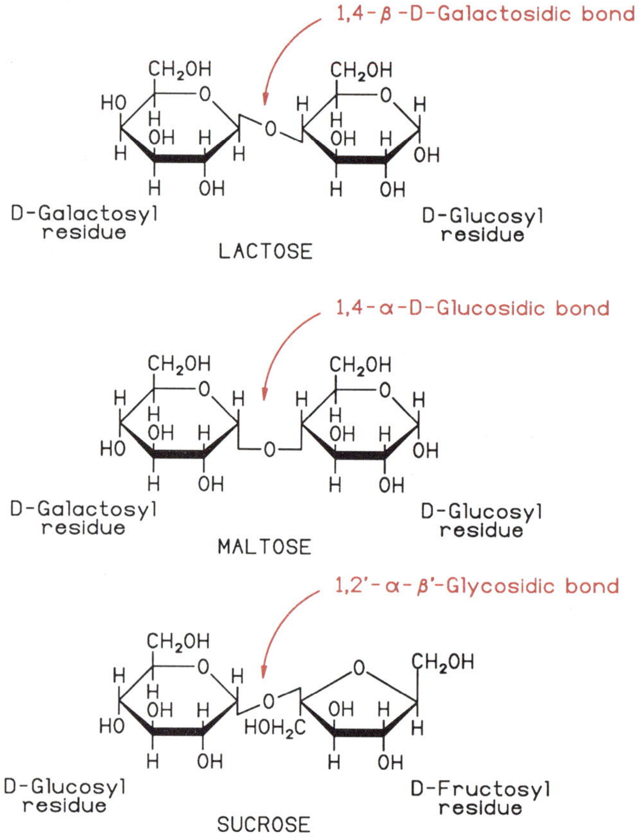

Figure 2.17.
Structures of three disaccharides.

dation at carbon 6 gives rise to -**uronic acids** (e.g., glucuronic acid). The structures of these and other sugars, sugar derivatives, and polysaccharides will be described in Sections 2.8 and 2.10, where their biosynthesis and metabolism are discussed.

2.3 GLYCOLYSIS

The degradation of glucose to pyruvate or lactate is called **glycolysis**. It occurs in the cytoplasm and begins with the phosphorylation of glucose to **glucose 6-phosphate**, which is quickly isomerized to **fructose 6-phosphate** and then further phosphorylated to **fructose 1,6-bisphosphate**. Fructose 1,6-bisphosphate is subsequently cleaved by the enzyme **aldolase** into two three-carbon-sugar phosphates, **dihydroxyacetone phosphate**, and **glyceraldehyde 3-phosphate**.

Under aerobic conditions, the glycolytic degradation of glucose proceeds to pyruvate (Figure 2.18), which enters the mitochondrion for further oxidation to CO_2 and water in the citric acid cycle. Under anaerobic conditions, the end product of glycolysis is lactate (Figure 2.19).

The glycolytic pathway, also called the **Embden-Meyerhof pathway**, consists of 10 enzymatic steps, all of which take place in the cytoplasm of the cell. Note that all intermediates consist of phosphorylated compounds that, because of their charge, cannot cross the cell membrane and hence cannot escape from the cell. The individual reactions of the pathway are listed in Table 2.3.

The overall process requires two ATP molecules per glucose for the initial phosphorylations and generates four ATP molecules (there are two glyceraldehyde-3-phosphates formed from each glucose). The net

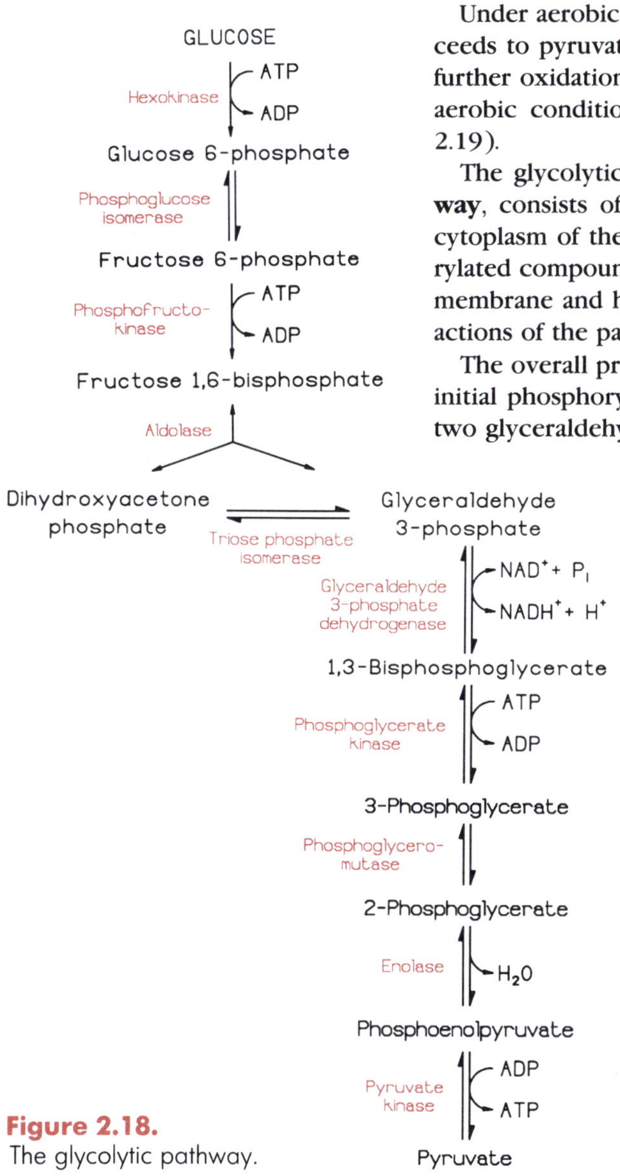

Figure 2.18.
The glycolytic pathway.

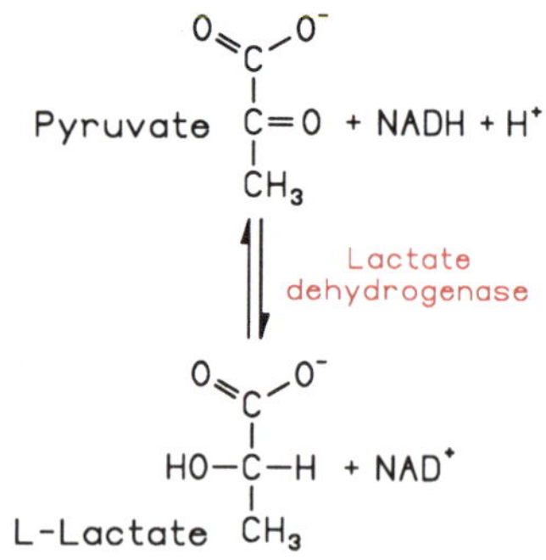

Figure 2.19.
Formation of lactate.

Table 2.3. Reactions of Glycolysis

Step	Enzyme	Reaction
1	Hexokinase	Glucose + ATP → glucose 6-phosphate + ADP + H^+
2	Phosphoglucose isomerase	Glucose 6-phosphate ⇌ fructose 6-phosphate
3	Phosphofructokinase	Fructose 6-phosphate + ATP → fructose 1,6-bisphosphate + ADP + H^+
4	Aldolase	Fructose 1,6-bisphosphate ⇌ dihydroxyacetone phosphate + glyceraldehyde 3-phosphate
5	Triose phosphate isomerase	Dihydroxyacetone phosphate ⇌ glyceraldehyde 3-phosphate
6	Glyceraldehyde 3-phosphate dehydrogenase	Glyceraldehyde 3-phosphate + P_i + NAD^+ ⇌ 1,3-bisphosphoglycerate + NADH + H^+
7	Phosphoglycerate kinase	1,3-Bisphosphoglycerate + ADP ⇌ 3-phosphoglycerate + ATP
8	Phosphoglycerate mutase	3-Phosphoglycerate ⇌ 2-phosphoglycerate
9	Enolase	2-Phosphoglycerate ⇌ phosphoenolpyruvate + H_2O
10	Pyruvate kinase	Phosphoenolpyruvate + ADP + H^+ → pyruvate + ATP

gain, therefore, is two ATPs per glucose molecule. The overall reaction converting one molecule of glucose into two molecules of pyruvate is

$$\text{Glucose} + 2\ P_i + 2\ \text{ADP} + 2\ \text{NAD}^+ \rightarrow 2\ \text{pyruvate} + 2\ \text{ATP} + 2\ \text{NADH} + 2\ H^+$$

In **anerobic glycolysis**, the two molecules of NADH that are generated during the oxidation of glyceraldehyde 3-phosphate to 1,3-bisphosphoglycerate transfer their electrons to O_2 through the electron-transport chain, thereby regenerating NAD^+. Under **anaerobic conditions**, NAD^+ is regenerated by the reduction of pyruvate to lactate. The enzyme responsible for this reaction, lactate dehydrogenase, has been discussed previously as an example of an isozyme (see Section 1.11).

Glycolysis occurs in all cells of the human body. Liver cells, however, differ from other cells by using the enzyme **glucokinase** instead of **hexokinase** for the initial step in glycolysis. Glucokinase has a much higher K_m value than hexokinase (20 m*M* versus 0.1 m*M*). Therefore, at prevailing glucose levels (5 m*M*), the liver is not competing with other tissues for glucose. Only if there is an excess of glucose, as observed after a carbohydrate meal, will the liver enzyme phosphorylate glucose. This glucose, however, is destined for glycogen synthesis and storage rather than glycolysis. Once glycogen stores are filled, excess glucose is converted to fat.

Glucokinase is induced by insulin (see Section 13.5). The enzyme is inhibited by a specific regulatory protein that together with fructose 6-phosphate, forms a complex with the enzyme. Fructose 1-phosphate acts competitively and thus prevents this inhibition. This accounts for the fact that fructose stimulates phosphorylation of glucose in isolated rat hepatocytes. (Glucokinase and hexokinase are examples of one-polypeptide-chain isoenzymes.)

2.3.1 Control of Glycolysis

The **committed step** in glycolysis (i.e., the reaction that irreversibly commits glucose to glycolytic degradation) is catalyzed by **phosphofructokinase**. This enzyme is allosterically regulated by ATP and citrate, both of which act as inhibitors, and AMP, which functions as an activator. This regulation explains the **Pasteur effect**, the phenomenon whereby the rate of utilization of glucose in muscle tissue is markedly lower under aerobic conditions than under anaerobic conditions. During the complete oxidation of glucose to CO_2 and water, 36 molecules of ATP are formed (see Table 2.4), as compared with only two molecules of ATP formed during anaerobic metabolism (limited to glycolysis). This much more efficient generation of ATP in aerobic conditions results in high levels of ATP in the cell that inhibit phosphofructokinase, thereby lowering the overall rate of utilization of glucose.

Another important activator of phosphofructokinase if **fructose 2,6-bisphosphate (F-2,6-BP)**. It is found in most tissues with the exception of erythrocytes. F-2,6-BP is formed from fructose 6-phosphate by **phosphofructokinase 2**. (The 2 distinguishes it from the enzyme of glycolysis, which may be called phosphofructokinase 1.) This enzyme can be inactivated through phosphorylation by a cAMP-dependent protein kinase (see glucagon, Chapter 13). In its phosphorylated form, the enzyme becomes a phosphatase, called **fructose 2,6-bisphosphatase 2**, that converts F-2,6-BP back to fructose 6-phosphate (Figure 2.20). In contrast to the phosphofructokinase 2 found in liver, the muscle enzyme appears to be stimulated by cAMP-dependent phosphorylation.

Two additional control points exist at the level of hexokinase and pyruvate kinase. **Hexokinase** is inhibited by glucose 6-phosphate. Thus, when phosphofructokinase is inhibited, the rising level of glucose 6-phosphate, which is in equilibrium with fructose 6-phosphate, shuts down hexokinase as well. **Pyruvate kinase** exists in the form of three

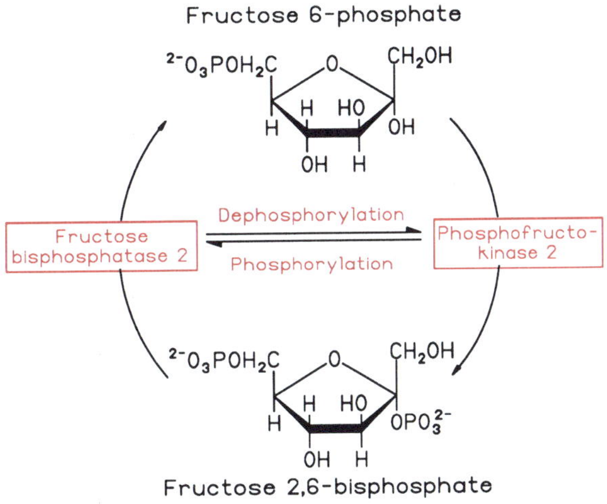

Figure 2.20.
Regulation of fructose 2,6-bisphosphate levels.

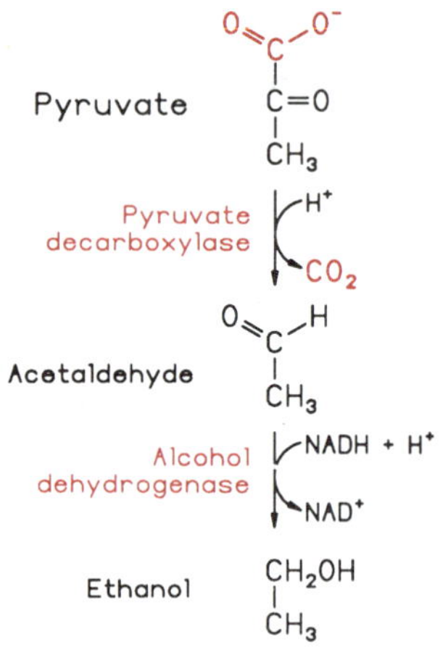

Figure 2.21.
Alcohol fermentation.

isozymes: the L form (liver), the M form (muscle and brain), and the A form (other tissues). The L form of the enzyme is allosterically inhibited by ATP (high energy charge) and activated by fructose 1,6-bisphosphate. In addition, a cAMP-dependent protein kinase phosphorylates the enzyme, thereby inactivating it. The M enzyme is not phosphorylated.

2.3.2 Anaerobic Glycolysis and Fermentation

The conversion of glucose to pyruvate generates two molecules of NADH that must be reoxidized to NAD^+ for glycolysis to continue. Under aerobic conditions, that is, when the tissue is well oxygenated, this reoxidation of NADH occurs in the mitochondrion in the electron-transport chain. Under anaerobic conditions, for example, in heavily exercising muscle, the NADH is reoxidized by the enzyme **lactate dehydrogenase**. In the process, pyruvate is reduced to L-lactate.

$$\text{Pyruvate} + \text{NADH} + H^+ \rightleftharpoons \text{L-lactate} + NAD^+$$

The equilibrium of this reaction strongly favors the formation of lactate. The lactate is then transported to the liver, where it is reoxidized to pyruvate. After conversion to glucose by the process of gluconeogenesis, it is recycled to the muscle tissue.

Other tissues that may produce lactate as the end product of glycolysis are the erythrocytes, the kidney, and the adrenal glands. Recently it has been shown that even in well-oxygenated muscle, as much as 50% of the glucose that enters glycolysis is converted to lactate and recycled to other tissues for reoxidation.

Anaerobic glycolysis is similar to the process of **fermentation** that occurs in many microorganisms under anaerobic conditions. In yeast, for example, the fermentation of glucose to alcohol follows essentially the glycolytic pathway, except that here pyruvate is decarboxylated to acetaldehyde, which is subsequently reduced to ethanol by **alcohol dehydrogenase** (Figure 2.21). This last reaction regenerates the NAD^+ used up before this step by glyceraldehyde 3-phosphate dehydrogenase. (This is analogous to the formation of lactate in anaerobic muscle.)

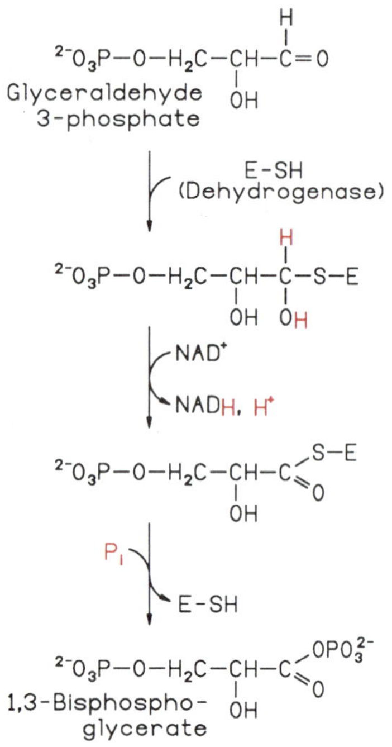

Figure 2.22.
Mechanism of oxidation of glyceraldehyde 3-phosphate to 1,3-bisphosphoglycerate. E = enzyme.

2.3.3 Discussion of Some Glycolytic Reactions

The oxidation of glyceraldehyde 3-phosphate to 1,3-bisphosphoglycerate represents an example of **substrate-level oxidative phosphorylation**, as distinguished from the oxidative phosphorylation that is linked to the respiratory chain (see Section 2.6). During this process a high-energy thioester between 3-phosphoglyceric acid and an SH group of glyceraldehyde 3-phosphate dehydrogenase is formed. The thioester is then converted into an acyl phosphate (1,3-bisphosphoglycerate) by phosphorolysis (Figure 2.22). The enzyme phosphoglycerate kinase transfers the newly created high-energy phosphate to ADP, thus generating ATP (Figure 2.23).

Phosphoglyceromutase converts 3-phosphoglycerate into 2-phos-

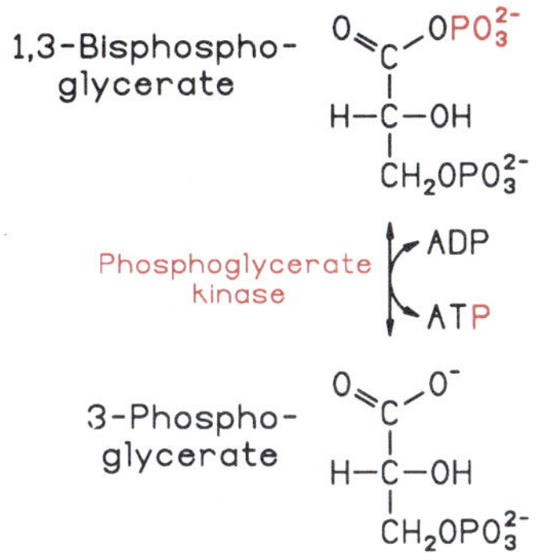

Figure 2.23.
ATP generation by phosphoglycerate kinase.

phoglycerate. The enzyme requires 2,3-bisphosphoglycerate as cofactor (or reaction intermediate). 2-Phosphoglycerate is then transformed to phosphoenolpyruvate by the enzyme enolase (Figure 2.24). In the last reaction of the glycolytic pathway, pyruvate kinase generates another ATP molecule (Figure 2.25).

Note that most reactions of glycolysis are reversible under physiologic conditions. The exceptions are the reactions catalyzed by hexokinase (glucokinase), phosphofructokinase, and pyruvate kinase.

The overall change in free energy during glycolysis is negative ($\Delta G^{0\prime} = -8.6$ kcal/mol), even though individual reactions show positive $\Delta G^{0\prime}$ values (e.g., aldolase: $\Delta G^{0\prime} = +5.7$ kcal/mol). The energetically unfavorable reactions are driven by the energetically more favorable reactions, which remove the reaction products from the reaction mixture and thus shift the equilibrium.

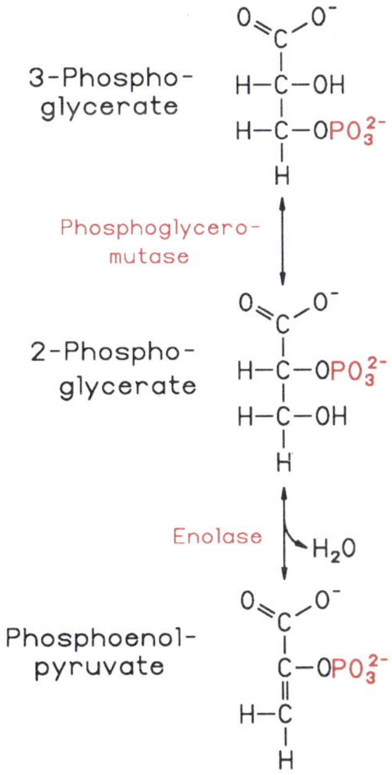

Figure 2.24.
Mechanism of conversion of 3-phosphoglycerate into phosphoenolpyruvate.

2.3.4 Transport of Reducing Equivalents into the Mitochondrion

The cytoplasmic NADH created by the oxidation of glyceraldehyde 3-phosphate cannot cross the mitochondrial membrane for reoxidation. Instead, it transfers only its electrons into the mitochondrion via the **glycerol phosphate shuttle** mechanism (Figure 2.26).

L-Glycerol 3-phosphate, created from dihydroxyacetone phosphate in the cytoplasm, is capable of crossing the outer mitochondrial membrane. It is reoxidized to dihydroxyacetone phosphate by **glycerol phosphate dehydrogenase** located on the inner mitochondrial membrane. This enzyme is covalently linked to FAD, which transfers its accepted electrons directly into the respiratory chain (see Section 2.5). During its oxidation, glycerol phosphate remains outside the inner mitochondrial membrane, and the regenerated dihydroxyacetone phosphate reenters the cytoplasm to complete the cycle. The glycerol phosphate shuttle operates in most cells. It can move reducing equivalents into the mitochondrion against an NADH concentration gradient.

In heart, muscle, and liver cells, an additional, more complex shuttle mechanism operates, the **malate-aspartate shuttle** (Figure 2.27). In contrast to the glycerol phosphate shuttle, the malate-aspartate shuttle is reversible and thus capable of transporting reducing equivalents into and out of the mitochondrion. It is also more efficient, since it recreates NADH on the other side of the mitochondrial membrane, whereas the glycerol phosphate shuttle produces $FADH_2$, which generates only two ATPs during its oxidation in the respiratory chain (as compared with three ATPs by NADH; see Section 2.6).

The inner mitochondrial membrane contains specific **translocases** that function in the shuttling of different molecules in and out of the mitochondrion. Most of these translocases are **antiporters** in nature (i.e., they transport one substrate in one direction while transporting another in the opposite direction). For example, cytoplasmic pyruvate is specifically exchanged for mitochondrial hydroxyl ions, cytoplasmic malate for mitochondrial α-ketoglutarate, aspartate (mitochondria) for glutamate (cytoplasm), citrate (mitochondria) for malate (cytoplasm), ATP (mitochondria) for ADP (cytoplasm), and so on.

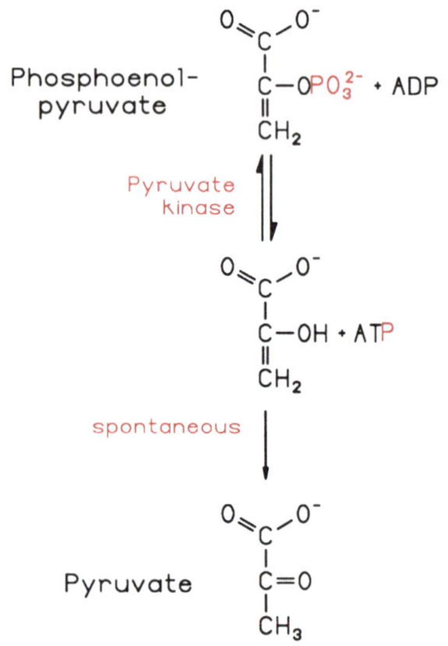

Figure 2.25.
ATP generation by pyruvate kinase.

2.3.5 Entry of Other Sugars into the Glycolytic Pathway

Two other important sugars besides glucose, namely fructose (derived from sucrose) and galactose (derived from lactose), are normal components of the human diet.

Fructose enters the glycolytic pathway after being phosphorylated to **fructose 1-phosphate** by **fructokinase**. **Fructose 1-phosphate aldolase** then cleaves the molecule into glyceraldehyde and dihydroxyacetone phosphate. **Glyceraldehyde** is phosphorylated by **triose kinase**. These reactions occur almost exclusively in the liver.

In contrast to phosphofructokinase, fructokinase is not regulated by the energy charge of the system. Although excess dietary glucose is converted into glycogen, excess fructose is more likely to be degraded to acetyl CoA, which is used for fatty acid or cholesterol synthesis. The possible contribution of excess dietary intake of fructose, either in the form of sucrose or as the powerful soft drink sweetener fructose, to the development of atherosclerosis is still being studied.

Hereditary fructose intolerance is due to a defective fructose 1-phosphate aldolase and is characterized by hypoglycemia, vomiting, and a general failure of the infant to thrive. The symptoms can be avoided by strict adherence to a fructose-free diet.

Galactose is phosphorylated to galactose 1-phosphate by **galactokinase**. Galactose 1-phosphate then is converted into UDPG, a reaction that will be discussed in more detail in Section 2.10.

2.4 THE CITRIC ACID CYCLE

The final oxidation of food to CO_2 and water occurs in the citric acid cycle, also called the **Krebs cycle** or tricarboxylic acid cycle (TCA cycle). The degradation products of carbohydrates, lipids, and some amino acids enter the cycle in the form of acetyl CoA and are completely oxidized to CO_2 and water.

2.4.1 Acetyl CoA Synthesis

The link between glycolysis and the citric acid cycle is the oxidative decarboxylation of pyruvate to acetyl CoA. This reaction occurs inside

Figure 2.26.
The glycerol phosphate shuttle.

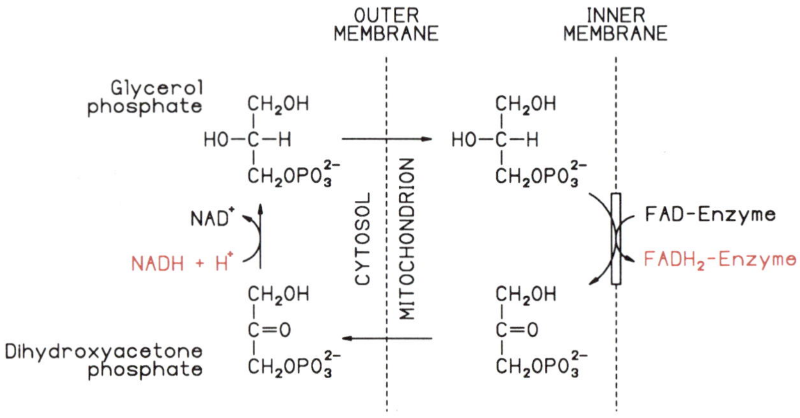

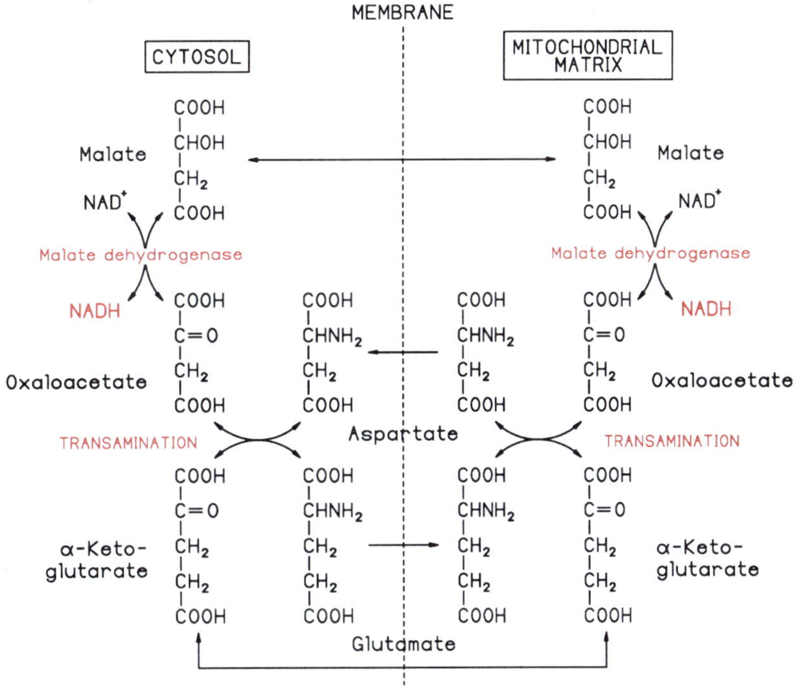

Figure 2.27.
The malate-aspartate shuttle.

the mitochondrion by the pyruvate dehydrogenase enzyme complex. Pyruvate is transported into the mitochondrion in exchange for hydroxyl ions. The overall reaction equation is

$$\text{Pyruvate} + \text{CoA} + \text{NAD}^+ \rightarrow \text{acetyl CoA} + \text{CO}_2 + \text{NADH} + \text{H}^+$$

Pyruvate dehydrogenase exists as a multienzyme complex with a molecular weight of over 7000 kd containing four different enzymes:

1. Pyruvate decarboxylase with thiamine pyrophosphate as coenzyme
2. Dihydrolipoyl transacetylase with lipoamide
3. Dihydrolipoyl dehydrogenase with FAD
4. Pyruvate dehydrogenase kinase

A fifth enzyme, pyruvate dehydrogenase phosphatase, is not part of the complex; together with pyruvate dehydrogenase kinase, it functions in the regulation of enzyme activity.

There are four steps to the overall process. Only the first three enzymes are directly involved in the formation of acetyl CoA. The coenzymes are covalently linked to enzymes 1, 2, and 3.

In the first step, pyruvate is added to the acidic carbon atom of thiamine pyrophosphate (TPP; see Figure 2.10), bound to E_1, the first enzyme of the pyruvate dehydrogenase complex, and the adduct is decarboxylated (Figure 2.28).

$$\text{CH}_3\text{—CO—COO}^- + \text{TPP—E}_1 \rightarrow \text{CH}_3\text{—CH(OH)—TPP—E}_1 + \text{CO}_2$$

In the second step, the newly formed hydroxyethyl group is oxidized and transferred from TPP to the proximal sulfur atom of lipoamide

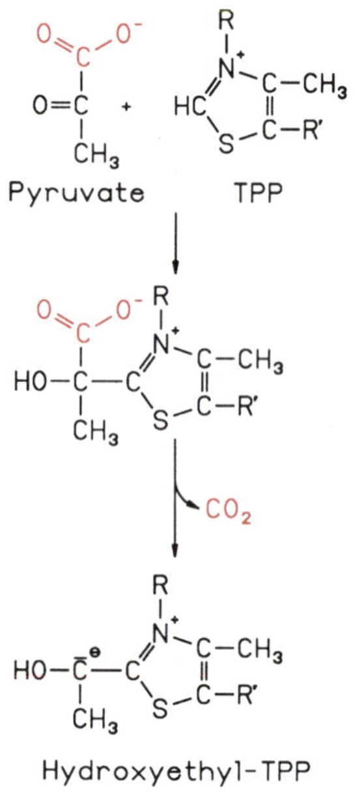

Figure 2.28.
Decarboxylation of pyruvate to hydroxyethyl thiamine pyrophosphate.

bound to enzyme E_2. The lipoamide acts as the oxidizing agent and is converted to dihydrolipoamide, its sulfhydryl form. (The hydrogen is transferred to the distal sulfur atom of lipoamide.)

$$CH_3—CH(OH)—TPP—E_1 + \text{lipoamide}—E_2 \rightarrow TPP—E_1 + CH_3—CO—S\text{-lipoamide}—E_2$$

In the third step, catalyzed by dihydrolipoyl transacetylase, the acetyl group is transferred to coenzyme A to form acetyl CoA. Simultaneously, the dihydrolipoamide on enzyme E_2 is reoxidized by FAD, bound to enzyme E_3.

$$CH_3—CO—S—\text{lipoamide}—E_2 + CoA + E_3—FAD \rightarrow CH_3—CO—S—CoA + \text{lipoamide}—E_2 + E_3—FADH_2$$

In the final step, catalyzed by dihydrolipoyl dehydrogenase, $FADH_2$ on E_3 is reoxidized to FAD by NAD^+.

The overall process is highly exergonic and irreversible. In humans, there is no mechanism by which acetyl CoA can be converted back into pyruvate. Therefore, the fatty acid component of lipids, which is metabolized to acetyl CoA, cannot be converted into carbohydrates (only the glycerol portion of lipids and the propionate, derived from the degradation of odd-numbered fatty acids, are glucogenic).

Regulation of Pyruvate Dehydrogenase

The pyruvate dehydrogenase complex is regulated by several factors:

1. The end products acetyl CoA and NADH both inhibit the complex.
2. In the presence of ATP, pyruvate dehydrogenase kinase phosphorylates, and thereby inactivates, pyruvate dehydrogenase.

The activity of the pyruvate dehydrogenase complex is regulated by phosphorylation and dephosphorylation. This is accomplished by a **protein kinase** that is tightly bound to the E_2 subunits of the complex and a **phosphatase** that is only loosely associated with the complex.

In the presence of high concentrations of ATP (as well as guanosine triphosphate [GTP]), the enzymatic activity of the complex is inhibited by phosphorylation (AMP activates the complex). Substrate and products also affect the kinase and phosphatase activities in a complex manner (Figure 2.29). In addition, acetyl CoA inhibits the transacetylation reaction, whereas NADH inhibits the dihydrolipoyl dehydrogenase. Thus, the complex is capable of regulating the metabolism of pyruvate, through the citric acid cycle, according to the needs of the cell as reflected in the levels of ATP and reducing power in the form of NADH.

Subunit Homology with Other Enzymes

Oxidative decarboxylation of α-keto acids such as pyruvate is a very common metabolic reaction. Another example of this type of reaction is the conversion of α-ketoglutarate to succinyl CoA and CO_2, one of

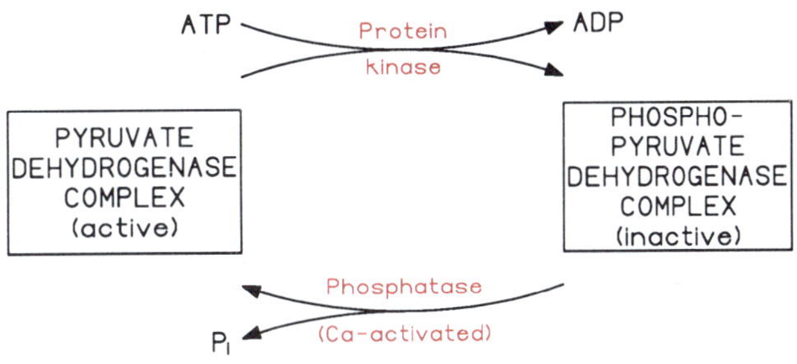

Figure 2.29. Regulation of pyruvate dehydrogenase by phosphorylation and dephosphorylation.

the two decarboxylation reactions of the citric acid cycle. The reaction sequence is identical to that described for pyruvate dehydrogenase, the only difference being that the methyl group in pyruvate is replaced by a carboxymethyl group.

The α-ketoglutarate dehydrogenase complex shows the same basic architecture as the pyruvate dehydrogenase complex:

1. A decarboxylase (for α-ketoglutarate instead of pyruvate).
2. A core transacylase (transsuccinylase instead of transacetylase).
3. A dihydrolipoyl dehydrogenase (identical reaction).

Dissociation of the three catalytic components from the complex and reassociation of hybrid complexes containing subunits of both α-ketoglutarate and pyruvate dehydrogenases show that the dihydrolipoyl dehydrogenase (E_3) subunit is identical in both enzyme complexes. This suggests that extensive surface homologies exist between the two complexes for the precise recognition and binding of E_3.

2.4.2 The Reactions of the Citric Acid Cycle

Acetyl CoA enters the citric acid cycle by reacting with oxaloacetate to form citrate. During the completion of the cycle, one acetate is completely oxidized to CO_2, and oxaloacetate is regenerated (Figure 2.30).

All intermediates of the citric acid cycle consist of dicarboxylic or tricarboxylic acids. There are no phosphorylated intermediates. The entire process takes place in the mitochondrion (Figure 2.31). All enzymes involved are located in the mitochondrial matrix except succinate dehydrogenase, which is attached to the inner mitochondrial membrane together with the components of the electron-transport chain. The individual reactions of the citric acid cycle are listed in Table 2.4.

After isomerization of citrate, via *cis*-aconitate, to isocitrate (C_6), two successive oxidative decarboxylation steps produce first α-ketoglutarate (C_5) and then succinyl CoA (C_4). Two molecules of NADH are generated during these steps, yielding three molecules of ATP each during the transfer of their electrons to oxygen in the respiratory chain (see Section 2.6). The cleavage of the thioester bond of succinyl CoA yields another high-energy phosphate bond in the form of GTP. Oxidation of succinate to fumarate generates $FADH_2$, which yields two ATP molecules in the respiratory chain. Overall, during the four oxidation reactions of the cycle, three pairs of electrons are transferred to NAD^+

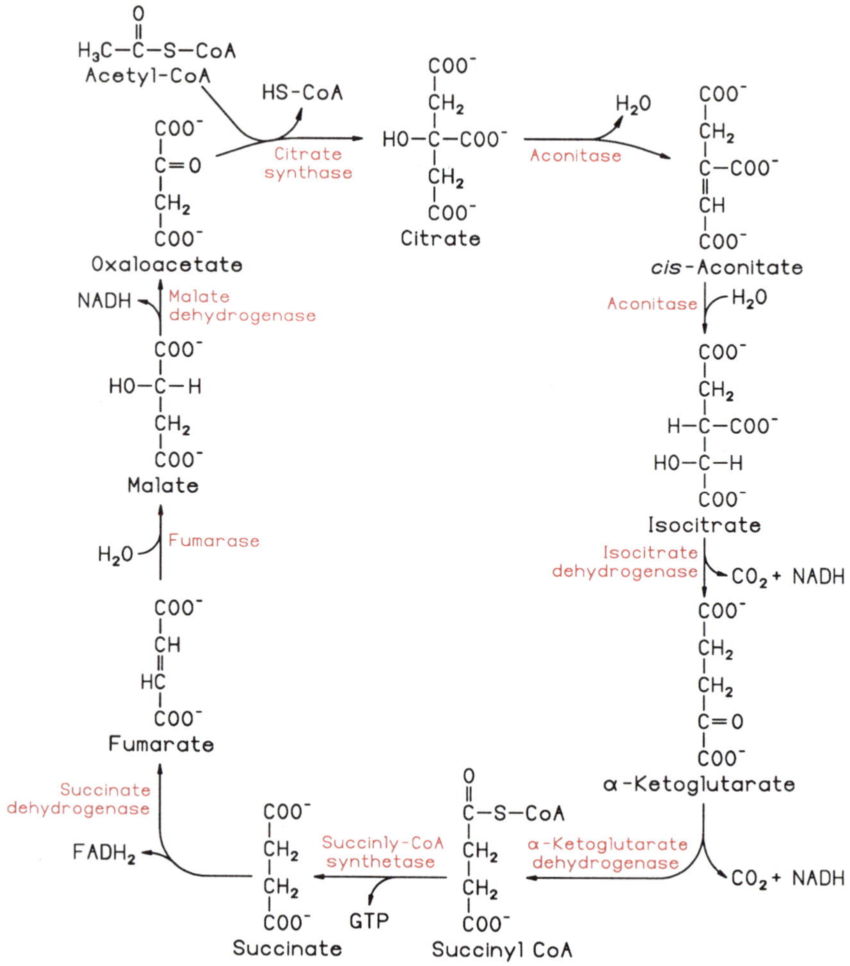

Figure 2.30.
The citric acid cycle.

Table 2.4. Reactions of the Citric Acid Cycle

Step	Enzyme	Reaction
1	Citrate synthase	Acetyl CoA + oxaloacetate + H_2O → citrate + CoA + H^+
2	Aconitase	Citrate ⇌ *cis*-aconitate + H_2O ⇌ isocitrate
3	Isocitrate dehydrogenase	Isocitrate + NAD^+ → α-ketoglutarate + CO_2 + NADH
4	α-Ketoglutarate dehydrogenase complex	α-Ketoglutarate + NAD^+ + CoA → succinyl CoA + CO_2 + NADH
5	Succinyl CoA synthetase	Succinyl CoA + P_i + GDP ⇌ succinate + GTP + CoA
6	Succinate dehydrogenase	Succinate + FAD (enzyme bound) ⇌ fumarate + $FADH_2$ (enzyme bound)
7	Fumarase	Fumarate + H_2O ⇌ L-malate
8	Malate dehydrogenase	L-Malate + NAD^+ ⇌ oxaloacetate + NADH + H^+

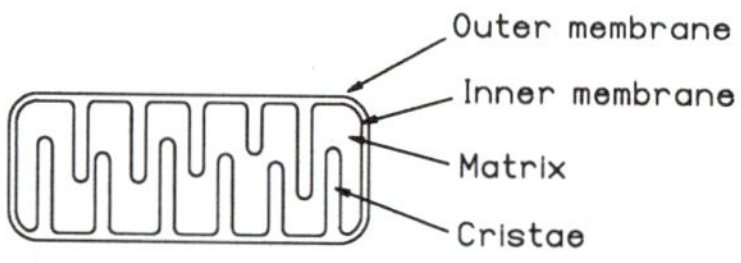

Figure 2.31.
Schematic drawing of a mitochondrion.

and one to FAD, yielding a total of 11 ATP molecules. Together with the GTP, the oxidation of a two-carbon fragment (in the form of acetyl CoA) to CO_2 and H_2O generates 12 high-energy phosphate bonds.

$$\text{Acetyl CoA} + 3\ NAD^+ + FAD + GTP + P_i + 2\ H_2O \rightarrow 2\ CO_2 + 3\ NADH + FADH_2 + GTP + 2\ H^+ + CoA$$

The overall $\Delta G^{0\prime}$ of the cycle is -23 kcal/mol; therefore, the cycle is unidirectional.

The carbon atoms of the acetyl group that enters the citric acid cycle do not appear in CO_2 during the first turn of the cycle. This indicates that aconitase, the enzyme that isomerizes citrate to isocitrate, is capable of distinguishing between the two $-CH_2-COOH$ groups in citrate. The two substituents are not sterically identical in relation to the two other substituents $-H$ and $-COOH$, even though the mirror images of citric acid are identical and, as a consequence, the molecule is optically inactive. (See the discussion of prochiral centers in Section 1.11.)

The three reactions converting succinate into fumarate (dehydrogenation), malate (hydration), and then finally into oxaloacetate (oxidation) are identical in their chemistry ($-CH_2-CH_2-$ to $-CO-CH_2-$) to the reactions occurring in the β oxidation of fatty acids (see Chapter 3).

α-Ketoglutarate dehydrogenase is similar to pyruvate dehydrogenase in that it also functions in a multienzyme complex using the same cofactors: thiamine pyrophosphate, lipoamide, CoA, FAD, and NAD^+.

The citric acid cycle operates only under aerobic conditions because of its requirement of NAD^+ and FAD. The oxidized forms of these cofactors must be regenerated by the respiratory chain in a process coupled to oxidative phosphorylation.

2.4.3 Regulation of the Citric Acid Cycle

The energy charge of the cell directly controls the activity of the citric acid cycle. The first enzyme in the cycle, **citrate synthase**, is allosterically inhibited by ATP. Furthermore, the availability of oxaloacetate regulates the citric acid cycle. For example, during a diet that is low in carbohydrate and protein, oxaloacetate levels are reduced because oxaloacetate participates in gluconeogenesis (see Section 2.9). As a consequence, the entry of acetyl CoA into the cycle is reduced, and most coenzyme A is tied up in acetyl CoA. In order to regenerate free CoA for further fatty acid degradation, two acetyl CoA molecules can condense to form acetoacetyl CoA.

$$\text{Acetyl CoA} + \text{acetyl CoA} \rightarrow \text{acetoacetyl CoA} + \text{CoA}$$

Acetoacetyl CoA is further converted into acetoacetate, β-hydroxybutyrate, and acetone, which are collectively called **ketone bodies** (see Chapter 3). This also occurs during starvation and in diabetes mellitus. This condition of **ketosis** leads to acidosis and dehydration (see Chapter 12).

Two other enzymes in the cycle are control points:

1. Isocitrate dehydrogenase is inhibited by ATP and NADH and stimulated by ADP.
2. α-Ketoglutarate dehydrogenase is inhibited by succinyl CoA and NADH.

The inhibition of isocitrate dehydrogenase increases intramitochondrial citrate levels, thus facilitating citrate export into the cytoplasm for biosynthetic purposes (see below). The regulation of α-ketoglutarate dehydrogenase may serve a similar function.

2.4.4 Anaplerotic Reactions

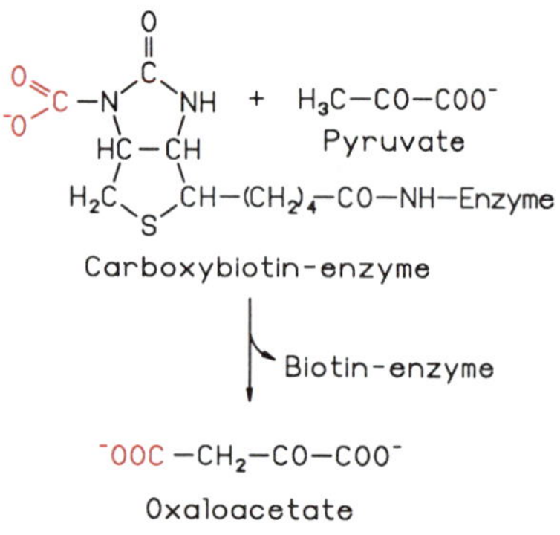

Figure 2.32.
Carboxylation of pyruvate to oxaloacetate.

When intermediates of the citric acid cycle are removed by side reactions, for example, the use of oxaloacetate in gluconeogenesis, they have to be replenished in an **anaplerotic reaction** (i.e., fill-up reaction) in order to maintain the cycle. The most important anaplerotic reaction is the synthesis of oxaloacetate from pyruvate; it is catalyzed by **pyruvate carboxylase**. The enzyme contains the cofactor **biotin** (see Figure 2.10), to which CO_2 becomes covalently linked in a reaction driven by the hydrolysis of ATP. The **carboxybiotin** then transfers the CO_2 to pyruvate to form oxaloacetate (Figure 2.32).

Pyruvate carboxylase is totally dependent on acetyl CoA for activity. When acetyl CoA accumulates because of insufficient levels of oxaloacetate, the anaplerotic synthesis of oxaloacetate is activated. Of course, in starvation or diabetes mellitus, the availability of pyruvate or citric acid cycle intermediates is diminished, and the reestablishment of normal oxaloacetate levels is impossible.

Oxidative degradation of all 20 amino acids produces intermediates that feed into the citric acid cycle at the following entry points: α-ketoglutarate, succinyl CoA, fumarate, and oxaloacetate. These can be oxidized in the cycle, but since oxaloacetate is regenerated at the end of the cycle, no net oxidation can take place unless oxaloacetate can be removed. This is accomplished by **phosphoenolpyruvate carboxykinase**, which converts oxaloacetate in the cytoplasm into phosphoenolpyruvate, the starting point of gluconeogenesis (see Section 2.9).

The citric acid cycle is **amphibolic** in nature: it functions not only in catabolism but also produces precursors for many anabolic reactions. For example, α-ketoglutarate and oxaloacetate are easily converted, by transamination, into glutamate and aspartate, respectively (see Section 4.4). Citrate serves as the cytoplasmic precursor of acetyl CoA for fatty acid biosynthesis, since acetyl CoA cannot cross the mitochondrial membrane (see Chapter 3).

2.5 THE ELECTRON-TRANSPORT CHAIN

Oxidation of NADH and $FADH_2$, generated in the citric acid cycle or in glycolysis, occurs in the mitochondrial electron-transport chain, also called the respiratory chain, where hydrogen ions and electrons are transferred to molecular oxygen to form water. Before discussing the molecular reactions of this process, however, we have to understand the general nature of oxidation-reduction reactions.

2.5.1 Oxidation-Reduction Reactions

An **oxidation** of a molecule is defined as the **loss of electrons** by the molecule; a **reduction** is the **gain of electrons**. The oxidation of one molecule is always coupled with the reduction of another molecule that takes up the electrons from the oxidized molecule.

In analogy to the standard free energy $G^{0\prime}$, which is a measure of the reactive potential of a molecule, the standard **redox potential** E_0' of a molecule describes its potential to accept electrons (at pH 7.0, 25°C). By definition, the redox potential E_0' of the reaction

$$H^+ + e^- \rightleftharpoons \tfrac{1}{2} H_2$$

measured under standard conditions is zero.

Molecules that have a lower affinity for electrons than hydrogen have a negative redox potential; molecules with a higher affinity for electrons than hydrogen have a positive redox potential.

The standard redox potential of NAD^+ is

$$NAD^+ + H^+ + 2e^- \rightleftharpoons NADH \qquad E_0' = -0.32\ V$$

and that of oxygen is

$$\tfrac{1}{2}O_2 + 2H^+ + 2e^- \rightleftharpoons H_2O \qquad E_0' = +0.82\ V$$

The combination of these two half reactions describes the oxidation of NADH in the electron-transport chain:

$$NADH + \tfrac{1}{2}O_2 + H^+ \rightleftharpoons NAD^+ + H_2O \qquad \Delta E_0' = 1.14\ V$$

The relationship between the standard free energy change, $\Delta G^{0\prime}$, and the change in redox potential, $\Delta E_0'$, is

$$\Delta G^{0\prime} = -nF\Delta E_0' \qquad (2.6)$$

where n is the number of electrons transferred, and F is the Faraday constant (23.062 kcal/V).

The free energy change in the electron-transport chain reaction thus is

$$\Delta G^{0\prime} = -2 \times 23.062 \times 1.14 = -52.6\ \text{kcal/mol}$$

This energy differential is large enough to create three high-energy phosphate bonds in the form of three ATP molecules. (Note that a positive $\Delta E_0'$ indicates an exergonic reaction.)

The standard redox potentials of some important metabolic reactions are compared in Table 2.5.

Table 2.5. Standard Oxidation-Reduction Potentials of Some Metabolic Reactions

Reductant	Oxidant	*n*	E_0' (V)
Succinate + CO_2	α-Ketoglutarate	2	−0.67
H_2	2 H^+	2	−0.42
NADH + H^+	NAD^+	2	−0.32
NADPH + H^+	$NADP^+$	2	−0.32
Dihydrolipoamide	Lipoamide	2	−0.29
Glutathione (reduced)	Glutathione (oxidized)	2	−0.23
$FADH_2$	FAD	2	−.021
Ethanol	Acetaldehyde	2	−0.20
Lactate	Pyruvate	2	−0.19
Succinate	Fumarate	2	+0.03
Cytochrome *b* (Fe^{2+})	Cytochrome *b* (Fe^{3+})	1	+0.07
Ascorbate	Dehydroascorbate	2	+0.08
Ubiquinone (reduced)	Ubiquinone (oxidized)	2	+0.10
Cytochrome *c* (Fe^{2+})	Cytochrome *c* (Fe^{3+})	1	+0.24
Fe^{2+}	Fe^{3+}	1	+0.77
H_2O	$\frac{1}{2}O_2$	2	+0.82

2.5.2 The Components of the Electron-Transport Chain

The components of the electron-transport chain are located on the inner mitochondrial membrane and on its projections into the matrix, the cristae. They can be separated into individual enzyme complexes, each exhibiting properties of an oxidoreductase (Figure 2.33). The first three complexes are proton pumps that generate a pH gradient across the inner mitochondrial membrane, which is used for ATP generation (see below).

Complex I is an **NADH–coenzyme Q (CoQ) reductase** (NADH dehydrogenase), which contains flavin mononucleotide (FMN) bound to the enzyme.

FMN is structurally similar to FAD, but without the AMP residue, and functions identically to FAD (Figure 2.34). In addition, complex I contains several nonheme iron-sulfur proteins, in which the iron is bound to the sulfur atoms of cysteinyl residues as well as to free sulfide. There appear to be three types of iron-sulfur protein containing Fe_2S_2 or Fe_4S_4 centers. An example of a protein with an Fe_2S_2 center is shown in Figure 2.35. The redox potentials of these iron-sulfur proteins can vary by more than 800 mV. Iron-sulfur proteins participate in the transfer of

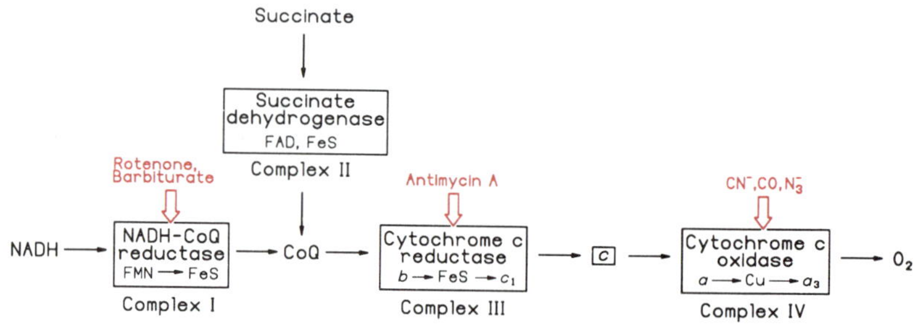

Figure 2.33. Schematic drawing of the electron-transport chain.

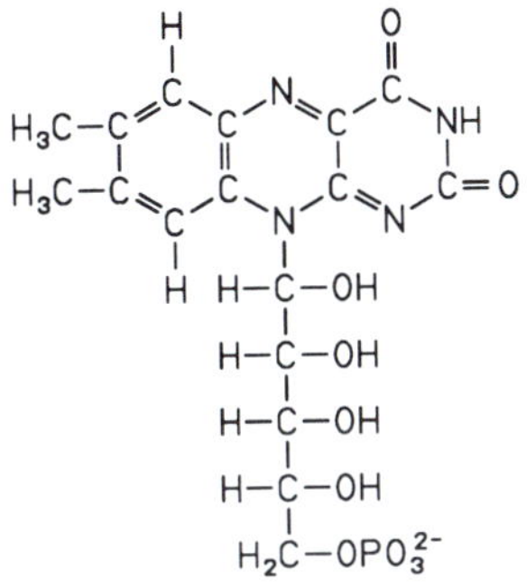

Figure 2.34.
Flavin mononucleotide (FMN).

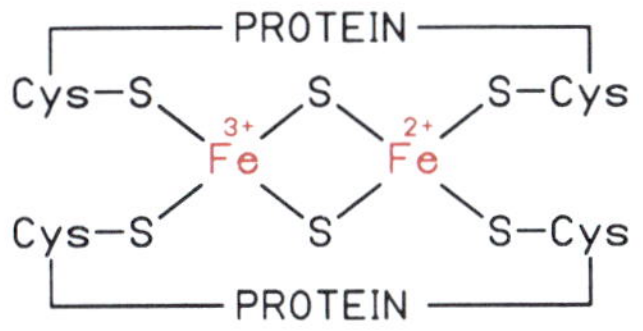

Figure 2.35.
Structure of an iron-sulfur protein (reduced form).

electrons by changing the oxidation state of their Fe ion:

$$Fe^{3+} + e^- \rightleftharpoons Fe^{2+}$$

The overall reaction catalyzed by complex I is

$$NADH + H^+ + CoQ \rightleftharpoons NAD^+ + CoQH_2$$

In reality, the reaction is more complex because it generates a proton gradient across the mitochondrial membrane that is used to fuel the synthesis of ATP (see Section 2.6).

$$NADH + H^+_{inside} + 2Fe^{3+} \rightarrow 2Fe^{2+} + NAD^+ + 2H^+_{outside}$$
$$2CoQ + 2H^+_{inside} + 2Fe^{2+} \rightarrow 2Fe^{3+} + CoQH_2$$

Thus, three protons disappear from the inside of the mitochondrion and two protons are generated on the outside.

Complex III is a **$CoQH_2$–cytochrome *c* reductase**. It contains a nonheme iron protein as well as cytochromes *b* and c_1. The cytochromes are proteins that contain a heme group, as in hemoglobin. However, in contrast to the iron in hemoglobin, which always is in the Fe^{2+} state, the iron in cytochromes functions as an electron acceptor and donor. Complex III catalyzes the reaction

$$CoQH_2 + 2 \text{ cytochrome } c_1 (Fe^{3+}) \rightleftharpoons CoQ + 2 \text{ cytochrome } c_1 (Fe^{3+}) + 2 H^+$$

Cytochrome *c* is a soluble, crystallizable, low-molecular-weight heme protein (Figure 2.36). The protein is covalently linked to the vinyl groups of heme (protoporphyrin IX) by thioether bonds with cysteinyl residues. In contrast, cytochromes *a* and *b* are insoluble, and their heme groups are bound noncovalently.

Complex IV is a **cytochrome *c* oxidase**. It contains cytochromes *a* and a_3. Cytochrome *c* oxidase contains two copper atoms, one of which is closely associated with the Fe of cytochrome a_3, that participate in the electron transfer by changing their oxidation states between Cu^+ and Cu^{2+}. Complex IV catalyzes the reaction

$$2 \text{ Cytochrome } c (Fe^{2+}) + \tfrac{1}{2} O_2 + 2H^+ \rightleftharpoons 2 \text{ cytochrome } c (Fe^{3+}) + \tfrac{1}{2} H_2O$$

Figure 2.36.
Cytochrome *c*.

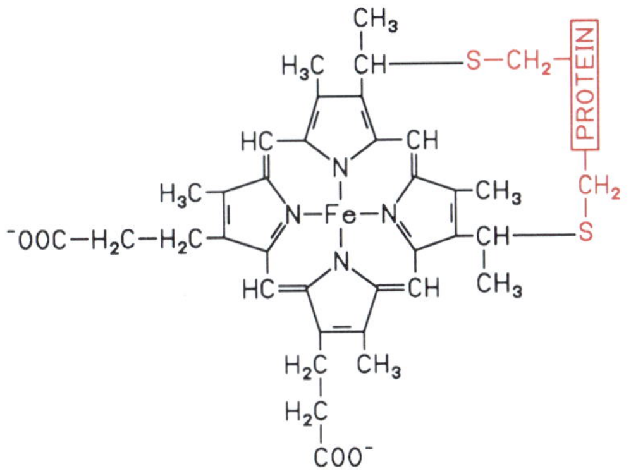

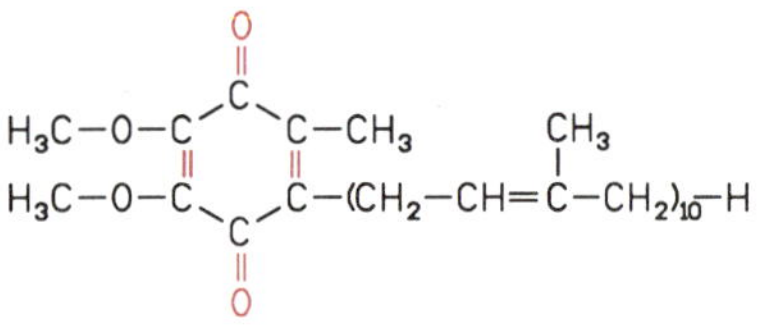

Figure 2.37.
Coenzyme Q (ubiquinone; oxidized form).

As indicated in Figure 2.33, complexes I, III, and IV act in sequence to transfer electrons from NADH to oxygen. A fourth complex (complex II) couples $FADH_2$ oxidation to the main electron-transport chain. $FADH_2$ arises from succinate dehydrogenase and from intramitochondrial glycerol phosphate dehydrogenase (shuttle enzyme) as well as from acetyl CoA dehydrogenase (involved in fatty acid degradation).

The fourth complex (complex II) is **succinate-CoQ reductase** (succinate dehydrogenase), the same enzyme that participates in the citric acid cycle. It contains bound FAD and an iron-sulfur protein and catalyzes the reaction

$$\text{Succinate} + \text{CoQ} \rightarrow \text{fumarate} + \text{CoQH}_2$$

The intermediary electron carrier **coenzyme Q (ubiquinone)** is a quinone with a polyisoprene side chain (Figure 2.37). It can accept either two electrons or a single electron (in the latter case it forms a semiquinone) and is thus capable of coupling two-electron donors, such as NADH dehydrogenase (complex I) and succinate dehydrogenase (complex IV), with one-electron acceptors, the cytochromes (Figure 2.38).

Figure 2.38.
Electron transfer between quinone, semiquinone, and hydroquinone.

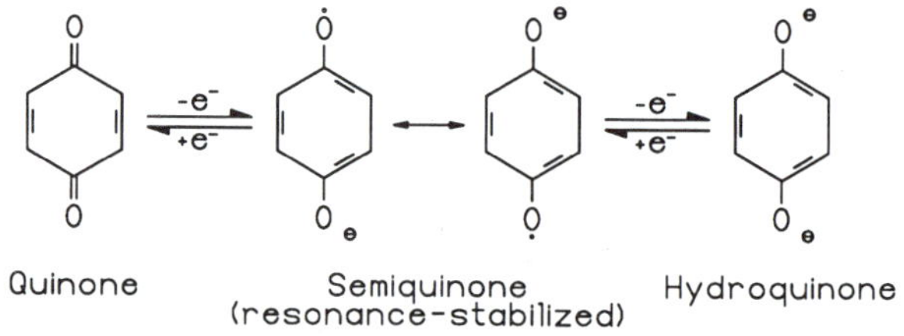

2.6 OXIDATIVE PHOSPHORYLATION

Electron transport is tightly coupled to oxidative phosphorylation, the process by which the energy generated during the electron transport is used to synthesize ATP.

ATP synthesis is driven by a **pH gradient** and the ensuing electrical potential difference that is created across the inner mitochondrial membrane during the electron-transport reactions. During the operation of each of the complexes I, II, and III, protons are taken up from the matrix and released on the outside of the mitochondrial membrane. This **chemiosmotic mechanism** of ATP generation, first postulated by **Peter Mitchell**, is now supported by extensive experimental evidence. For example, in vitro ATP generation can be observed only in vesicles with intact membranes. Furthermore, all known **uncouplers** (i.e., compounds that inhibit oxidative phosphorylation without affecting electron transport) are cation **ionophores**, compounds that allow ions to pass through the cell membrane. **Gramicidin A**, for example, interacts with the cell membrane and thereby creates channels for the passage of monovalent cations. **Valinomycin**, on the other hand, functions as a cation carrier that passes through the membrane.

Both **antimycin A** and **oligomycin** inhibit electron transport. Uncouplers relieve the inhibition caused by oligomycin but not that caused by antimycin A. Apparently, if oxidative phosphorylation is blocked by

Table 2.6. Inhibitors of the Electron-Transport Chain

Inhibitor	Site of Inhibition
Electron transport	
Barbiturate	Complex I
Rotenone	Complex I
Antimycin A	Complex III
Cyanide	Complex IV
Sulfide	Complex IV
Azide	Complex IV
Carbon monoxide	Complex IV
Phosphorylation	
Oligomycin	ATP synthesis
ATP transport	
Atractyloside	ATP/ADP antiport system
Bonkrekic acid	ATP/ADP antiport system
Cation ionophores	
Dicoumarol (H^+)	Uncoupler
Dinitrophenol (H^+)	Uncoupler
Valinomycin (K^+)	Uncoupler
Gramicidin A (Na^+, K^+, H^+)	Uncoupler
Nigericin (K^+, H^+)	Uncoupler

oligomycin, the pH gradient across the mitochondrial membrane, reaches a level that inhibits further electron transport. Upon addition of uncoupler, new "channels" for the release of excess H^+ ions are provided so that electron transport can continue. Table 2.6 lists some of the inhibitors of oxidative phosphorylation that were used to study the individual steps of the reaction.

ATP synthase is an enzyme complex of about 360 kd. It consists of two multi-subunit factors, F_1 and F_O. F_1 in solubilized form is an ATPase that is not inhibited by oligomycin. In combination with F_O, an integral membrane protein of six identical subunits, its ATPase activity becomes sensitive to oligomycin. When both factors are inserted into liposomes and a pH gradient is artificially created across the liposomal membrane, the enzyme complex synthesizes ATP when supplied with ADP and inorganic phosphate. Actually, ATP is formed from ADP and P_i even in the absence of a proton motive force. However, the ATP remains bound to the enzyme until the movement of H^+ through the transmembrane channel created by the six F_O subunits causes a conformational change in F_1 that releases ATP so that another molecule of ADP can bind and be phosphorylated.

The P/O ratio indicates how many molecules of ATP are formed per atom of oxygen that is consumed. For malate, the P/O ratio is 3; for succinate, it is only 2, since complex I is bypassed. Reduced ascorbate, which reduces cytochrome *c* directly, exhibits a P/O ratio of 1. In the presence of the uncoupler dinitrophenol, all the above substances show P/O ratios of zero.

Normally, electron transport is controlled by the availability of ADP. ADP is transported into the mitochondrion in exchange for ATP. Similarly, inorganic phosphate enters the mitochondrion in exchange for

Table 2.7. Recapitulation of the Energy-Producing Reactions Involved in the Complete Oxidation of Glucose

Reaction	Net Moles of ATP Generated per Mole of Glucose
Glycolysis (phosphoglycerate kinase, pyruvate kinase;. 2 ATPs are expended)	2
Glycerol phosphate shuttle (glyceraldehyde 3-phosphate → 1,3-bisphosphoglycerate + NADH → $FADH_2$)	4
Pyruvate dehydrogenase (NADH)	6
Succinyl CoA synthetase (succinate → fumarate + $FADH_2$)	4
Other citric acid cycle reactions (isocitrate → α-ketoglutarate, α-ketoglutarate → succinyl CoA, malate → oxaloacetate; total of 3 NADH generated)	18
Total	36

hydroxyl ions. This antiport system does not require energy; it is also referred to as facilitated exchange diffusion.

Uncoupling ATP formation from electron transport causes release of energy in the form of heat, a mechanism that appears to be utilized by some hibernating mammals.

Table 2.7 summarizes the steps that generate ATP during the complete oxidation of one molecule of glucose.

2.7 DEHYDROGENASES, OXIDASES, AND OXYGENASES

In addition to the enzymes of the electron transport chain in the mitochondrion, there are three main classes of oxidizing enzymes in the cell: **dehydrogenases**, **oxidases**, and **oxygenases**. Dehydrogenases transfer electrons to cofactors, whereas the reactions catalyzed by oxidases and oxygenases involve the participation of molecular oxygen.

2.7.1 Dehydrogenases

Dehydrogenases typically transfer electrons from one substrate to another. Most dehydrogenases employ NAD^+, FAD, or FMN as cofactors (cosubstrates). The NAD^+ may be separate from the enzyme (e.g., lactate dehydrogenase) or tightly bound (e.g., glyceraldehyde 3-phosphate dehydrogenase). Flavin nucleotides are always tightly bound, in some cases even covalently linked, to the enzyme.

2.7.2 Oxidases

Oxidases transfer electrons to oxygen, forming either **peroxide** or **water** in the process. The intermediary electron carriers may be metal ions, such as Fe^{3+} or Cu^{2+}, or flavin nucleotides. **Xanthine oxidase,** an enzyme involved in purine catabolism, for example, employs FAD as well as Fe^{3+} and Mo^{6+} ions. An example of a peroxide-forming enzyme is **amino acid oxidase**, which oxidizes amino acids to keto acids. Since peroxide is rather toxic, organisms have evolved detoxification mechanisms, such as the heme-containing enzyme **catalase,** which decomposes H_2O_2 to oxygen and water. Peroxide may also be inactivated by **peroxidases**, enzymes that use peroxide to oxidize another substrate. An example is the **glutathione peroxidase** found in erythrocytes. It is a selenium-containing enzyme that uses glutathione to reduce H_2O_2.

2.7.3 Oxygenases

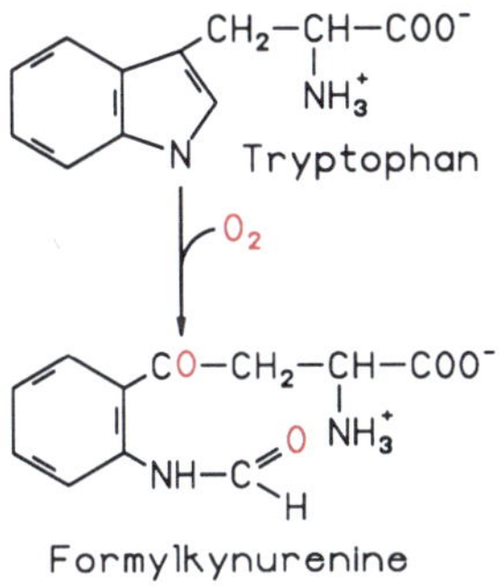

Figure 2.39.
Oxidation of tryptophan to formylkynurenine by a dioxygenase.

Oxygenases incorporate molecular oxygen into organic substrates. One distinguishes between monooxygenases (see hydroxylases below) and dioxygenases, depending on whether only one or both atoms of molecular oxygen are incorporated into the substrate. Most dioxygenases contain Fe^{2+} ions (heme or nonheme). **Tryptophan 2,3-dioxygenase** (a heme enzyme) is an example of a dioxygenase; it functions in the catabolism of tryptophan (Figure 2.39).

Monooxygenases are also called **hydroxylases**; they occur in mitochondria or in the membranes of the smooth endoplasmic reticulum (SER). The latter enzyme systems are very important in the **detoxification of drugs in the liver**. They are inducible; that is, their levels increase in response to the intake of drugs. They are also known as **microsomal hydroxylases** because the SER is obtained experimentally in the form of small vesicles called microsomes.

Hydroxylases require the participation of a second reducing substrate, such as NADPH, for the reduction of the oxygen atom that is not incorporated into the substrate:

$$R{-}H + O_2 + NADPH + H^+ \rightarrow R{-}OH + H_2O + NADP^+$$

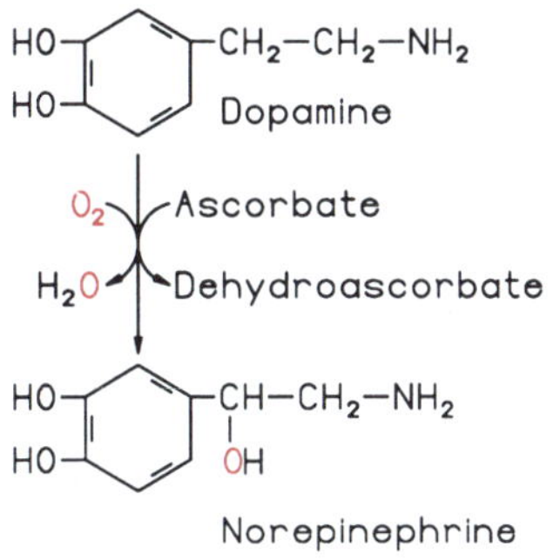

Figure 2.40.
Hydroxylation of dopamine to norepinephrine by a dioxygenase.

The dual function of these enzymes in oxidizing both the substrate and NADPH, while reducing oxygen, led to their description as **mixed-function oxidases**. A typical microsomal mixed-function oxidase (hydroxylase) consists of a FAD-containing protein and **cytochrome P_{450}**. This cytochrome is a heme-protein that derives its name from the characteristic light absorption maximum at 450 nm of the carbon monoxide complex of the reduced form. It will be discussed in more detail in Section 10.2

Dopamine β-hydroxylase uses ascorbic acid (instead of NADPH) as cosubstrate in the hydroxylation of dopamine to norepinephrine (Figure 2.40).

2.8 PENTOSE PHOSPHATE PATHWAY

The pentose phosphate pathway, also called **hexose monophosphate shunt**, serves two important functions: (1) the generation of NADPH, and (2) the synthesis of ribose 5-phosphate. NADPH is required for

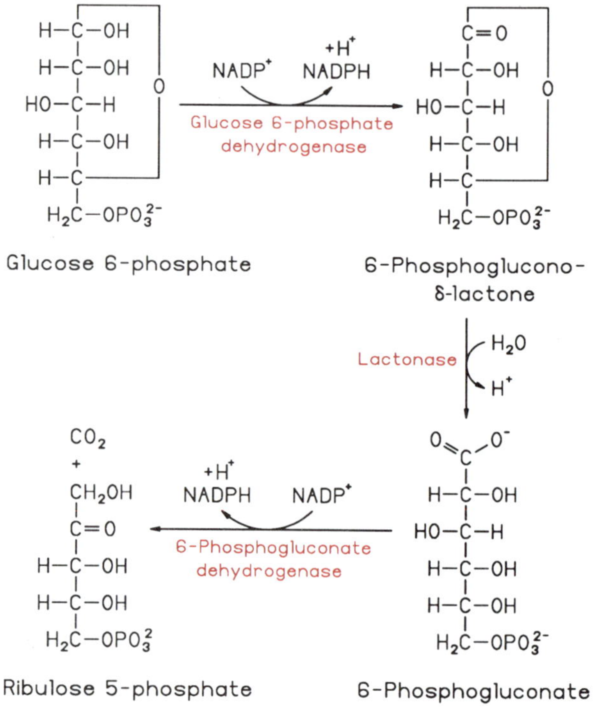

Figure 2.41.
The oxidative arm of the pentose phosphate pathway.

many biosynthetic pathways, and ribose 5-phosphate is a precursor of nucleotides (see Chapter 5).

The pathway consists of two parts, the oxidative arm, in which glucose 6-phosphate is converted to ribulose 5-phosphate and two molecules of NADPH are produced (Figure 2.41), and the nonoxidative arm, which consists of interconversions of various three-, four-, five-, six-, and seven-carbon sugars (see Figures 2.42–2.45).

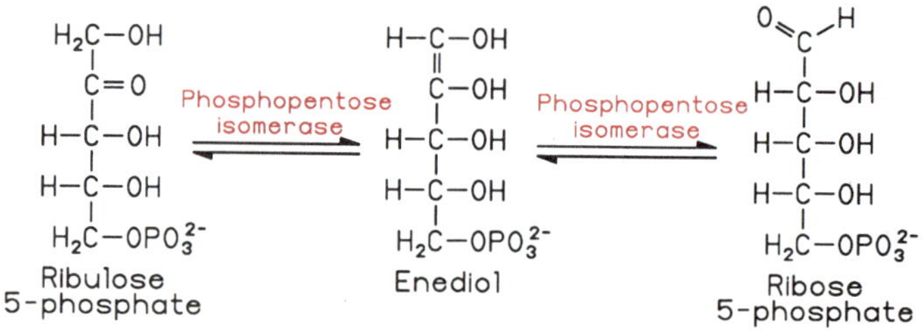

Figure 2.42.
Isomerization of ribulose 5-phosphate to ribose 5-phosphate.

2.8.1 The Oxidative Arm

The oxidative arm of the pentose phosphate pathway is irreversible and occurs only in cells that require a supply of NADPH for reductive synthesis, such as liver, adipose tissue, lactating mammary gland, kidney, and erythrocyte. It does not function in skeletal muscle, heart, or brain.

The three reactions are catalyzed by **glucose 6-phosphate dehydrogenase**, **lactonase**, and **6-phosphogluconate dehydrogenase** (see Figure 2.41). No ATP is generated in this pathway.

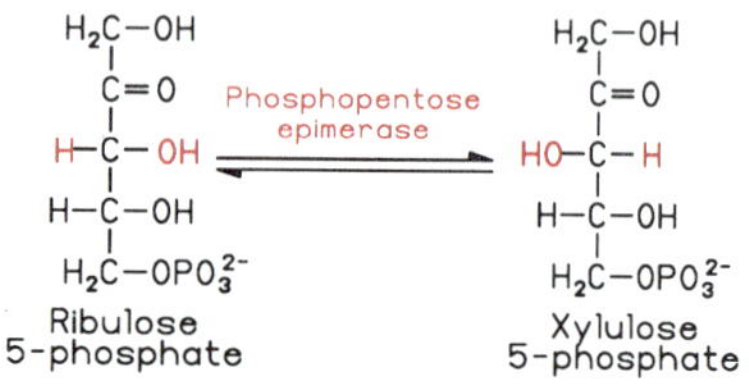

Figure 2.43.
Epimerization of ribulose 5-phosphate to xylulose 5-phosphate.

2.8.2 The Nonoxidative Arm

In contrast to the oxidative arm, the nonoxidative arm of the pentose phosphate pathway is completely **reversible** and occurs in most cells.

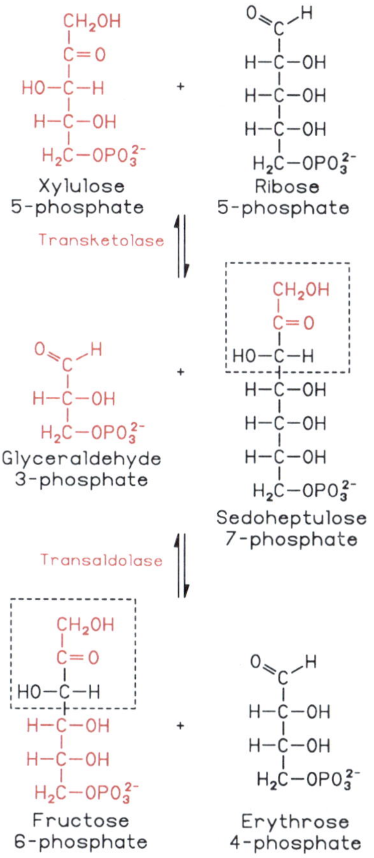

Figure 2.44.
Isomerization reactions catalyzed by transketolase and transaldolase.

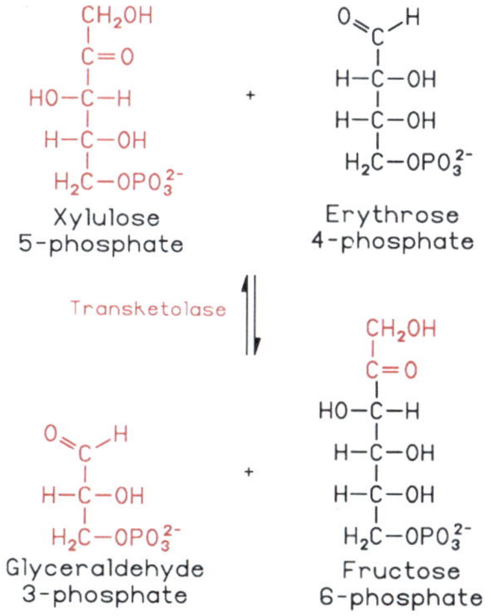

Figure 2.45.
Isomerization reactions catalyzed by transketolase.

First, ribulose 5-phosphate is isomerized to ribose 5-phosphate. Similar to other **ketose-aldose isomerization** reactions encountered in glycolysis (e.g., glucose 6-phosphate → fructose 6-phosphate; dihydroxyacetone phosphate → glyceraldehyde 3-phosphate), an **enediol intermediate** is formed (Figure 2.42.)

A different isomer of ribulose 5-phosphate is produced by an **epimerase**. Epimers are isomers of a sugar in which an *H—C—OH* group other than the anomeric (see the definition in Section 2.2) carbon atom is inverted (Figure 2.43). [Note: all sugars ending with "ulose" are ketoses. Fructose was originally called levulose, as opposed to dextrose (glucose).]

The interconversions of these pentose phosphate isomers into other sugars are catalyzed by two enzymes: **transaldolase** and **transketolase**. The reactions involve the transfer of carbon fragments from one sugar to another. The donor substrate always is a ketose, and the acceptor is always an aldose. In the process, the donor ketose becomes an aldose, and the aldose receptor becomes a ketose. Transaldolase transfers the three-carbon fragment (CH_2OH—*CO*—*CHOH*—), and transketolase transfers the two-carbon fragment (CH_2OH—*CO*—); Figures 2.44 and 2.45).

Transaldolase forms a covalent intermediate with the substrate, a Schiff base between the carbonyl group of the sugar and the ϵ amino group of a lysine residue in the protein. This destabilizes the neighboring carbon-carbon bond, thus facilitating its disruption (Figure 2.46).

Transketolase, on the other hand, uses thiamine pyrophosphate as the prosthetic group. The acidic C-2 atom of TPP reacts as carbanion (negatively charged carbon) with the carbonyl group of the substrate (Figure 2.47).

The actions of these two enzymes interconnect the glycolytic and pentose phosphate pathways:

$$C_5 + C_5 \rightleftharpoons C_3 + C_7 \rightleftharpoons C_4 + C_6$$

and

$$C_4 + C_5 \rightleftharpoons C_6 + C_3$$

Even in cells where the oxidative arm of the pentose phosphate pathway is not functional, ribose 5-phosphate can be synthesized by the reversal of the above reactions.

In cells that require large amounts of NADPH for synthetic purposes, the fructose 6-phosphate generated in the nonoxidative arm of the pathway is isomerized to glucose 6-phosphate and recycled through the oxidative arm. This recycling allows, at least theoretically, a complete oxidation of glucose 6-phosphate to CO_2.

$$\text{Glucose 6-P} + 12\ NADP^+ + 7\ H_2O \rightarrow 6\ CO_2 + 12\ NADPH + 12\ H^+\ P_i$$

It is doubtful, however, whether the pathway operates as a cycle in this way. All enzymes of the pentose phosphate pathway are located in the cytoplasm of the cell. The reactions of the pathway are summarized

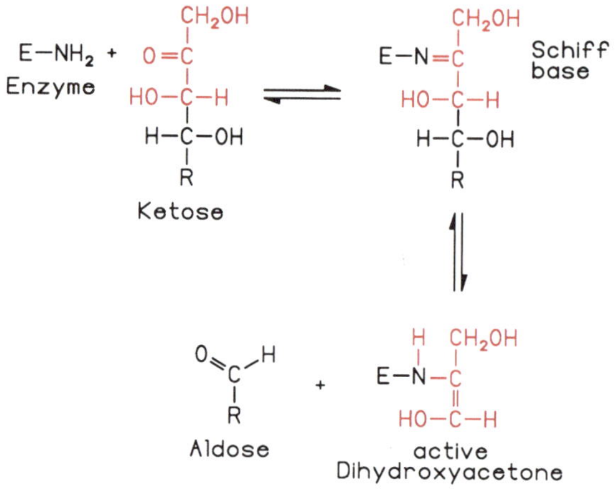

Figure 2.46.
Reaction mechanism of transaldolase.

in Table 2.8. Note that all intermediates are sugar phosphates and that nowhere is ATP used or generated.

2.8.3 Regulation of the Pentose Phosphate Pathway

The level of $NADP^+$ directly controls the first step of the pathway, the oxidation of glucose 6-phosphate to 6-phosphoglucono-δ-lactone, which is also the rate-limiting step of the pathway.

Table 2.8. Reactions of the Pentose Phosphate Pathway

	Step	Enzyme	Reaction
Oxidative arm	1	Glucose 6-phosphate dehydrogenase	Glucose 6-phosphate + $NADP^+$ ⇌ 6-phosphoglucono-δ-lactone + NADH + H^+
	2	Lactonase	6-Phosphoglucono-δ-lactone + H_2O → 6-phosphogluconate + H^+
	3	6-Phosphogluconate dehydrogenase	6-Phosphogluconate + $NADP^+$ → ribulose 5-phosphate + CO_2 + NADPH
Nonoxidative arm	4	Phosphopentose isomerase	Ribulose 5-phosphate ⇌ ribose 5-phosphate
	5	Phosphopentose epimerase	Ribulose 5-phosphate ⇌ xylulose 5-phosphate
	6	Transketolase	Xylulose 5-phosphate + ribose 5-phosphate ⇌ sedoheptulose 7-phosphate + glyceraldehyde 3-phosphate
	7	Transaldolase	Sedoheptulose 7-phosphate + glyceraldehyde 3-phosphate ⇌ fructose 6-phosphate + erythrose 4-phosphate
	8	Transketolase	Xylulose 5-phosphate + erythrose 4-phosphate ⇌ fructose 6-phosphate + glyceraldehyde 3-phosphate

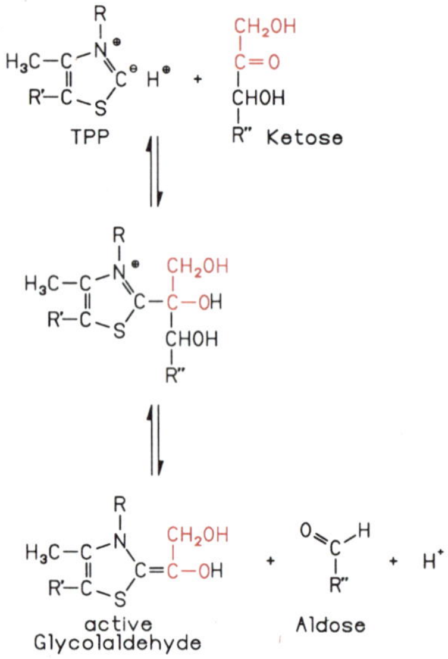

Figure 2.47.
Reaction mechanism of transketolase.

The reversible nonoxidative arm of the pathway is controlled mainly by the relative cytoplasmic concentrations of its intermediates.

2.9 GLUCONEOGENESIS

Gluconeogenesis is the synthesis of glucose from noncarbohydrate precursors such as amino acids, glycerol, or lactate. Over 90% of the gluconeogenesis normally occurs in the liver, the remainder in the kidney. It serves to maintain a constant level of glucose in the circulation even during periods of dietary restrictions, thereby fulfilling the demands of the central nervous system and a few other cells, such as red blood cells, that depend on glucose as their sole source of energy. Only after prolonged periods of starvation can the brain adapt to metabolizing ketone bodies.

The main **point of entry** into the gluconeogenesis pathway is via **pyruvate**. The reactions from pyruvate to glucose, however, do not represent a simple reversal of glycolysis, since glycolysis is an exergonic process including several irreversible steps. Gluconeogenesis from lactate must overcome the energy barrier of −47 kcal/mol that are released during the conversion of glucose into lactate. This is accomplished through the expenditure of six ATP molecules for the synthesis of one molecule of glucose:

$$2 \text{ Lactate} + 6 \text{ ATP} \rightarrow \text{glucose} + 6 \text{ ADP} + 6 \text{ P}_i$$

Gluconeogenesis utilizes several reversible steps of glycolysis. In addition, it employs four reactions that are different from glycolysis:

1. **Pyruvate carboxylase** converts pyruvate into oxaloacetate (see anaplerotic reactions of the citric acid cycle). This reaction occurs in the mitochondrion.
2. **Phosphoenolpyruvate carboxykinase** converts oxaloacetate into phosphoenolpyruvate (PEP). In this cytoplasmic reaction, oxaloacetate is decarboxylated and phosphorylated by GTP (Figure 2.48). The $\Delta G^{0\prime}$ of this reaction is slightly endergonic (+0.2 kcal/mol), but the removal of CO_2 shifts the equilibrium in favor of PEP.

   Since oxaloacetate cannot cross the mitochondrial membrane, it is converted into malate by **malate dehydrogenase**. The malate then enters the cytoplasm, where it is reoxidized to oxaloacetate by cytoplasmic malate dehydrogenase. Both mitochondrial and cytoplasmic malate dehydrogenases require NADH or NAD^+, respectively. This represents a "reverse shuttle" for the transfer of NADH out of the mitochondrion (see the discussion of the glycerol phosphate shuttle in Section 2.3). The cytoplasmic NADH is later used for the conversion of 1,3-diphosphoglycerate into glyceraldehyde 3-phosphate in a reversal of the glycolytic reaction.
3. **Fructose 1,6-bisphosphatase** converts fructose 1,6-bisphosphate to fructose 6-phosphate.

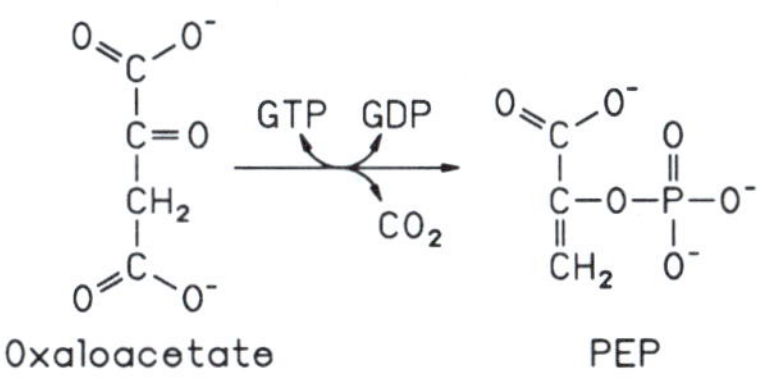

Figure 2.48.
Conversion of oxaloacetate into phosphoenolpyruvate by PEP carboxykinase.

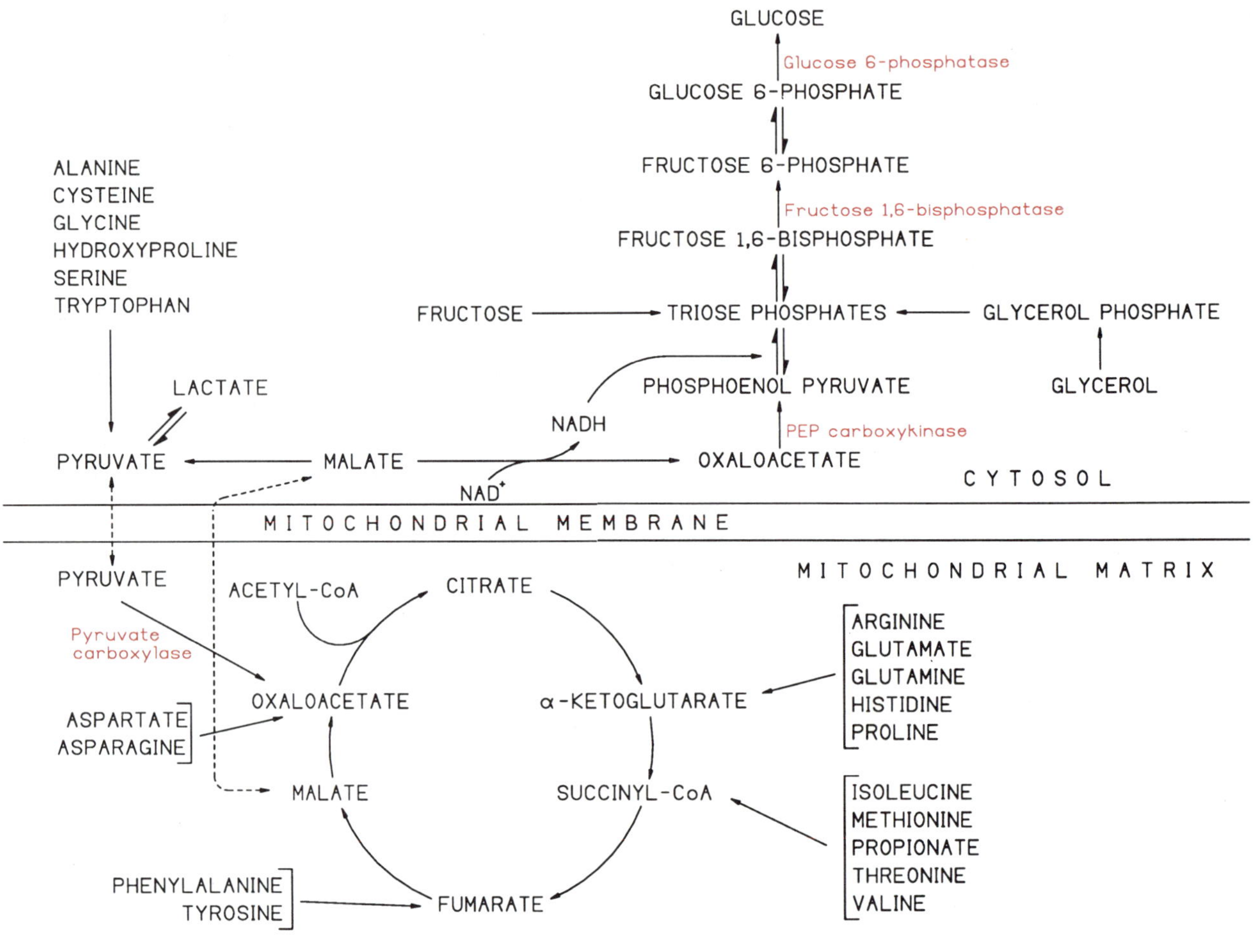

Figure 2.49.
Overview of gluconeogenesis. Key enzymes of gluconeogenesis are indicated in blue. Note the different entry points for the various amino acids into the gluconeogenic pathway. (Modified from Whelan WJ. Biochemistry lecture notes. Miami, FL: University of Miami; 1982, mimeographed.)

4. **Glucose 6-phosphatase** hydrolyzes glucose 6-phosphate to glucose and inorganic phosphate, allowing the unphosphorylated glucose to cross the cell membrane and enter the bloodstream for transport to the tissues where it is needed.

The complete pathway of gluconeogenesis is illustrated in Figure 2.49.

2.9.1 Regulation of Gluconeogenesis

Gluconeogenesis is regulated at several points. The first control point is the carboxylation of pyruvate to oxaloacetate. Pyruvate carboxylase is activated by acetyl CoA and inhibited by ADP. If the cell has an excess of ATP, more oxaloacetate is synthesized and used for gluconeogenesis. When ATP levels fall, acetyl CoA combines with the oxaloacetate to enter the citric acid cycle. (Compare this with the regulation of pyruvate kinase.)

The reaction catalyzed by fructose 1,6-bisphosphatase represents another control point. The enzyme is activated by ATP and by citrate and is inhibited by AMP and fructose 2,6-bisphosphate. This control is the opposite of the regulation of phosphofructokinase, which is activated by AMP and inhibited by ATP. This prevents futile cycling (i.e., the consumption of ATP for no purpose other than to generate heat).

Insulin represses the biosynthesis of the four key enzymes of gluconeogenesis: pyruvate carboxylase, PEP-carboxykinase, fructose 1,6-bisphosphatase, and glucose 6-phosphatase. This appears understandable, since insulin is released when blood glucose levels are high. Cortisol or glucagon, on the other hand, both stimulate de novo synthesis of these enzymes (see Chapter 13).

In contrast to the liver, where gluconeogenesis mainly serves to maintain glucose homeostasis, the kidney uses gluconeogenesis as a means of producing ammonia from amino acids in order to neutralize urinary acids (see Chapter 12). The resulting α-keto acids, as well as the reabsorbed lactate, are converted to glucose. Gluconeogenesis in the kidney is not influenced by insulin or glucagon. Instead, it is stimulated by parathyroid hormone, whose level rises in response to metabolic acidosis.

Some of the lactate that is produced in muscle during strenuous exercise (anaerobic conditions) is reconverted into glucose in the liver and then resupplied to the muscle via the bloodstream. This cyclic interconversion of lactate and glucose is sometimes referred to as the **Cori cycle**. A true Cori cycle probably operates only during fasting when the liver engages in gluconeogenesis. Normally, lactate is being utilized directly by many other tissues.

2.10 STRUCTURE AND METABOLISM OF COMPLEX CARBOHYDRATES

In humans, relatively few types of carbohydrates are encountered in the form of monosaccharides. However, a variety of different sugars are found as monomeric units in polysaccharides. With the exception of **glycogen**, which is a **homopolymer** of glucose residues, all polysaccharides found in human tissues are **heteropolymers**; that is, they contain more than one type of monomeric subunit.

Most of these polysaccharide chains are covalently linked to some protein or lipid. Compounds with a predominantly protein character are called **glycoproteins**, those with predominantly carbohydrate character are called **proteoglycans** or **mucopolysaccharides**. **Glycolipids** have relatively short carbohydrate chains. The structures of some monosaccharide units found in mammalian oligosaccharides and polysaccharides are shown in Figure 2.50.

Proteoglycans consist of multiple polysaccharide chains attached to a protein backbone. They serve mostly structural functions in skin, connective tissue, cartilage, and bone. The structures of these polysaccharides are discussed in Section 11.2. This discussion also includes two pure polysaccharides: hyaluronic acid, which forms very viscous solutions and functions as a lubricant in joints, and heparin, an important anticoagulant in blood.

Glycoproteins contain oligosaccharides and have both structural and informational functions. They are found in cell membranes and appear to be responsible for **cell-cell recognition**. For example, **N-CAMs** (neural cell adhesion molecules) function in cell-cell adhesion

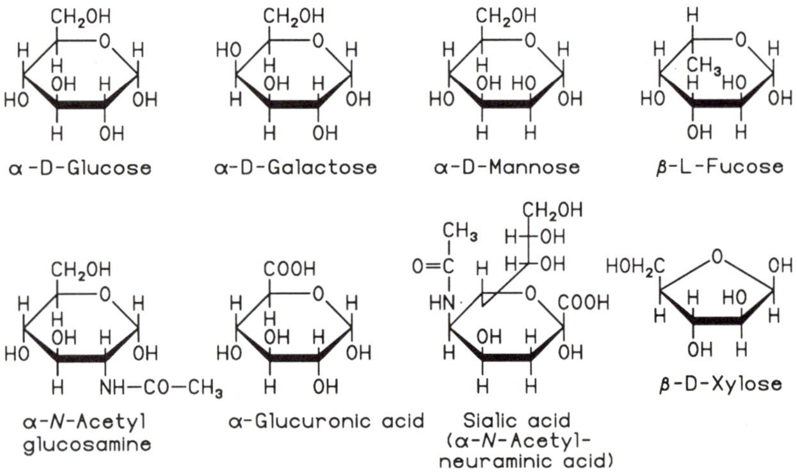

Figure 2.50. Structures of some monosaccharide units found in human oligosaccharides and polysaccharides.

during neural development. Secreted proteins such as extracellular enzymes and antibodies also are mostly glycoproteins. All glycoproteins are synthesized in the endoplasmic reticulum, where part of the oligosaccharide side chains are attached. Additional sugars are attached in the Golgi apparatus. (The synthesis of glycoproteins will be discussed in detail in Section 6.6.)

The oligosaccharides are linked to the protein at their reducing end. Depending on the amino acid to which the oligosaccharides are attached, one distinguishes two major classes of glycoproteins: O-linked glycoproteins and N-linked glycoproteins.

In **O-linked glycoproteins**, the first sugar, usually a **D-xylose** or **D-*N*-acetylgalactosamine**, is linked to a **serine** or **threonine** residue of the protein. Examples of O-linked glycoproteins are the ABO blood group antigens expressed on cell surfaces. In collagen, the first sugar attached to hydroxylysine is D-galactose. The biosynthesis of the oligosaccharide side chains is catalyzed by a large number of different enzymes, each specific for the attachment of only one kind of monosaccharide.

N-linked glycoproteins have their oligosaccharides attached to the amide group of asparagine residues in the sequence Asn-X-Ser/Thr, where X may be any amino acid except Pro and Asn. There are three major classes of N-linked glycoproteins: complex, hybrid, and high mannose. All have the same branched pentasaccharide structure attached through a **D-*N*-acetylglucosamine** residue to **asparagine** but differ in the outer monosaccharide units (Figure 2.51). Sugars that occupy other positions in the oligosaccharide side chains include D-*N*-acetylglucosamine, D-*N*-acetylgalactosamine, D-galactose, D-glucose (mainly in collagen), D-mannose, L-fucose, and D-*N*-acetylneuraminic acid (sialic acid). They are interconnected by a variety of different glycosidic bonds.

Terminal **sialic acid** residues protect circulating glycoproteins from binding to specific asialoglycoprotein receptors located in the endothelium of blood vessels. However, circulating sialylases tend to remove these sialic acid residues. Once attached to these receptors, the glycoproteins are internalized by endocytosis and degraded. Thus, the lifetime of a glycoprotein is determined by its sensitivity to sialylases, which, in turn, depends on its structure.

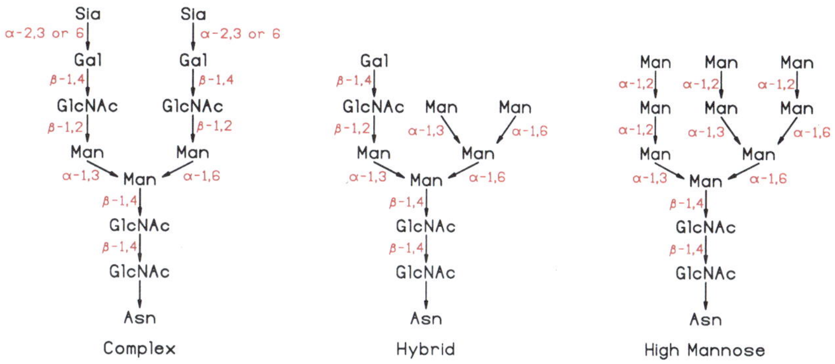

Figure 2.51.
Carbohydrate side chains in N-linked glycoproteins.

The distribution of glycoproteins in cellular membranes has been studied with the help of **lectins**, proteins that specifically bind to certain sugar residues. Although found throughout nature (e.g., the asialoglycoprotein receptors can be viewed as lectins), plant lectins have been described most extensively. For example, **concanavalin A** (jackbean) binds to mannose residues, and **wheat germ agglutinin** binds to *N*-acetylglucosamine trisaccharides.

2.10.1 Biosynthesis of Polysaccharides

The biosynthesis of polysaccharides in humans utilizes nucleoside diphosphate sugars as precursors. These precursors are generated from a monosaccharide 1-phosphate and nucleoside triphosphate. For example, **uridine diphosphate glucose** (Figure 2.52) is synthesized from glucose 1-phosphate (G-1-P) and uridine triphosphate (UTP).

$$\text{UTP} + \text{G-1-P} \rightarrow \text{UDPG} + \text{PP}_i$$

Polysaccharide synthesis requires a primer, a short piece of preformed polymer. The polymer then grows by sequential attachment of monomeric sugar units, derived from nucleoside diphosphate sugars, to the nonreducing end of the primer. The primer remains part of the polymer product. Figure 2.53 illustrates the elongation of a glycogen chain by a glucose residue provided by UDP-glucose.

When different monosaccharides are incorporated into the polymer, as is the case in glycoprotein synthesis, a different transglycosylase enzyme is required for the addition of each particular monosaccharide. In the absence of one of these enzymes, the synthesis of the mucopolysaccharide or glycoprotein is faulty. Similarly, the catabolism of these polysaccharides is impaired or blocked if one or another of the specific hydrolases that cleave the glycosidic bonds are absent. Genetic deficiencies of the hydrolytic enzymes are known as mucopolysaccharidoses; they are discussed in Section 11.2

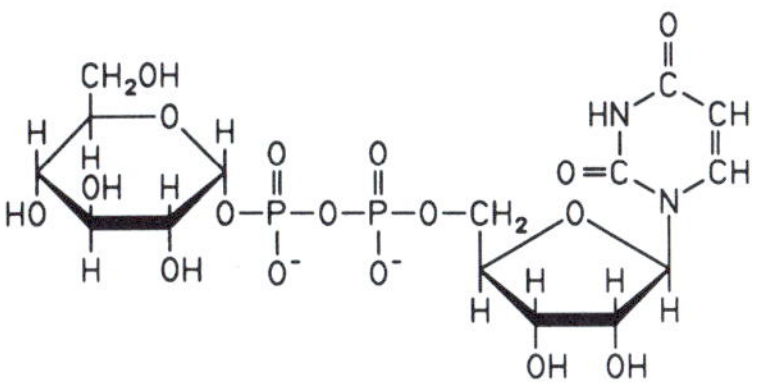

Figure 2.52.
Uridine diphosphate glucose (UDP-glucose).

The various monomeric precursors of polysaccharides are all derived from glucose. **D-Mannose**, for example, is the epimer of glucose at C-2. Its 6-phosphate is formed from fructose 6-phosphate by **phosphomannoisomerase**. **L-Fucose** is a 6-deoxyhexose that is formed at the nucleoside diphosphate sugar level. Similarly, UDP-galactose is formed from UDP-glucose by **UDP-galactose 4-epimerase**. This enzyme par-

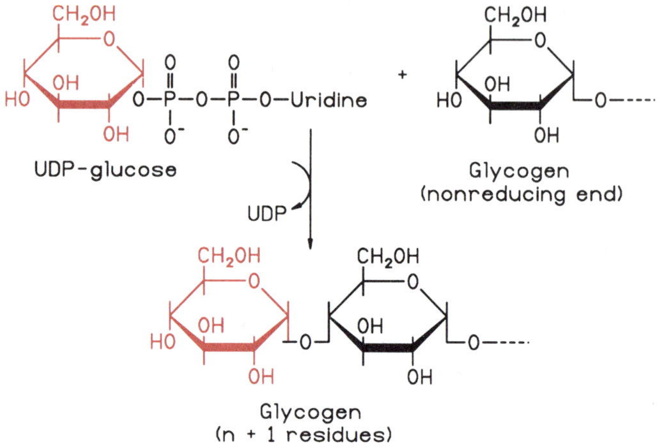

Figure 2.53.
Polymerization of glucose (from UDP-glucose) to glycogen.

ticipates in the conversion of dietary galactose into glucose. As mentioned earlier, galactose is first phosphorylated to galactose 1-phosphate, which then undergoes an exchange reaction with UDP-glucose to form UDP-galactose and glucose 1-phosphate:

$$\text{UDP-glucose} + \text{galactose-1-P} \rightleftharpoons \text{UDP-galactose} + \text{glucose-1-P}$$

A genetic deficiency of **galactose 1-phosphate uridyltransferase** causes **galactosemia**, a severe disease leading to mental retardation unless galactose is excluded from the diet. Patients can still synthesize UDP-galactose from UDP-glucose by using the epimerase and thus incorporate galactosyl residues into glycoproteins as needed.

Figure 2.54 represents an overview of the interconversions of monosaccharides that are used in polysaccharide synthesis in the human body.

Neuraminic acid is derived from UDP-*N*-acetylglucosamine via *N*-acetylmannosamine, which is first phosphorylated to *N*-acetylmannosamine 6-phosphate and then reacts with phosphoenolpyruvate to form *N*-acetylneuraminic acid (NANA) 9-phosphate (Figure 2.55). NANA 9-phosphate is activated with cytosine triphosphate (CTP) to CMP-NANA for subsequent incorporation into glycoproteins.

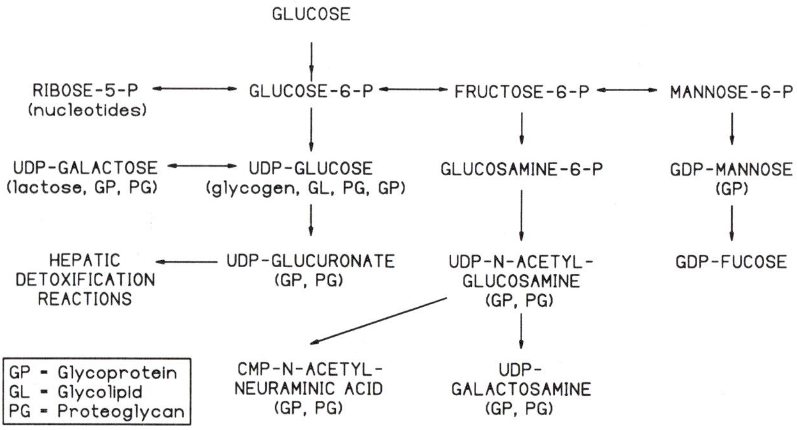

Figure 2.54.
Interconversions of monosaccharides used in polysaccharide synthesis. (Modified from Whelan WJ. Biochemistry lecture notes. Miami, FL: University of Miami; 1982, mimeographed.)

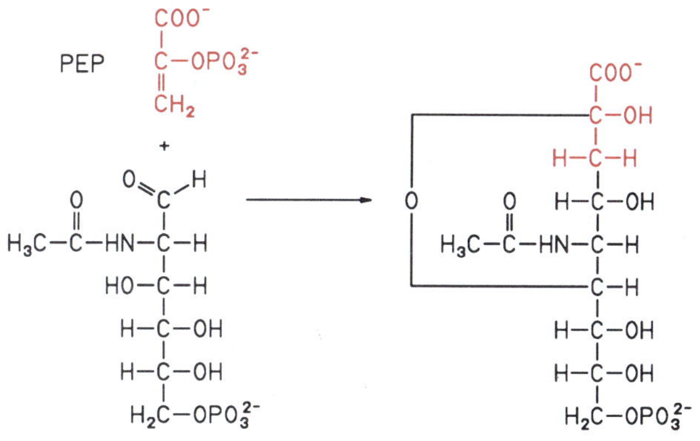

Figure 2.55.
Biosynthesis of *N*-acetylneuraminic acid (NANA) 9-phosphate.

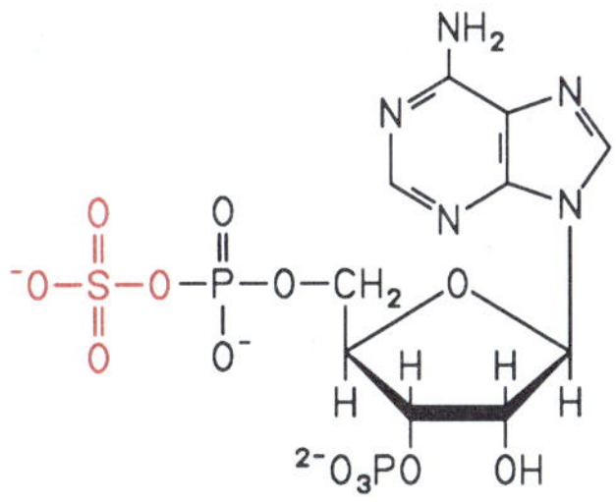

Figure 2.56.
3′-Phosphoadenosine 5′-phosphosulfate.

Glucosamine, as the 6-phosphate, is synthesized from fructose 6-phosphate. Glutamine is the amino donor. Other amino sugars are formed from glucosamine 6-phosphate.

Acetylation of glucosamine 6-phosphate is achieved with acetyl CoA. **Sulfation** of sugar residues occurs after the sugar has been incorporated into the polysaccharide. The sulfate donor is **3′-phosphoadenosine 5′-phosphosulfate (PAPS)** Figure 2.56, a compound that is synthesized from ATP and inorganic sulfate in a two-step reaction.

Those oligosaccharide chains that are to be attached to asparagine are synthesized on lipid carrier molecules before they are incorporated into glycoproteins. The lipid carriers are polyisoprene phosphates, called **dolichol phosphates**, containing 15–20 isoprene units (Figure 2.57). They are located in the membrane of the endoplasmic reticulum (ER).

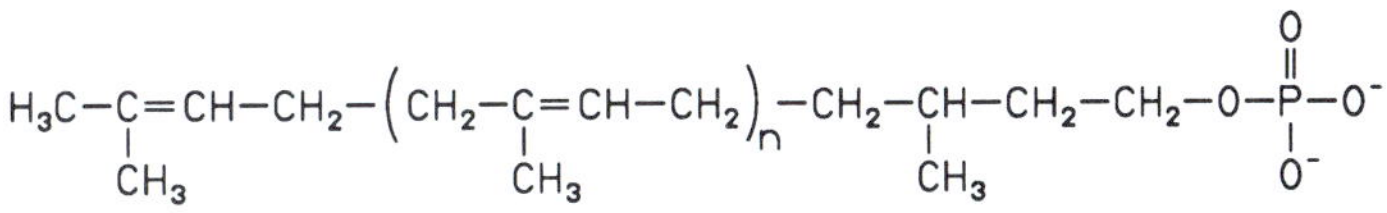

Figure 2.57.
Dolichol phosphate (n = 16–20).

Dolichol phosphate (Dol-P) reacts with *N*-acetylglucosamine (GlcNAc) to form a glycosyl diphosphodolichol:

$$\text{UDP—GlcNac} + \text{Dol-P} \rightarrow \text{Dol-P—P—GlcN Ac} + \text{UMP}$$

It appears that this reaction occurs on the cytoplasmic side of the ER membrane. Additional sugar residues are then added to yield a common dolichol phosphate–linked oligosaccharide precursor for all N-linked glycoproteins that contains 14 monosaccharide units (Figure 2.58). There is evidence that the *N*-acetylglucosamine and from three to five mannose residues are derived from UDP-GlcNAc and GDP-Man, respectively, and added on the cytoplasmic side of the ER. The growing oligosaccharide chain is then flipped to the lumenal side of the ER where the remaining sugars, derived from Dol-P-Man and Dol-P-Glc, are added. The completed Dol-P–linked oligosaccharide is then transferred to an asparagine residue of the protein. Most of the sugar residues of the

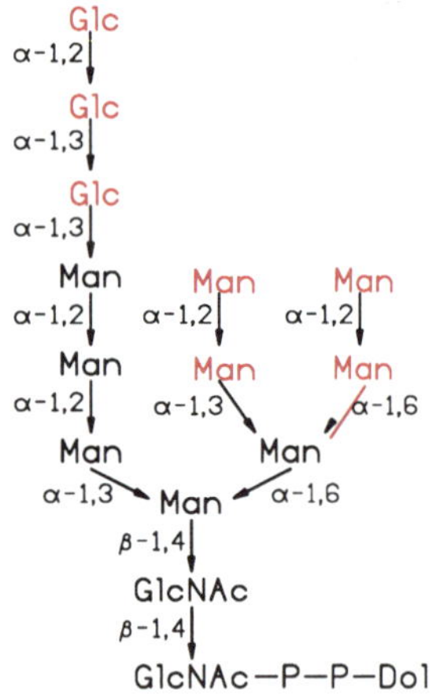

Figure 2.58.
Common oligosaccharide chain synthesized on the dolichol phosphate carrier. The sugar units added in the cytoplasm are indicated in black, those added in the endoplasmic reticulum are indicated in color.

oligosaccharide are then removed, first in the ER and subsequently in the Golgi apparatus. When only the core pentasaccharide (Figure 2.51) is left, new sugar residues are added in the Golgi apparatus.

2.11 SYNTHESIS OF LACTOSE AND FRUCTOSE

In addition to glucose, the human body is capable of synthesizing two sugars: lactose in the mammary gland, and fructose in male seminal vesicles.

Lactose is a disaccharide containing galactose and glucose linked by a β1,4-glycosidic bond. Its synthesis is catalyzed by **lactose synthase**, an enzyme that consists of two subunits, **galactosyl transferase** and **α-lactalbumin**. Galactosyl transferase, present in many tissues, normally transfers a galactosyl residue from UDP-galactose to an *N*-acetylglucosamine (GlcNAc) residue in a glycoprotein.

$$\text{UDP-Gal} + \text{GlcNAc}\cdots\text{protein} \rightarrow \text{Gal}(\beta 1{,}4)\text{-GlcNAc}\cdots\text{protein} + \text{UDP}$$

In the lactating mammary gland, however, the membrane-bound galactosyl transferase combines with α-lactalbumin, a milk protein. This interaction alters the substrate specificity of galactosyl transferase so that glucose, instead of *N*-acetylglucosamine, becomes the acceptor substrate.

$$\text{UDP-Gal} + \text{glucose} \rightarrow \text{galactose-}(\beta 1{,}4)\text{-glucose} + \text{UDP}$$

Fructose is synthesized in seminal vesicles from D-glucose via D-sorbitol (Figure 2.59).

Figure 2.59.
The biosynthesis of fructose.

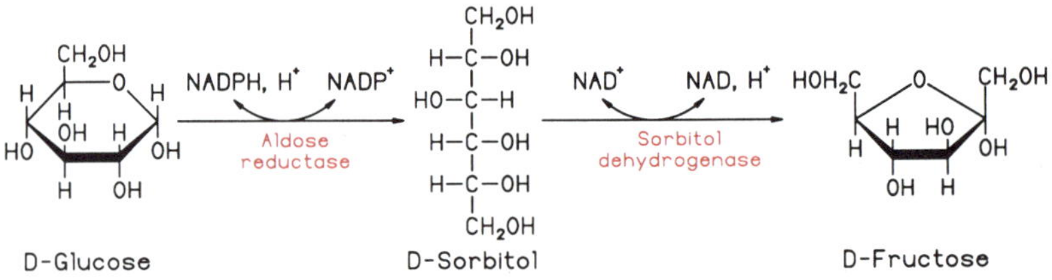

2.12 GLYCOGEN METABOLISM

Glycogen is the major storage form of glucose in humans. It is a very large macromolecule containing up to 1 million glucose residues linked together by α-1,4-glucosidic bonds. The molecule is highly branched through additional α-1,6-glucosidic bonds (7%; Figure 2.60). Because of the branched nature of glycogen, about 50% of the glucose residues of the molecule are located on the surface of the molecule and 10% at chain ends. This permits rapid degradation by **phosphorylase**, an enzyme that degrades α-1,4-glucan (polyglucose) chains sequentially

Figure 2.60.
Structure of glycogen. (Modified from Cori GT. *The Harvey Lectures.* 1954; 48:145–171.)

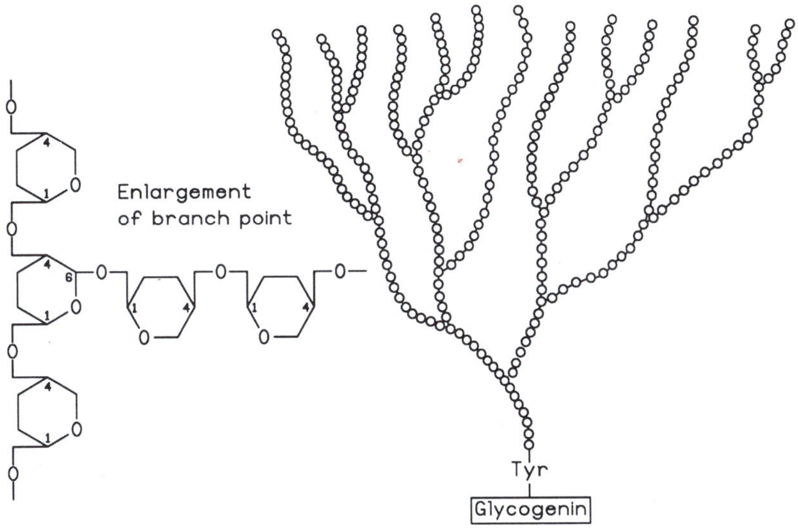

from their nonreducing ends. Glycogen is stored in cytoplasmic granules predominantly found in liver and muscle cells.

2.12.1 Glycogen Synthesis

Glycogen synthesis proceeds from a primer that is linked to the tyrosine hydroxyl group of a cellular protein, **glycogenin**. **Glycogen synthase**, using UDP-glucose as a precursor, adds glucose units to the nonreducing end of the primer, joining them by α-1,4 linkages (see Figure 2.53). An additional enzyme, called **branching enzyme,** randomly breaks α-1,4 bonds and reattaches the severed fragments to other parts of the molecule, creating a new α-1,6 bond and thus generating branch points (Figure 2.61). The average number of glucose units in each linear sequence of α-1,4 bonds lies between 10 and 14 (Figure 2.60).

2.12.2 Glycogen Degradation

Glycogen degradation for energy production proceeds by the combined actions of **phosphorylase** and the **debranching system**. Phosphorylase removes glucosyl residues from the nonreducing ends and trans-

Figure 2.61.
Action of branching enzyme during glycogen synthesis.

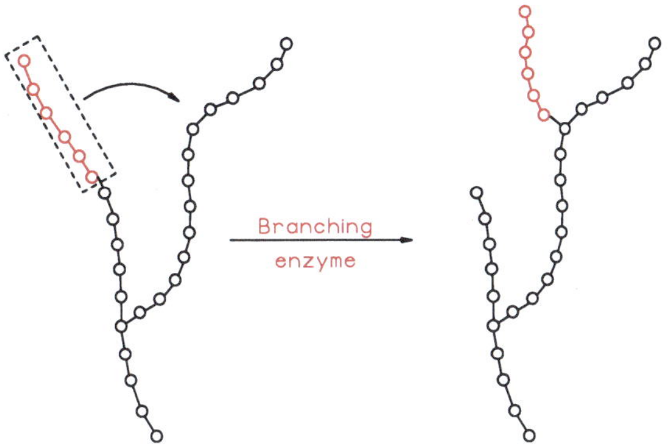

fers them to inorganic phosphate, forming glucose 1-phosphate in the process.

$$(\text{Glucose})_n + P_i \rightleftharpoons \text{glucose-1-P} + (\text{glucose})_{n-1}$$

This phosphorolytic process continues until there are four glucosyl residues left before a branch point, then the debranching enzyme system comes into action. A **glucan transferase** transfers a triglucosyl segment to the nonreducing end of another chain. The remaining glucosyl residue is then hydrolytically cleaved by amylo-1,6-glucosidase. The two enzyme activities appear to be expressed by a single polypeptide chain. Glycogenolysis thus will yield about one free glucose for every 10 glucose 1-phosphate molecules.

Glycogen is also hydrolyzed by **α-amylase** and **α-glucosidase**, cytoplasmic enzymes whose physiologic significance is not understood. Some glycogen is also degraded in lysosomes by a **lysosomal** α-**glucosidase**. In **Pompe's disease**, this enzyme is defective, and glycogen accumulates in the lysosomes, causing early death. The lysosomal α-glucosidase can hydrolyze both α-1,4-glucosidase bonds and α-1,6-glucosidic bonds.

In humans, several genetic deficiencies of enzymes involved in glycogen metabolism are known; they are called **glycogen storage diseases** (Table 2.9).

2.12.3 Control of Glycogen Metabolism

Both the synthesis and degradation of glycogen are hormonally controlled by a complex cascade mechanism (Figure 2.62).

Glycogen degradation in the liver is induced by **glucagon** (released in response to low blood glucose) or **epinephrine** (released in response to exercise or stress). Glucagon acts on the cell membrane, where it activates **adenylate cyclase** (for the detailed mechanism, see Section 13.5). This enzyme converts ATP into cyclic AMP (cAMP), the intracellular second messenger for many hormones (Figure 2.63).

The cAMP, in turn, activates a **protein kinase,** an enzyme capable of phosphorylating several different proteins using ATP as a phosphate

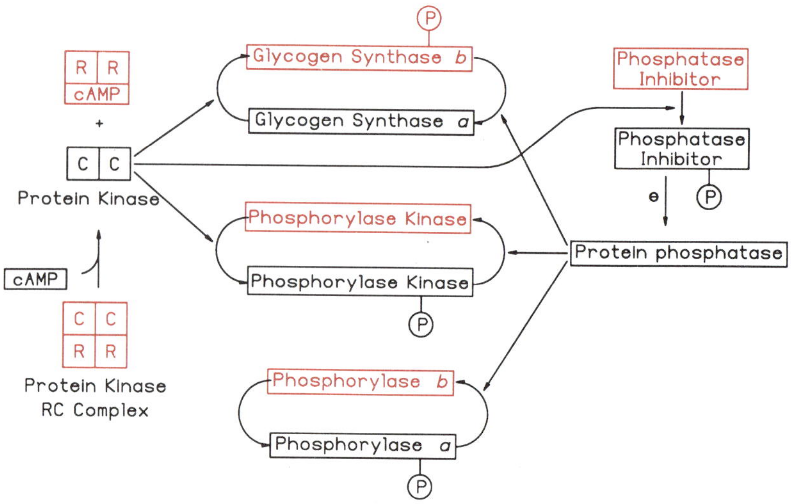

Figure 2.62.
Control of glycogen metabolism by phosphorylation and dephosphorylation of glycogen synthase and phosphorylase. Inactive enzymes are shown in color.

Table 2.9. Glycogen Storage Diseases

Type	Name	Enzyme Defect	Affected Organ	Glycogen Content, Structure	Symptoms
I	von Gierke's disease	Glucose 6-phosphatase	Liver and kidney	Increased, normal structure	Enlargement of the liver, hypoglycemia, hyperlipemia, gout
II	Pompe's disease	Lysosomal α-1,4-glucosidase	Liver, heart, muscle	Increased, normal structure	Cardiorespiratory failure at early age, muscular weakness
III	Cori's disease	Amylo-1,6-glucosidase (debranching enzyme)	Muscle, heart, liver	Increased, short outer branches	Mild liver enlargement, ketosis, muscle weakness, cramps
IV	Andersen's disease	Branching enzyme	Liver and spleen	Normal, very long outer branches	Cirrhosis of the liver, muscular weakness, early death
V	McArdle's disease	Phosphorylase	Muscle	Moderately increased, normal structure	Muscle cramps, but only after strenuous exercise
VI	Hers' disease	Phosphorylase	Liver	Increased	Similar to type III
VII		Phosphofructokinase	Muscle	Increased, normal structure	Similar to type V
VIII		Phosphorylase kinase	Liver	Increased, normal structure	Liver enlargement (relatively benign)
IX		Glycogen synthase	Liver	Decreased, normal structure	Hypoglycemia, slow growth

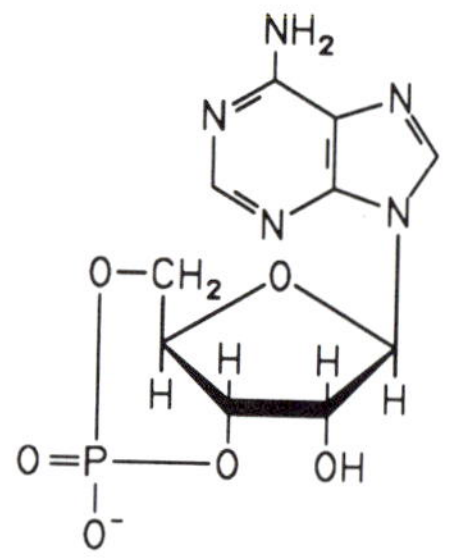

Figure 2.63.
Cyclic AMP (cAMP).

donor. Cyclic AMP-dependent protein kinase consists of four subunits, two catalytic subunits C and two regulatory subunits R. Cyclic AMP combines with the R subunits, making them dissociate from the C subunits. As a result, the C subunits dissociate into monomers and become activated. This active protein kinase now phosphorylates the enzyme **phosphorylase kinase**, thereby activating it. Active phosphorylase kinase finally phosphorylates the inactive phosphorylase b, thus creating active phosphorylase a. The purpose of this cascade of reactions appears to be an amplification of a small initial stimulus into a large response.

Phosphorylase kinase is a tetrameric protein with the subunit structure $(\alpha\beta\gamma\delta)_4$. The γ subunit is the catalytic subunit, the α and β subunits are regulatory in nature and can be phosphorylated, and the δ subunit is **calmodulin**. The affinity of calmodulin for Ca^{2+} ions is increased when the regulatory subunits are phosphorylated. The binding of Ca^{2+} ions is required for kinase activity.

The mechanism of epinephrine action on liver cells is different. **Epinephrine** combines with **α-adrenergic receptors** in the cell membranes. This interaction releases intracellular Ca^{2+} ions that combine with calmodulin to activate phosphorylase kinase. The activated phosphorylase kinase then activates glycogen phosphorylase as described above. (Note that the mechanism of epinephrine action on muscle cells is different from that described for liver cells; see below.)

The resulting phosphorolytic degradation of glycogen can be stopped by the action of a **protein phosphatase 1**, which converts phosphorylase back into its inactive form b. The activity of phosphatase 1 is regulated by **inhibitor 1**, a protein that in phosphorylated form inhibits the enzyme but that in nonphosphorylated form has no effect.

The same protein kinase that initiates glycogen degradation, through activation of phosphorylase, inhibits the biosynthesis of glycogen by inactivating glycogen synthase a through phosphorylation. Note that glycogen synthase is inactive when phosphorylated (b form), whereas phosphorylase is active when phosphorylated (a form).

In skeletal muscle, the principal activator of protein kinase appears to be Ca^{2+} ions (through the action of calmodulin), which are released from the sarcoplasmic reticulum during muscle contraction. Epinephrine and neural stimulation (norepinephrine) exert an additional effect through cAMP. This effect is comparable to the activation of adenylate cyclase by the glucagon-receptor complex. Cyclic AMP is hydrolyzed by cAMP phosphodiesterase. Caffeine inhibits this phosphodiesterase, thus prolonging the effects of cAMP.

In addition to this regulation via phosphorylation and dephosphorylation, glucose 6-phosphate allosterically activates glycogen synthase b, whereas AMP activates phosphorylase b. Note that enzyme regulation through phosphorylation is not always mediated by cAMP. The protein kinase that phosphorylates pyruvate dehydrogenase does not require cAMP.

For a more extensive discussion of carbohydrate metabolism and energy supply, consult the following textbooks:

Mathews et al. (1990): Chapters 3, 8, 12–16

Stryer (1988): Chapters 13–19

Further References

Chock PB, Rhee SG, Stadtman ER. Interconvertible enzyme cascades in cellular regulation. *Annu Rev Biochem.* 1980; 49:813–843. (Review of glycogen synthesis and degradation mechanisms.)

Elbein AD. Inhibitors of the biosynthesis and processing of N-linked oligosaccharide chains. *Annu Rev Biochem.* 1987; 56:497–534.

Hatefi Y. The mitochondrial electron transport and oxidative phosphorylation system. *Annu Rev Biochem.* 1985; 54:1015–1069.

Leloir LF. Two decades of research on the biosynthesis of saccharides. *Science* 1971; 172:1299–1302.

Malmstrom B. Enzymology of oxygen. *Annu Rev Biochem.* 1982; 51:21–59.

McCormack JG, Denton RM. Ca^{2+} as a second messenger within mitochondria. *Trends Biochem Sci.* 1986; 11:258–262.

Koshland DE. Control of enzyme activity and metabolic pathways. *Trends Biochem Sci.* 1984; 9:155–159.

Krebs HA. The history of the tricarboxylic acid cycle. *Perspect Biol Med.* 1970; 14:154–170.

Mitchell P. Keilin's respiratory chain concept and its chemiosmotic consequences. *Science.* 1979; 206:1148–1159.

Nishizuka Y. Protein kinases in signal transduction. *Trends Biochem Sci.* 1984; 9:163–166.

Ovádi J. Old pathway–new concept. Control of glycolysis by metabolite-modulated dynamic enzyme associations. *Trends Biochem Sci.* 1988; 13:486–490.

Pilkis SJ, El-Maghrabi MR, Claus TH. Hormonal regulation of hepatic gluconeogenesis and glycolysis. *Annu Rev Biochem.* 1988; 57:755–784.

Prince RC. The proton pump of cytochrome oxidase. *Trends Biochem Sci.* 1988; 13:159–160.

Sharon N. Glycoproteins. *Trends Biochem Sci.* 1984; 9:199–201.

Srere PA. Complexes of sequential metabolic enzymes. *Annu Rev Biochem.* 1987; 56:89–124.

CHAPTER

3 Lipid Metabolism

OBJECTIVES

After studying this chapter, the student should be able to

1. Give an example for each of the naturally occurring lipid classes and describe its principal function.
2. Understand the difference between the Δ system and the n system for describing the structures of fatty acids and explain how the two double bonds of linoleic acid (18:2Δ9,12) are retained when it is converted to arachidonic acid (20:4Δ5,8,11,14).
3. Describe the structures of the different types of phospholipids and sphingolipids.
4. Explain the asymmetry of biological membranes as a result of their biosynthesis.
5. Explain the terms chylomicron, VLDL, LDL, and HDL and outline their function in the transport of lipids in the bloodstream.
6. Understand the significance of HDL and LDL in cholesterol transport and metabolism.
7. Describe the functions of hormone-sensitive lipase, lipoprotein lipase, ACAT, and LCAT.
8. Explain why the genetic deficiency of cell surface receptors for apolipoprotein B causes hypercholesterolemia.
9. Describe the three mechanisms for the oxidation of fatty acids and their relative importance.
10. Explain how acetyl CoA is shuttled into and out of the mitochondrion.
11. Understand why starvation increases the production of ketone bodies.
12. Describe the functioning of the fatty acid synthase complex.
13. Explain how longer-chain fatty acids (C_{18} and higher) are synthesized and how double bonds can be introduced into the molecule.
14. Compare the de novo and salvage pathways for the synthesis of phospholipids.
15. Describe the synthesis of mevalonic acid from acetyl CoA and its regulation.

Lipids are a highly heterogeneous class of biological compounds. They are insoluble in water but variously soluble in ether, chloroform, and other organic solvents. Their hydrophobic nature is due to their large hydrocarbon content.

Neutral fats, consisting of fatty acids esterified to glycerol, serve as the quantitatively most important store of metabolic energy. These **triacylglycerols**, present in large quantities in adipose tissue, contain almost no water and yield 8–9 kcal of energy per gram of stored fat. In contrast, glycogen and protein contain at least 75% water and yield only about 1 kcal/g of stored material. Fat deposits turn over rapidly; about 80% of the caloric yield from a mixed diet is first deposited in the form of triacylglycerides and then mobilized again.

Other lipids include the **phospholipids**, the major constituents of cellular membranes, which have a hydrophilic, polar head and a hydrophobic tail, the **sterols**, another important membrane component, the **steroids**, which function as hormones and vitamins, the **prostaglandins**, a class of locally acting hormones, the **dolichol phosphates**, which serve as carriers of oligosaccharides in glycoprotein synthesis, and a group of **fat-soluble vitamins**.

3.1 STRUCTURE AND NOMENCLATURE

3.1.1 Fatty Acids

Fatty acids are long aliphatic hydrocarbon chains with a carboxyl group at one end. There are both saturated and unsaturated fatty acids (containing double bonds), and they may have an even or an odd number of carbon atoms. Most naturally occurring fatty acids are even-numbered, containing 16 or 18 carbon atoms. However, longer chain lengths of up to 24 carbon atoms can occur in humans. Naturally occurring unsaturated fatty acids contain nonconjugated double bonds (i.e., double bonds that do not alternate with single bonds). All natural double bonds are in the *cis*-configuration (Figure 3.1).

There are two major systems for numbering the carbon atoms in fatty acids. The **Δ system** counts the carbon atoms beginning from the carboxyl end. For example, linoleic acid, a C_{18} fatty acid, is written as 18:2Δ9,12, the number before the Δ (2) indicating the number of double bonds in the molecule, and the numbers after the Δ (9,12) marking their positions.

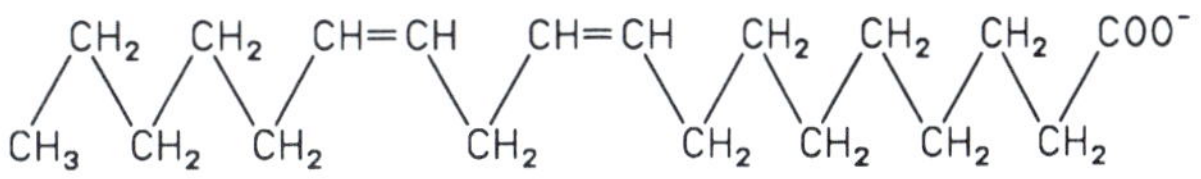

Figure 3.1
Linoleic acid (18:2Δ9,12).

Table 3.1. Some Important Fatty Acids

Common Name	Number of Carbon Atoms	Number of Double Bonds	Position of Double Bonds (Δ system)	Structure
Palmitate	16	0		
Stearate	18	0		
Oleate	18	1	9	
Linoleate	18	2	9, 12	
Linolenate	18	3	9, 12, 15	
Arachidonate	20	4	5, 8, 11, 14	

In contrast, the **n system** begins counting from the methyl end of the molecule. Linoleic acid would thus be designated as 18:2n-6,9. The n system has the advantage that the position of the double bond remains unchanged if the fatty acid is elongated or shortened at its carboxyl end, as happens in metabolism. Some metabolically important fatty acids are listed in Table 3.1.

Arachidonic acid is the precursor of prostaglandins and related compounds (see Chapter 13). The human body can synthesize arachidonic acid from **linoleic acid**, an essential fatty acid that must be supplied in the diet. **Linolenic acid**, a derivative of which is found in the retinal photoreceptor membrane, cannot be made from linoleic acid. Thus, it is an essential fatty acid as well.

At physiologic pH, fatty acids are ionized and possess **amphipathic** character (i.e., their hydrocarbon tail is hydrophobic whereas the carboxyl group or head is hydrophilic). Amphipathic compounds behave as **detergents**.

3.1.2 Neutral Fats and Waxes

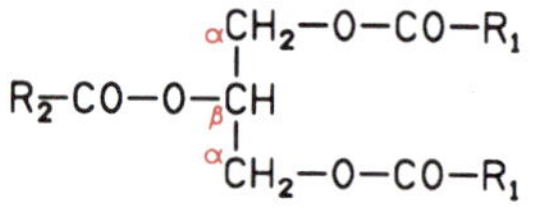

Figure 3.2.
Triacylglycerol.

Triacylglycerides, or neutral fats, have the general structure shown in Figure 3.2. At least 10% of the body weight (even in slim persons) consists of neutral fat. Human fat contains between 30% and 50% unsaturated fatty acids. Cottonseed and corn oil contain mostly polyunsaturated fatty acids (i.e., fatty acids containing more than one double bond).

If one or two of the three hydroxyl groups of glycerol are not esterified, the compound has amphipathic character. Strong alkali will hydrolyze all ester bonds, a process called **saponification**, releasing free glycerol and salts of fatty acids, **soaps**.

Waxes are esters of long-chain primary alcohols (C_{14}–C_{18}) with fatty acids. They are primarily found in skin lipids.

3.1.3 Sterols

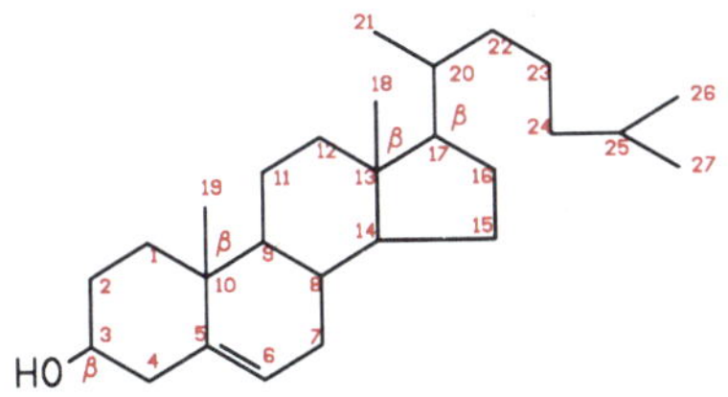

Figure 3.3.
The structure of cholesterol and the numbering convention of its carbon atoms.

Sterols are alcohols containing four condensed carbon rings with various side chains. Depending on whether the side chains project below or above the three-dimensional steroid ring system, they are given the designation α or β, respectively. The most abundant sterol in mammals is **cholesterol** (Figure 3.3).

The 3β-hydroxyl group in cholesterol is frequently esterified with polyunsaturated fatty acids. Cholesterol is a vital component of cell membranes. It is a precursor of **bile acids**, which are required for lipid digestion. Cholesterol is also the precursor of the **steroid hormones** and **vitamin D_3**.

3.1.4 Lipid Components of Biological Membranes

Phospholipids

The major lipid component of biological membranes are the **phospholipids**. They are derivatives of **phosphatidic acid**, a 1,2-diacylglycerol in which the third hydroxyl group is esterified to phosphoric acid. The two acyl groups in the phosphatidic acid component of the phospholipids found in biological membranes are mostly unsaturated acids. Its phosphate group is esterified with **choline**, **ethanolamine**, **serine**, or **inositol** (Figure 3.4). **Cardiolipin**, a constituent of the mitochondrial membrane, consists of two molecules of phosphatidic acid connected by a glycerol bridge.

The removal of the middle (β) fatty acid from **phosphatidyl choline**, also called **lecithin**, gives rise to **lysolecithin**, a detergent that is highly disruptive to cell membranes. The toxic action of snake venom

Figure 3.4.
Phosphatidic acid and derivatives.

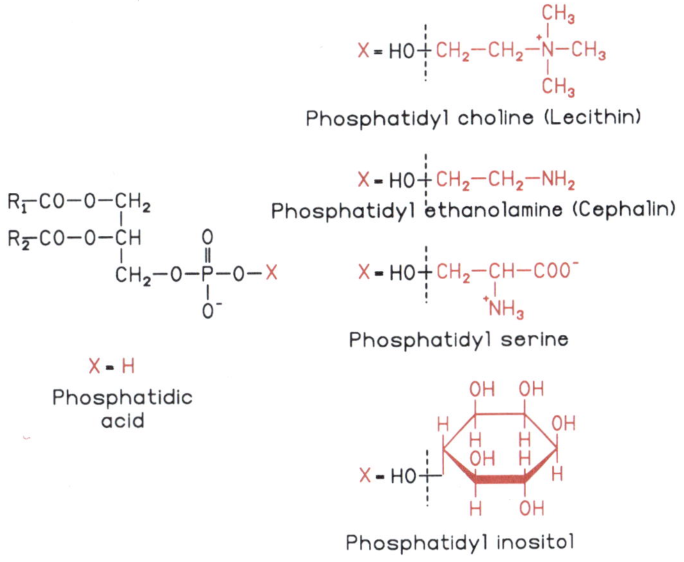

is partially due to its content of phospholipase A_2, an enzyme that removes the acyl group from position 2, thus converting lecithin into lysolecithin. (Phospholipase A_1 hydrolyzes the acyl group at position 1.)

Phospholipases, especially phospholipase A_2, play important roles in the repair of membrane phospholipids. Some of the unsaturated fatty acyl groups suffer oxidative damage (generation of peroxide groups in the fatty acid chain) and must be removed to be replaced by new acyl groups.

Dipalmityl lecithin, an unusual phospholipid because of its two saturated fatty acids, is a major component of the alveolar surfactant in the lung.

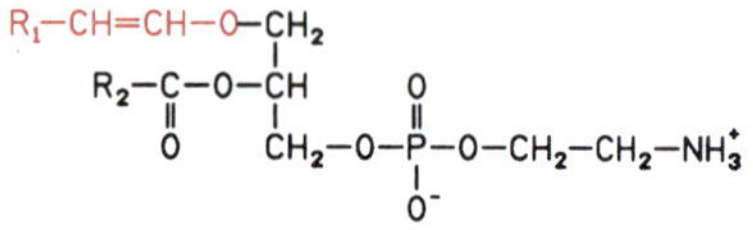

Figure 3.5.
Plasmalogen.

Plasmalogens are phospholipids in which an alkyl ether is attached at the α position of glycerol (Figure 3.5). Ethanolamine is found most often in this phospholipid. Plasmalogens are important membrane components of nervous tissue.

Another important ether phospholipid is the **platelet-activating factor**. It is a 1-alkyl-2-acetyl phosphatidyl choline and causes platelet aggregation at concentrations of only 0.1 n*M*.

Sphingolipids

Another important class of membrane lipids are the sphingolipids. They contain the amino alcohol **sphingosine** in place of glycerol. Its amino group forms amides with fatty acids, and the terminal hydroxyl group is esterified to a variety of compounds. All sphingolipids are membrane components; they include **sphingomyelin**, **cerebrosides**, and **gangliosides** (Figure 3.6).

If the 3′ hydroxyl group of the galactose residue in cerebroside is esterified to sulfuric acid, the compound is called a **sulfatide**. The carbohydrate part of gangliosides consists mainly of glucose, galactose, *N*-acetylgalactosamine, and *N*-acetylneuraminic acid.

3.1.5 Terpenes

Terpenes are polymers of isoprene units (C_5). Vitamin A, E, and K belong to this class of lipids, as do the dolichol phosphates (see Figure 2.56), which function in glycoprotein biosynthesis, and the farnesyl groups, which are attached to heme A, the lamins of the nuclear envelope, and the *ras* oncogene products (see Chapter 7). Cholesterol and vitamin D are also derived from a terpene. The structures of these compounds are discussed elsewhere in this book.

Figure 3.6.
Structures of sphingolipids.

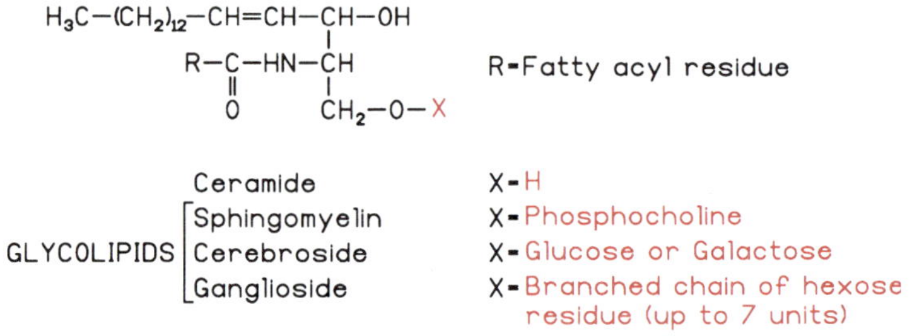

3.2 BIOLOGICAL MEMBRANES

Phospholipids and sphingolipids are strongly amphipathic, having a highly polar head and a nonpolar hydrocarbon tail. In an aqueous environment, these molecules have a tendency to orient themselves so that they minimize exposure of their hydrophobic regions to water. This is accomplished by the formation of either **micelles**, spherical droplets with the polar heads on the outside and the hydrophobic tails on the inside, or by the formation of a **lipid bilayer**, the characteristic structure of biological membranes (Figure 3.7).

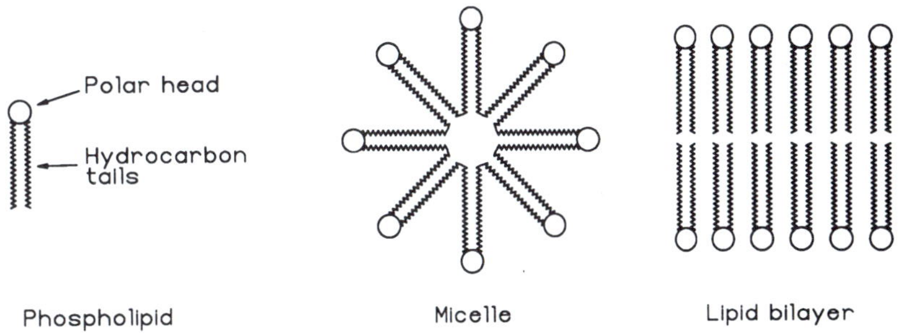

Figure 3.7. Aggregation of phospholipids in water.

The driving force for the spontaneous formation of these structures is **hydrophobic interaction**. Molecules in a water solution arrange themselves so that the water molecules can form the maximum number of hydrogen bonds with the solute as well as with other water molecules. A hydrophobic hydrocarbon chain disrupts potential water-water hydrogen bonding. As a result of these forces, phospholipids and glycolipids will spontaneously form membranelike structures when brought into an aqueous environment. Cholesterol possesses only a weakly polar hydroxyl group. It will arrange itself between the hydrophobic tails of phospholipids and sphingolipids in the bilayer, with its hydroxyl group near the polar surface.

Biological membranes consist of phospholipids, sphingolipids, cholesterol, and proteins. The ratio of lipids to proteins in a plasma membrane is about 2:1 on a weight basis. Plasma membranes also contain carbohydrates ($<10\%$), largely in the form of glycolipids and glycoproteins. The carbohydrate residues are located exclusively on the extracellular side of the plasma membrane (Figure 3.8).

Biological membranes are **asymmetrical** with regard to their inside and outside composition. The asymmetry arises as a consequence of the **biosynthesis** and **assembly** of the membrane. The major cellular membrane phospholipids are synthesized in the cytosol and are inserted into the cytosolic side of the endoplasmic reticulum. The movement of components from one side of the membrane to the other appears to be catalyzed by **flippases**, enzymes that cause the transfer of phospholipids from one side of the membrane to the other. Spontaneous flip-flop of membrane phospholipids occurs only very slowly.

Membrane phospholipids are constantly being synthesized and degraded. Three types of **phospholipid-transfer proteins** are known:

1. Carriers for phosphatidyl choline
2. Carriers for phosphatidyl choline and phosphatidyl inositol
3. Nonspecific carriers

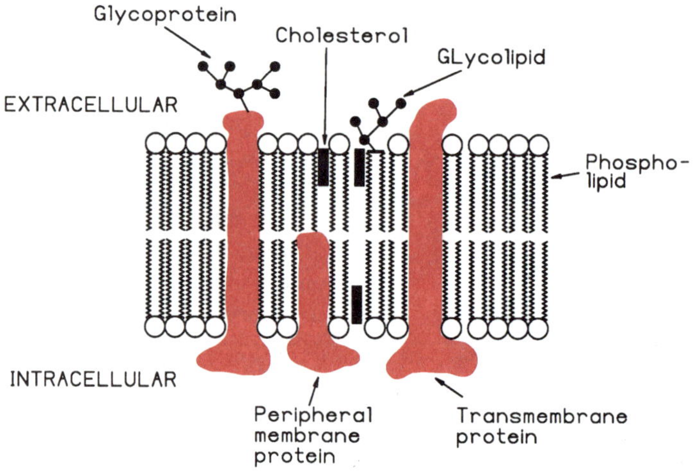

Figure 3.8.
Architecture of biological membranes.

In yeast, a temperature-sensitive mutant for the type 2 carrier was shown to be nonviable at the restrictive temperature. It was shown that phospholipids accumulated in the Golgi apparatus, whereas the transport from endoplasmic reticulum (ER) to Golgi was not affected. This result suggests that the type 2 carrier is involved in the transport of phospholipids between the Golgi and the plasma membrane.

Many membrane proteins are synthesized on the rough endoplasmic reticulum (RER) and inserted into its membrane by means of specific **signal peptide** sequences within the nascent protein. The endoplasmic reticulum then matures into the smooth endoplasmic reticulum (SER), from which small shuttle vesicles separate, carrying in their membranes the inserted proteins. These shuttle vesicles fuse with the membranes of the Golgi apparatus, where additional glycosylation of the membrane proteins occurs. (Note that the carbohydrate moiety of the membrane proteins is located on the inside of the ER or Golgi membrane.) After final posttranslational modification, shuttle vesicles again separate, carrying the glycosylated proteins into the plasma membrane. Fusion with this membrane brings the glycosylated part of the membrane proteins to the outside of the plasma membrane (Figure 3.9).

In contrast to the strict limitation of membrane components to one side or the other of the lipid bilayer, horizontal movement is essentially unrestricted and occurs continuously. Membranes are fluid in the sense that most components can easily diffuse over the entire surface of the membrane as if they were floating on a thick liquid.

Liposomes are artificial phospholipid membrane vesicles that are

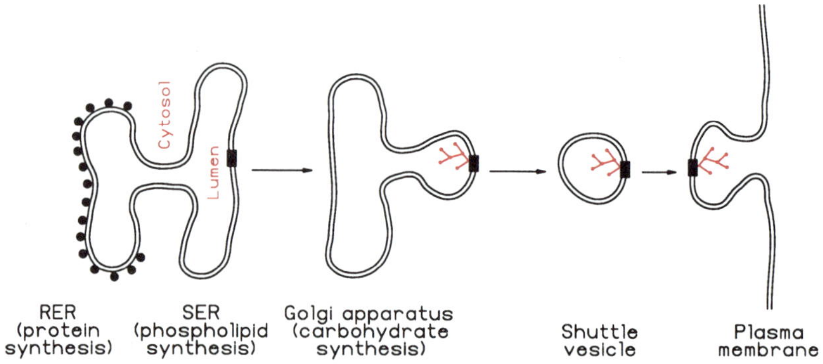

Figure 3.9.
Maturation of membrane proteins.

easily prepared by removing the organic solvent from an aqueous emulsion of phospholipid droplets dissolved in chloroform or ether. Water-soluble substances, present in the aqueous phase, become encapsulated in the membrane vesicles that form as the organic solvent is removed. Since liposomes readily fuse with cell membranes (catalyzed by Ca^{2+} ions), the contents of a liposome, such as drugs, enzymes, or nucleic acids, can be introduced into a living cell.

3.3 ABSORPTION OF LIPIDS

Dietary fat consists mainly of triacylglycerides with small amounts of cholesterol, cholesterol esters, and phospholipids, depending on the type of food. Intestinal absorption of lipids requires partial degradation by pancreatic enzymes. This degradation is facilitated by **emulsification** with **bile salts** in the aqueous chyme of the duodenum. As amphipathic molecules, bile salts, together with lecithin and other still undigested phospholipids, promote the formation of **micelles**. Neutral fat and cholesterol become components of these micelles (mixed micelles), and are thus made more accessible to the pancreatic enzymes **lipase (L)**, **cholesteryl esterase (C)**, and **phospholipase A_2 (P)**.

$$\text{Triacylglycerol} + 2\,H_2O \xrightarrow{L} \beta\text{-monoacylglycerol} + \text{fatty acids}$$

$$\text{Cholesteryl ester} + H_2O \xrightarrow{C} \text{cholesterol} + \text{fatty acid}$$

$$\text{Phosphoglyceride} + H_2O \xrightarrow{P} \alpha\text{-acyl-lysophosphoglyceride} + \text{fatty acid}$$

The products of this limited digestion, monoacylglycerol, fatty acids, cholesterol, and lysophospholipids, slowly enter the intestinal epithelial cells by passive diffusion. There, fatty acids are reconverted to triacylglycerols, which are again assembled, together with phospholipids, cholesterol, and a small amount of protein (apolipoprotein, see below), into large complexes called **chylomicrons**. Chylomicrons are secreted into the lymphatic system, from where they enter the bloodstream by way of the thoracic duct. In the blood, fat is carried via chylomicrons to the various tissues for storage (adipose tissue), modification (liver), or oxidation and energy generation (liver, muscle, heart, etc.).

3.4 PLASMA TRANSPORT AND DEPOSITION OF LIPIDS

Lipids are transported in the bloodstream in the form of multimolecular complexes called **lipoproteins**. The lipoproteins are divided into four major classes of increasing density that are determined by their relative contents of lipids and proteins: **chylomicrons**, **very-low-density lipoproteins (VLDL)**, **low-density lipoproteins (LDL)**, and **high-density lipoproteins (HDL)**. There are no sharp boundaries between these classes, since components are constantly being exchanged between different lipoproteins and also between lipoproteins and cells. The protein component of these complexes is called **apolipoprotein (apo)**. The major apolipoproteins are apo A-I, apo A-II, apo B-48, apo B-100, apo C-I, apo C-II, apo C-III, apo D, and apo E. The lipid and protein compositions of the major lipoproteins are summarized in Table 3.2.

As discussed above, triacylglycerides of dietary origin enter the blood in the form of chylomicrons. Triacylglycerides, synthesized in the liver, are secreted via the Golgi apparatus as part of a pseudomicellar complex called **VLDL** (see above).

Both chylomicrons and VLDL are acted upon by **lipoprotein lipase** present on the capillary endothelium of many tissues. This enzyme will only attack triacylglycerols when contained in a lipoprotein complex because it is specifically activated by apo C-II, a protein component of both chylomicrons and VLDL.

The synthesis of lipoprotein lipase in adipocytes occurs in response to the hormone insulin. The enzyme is released into the extracellular fluid. It then becomes locally bound to the endothelial surface. In uncontrolled **diabetes mellitus**, that is, in the absence of insulin, there is no lipoprotein lipase synthesized. Therefore, chylomicrons and VLDL accumulate in the plasma, a situation that is called **lipemia**.

Lipoprotein lipase has only a short half-life. After a meal, when insulin is released in response to rising glucose levels, the lipoprotein lipase in adipose tissue is induced. This leads to uptake of free fatty acids, released by the enzyme from the lipoproteins, into the adipose cells. As soon as insulin levels drop, no new lipoprotein lipase is synthesized, and enzyme levels drop rapidly. After a long fasting period, that is, when insulin levels are very low, lipoprotein lipase is essentially absent from adipose tissue but is greatly increased in heart muscle and other muscle tissue. There, the synthesis of lipoprotein lipase appears to be under the control of glucocorticoids. The released free fatty acids are taken up by muscle cells for subsequent metabolism. Thus, there exists a **reciprocal regulation** of lipoprotein lipase synthesis in adipose tissue as compared with muscle and other tissues. When there is excess lipid, it is stored in adipose tissue. On the other hand, when energy is required

Table 3.2. Serum Lipoproteins

Lipoprotein	Particle: Diameter (nm)	Particle: Density (g/ml)	Protein Content; Components	Lipid Composition		Function
Chylomicrons	100–1000	0.900–0.950	1%; A-I, A-II, B-48, C-I, C-II, C-III, E	TG PL CH	90% 4% 5%	Transport of triacylglycerol from intestinal mucosa to tissue
VLDL	30–50	0.950–1.006	10%; B-100, C-I, C-II, C-III, E	TG PL CH	60% 15% 15%	Transport of triacylglycerol from liver to tissue
LDL	20–25	1.006–1.063	26%; B-100	TG PL CH	10% 22% 42%	Transport of phospholipids and cholesterol for membrane synthesis
HDL	8–10	1.063–1.210	45%; A-I, A-II, C-I, C-II, C-III, D, E	TG PL CH	5% 30% 20%	Transport of phospholipids and cholesterol, reservoir for apolipoprotein Cs.

Note: A, B, C, D, E = classes of apolipoproteins, TG = triacylglycerides, PL = phospholipids, CH = cholesterol and esters.

in muscle cells, as for example in a marathon runner, fatty acids are removed from circulating lipoproteins in muscle capillaries. In addition, free fatty acids are transported from adipose tissue to muscle for oxidation and energy production. (The hormone-controlled release of fatty acids from fats stored in adipose tissue is discussed below.)

Hydrolysis of triacylglycerides in chylomicrons and VLDL leaves lipoprotein complexes that are enriched for cholesterol, apolipoprotein, and phospholipids. Some of the phospholipids, unesterified cholesterol, and apolipoproteins are transferred to **HDL**. The remaining triacylglyceride-poor particles are called **remnants** or **LDL**, depending on their origin from chylomicrons or VLDL, respectively. LDL serves as a major source of cholesterol and phospholipids for peripheral tissues. Cells take up LDL by endocytosis aided by specific **LDL receptors** that recognize the apolipoprotein B-100 present in LDL. LDLs are also recycled to the liver. Remnants are returned to the liver, where they bind to receptors that recognize their apolipoprotein E component.

The metabolism and function of **HDL** are complex and still only partly understood. The major source of HDL is the liver that releases it in the form of disc-shaped **nascent HDL particles**. They are also formed indirectly during the conversion of VLDL or chylomicrons to LDL and remnants, respectively. Nascent HDL particles consist mainly of phospholipids and apolipoproteins. They are transformed in the plasma into spherical HDL particles by exchange reactions with other lipoproteins. During these exchange reactions, HDL takes up cholesterol and other lipids from peripheral tissues as well as from other lipoproteins. The cholesterol is converted to cholesteryl esters by the plasma enzyme **lecithin-cholesterol-acyl transferase** (**LCAT**), which transfers the β-acyl group from lecithin to cholesterol. LCAT is activated by apo A-I. Some of the cholesteryl esters are transferred from HDL back to LDL.

The excess of cholesteryl esters in HDL is carried directly to the liver for excretion in the bile. Thus, HDL is a very important carrier for the removal of excess cholesterol from peripheral tissue. The incidence of coronary heart disease is inversely correlated with the level of plasma HDL (see below). Figure 3.10 gives an overview of lipid transport and the metabolism of lipoproteins. The indicated numbered reactions are defective in certain **lipoproteinemias**.

In addition to the cellular uptake of lipoproteins via specific receptors, there is also a **scavenger pathway** that operates in macrophages and other scavenger cells of the reticuloendothelial system. In vivo, about 45% of the LDL pool turns over every day, with 30% being removed by the receptor pathway and 15% by the scavenger pathway. The percentage rate of LDL removal by the scavenger pathway increases with increased plasma concentrations of LDL. It is thought that the scavenger pathway is closely related to **atherosclerosis**, the process of formation of cholesterol plaques on the endothelial surface of blood vessels.

Recently, a new type of lipoprotein, **lipoprotein (a)** was discovered. This protein is not a real lipoprotein because it is water-soluble. When found, it is associated with LDL particles. There appears to be a direct correlation between levels of lipoprotein (a) and an increased risk of development of coronary heart disease. It has been speculated that this correlation is the result of close sequence similarities between lipoprotein (a) and plasminogen.

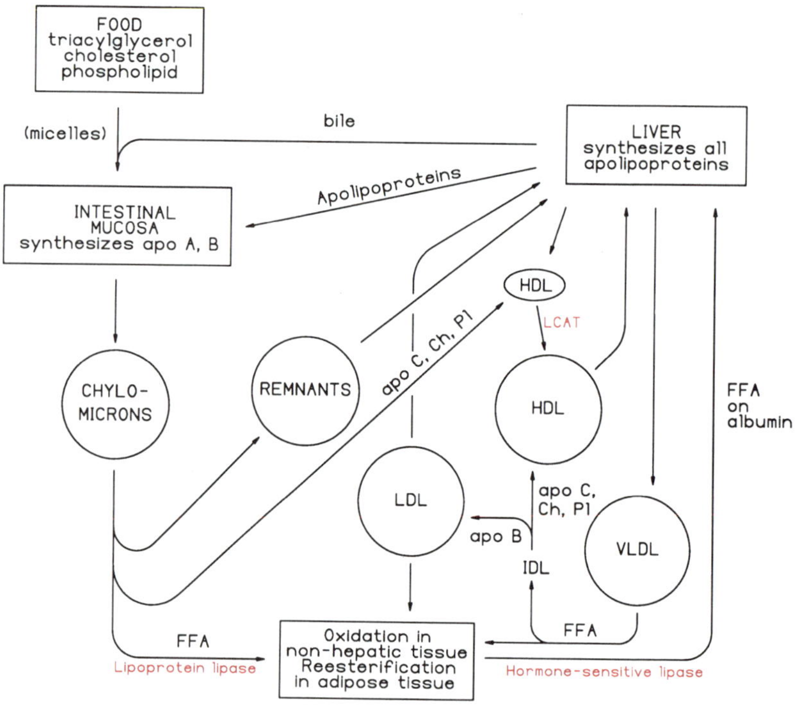

Figure 3.10.
Lipid transport and interactions between lipoproteins. The numbered pathways are defective in certain lipoproteinemias.

3.5 TRANSPORT AND METABOLISM OF CHOLESTEROL

In intestinal cells, a substantial portion of the absorbed cholesterol is reesterified by **acyl CoA-cholesterol acyl transferase** (**ACAT**). Both free and esterified cholesterol are packaged into chylomicrons. Unesterified cholesterol, located on the particle surface, freely exchanges with other plasma lipoproteins as well as with cellular membranes. Esterified cholesterol, buried in the hydrophobic core of the lipoprotein particle, is more restricted in its exchange.

As discussed above, esterification of cholesterol contained in VLDL and chylomicrons occurs after its transfer to HDL and is catalyzed by LCAT. Plasma lecithin contains mainly polyunsaturated fatty acids in its β position. As a result, 50% of the circulating cholesterol esters contain linoleic acid. Part of the esterified cholesterol is transferred back to LDL, which is subsequently taken up by many different cells. The cholesterol esters remaining in HDL are carried to the liver.

In the liver, cholesterol is converted to bile acids that are secreted via the bile duct, partly reabsorbed in the intestine, and returned to the liver by the portal circulation (enterohepatic circulation). The only way cholesterol is normally removed from the body is in the form of bile salts. A substantial amount of free cholesterol is secreted into the bile in the form of mixed micelles made of lecithin, bile acids, and cholesterol. The ratio of these components is critical to prevent precipitation (gall bladder stones).

De novo synthesis of cholesterol in the liver (see Section 3.12) is controlled mainly by dietary cholesterol, which reaches the liver in the form of chylomicron remnants. Nonhepatic cholesterol biosynthesis is regulated by dietary or endogenous cholesterol contained in LDL. Cell surface receptors recognize LDL and promote its uptake by endocytosis. Within the cells, LDL is disrupted by lysosomal enzymes, and cholesteryl esters are cleaved. Free cholesterol (or its oxidation products) inhibits **3-hydroxy-3-methylglutaryl coenzyme A (HMG-CoA) reductase** (see Section 3.12) and activates **ACAT**, which is responsible for reesterification of free cholesterol.

3.6 MOBILIZATION OF LIPIDS

When energy supplies are low, triacylglycerol, deposited in adipose tissue, is hydrolyzed, and free fatty acids are released into the bloodstream. The process is controlled by an intracellular **hormone-sensitive lipase**, whose activity is determined by the intracellular level of cyclic adenosine monophosphate (cAMP). In response to norepinephrine, epinephrine, adrenocorticotropic hormone (ACTH), or glucagon, membrane-bound **adenylate cyclase** converts adenosine triphosphate (ATP) into cAMP, which activates a cytoplasmic **protein kinase**. This enzyme, in turn, activates an inactive hormone-sensitive lipase by phosphorylation (Figure 3.11). The active hormone-sensitive lipase hydrolyzes only one α-acyl group of triacylglycerol. The resulting diacylglycerol is then further hydrolyzed by another lipase not sensitive to hormone action. The free fatty acids (FFA) are released into the bloodstream, where they are transported, bound to serum albumin, to other tissues for oxidation.

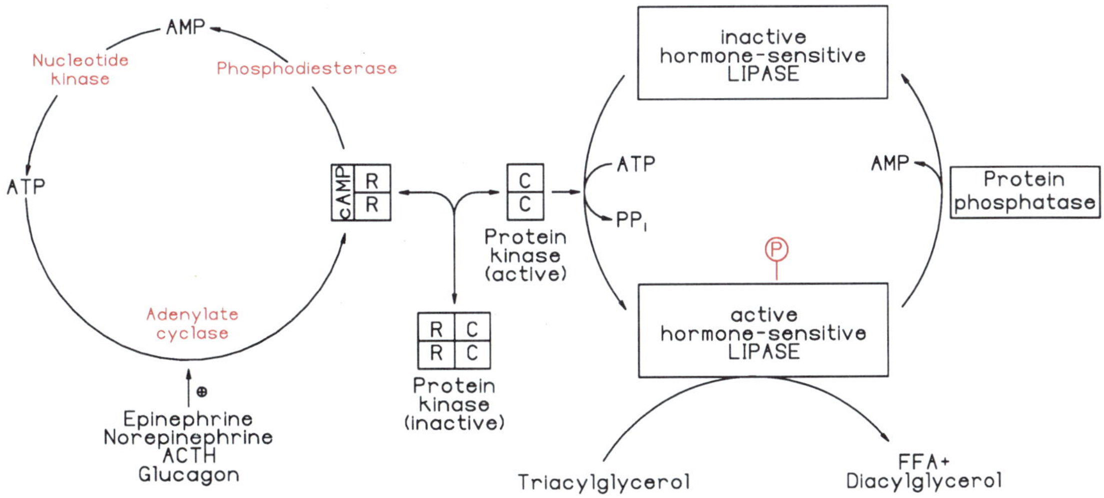

Figure 3.11.
Control of hormone-sensitive lipase activity.

High concentrations of glucose in blood reduce intracellular levels of cAMP and thus stimulate fat deposition rather than mobilization in the adipocyte (low glucose concentrations have the opposite effect). Cyclic AMP is constantly being hydrolyzed by phosphodiesterase. If this enzyme is inhibited by methylxanthines, such as caffeine or theophylline, cAMP levels rise and triacylglycerol hydrolysis is increased. Thyroxine, growth hormone, or glucocorticoids also enhance fat mobilization, but these hormones do not directly activate the hormone-sensitive lipase (see Chapter 13). Insulin inactivates hormone-sensitive lipase.

3.7 DISORDERS IN LIPID TRANSPORT AND METABOLISM

3.7.1 Abnormal Lipid Transport Caused by Genetic Deficiencies of Specific Apolipoproteins

Tangier disease is a genetic disorder characterized by a **deficiency in apolipoprotein A**. The blood contains high levels of triacylglycerides but low concentrations of cholesterol. Because of the almost complete absence of HDL, cholesterol cannot be removed from peripheral tissues, leading to deposits of cholesterol and cholesteryl esters (**xanthomas**). The VLDL is deficient in apo Cs, which includes the activator for lipoprotein lipase, C-II, that is normally obtained by exchange from the now absent HDL. This may explain the high levels of triglycerides and the failure of VLDL to mature into LDL, resulting in low plasma levels of cholesterol.

Abetalipoproteinemia is a rare genetic disease in which no chylomicrons, VLDL, or LDL lipoproteins are formed due to a **deficiency in apolipoprotein B**. As a consequence, fat absorption in the gut is severely inhibited, leading to steatorrhea (fatty stool), retinitis from malabsorption of carotinoids, and neuromuscular degeneration. The cells of the intestinal mucosa become stuffed with fat droplets, and the erythrocytes appear prickled because of the lack of cholesterol in their plasma membranes.

Apolipoprotein CVII deficiency causes marked **hypertriglyceridemia** (>1000 mg/dl) because apo CVII is normally required for the activation of lipoprotein lipase. This rare genetic disease causes recurrent pancreatitis, lipemia retinalis, and xanthomas.

Apolipoprotein E deficiency causes **dysbetalipoproteinemia**. The cholesteryl ester-rich chylomicron remnants cannot bind to the apo E-directed receptors in the liver and as a result accumulate in the blood. This leads to the deposition of lipid in the skin and arterial walls, called **xanthomas** and **atheromas**, respectively.

3.7.2 Abnormal Lipid Metabolism Caused by Genetic Deficiencies of Specific Enzymes

Lipoprotein lipase deficiency, also called **hyperlipoproteinemia type I**, is a recessive disorder marked by massive hyperchylomicronemia, which causes high concentrations of plasma triacylglycerides and lesser increases in cholesterol.

Wolman disease is characterized by an abnormal lipoprotein pattern because of a **deficiency of lysosomal cholesteryl esterase**. This deficiency causes accumulation of triacylglycerides and cholesteryl esters in many tissues. The disorder is fatal in infancy if the deficiency is complete.

LCAT deficiency causes abnormal cholesterol and lipoprotein metabolism. The normal transfer of cholesterol as cholesteryl esters between lipoprotein particles does not occur. As a result, nascent HDL cannot accept cholesteryl esters and thus does not mature into the spherical HDL. Opacities in the cornea appear during childhood, with anemia and kidney damage developing later.

3.7.3 LDL-Receptor Deficiency

Familial hypercholesterolemia, classified as **hyperlipoproteinemia type IIa**, is an autosomal dominant condition in which LDL cannot be removed from the circulation because of the deficiency or absence of LDL receptors. Heterozygotes have total serum cholesterol concentrations between 275 and 550 mg/dl, whereas homozygotes have levels of 650–1000 mg/dl. The normal range is 120–200 mg/dl. Half of the male heterozygotes suffer heart attacks before age 50 and half the females before age 60. The homozygotes develop cardiovascular disease in childhood.

3.7.4 Other Inherited Hyperlipoproteinemias

Several other types of hyperlipoproteinemias are known, some of which are much more common than the ones mentioned above. Most are characterized by elevated levels of LDL. The exact nature of the mutations that cause these different hyperlipoproteinemias (types IIb, IV, and V) is not known. Hyperlipoproteinemia type III seems to be caused by a defect in the mechanism that converts VLDL into LDL. As a consequence, an **intermediate density lipoprotein** (**IDL**) accumulates. Most of these conditions are associated with an increased risk of early heart attacks.

3.7.5 Acquired Hypercholesterolemia

Many people in our Western societies suffer heart attacks or strokes without having a genetic defect in their metabolism of lipoproteins or their LDL receptors. The **LDL receptor hypothesis** states that most of the atherosclerosis in the general population is caused by high blood levels of LDL resulting from inadequate numbers of LDL receptors. It is proposed that the number of LDL receptors is reduced by environmental factors such as high dietary intake of cholesterol and/or saturated fats. The dietary cholesterol accumulates in the liver and suppresses the synthesis of LDL receptors. This could lead to an increase in the level of circulating LDL.

3.7.6 Cholesterol and the Risk of Coronary Heart Disease

Extensive studies indicate a close relationship between serum cholesterol levels and the development of coronary heart disease. **Atherosclerosis** begins in childhood with the accumulation of lipids, principally cholesterol and its esters, in the intima of the large arteries in the form of fatty streaks. These streaks then thicken by accumulation of further lipid, smooth muscle cells, and connective tissue to form fibrous plaques. Later in life these plaques increase in size and undergo calcification. Eventually, this causes occlusion of the arterial lumen, leading to ulceration, thrombosis, and sudden blockage of blood flow. In the coronary arteries, this causes myocardial infarction; in the carotid arteries, it may lead to stroke.

Atherosclerosis is accelerated by hypertension and cigarette smoking. The risk of coronary heart disease is often assessed by measuring **total plasma cholesterol**. Individuals with values of 200 mg/dl and higher are considered at risk. However, a better predictor for development of coronary heart disease is not total serum cholesterol but the ratio of total cholesterol to cholesterol contained in HDL. The risk is inversely related to the level of HDL cholesterol. The reason for this beneficial action of HDL is not completely understood. It is thought that HDL facilitates the transport of cholesterol and its esters from peripheral tissues to the liver, where cholesterol is converted into bile acids and excreted.

3.8 FATTY ACID OXIDATION

3.8.1 β-Oxidation

The major pathway of fatty acid degradation is **β oxidation**; it occurs inside the mitochondrion and requires the transport of fatty acids across the inner mitochondrial membrane. First, free fatty acids are activated by coenzyme A and ATP to fatty acyl CoA. The enzyme responsible for this reaction, **acyl CoA synthetase**, also called **thiokinase**, is located on the outer mitochondrial membrane. Since CoA esters of fatty acids containing more than 10 carbon atoms cannot penetrate the inner mitochondrial membrane, they must first be converted to **acyl carnitine** by the enzyme **carnitine acyltransferase I** located on the outer surface of the inner mitochondrial membrane. An **acyl carnitine translocase** then brings the acyl carnitine to the inner side of the membrane, where a second acyl transferase, **carnitine acyltransferase II**, reforms acyl coenzyme A, releasing free carnitine that exits the mitochondrion to be used for the transport of another fatty acyl group (Figure 3.12).

β oxidation begins with the oxidation of the third carbon atom from the carboxyl end of the molecule. It occurs in three steps (analogous to the oxidation of succinate to oxaloacetate in the citric acid cycle) and yields a β-keto acid. β-Keto acyl CoA is then cleaved by **thiolase**, in a CoA-requiring step, into acetyl CoA and a fatty acyl CoA ester that is shortened by a two-carbon fragment. The process is repeated until the complete fatty acid chain has been converted to acetyl CoA (Figure 3.13).

The overall energy yield of the process in the form of ATP can be calculated as follows. The first oxidation involves the formation of reduced flavin adenine dinucleotide ($FADH_2$), which gives rise to 2 ATPs in the respiratory chain. The second oxidation results in the production of reduced nicotinamide adenine dinucleotide (NADH) + H^+, thus generating 3 ATPs. Further oxidation of the released acetyl CoA in the

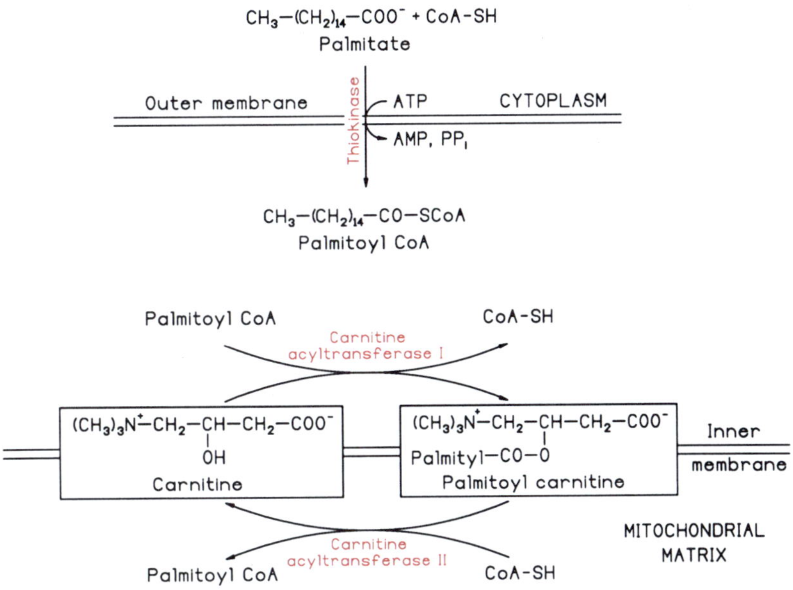

Figure 3.12.
The activation of palmitate to palmitoyl CoA and its transport into the mitochondrion as carnitine ester.

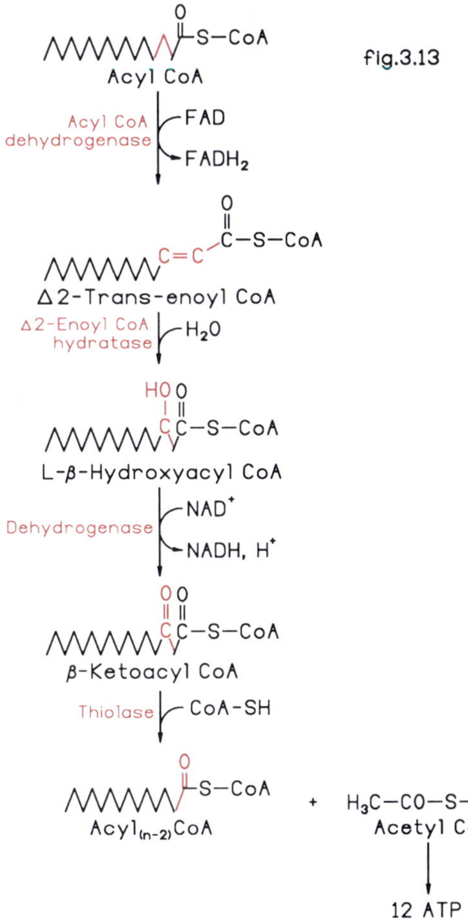

Figure 3.13.
β oxidation of fatty acids.

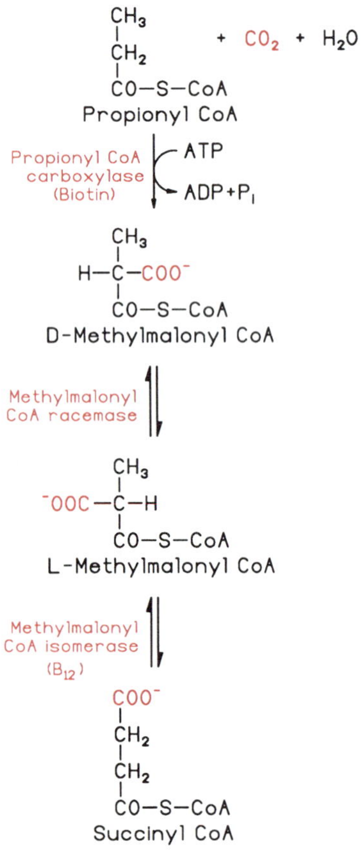

Figure 3.14.
Conversion of propionyl CoA to succinyl CoA.

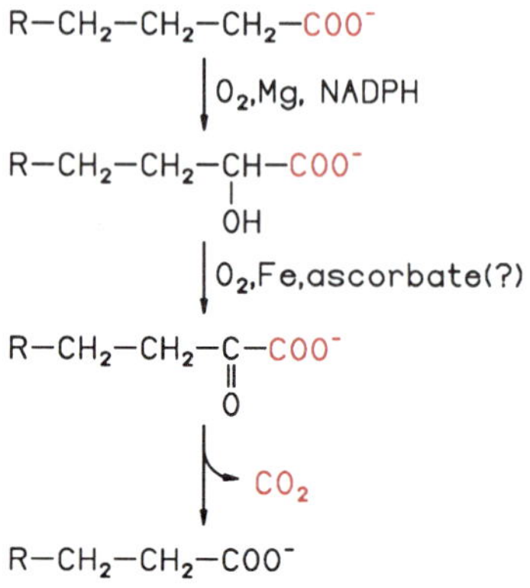

Figure 3.15.
α oxidation of fatty acids.

citric acid cycle yields another 12 ATP molecules. Since two high-energy bonds are expended in the initial fatty acid activation [note that ATP is hydrolyzed to adenosine monophosphate (AMP) and two P_is], the total yield from the removal and complete oxidation of the first two-carbon unit amounts to 15 ATPs. Oxidation of each subsequent two-carbon unit produces 17 ATPs, and the last two-carbon unit produces 12 ATPs. The overall energy yield from the oxidation of palmitate, a 16-carbon fatty acid, is 129 ATP/mol. This corresponds to a 40% recovery of the free energy that is released upon complete combustion of palmitate.

Odd-numbered fatty acid chains are degraded by the same mechanism. The final fragment, **propionyl CoA**, however, is not susceptible to β oxidation. It is carboxylated to **methylmalonyl CoA**, which is enzymatically rearranged to **succinyl CoA**, an intermediate of the citric acid cycle. This pathway requires two cofactors, biotin and vitamin B_{12}. Vitamin B_{12} deficiency causes urinary excretion of propionate and methylmalonate (Figure 3.14).

Unsaturated fatty acids may require rearrangement of their double bonds to allow β oxidation to proceed. This is accomplished by **enoyl CoA isomerase**, an enzyme that converts a Δ3-*cis*-enoyl CoA into Δ2-*trans*-enoyl CoA, the proper substrate for the enoyl CoA hydratase. When β oxidation of an unsaturated fatty acid produces a *cis* double bond at the Δ2 position, the **enoyl CoA hydratase** will hydrate this substrate to the D-hydroxyacyl intermediate. **β-Hydroxyacyl CoA dehydrogenase**, however, cannot utilize the D isomer directly. It must be converted to the L isomer by **β-hydroxyacyl CoA epimerase** before oxidation can proceed.

Although natural unsaturated fatty acids contain only *cis* double bonds, some *trans* unsaturated fatty acids are generated during the preparation of margarine. They are oxidized just as easily as *cis* unsaturated fatty acids.

3.8.2 Other Types of Fatty Acid Oxidation

There are two other mechanisms of fatty acid degradation, α oxidation and ω oxidation. In **α oxidation**, only one carbon atom at a time is removed from the carboxyl end. α oxidation takes place in the **endoplasmic reticulum**. Free fatty acids are first oxidized to α-hydroxy acids by a mixed-function oxidase. (α-Hydroxy fatty acids are important constituents of brain lipids.) Further oxidation gives rise to α-keto acids, which are subsequently decarboxylated. There are indications that ascorbic acid may participate in this process; Figure 3.15).

α oxidation is important for the degradation of **branched fatty acids**, such as **phytanic acid**, which is derived from the phytol side chain of chlorophyll. Phytanic acid is resistant to β oxidation until the first carbon atom is removed by α oxidation (Figure 3.16).

Patients with **Refsum disease** cannot degrade phytanic acid. The acid accumulates in neural tissues with serious neurologic consequences. The enzymatic defect must lie in the α hydroxylation step, since α-hydroxyphytanic acid is metabolized normally.

In **ω-oxidation**, the ω carbon atom (farthest removed from the carboxyl group) is oxidized to a carboxyl group, resulting in the creation of a dicarboxylic acid. This acid can then be degraded from either end

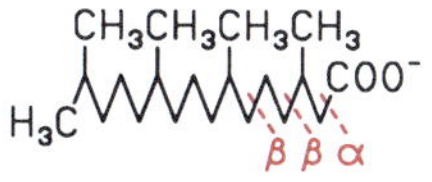

Figure 3.16.
Phytanic acid. The dashed lines indicate points of α or β cleavage.

by β oxidation. ω oxidation is catalyzed by a **microsomal mixed-function oxidase** that contains **cytochrome P_{450}**. Compared with β oxidation, ω oxidation represents only a minor pathway.

3.9 KETONE BODY FORMATION

In **starvation**, or in **diabetes mellitus**, the carbohydrate uptake by the liver is reduced, and fat is mobilized from adipose tissue. This occurs despite the high blood glucose levels that are observed in diabetes mellitus because of excessive catecholamine synthesis. These hormones stimulate hormone-sensitive lipase via cAMP signalling. The released fatty acids are taken up by the liver in large amounts, generating acetyl CoA in the process. In normal conditions, acetyl CoA would enter the citric acid cycle. In starvation, however, the lack of carbohydrates in the liver reduces the level of oxaloacetate below the level required for entry of acetyl CoA into the citric acid cycle. The excess acetyl CoA is metabolized to **acetoacetate** via the intermediate formation of 3-hydroxy-3-methylglutaryl CoA. Acetoacetate can be reduced to **3-hydroxybutyrate**. Alternatively, it decomposes spontaneously to **acetone** (Figure 3.17). Acetoacetate and its reaction products are known collectively as **ketone bodies**.

Ketone bodies are normal liver metabolites and are usually present in the blood at low levels. If ketone body production is increased, however, ketone bodies are excreted in the urine. Some of the acetone is exhaled through the lungs (typical acetone smell). Acetoacetic acid and hydroxybutyric acid are strong acids and cause a drop in blood pH

Figure 3.17.
Biosynthesis of ketone bodies.

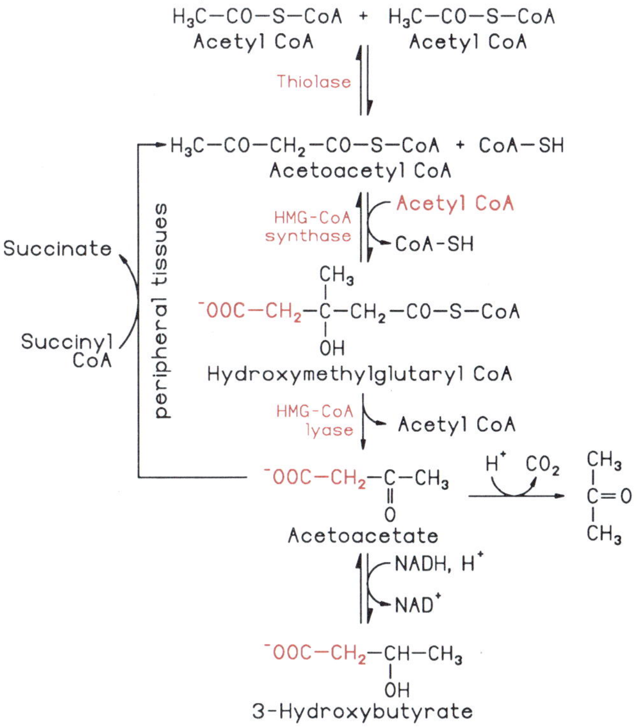

(acidosis). Their excretion in the urine is accompanied by excretion of sodium ions. This results in the loss of water (dehydration).

Most other tissues can utilize ketone bodies for energy production. The ketone bodies are simply converted back to acetyl CoA since the reactions in Figure 3.17 are reversible. Only acetoacetate must be reactivated. This occurs by transfer of CoA from succinyl CoA.

$$\text{Acetoacetate} + \text{succinyl CoA} \rightarrow \text{acetoacetyl CoA} + \text{succinate}$$

This reaction does not occur in the liver.

The brain is normally dependent on glucose as its sole source of energy and cannot oxidize either fatty acids or ketone bodies. In prolonged starvation, however, part of the brain can adapt to utilizing ketone bodies.

No net conversion of fat to carbohydrate is possible in humans. The synthesis of carbohydrate from smaller carbon fragments (gluconeogenesis) begins with oxaloacetate. Although oxaloacetate is an intermediate in the citric acid cycle, no net conversion of acetyl CoA to oxaloacetate takes place because acetyl CoA is completely oxidized to CO_2 and water while passing through the cycle. Only the glycerol part of neutral fats and the terminal propionate residue, resulting from the degradation of odd-numbered fatty acids, can be converted to carbohydrate.

3.10 FATTY ACID BIOSYNTHESIS

Fatty acid biosynthesis occurs primarily in the cytoplasm of liver cells and adipocytes. Acetyl CoA, the precursor of fatty acids, is formed in the mitochondrion but cannot cross the mitochondrial membrane. Instead it passes into the cytoplasm in the form of citrate, to which the mitochondrial membrane is permeable. In the cytoplasm, citrate is then cleaved to acetyl CoA and oxaloacetate, which is further metabolized via malate to pyruvate. Pyruvate crosses back into the mitochondrion, where it is carboxylated to oxaloacetate, which subsequently combines with acetyl CoA to regenerate citrate (Figure 3.18).

In this **citrate-malate-pyruvate shuttle**, one NADH is converted to NADPH, which is subsequently used for fatty acid synthesis. For each acetyl CoA transported across the mitochondrial membrane, two molecules of ATP are cleaved.

The first, and rate-limiting, step in fatty acid synthesis is the carboxylation of acetyl CoA to **malonyl CoA**. The reaction requires ATP and carbon dioxide (Figure 3.19). The enzyme **acetyl CoA carboxylase** carries as a prosthetic group the coenzyme biotin, covalently linked to the ϵ amino group of a lysine residue. (Remember that biotin also functions in the carboxylation of pyruvate and propionyl CoA.)

Acetyl CoA carboxylase is a multimeric enzyme, containing 15–20 identical monomeric polypeptides. Citrate promotes the polymerization of the inactive monomers, thereby activating the enzyme, while palmitoyl CoA causes depolymerization of the enzyme and thus inactivation. Citrate concentrations are high when acetyl CoA and ATP are abundant.

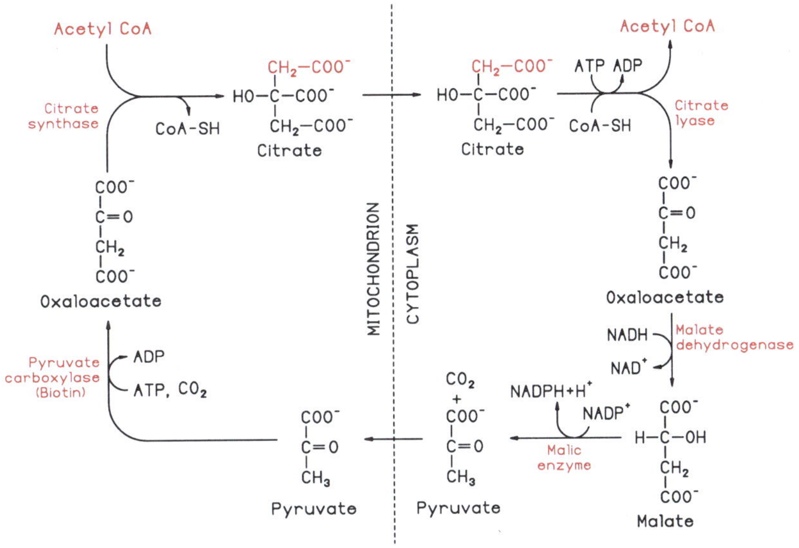

Figure 3.18.
The citrate-malate-pyruvate shuttle.

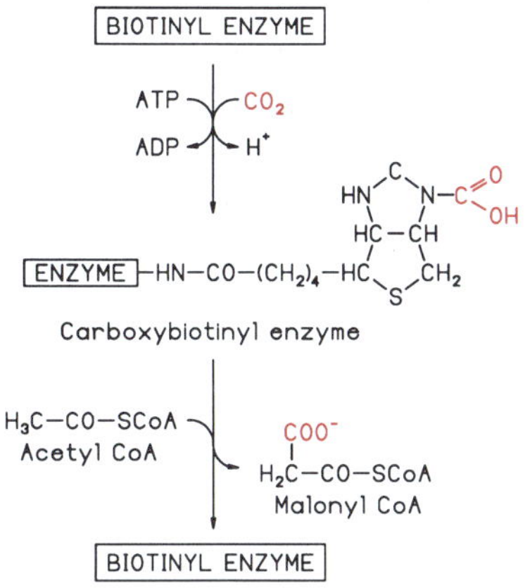

Figure 3.19.
Carboxylation of acetyl CoA to malonyl CoA by acetyl CoA carboxylase.

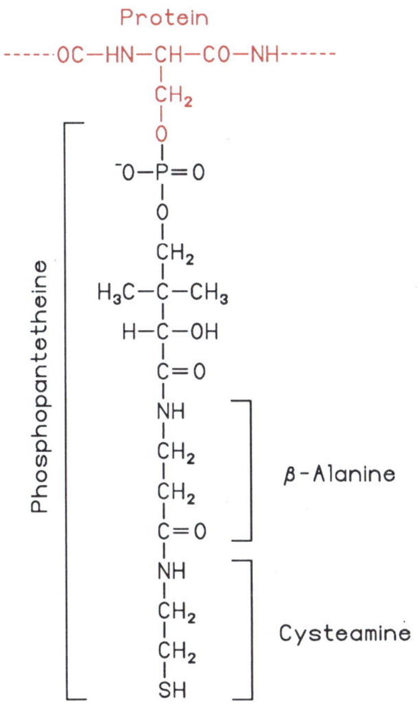

Figure 3.20.
Acyl carrier protein (ACP).

The activity of acetyl CoA carboxylase is also regulated by reversible phosphorylation. When blood glucose concentration is low, rising glucagon levels activate intracellular adenylate cyclase. This raises intracellular cAMP levels, thereby activating protein kinase A, which then phosphorylates acetyl CoA carboxylase. This results in the inactivation of acetyl CoA carboxylase. On the other hand, when blood glucose levels are high, glucagon levels fall (insulin rises), and acetyl CoA carboyxlase is activated by dephosphorylation. This causes malonyl CoA levels to rise, which in turn inhibits carnitine acyl transferase and thus prevents the transport of fatty acids into the mitochondrion for degradation.

The polymerization of malonyl CoA into a fatty acid chain occurs on a multienzyme complex called **fatty acid synthase** (**FAS**). The complex contains two functional SH groups, one of which is central (C) while the other is peripheral (P). In eukaryotes, the central SH group is part of the **phosphopantetheine** group (also found in CoA; Figure 3.20) attached to a serine residue of one of two polypeptide chains with multiple functional sites. [In bacteria, phosphopantetheine is attached to a separate **acyl carrier protein** (**ACP**).] The peripheral SH group belongs to a cysteine residue of the **condensing enzyme**, which in eukaryotes is part of one of the two polypeptide chains.

The first step catalyzed by fatty acid synthase is the transfer of an acetyl group from acetyl CoA to the central SH group (corresponding to ACP in bacteria). From there the acetyl group is temporarily transferred to the peripheral SH site, and a malonyl group is transferred from malonyl CoA to the central site. The condensing enzyme then catalyzes the condensation of the acetyl and malonyl groups, resulting in the formation of an acetoacetyl group, still linked to the central site, and the release of CO_2. Acetoacetyl, still attached to the enzyme, is reduced to β-hydroxybutyryl, dehydrated to crotonyl, and finally reduced to butyryl. Addition of the next two-carbon unit to the growing fatty acid chain occurs after the temporary transfer of the butyryl residue to the peripheral SH site and attachment of a new malonyl residue to the central site. The sequence of events is repeated until a chain length of

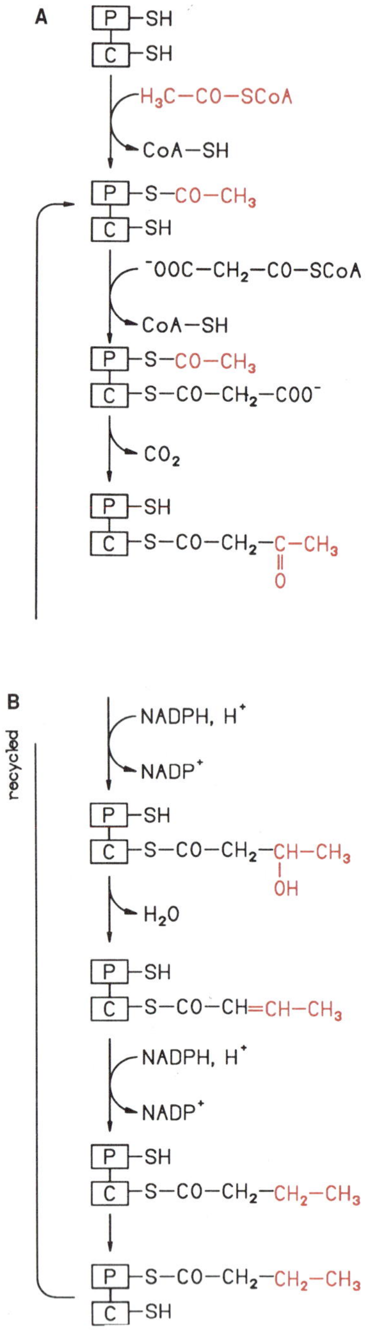

Figure 3.21.
Polymerization of acetyl units on the fatty acid synthase complex. P refers to the peripheral SH group and C to the central SH group of the enzyme.

16 carbon atoms is reached (Figure 3.21). A **deacylase** finally removes the completed fatty acid from the central site.

Each addition of a two-carbon unit to the growing fatty acid chain requires two NADPH molecules for reduction. The NADPH is derived in part (40%) from the pentose phosphate pathway and in part (60%) from the citrate-malate-pyruvate shuttle.

Saturated and unsaturated fatty acids with more than 16 carbon atoms are synthesized by **elongation** of preformed fatty acids in the mitochondrion or in the endoplasmic reticulum. The mitochondrial process requires acetyl CoA, the ER process requires malonyl CoA. In either case, the fatty acid chain is elongated by a two-carbon unit at the carboxyl end. For example, linoleic acid (18:2Δ9,12) can be converted to arachidonic acid (20:4Δ5,8,11,14) via chain elongation and introduction of two more double bonds (Figure 3.22).

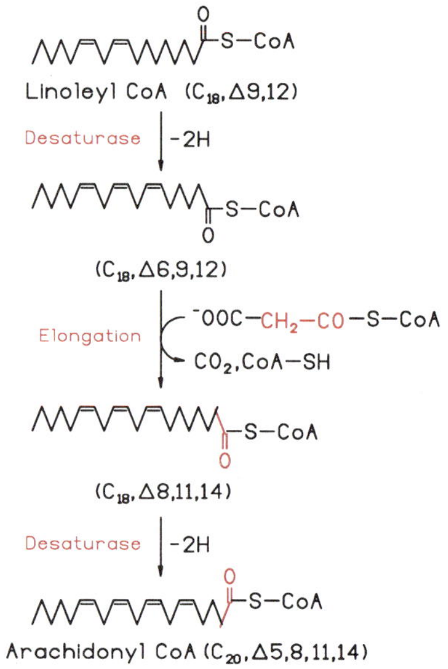

Figure 3.22.
Synthesis of arachidonic acid from linoleic acid.

The introduction of a double bond, called **desaturation**, is catalyzed by a microsomal [smooth ER (SER)] **desaturase** and requires oxygen and NADH. The reaction mechanism probably involves hydroxylation and subsequent dehydration. Desaturation, in humans, can occur only from carbon atom 9 in the direction of the carboxyl group. This is why linoleic acid and linolenic acid are essential fatty acids.

3.11 BIOSYNTHESIS OF TRIACYLGLYCEROL AND PHOSPHOLIPIDS

The synthesis of triacylglycerol begins with glycerol 3-phosphate, which is either derived from glycolysis or synthesized from glycerol and ATP by the enzyme **glycerol kinase**. The site of triacylglycerol

(TAG) biosynthesis is the smooth endoplasmic reticulum; the reaction is catalyzed by **acyltransferases** that transfer fatty acyl residues from acyl CoA to the 1 and 2 positions of glycerol 3-phosphate, thus giving rise to **phosphatidic acid**. The third hydroxyl group of glycerol is esterified after removal of the phosphate group from phosphatidic acid, again deriving the fatty acyl residues from fatty acyl CoA (Figure 3.23).

The predominant pathway of de novo synthesis of **phospholipids**

Figure 3.23.
Biosynthesis of triacylglycerol.

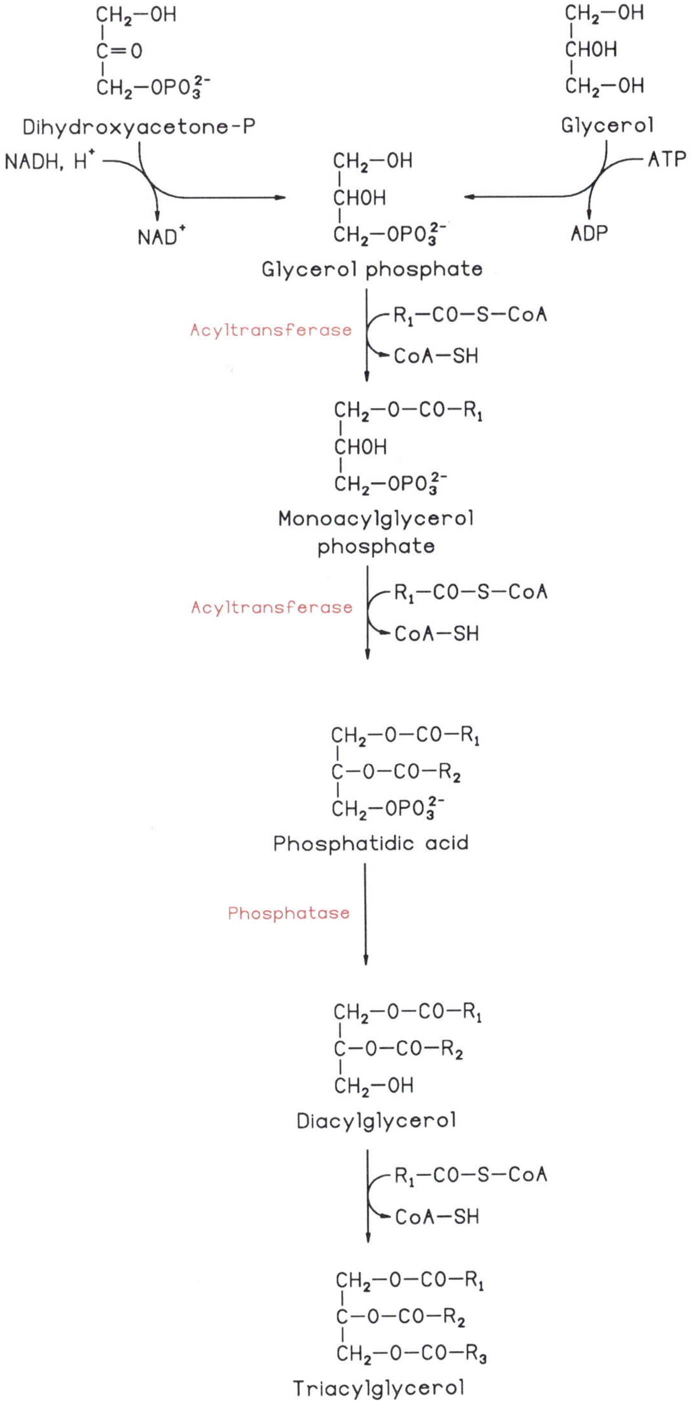

Figure 3.24.
Biosynthesis of phospholipids in mammalian cells.

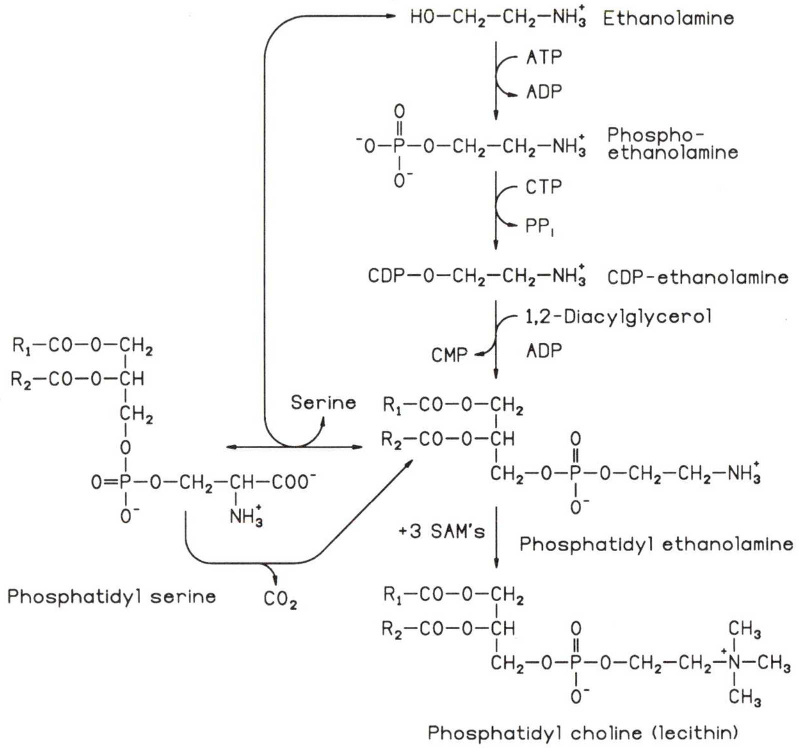

Figure 3.25.
Biosynthesis of sphingolipids.

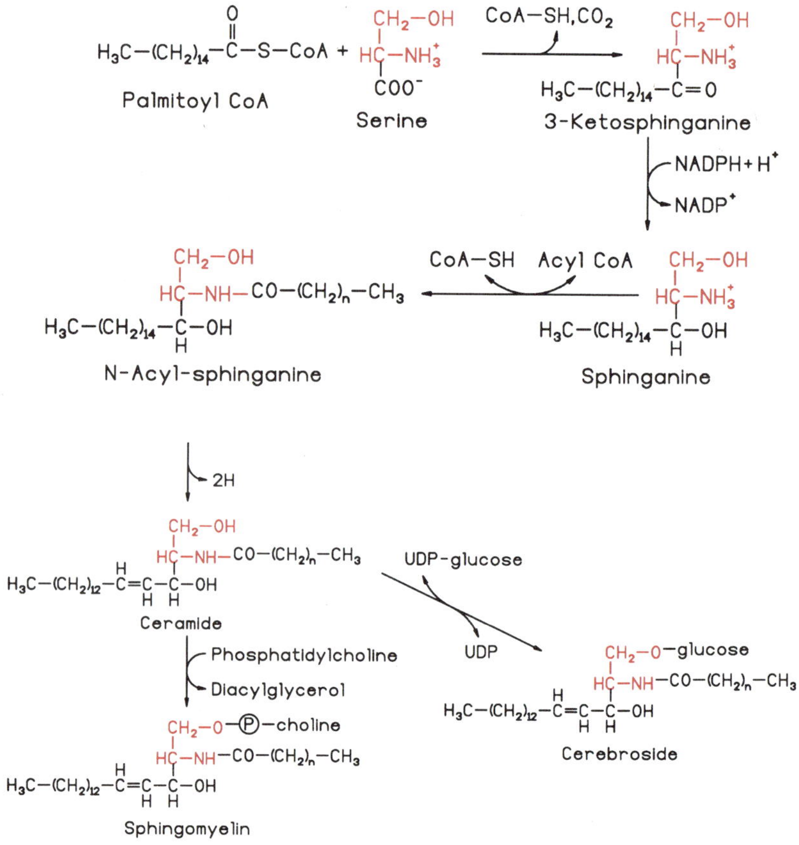

in humans begins with the activation of ethanolamine by cytidine triphosphate (CTP) to cytidine diphosphate-ethanolamine (CDP-ethanolamine). CDP-ethanolamine then reacts with 1,2-diacylglycerol (derived from the triacylglycerol synthetic pathway) to form **phosphatidyl ethanolamine**, which is methylated by S-adenosylmethionine (AdoMet) to **phosphatidyl choline (lecithin)**. In a reversible reaction, the ethanolamine moiety of phosphatidyl ethanolamine can be exchanged with serine to yield **phosphatidyl serine**. Decarboxylation of phosphatidyl serine to phosphatidyl ethanolamine provides an additional mechanism for the synthesis (Figure 3.24). **Phosphatidyl inositol** is synthesized from CDP-diacylglycerol by reaction with inositol.

In a second pathway of phospholipid biosynthesis, the first acylation occurs on dihydroxyacetone phosphate, which is then reduced to 1-acylglycerol 3-phosphate before the second acylation occurs.

In addition to the de novo pathways of phospholipid biosynthesis, there is a **salvage pathway** that allows the utilization of dietary choline. Choline is activated by ATP to phosphocholine, which combines with CTP to yield **CDP-choline**. CDP-choline then reacts with 1,2-diacylglycerol to form lecithin.

3.11.1 Metabolism of Sphingolipids

The biosynthesis of **sphingolipids** begins with the condensation of palmitoyl CoA with serine followed by decarboxylation and desaturation to yield **sphinganine**. In humans, sphinganine is directly acylated, desaturated to **ceramide**, and then reacted with phosphatidyl choline to yield **sphingomyelin** (Figure 3.25).

Cerebrosides and **gangliosides** arise from ceramide by glycosyla-

Figure 3.26.
Lysosomal degradation of gangliosides. Glc = glucose; Gal = galactose; GalNac = *N*-acetylgalactose; CER = ceramide. Genetic deficiencies of key enzymes are indicated by numbers: 1 = hexosaminidase A and B, Sandhoff disease; 2 = β-galactosidase, GM_1-gangliosidosis; 3 = hexosaminidase A, Tay-Sachs disease; 4 = α-galactosidase, Fabry's disease; 5 = β-galactosidase, lactosyl ceramidosis; 6 = β-glucosidase, Gaucher's disease; 7 = sphingomyelinase, Niemann-Pick disease; 8 = galactosyl ceramide β-galactosyl hydrolase, Krabbe's leucodystrophy; 9 = sulfatase, Metachromatic leucodystrophy (sulfatidosis).

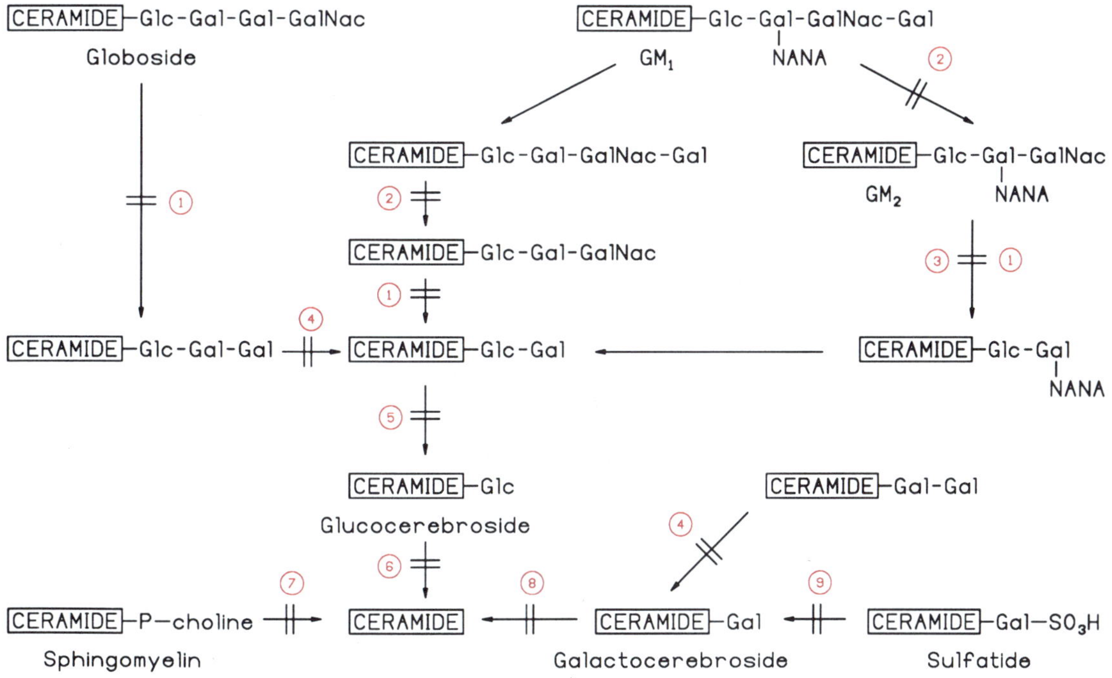

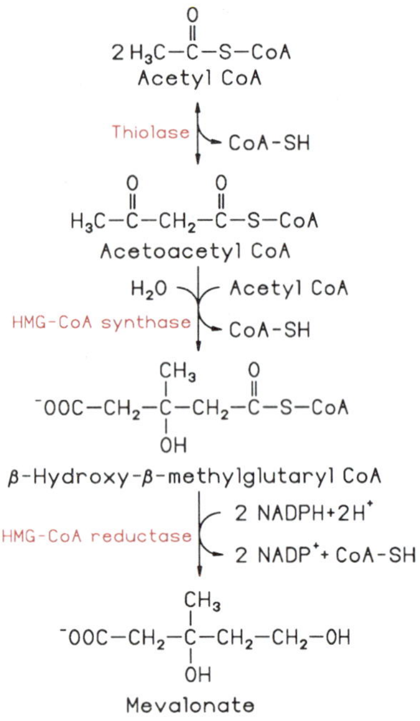

Figure 3.27.
Condensation of acetyl CoA to HMG-CoA and its reduction to mevalonic acid.

tion with UDP sugars and cytidine monophosphate (CMP)-*N*-acetylneuraminic acid. Sulfation of galactose in cerebrosides with 3′-phosphoadenosine 5′-phosphosulfate (PAPS) leads to **sulfatides**.

In several genetic diseases, called **sphingolipodystrophies**, the metabolism of sphingolipids is defective. The defect usually involves the absence of specific lysosomal hydrolases and leads to the accumulation of certain gangliosides in the brain. Often this causes demyelination and destruction of axons. Figure 3.26 illustrates the degradative pathways of gangliosides and the steps that are defective in genetic diseases.

3.12 BIOSYNTHESIS OF CHOLESTEROL AND BILE ACIDS

Although all cells of the human body are capable of cholesterol biosynthesis, most cholesterol is synthesized in the cytoplasm of hepatocytes and intestinal epithelium. All carbon atoms of cholesterol are derived from acetyl CoA. In a series of condensation steps, acetyl CoA is first converted to **hydroxymethylglutaryl CoA**, which is subsequently reduced to **mevalonate** (Figure 3.27). (The cytoplasmic HMG-CoA pool for cholesterol biosynthesis is independent of the intramitochondrial HMG-CoA pool that arises during ketogenesis.)

The reduction of HMG to mevalonate is the rate-limiting step in cholesterol biosynthesis. It is catalyzed by **HMG-CoA reductase**, an enzyme located on the smooth endoplasmic reticulum, and requires two NADPH. Cholesterol inhibits HMG-CoA reductase and thereby shuts down de novo biosynthesis of cholesterol when dietary cholesterol is abundant. In addition, cholesterol, as well as glucagon, inhibits the de novo synthesis of HMG-CoA reductase, whereas insulin stimulates the synthesis of this enzyme (Figure 3.28). Activation of HMG-CoA reductase involves phosphorylation by a cAMP-dependent protein kinase. [Patients with hypercholesterolemia are treated with inhibitors of HMG-CoA reductase (mevacor, lovastatin).]

Mevalonate is phosphorylated and decarboxylated to **isopentenyl pyrophosphate**, which is in equilibrium with its isomer **dimethylallyl pyrophosphate**. These **active isoprenoid units** condense to form first **geranyl pyrophosphate** (C_{10}), **farnesyl pyrophosphate** (C_{15}), and then **squalene** (C_{30}; Figure 3.29). In a series of complex reactions, squalene is finally converted to cholesterol (C_{27}). The cyclization of squalene and subsequent reactions take place in the membrane of the endoplasmic reticulum.

The **conversion of cholesterol to bile acids** involves the introduction of two additional hydroxyl groups and the shortening of the eight-carbon side chain to only five carbon atoms (Figure 3.30).

Bile is a fluid that is secreted by the liver and passes by way of the gall bladder and the bile duct into the duodenum, where it functions in the emulsification of dietary fats. It has an alkaline pH and contains salts of bile acids, phosphatidyl choline, and free cholesterol in the form of micelles. Essentially all of the bile acids are conjugated to either glycine or taurine (3 : 1; see Section 10.2). Intestinal bacteria hydrolyze conjugated bile acids and also remove the C_7 hydroxyl group to form

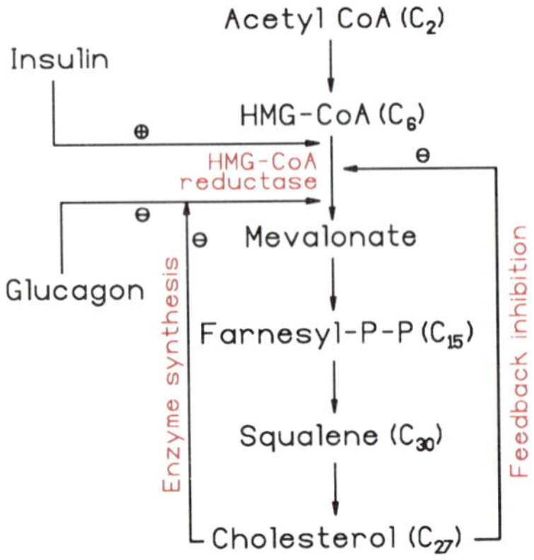

Figure 3.28.
Regulation of cholesterol biosynthesis.

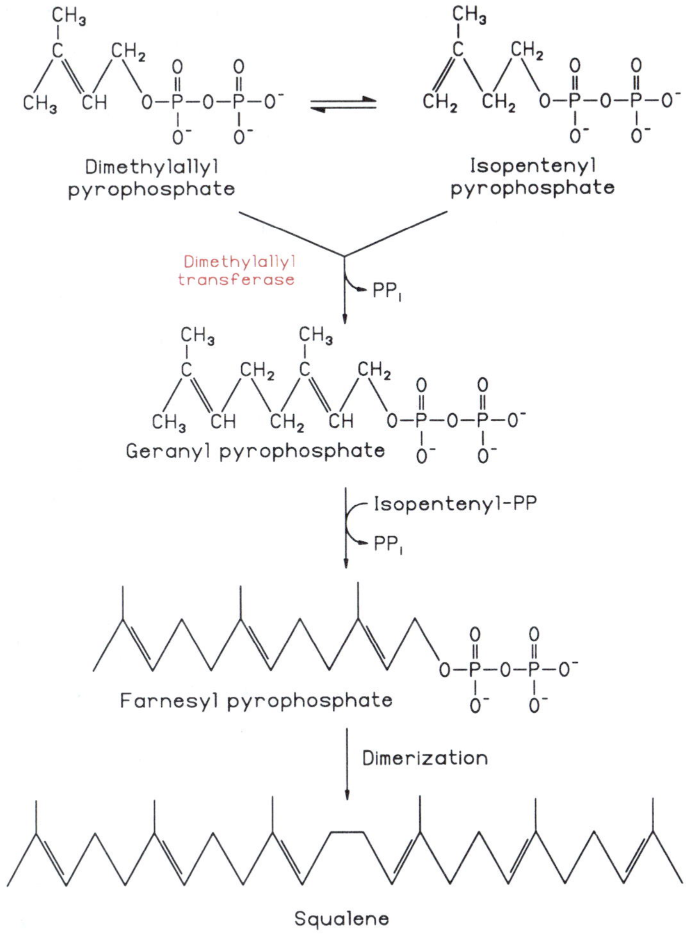

Figure 3.29.
Squalene biosynthesis.

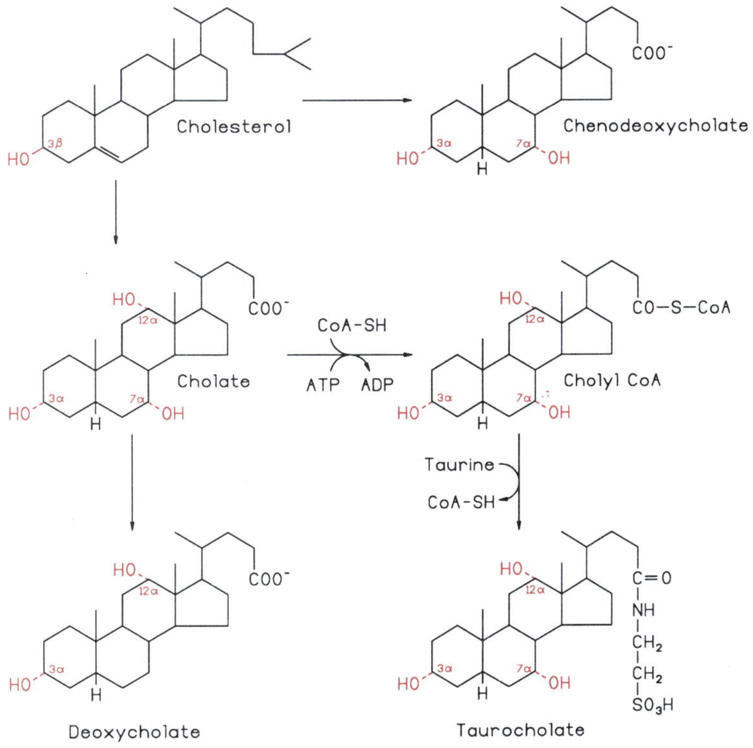

Figure 3.30.
Formation of bile acids from cholesterol.

the **secondary bile acids**, **deoxycholic acid**, and **lithocholic acid**. Both primary and secondary bile acids are reabsorbed in the large intestine. Only a small fraction (1%) is excreted.

In the bile, cholesterol is kept in solution by the presence of lecithin and bile acids. If the composition of bile changes, cholesterol may precipitate and form gallstones.

For a more extensive discussion of lipid metabolism, consult the following textbooks:

Mathews et al. (1990): Chapters, 9, 17, 18

Stryer (1988): Chapters 12, 20

Further References

Bremer J. Carnitine and its role in fatty acid metabolism. *Trends Biochem Sci.* 1977; 2:207–209.

Breslow J. Human apolipoprotein molecular biology and genetic variation. *Annu Rev Biochem.* 1985; 54:699–727.

Brown M, Goldstein JL. A receptor-mediated pathway for cholesterol homeostasis. *Science.* 1986; 232:34–47. (Nobel lecture on the discovery of the LDL receptor.)

McGarry JD, Foster DW. Regulation of hepatic fatty acid oxidation and ketone body production. *Annu Rev Biochem.* 1980; 49:395–420.

van Kujik FGJM, Sevanian A, Handelman GJ, Dratz EA. A new role for phospholipase A_2: protection of membranes from lipid peroxidase damage. *Trends Biochem Sci.* 1987; 12:31–34.

Volpe JJ, Vagelos PR. Mechanism and regulation of biosynthesis of saturated fatty acids. *Physiol Rev.* 1976; 56:339–417.

Wakil SJ, Stoops JK, Joshi VC. Fatty acid synthesis and its regulation. *Annu Rev Biochem.* 1983; 52:537–579.

CHAPTER

4 Metabolism of Amino Acids and Related Compounds

OBJECTIVES

After studying this chapter, the student should be able to

1. Name the nine essential amino acids.
2. Understand the concept of nitrogen balance.
3. Name two nonessential amino acids that are synthesized in the human body from amino acids.
4. Describe the principal pathway for the transport of amino acid nitrogen from peripheral tissues to the liver.
5. List the main sites of ammonia production in the human body.
6. Explain the mechanism of detoxification of ammonium ions.
7. Describe the reactions of the urea cycle and name their intracellular locations.
8. Remember which amino acids are degraded to α-ketoglutarate, succinyl CoA, fumarate, oxaloacetate, pyruvate, or acetoacetyl CoA.
9. Describe the different pathways of glycine metabolism.
10. Name the five amino acids that are both glucogenic and ketogenic and explain why.
11. List at least three reactions involving amino acids in which pyridoxal phosphate participates.
12. Give examples of activated one-carbon fragments in three different states of oxidation.
13. Explain how methyl groups can be shuttled from AdoMet via choline to N^{10}-formyl FH_4 and back to methionine.
14. Outline the pathways utilized in the biosynthesis of nonessential amino acids.
15. Describe how prolyl and lysyl residues in collagen are hydroxylated.
16. Explain the significance of creatine phosphate in muscle.
17. Explain the significance of MAO and COMT in the catabolism of catecholamines.
18. Describe the control of heme biosynthesis and explain how defective control leads to porphyria.

The primary function of amino acids is to provide the precursors for protein synthesis. In addition, amino acids serve as precursors of other important nitrogenous compounds such as biologically active amines (e.g., epinephrine, dopamine), purine and pyrimidine nucleotides, small peptide hormones, glutathione, carnitine, and many others. Finally, amino acids can be oxidized to provide metabolic energy. Figure 4.1 summarizes the metabolism of nitrogenous compounds.

4.1 THE AMINO ACID POOL

The amount of free amino acids in the human body, both extracellular and intracellular, is relatively small (about 40 g in a 70 kg man). This pool of free amino acids is not a mixable whole-body pool, since amino acids do not pass freely through cell membranes. The major flux of amino acids into and out of this pool occurs by **protein turnover.** Proteins vary in their half-lives. Intracellular enzymes, such as δ-amino levulinate synthetase (see heme biosynthesis, section 4.11), have short half-lives on the order of hours or minutes. In contrast, structural proteins, such as muscle proteins, have much longer half-lives, on the order of months.

Protein turnover occurs by other means as well. The cells of the **intestinal mucosa** constantly die and sluff off. The cellular proteins from these cells are digested by intestinal enzymes, and the amino acids are reabsorbed and returned to the pool. **Red blood cells** have a life span of about 120 days, after which time their proteins are degraded by the reticuloendothelial system. **Epidermal cells** (skin, hair, nails) are keratinized and sluff off, thus causing a continuous small loss from

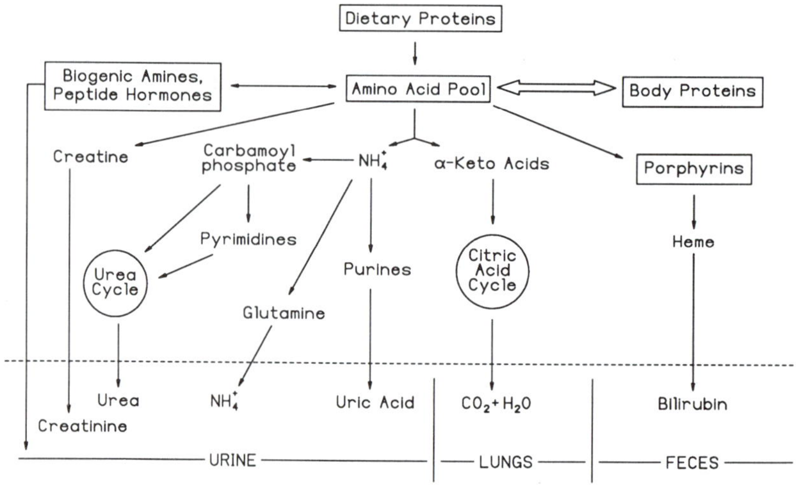

Figure 4.1.
Overview of nitrogen metabolism in the human body.

the total amino acid pool. Extracellular proteins, such as collagen (half-life on the order of years), albumin (half-life of a few days), and other plasma proteins also participate in protein turnover.

4.2 THE CONCEPT OF NITROGEN BALANCE

The healthy human adult usually exhibits **nitrogen balance,** which is defined as a steady state in which nitrogen intake is balanced by nitrogen loss. The minimum daily protein requirement for a 70 kg man has been determined to be about 20 g, equaling the obligatory amino acid losses. This value, however, was obtained under special experimental conditions and is probably not compatible with health under normal life conditions (see Section 8.2). No mechanism exists for storing excess amino acids. The excess amino acids derived from surplus dietary protein are metabolized, and the nitrogen is excreted in the form of urea (Figure 4.2).

Positive nitrogen balance (net protein synthesis) occurs during growth, pregnancy, or recuperation from a wasting disease. In contrast, malnutrition, surgery or severe trauma, long-term bed rest, or situations of excess corticosteroid secretion cause a **negative nitrogen balance** (nitrogen excretion exceeds intake).

During starvation, muscle protein provides the major source of amino acids for degradation and energy production. Once most fat stores are depleted, a starving human may lose as much as 6% of muscle mass per day. Plasma proteins, especially albumin, are depleted even faster. This causes a reduction in the osmotic pressure of the plasma, leading to **edema.** Children suffering from protein calorie malnutrition **(Kwashiorkor)** exhibit bulging abdomens caused by excess ascites (peritoneal fluid).

Nitrogen balance studies have been used to distinguish between **essential** and **nonessential amino acids.** Only nonessential amino acids can be synthesized by the human body; essential amino acids must be provided in the diet. The absence from the diet of a single essential amino acid leads to negative nitrogen balance because protein synthesis is reduced, whereas protein degradation continues.

The essential amino acids are **histidine, isoleucine, leucine, lysine, methionine, phenylalanine, threonine, tryptophan, and valine. Arginine,** although nonessential, can be in short supply (urea cycle); an additional dietary supply is considered essential for adequate

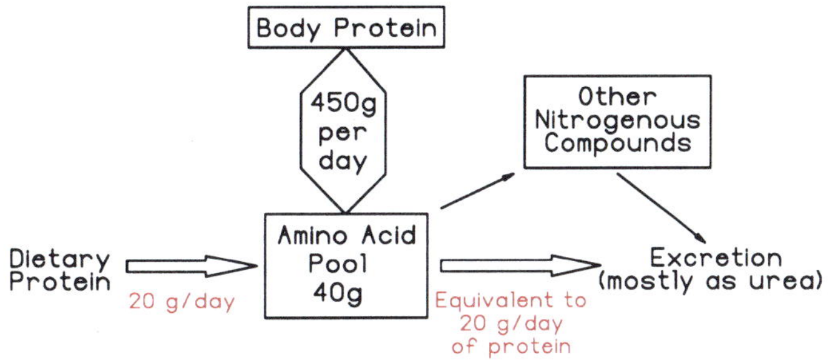

Figure 4.2.
Turnover of the amino acid pool under steady state conditions (nitrogen balance).

growth. **Tyrosine** is synthesized from phenylalanine (essential), whereas methionine is used for the synthesis of **cysteine.** Both of these amino acids, therefore, are not considered essential, although they are formed from essential amino acids.

4.3 ABSORPTION AND TRANSPORT OF AMINO ACIDS

Several amino acid **transport systems** are known in humans. Each system is responsible for a class of structurally similar amino acids. For example, there is a transport system for **acidic** amino acids, a system for **basic** amino acids, a **proline-glycine** system, and several systems for **neutral amino acids.** To a limited extent, amino acids can also be absorbed in the form of short oligopeptides. Little is known about this transport system.

Many of the amino acid transport systems are coupled with sodium transport. The amino acid crosses the cell membrane accompanied by a sodium ion (down a concentration gradient), which is subsequently pumped back out of the cell by the Na^+/K^+ exchange pump (see Section 11.1).

A different type of amino acid transport, the γ-glutamyl or **Meister cycle,** occurs in the renal tubular cells. Amino acids first bind to specific cell membrane receptors and are then transferred by the action of γ-glutamyl transpeptidase to the γ-glutamyl residue of **glutathione** (γ-glutamyl-cysteinyl-glycine), forming a γ-glutamyl amino acid and cysteinyl-glycine inside the cell. The internalized amino acid is then cleaved from the γ-glutamyl residue, which is released as **5-oxoproline**

Figure 4.3.
The γ-glutamyl or Meister cycle.

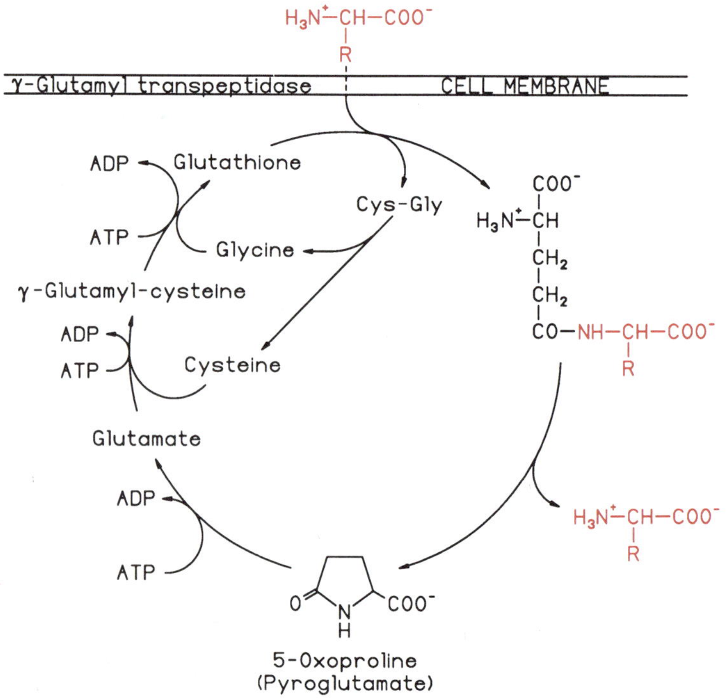

(pyroglutamate), the cyclized form of glutamate. Hydrolysis of 5-oxoproline requires one adenosine triphosphate (ATP), whereas resynthesis of glutathione requires two additional ATP molecules. The cycle is then repeated (Figure 4.3).

Oxoprolinuria is a genetic disease in which glutathione synthetase is defective. As a result, γ-glutamylcysteine, the substrate for the enzyme, is overproduced and subsequently converted to 5-oxoproline in an abortive cycle short circuit.

4.4 INTERCONVERSION AND DEAMINATION OF AMINO ACIDS

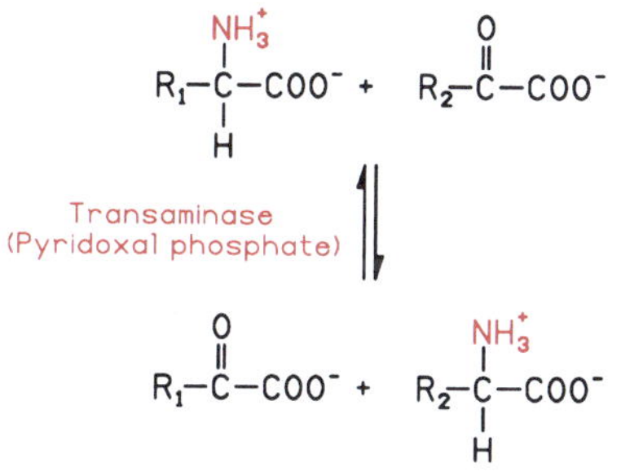

Figure 4.4.
Transamination between amino acid and α-keto acid.

In humans, the predominant pathway of nitrogen excretion is via urea formation. The amino groups of many amino acids, generated by intracellular proteolysis, are transferred primarily to α-ketoglutarate and, in skeletal muscle cells, subsequently to pyruvate. The resulting amino acids, glutamate and alanine, are transported via the bloodstream to the liver, where they are deaminated to form ammonia. The transfer of an amino group from an amino acid to an α-keto acid is called **transamination.** The reaction is catalyzed by **aminotransferases,** or simply called transaminases, enzymes that employ pyridoxal phosphate as cofactor (Figure 4.4).

All amino acids, with the exception of lysine and threonine, can participate in transamination. The most abundant aminotransferases are **alanine-glutamate aminotransferase** and **aspartate-glutamate aminotransferase.** After transamination or deamination (see below), the α-keto acid derived from the amino acids is oxidized by pathways that are discussed later in this chapter.

In the liver, glutamate is oxidatively deaminated by **glutamate dehydrogenase** (Figure 4.5). The enzyme is found both in the cytoplasm and in the mitochondria of liver cells. The equilibrium of the reaction favors the synthesis of glutamate. In the mitochondria, however, the ammonium ions are rapidly converted to carbamoyl phosphate, the first reaction of the urea cycle. This removal of ammonium ions favors the deamination of glutamate.

Glutamate dehydrogenase is inhibited by ATP and guanosine triphosphate (GTP) and is activated by adenosine diphosphate (AD) and guanosine diphosphate (GDP). Oxidation of glutamate, therefore, is increased when energy is required, as reflected in a lower ATP/ADP ratio. In the cytoplasm, an excess of the reduced form of nicotinamide adenine dinucleotide phosphate (NADPH) favors the conversion of ammonium ion into glutamate.

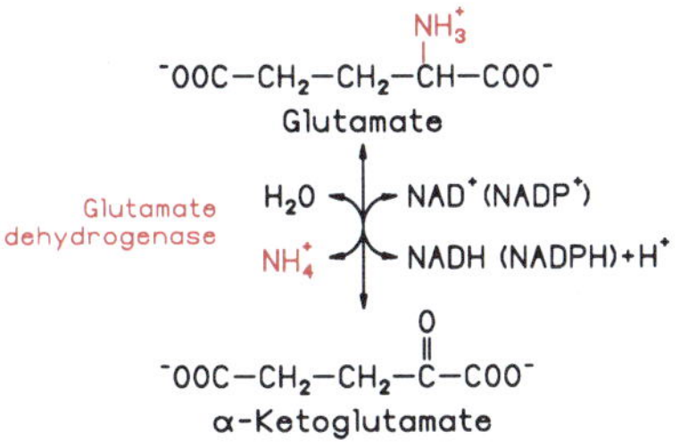

Figure 4.5.
Oxidative deamination of glutamate and de novo synthesis of glutamate.

In muscle, and probably also in liver, ammonia can also be produced by hydrolytic deamination of adenosine monophosphate (AMP). The amino group is transferred from aliphatic amino acids (valine, leucine, and isoleucine) to glutamate to aspartate to AMP, which is then deaminated to inosine monophosphate (IMP). This series of reactions is called the **IMP-AMP cycle** (Figure 4.6). During strenuous exercise, up to 10% of the energy production in the muscle can be derived from the oxidation of α-keto acids produced from muscle protein-derived amino acids via this pathway. Individuals with inherited **AMP deami-**

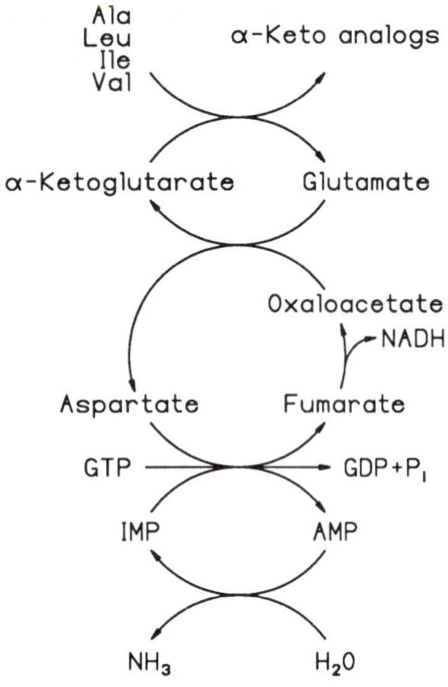

Figure 4.6.
The IMP-AMP cycle.

nase deficiency cannot deaminate AMP, so the cycle cannot be completed. When performing strenuous exercise, these people develop muscle cramps. Except for this condition, the defect is relatively harmless.

Other oxidative deamination reductions involve L and D **amino acid oxidases,** which are found in liver and kidney cells. Although their physiological significance is uncertain, these enzymes deaminate amino acids to α-keto acids with the concomitant production of hydrogen peroxide. The amino acid oxidases are located in **peroxisomes,** where the peroxide is disposed of by **catalase.**

Deamination of amino acids containing β-hydroxyl or β-sulfhydryl groups (serine, cysteine, threonine) is catalyzed by **dehydratases,** which use pyridoxal phosphate as cofactor. For example, **serine dehydratase** directly converts serine to pyruvate (Figure 4.7).

Histidine is deaminated by **histidase,** which is also called histidine ammonia lyase, an enzyme that employs pyridoxal phosphate as cofactor (Figure 4.8). Additional amino acids, such as methionine and threonine, also lose their α amino groups as NH_4^+. The ammonium ions produced by these reactions in extrahepatic tissues (where no urea cycle exists) are transported to the liver in the form of glutamate, glutamine, or alanine.

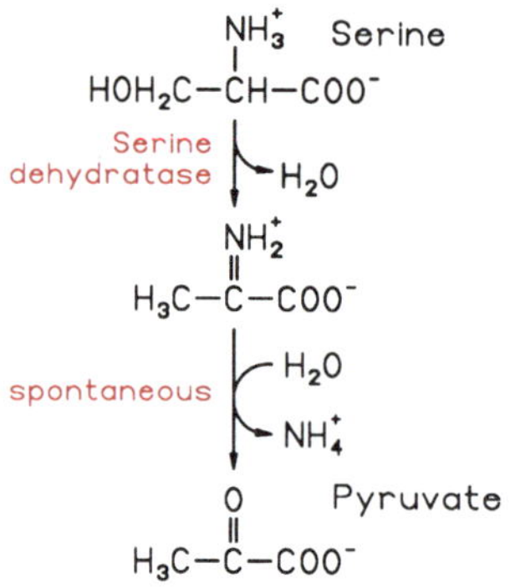

Figure 4.7.
Deamination of serine by serine dehydratase.

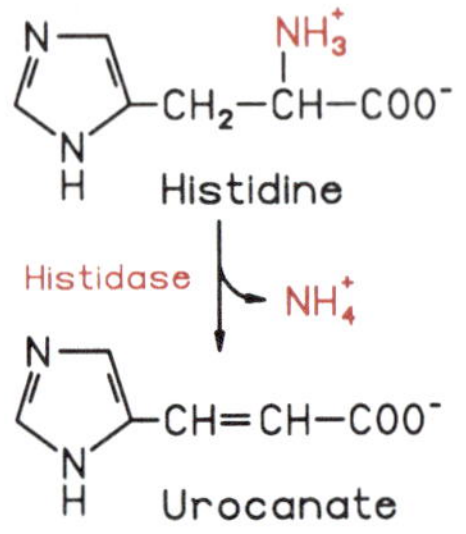

Figure 4.8.
Deamination of histidine.

4.5 THE UREA CYCLE

In the liver, the ammonia produced from glutamate, from AMP, or by direct deamination is converted to **urea** in a series of reactions called the urea cycle (Figure 4.9). As discussed before, the generation of ammonium ion (by glutamate dehydrogenase) occurs primarily in the mitochondrion, which is also the site of the first reaction of the urea cycle. Carbon dioxide and ammonium ion react with two molecules of ATP to form **carbamoyl phosphate.** The mitochondrial enzyme **carbamoyl phosphate synthetase** requires *N*-acetyl glutamate as an allosteric activator. (Note that this enzyme is different from the cytoplasmic carbamoyl phosphate synthetase involved in pyrimidine biosynthesis; see Chapter 5.) Carbamoyl phosphate reacts with **ornithine** to form **citrulline,** which moves out of the mitochondrion into the cytoplasm. There it reacts with aspartate to form **argininosuccinate,** which is subsequently cleaved to **arginine** and **fumarate.** Arginine is then hydrolyzed by **arginase** to urea and ornithine, which again enters the mitochondrion to initiate a new cycle.

The overall reaction from ammonium ion to urea requires energy in the form of four high-energy phosphate bonds (two ATPs are cleaved to two ADPs and one P_i in the synthesis of carbamoyl phosphate, and another ATP is cleaved to AMP and two P_is in the generation of argininosuccinate). Note that only one of the two nitrogen atoms of urea is derived from ammonium ion; the other originates from aspartate. The fumarate produced in this cytoplasmic process is converted to oxaloacetate, the direct precursor of phosphoenolpyruvate, thus funnelling amino acid–derived carbon atoms into gluconeogenesis.

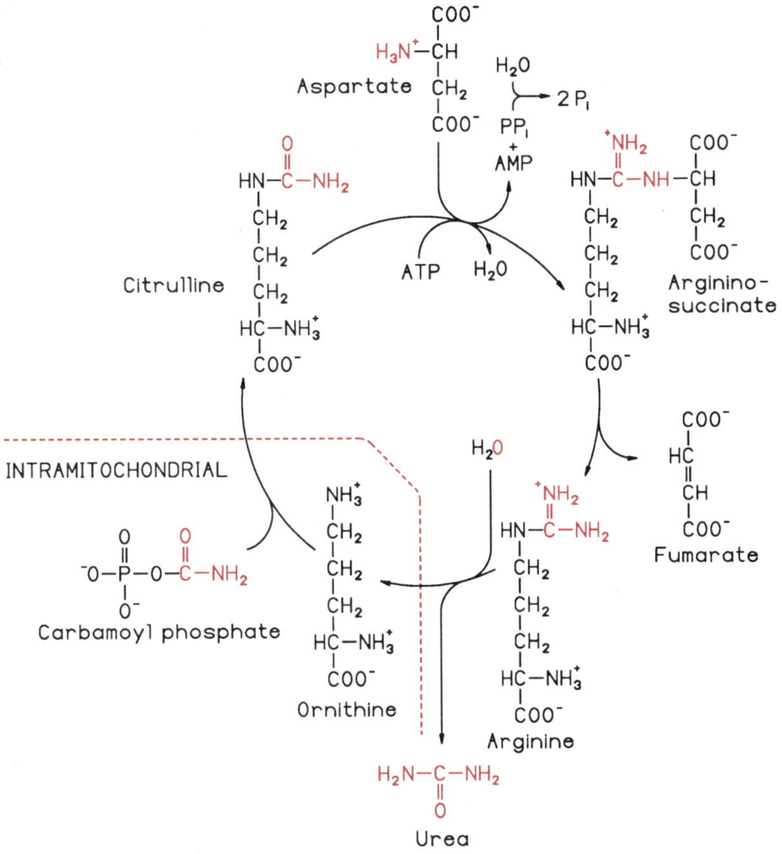

Figure 4.9.
The urea cycle.

4.6. CONTROL OF AMMONIUM ION CONCENTRATION

Although ammonium ion is a major metabolite, its steady state concentration in the body is kept at very low levels. Blood levels above 5 x $10^{-5}M$ are highly toxic, producing disorders of the central nervous system such as slurring of speech, blurring of vision, and eventually coma.

The principal sources of ammonium ions are

1. Intestinal cleavage of urea by bacterial urease. (The NH_4^+ ions enter the liver via the portal circulation and are removed there.)
2. Glutamate dehydrogenase.
3. Amino acid oxidases.
4. Amino acid dehydratases and histidase.
5. The cleavage of glycine by glycine synthase (see Section 4.7).
6. Deamination of threonine and homoserine (from methionine).
7. AMP deaminase.
8. Glutaminase and asparaginase.
9. Non-α-amino nitrogen of tryptophan and histidine.

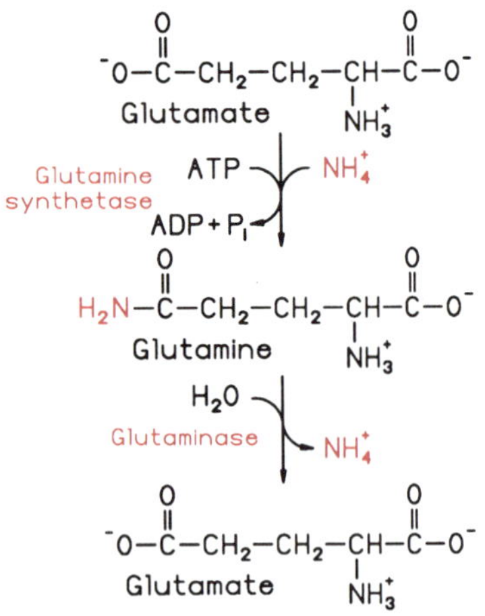

Figure 4.10.
Glutamine as a carrier of ammonium ions.

The last enzyme, **glutaminase,** is especially active in the kidney. Its main function is the production of ammonia for acid-base regulation (see Section 12.4). Most of the renal NH_4^+ ions are excreted in the urine; a small amount reenters the circulation and is eventually detoxified in the liver.

Free ammonium ions in nonhepatic tissue are primarily detoxified by **glutamine synthetase.** The resulting glutamine then carries the ammonium ion to the liver for final conversion to urea (Figure 4.10).

Glutamine synthetase is especially important in nervous tissue. However, the substrate for this enzyme, glutamate, cannot readily cross the blood-brain barrier and, therefore, has to be synthesized from α-ketoglutarate in the brain. One possible explanation why ammonium ions are so toxic to nervous tissue is that the conversion of α-ketoglutarate to glutamate depletes citric acid cycle intermediates and thus interferes with energy production in brain cells. Another, more recent, theory on ammonia toxicity proposes that ammonium ions indirectly promote tryptophan transport into the brain, thereby causing an overproduction of serotonin, an inhibitory neurotransmitter.

4.7 METABOLISM OF AMINO ACID CARBON CHAINS

An amino acid is **glucogenic** if its carbon skeleton can be metabolized to citric acid cycle intermediates or pyruvate, which can then be used for gluconeogenesis (for an overview, see Figure 2.49). An amino acid is **ketogenic** if its oxidative degradation leads to acetyl coenzyme A (acetyl CoA) or intermediates of fatty acid oxidation.

Most amino acids are glucogenic. Only one amino acid, **leucine,** is purely **ketogenic.** Five other amino acids, **isoleucine, lysine, phenylalanine, tyrosine,** and **tryptophan** are **mixed,** producing both pyruvate and acetyl CoA during their oxidative degradation. Five families of amino acids can be distinguished on the basis of their degradation patterns. The members of a family are not necessarily structurally related, and some amino acids belong to more than one family.

4.7.1 The α-Keloglutarate Family

All amino acids in this family, **glutamate, glutamine, proline, arginine (ornithine),** and **histidine,** contain five-carbon chains that are readily converted to α-ketoglutarate (Figure 4.11). (Please note that hydroxyproline, found in collagen, is degraded to pyruvate; see Figure 4.14.)

Histidinemia, a rare metabolic disorder, is caused by the absence of histidase and is characterized by elevated levels of histidine in blood and urine.

4.7.2 The Succinyl CoA Family

The amino acids **methionine, isoleucine,** and **valine** exhibit fairly complex degradative pathways. Their final common degradation product is **propionyl CoA,** which can be rearranged, after carboxylation

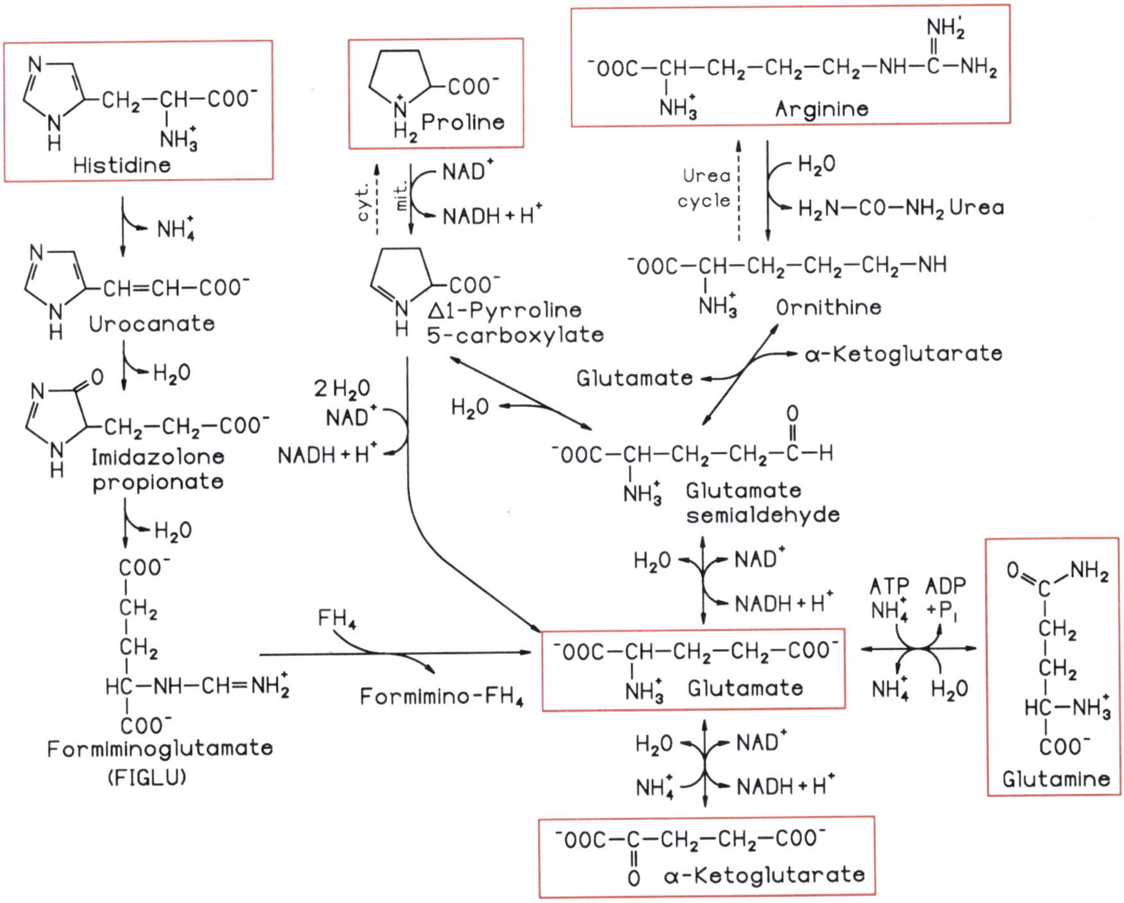

Figure 4.11.
Degradation of the amino acids of the α-glutamate family.

to **methylmalonyl CoA,** to **succinyl CoA** in a vitamin B_{12}–catalyzed reaction (Figure 4.12). Threonine also belongs to this family. It is first converted by threonine dehydratase to α-ketobutyrate, which is subsequently decarboxylated to propionyl CoA.

Methionine first reacts with ATP to form ***S*-adenosyl-methionine (AdoMet),** which subsequently loses its **active methyl group** by transmethylation; the resulting ***S*-adenosyl-homocysteine** is cleaved to adenosine and homocysteine. **Homocysteine** reacts with serine to yield cysteine and α-ketobutyrate plus NH_4^+. In this last reaction, **homoserine** is an intermediate but remains bound to the enzyme.

The degradation pathways for the branched-chain amino acids **isoleucine** and **valine** (as well as leucine, discussed later) use similar steps. Initial **transamination** is followed by **oxidative decarboxylation,** yielding the α-keto homologs as CoA esters. Further degradation occurs by β oxidation. Valine is degraded to **propionyl CoA,** which is carboxylated to **methylmalonyl CoA.** The degradation of isoleucine yields acetyl CoA in addition to propionyl CoA.

The oxidative carboxylation of the branched-chain α-keto acids derived from isoleucine, valine, and leucine is catalyzed by a multienzyme complex similar to the pyruvate dehydrogenase complex (see Section 1.11). A defective decarboxylase subunit (specific for branched-chain α-keto acids) in this complex causes the rare, genetic **maple syrup disease,** which is characterized by the presence of α-ketoisovalerate, α-keto-β-methylvalerate, and α-ketoisocaproate (the transamination

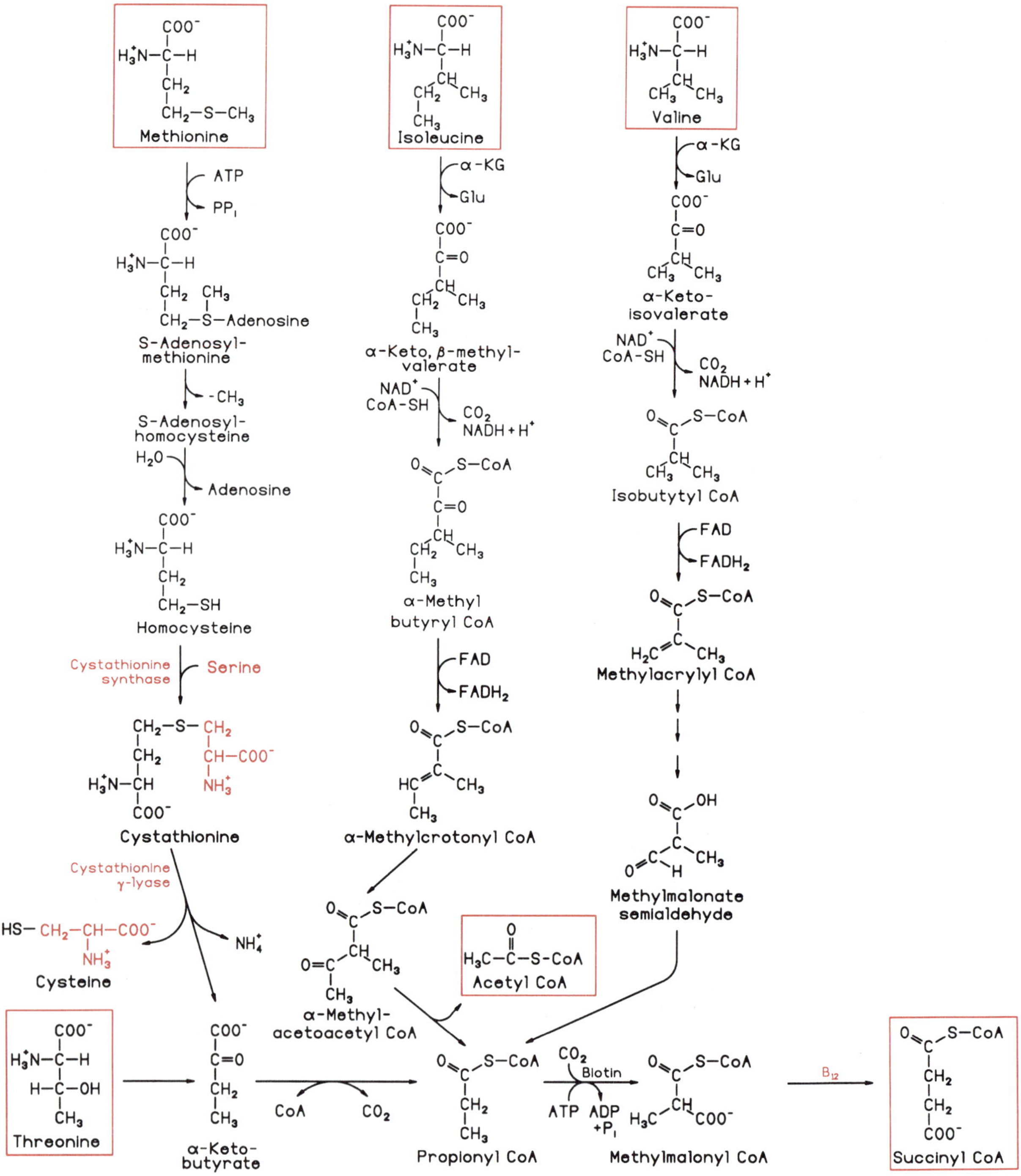

Figure 4.12.
Degradation of the amino acids of the succinyl CoA family.

products of valine, isoleucine, and leucine, respectively) in the urine. The characteristic smell of these keto acids is reminiscent of maple syrup. The disease is accompanied by severe mental retardation.

4.7.3 The Fumarate-Oxaloacetate Family

Aspartate and **asparagine** are predominantly transaminated to oxaloacetate but can be converted to fumarate via the IMP-AMP cycle (see Figure 4.6). **Phenylalanine** is hydroxylated to **tyrosine** by **phenylalanine hydroxylase,** a monooxygenase that uses tetrahydrobiopterin (see Figure 4.17) as cofactor. The absence of phenylalanine hydroxylase causes the genetic disease **phenylketonuria.** Dietary restriction of phenylalanine, especially during early childhood, prevents the mental retardation that otherwise results from this deficiency but must be continued into adult life.

After transamination of tyrosine to *p*-hydroxyphenylpyruvate, a dioxygenase introduces a second hydroxyl group into the phenyl ring in a reaction that is accompanied by decarboxylation, oxidation, and rearrangement of the side chain to **homogentisate.** The enzyme contains Cu^{2+} ions and requires ascorbate as cofactor. Homogentisate is then further oxidized under ring cleavage by **homogentisate 1,2-dioxygenase,** an enzyme that contains Fe^{2+} and requires both glutathione and ascorbate. (The genetic absence of this enzyme causes **alkaptonuria,** a condition characterized by brown-colored urine and abnormal pigmentation of connective tissue, caused by oxidation products of homogentisate.) The resulting eight-carbon acid is then further cleaved to acetoacetate and fumarate. Phenylalanine and tyrosine thus are partly ketogenic and partly glucogenic (Figure 4.13).

4.7.4 The Pyruvate Family

The conversion of **alanine** to pyruvate by transamination has been discussed in Section 4.4. **Serine** can be used as a source of one-carbon units and in the process is converted to glycine (see below); it can also be directly converted to pyruvate by the action of **serine dehydratase. Cysteine** undergoes a similar reaction accompanied by the release of H_2S and ammonia. This route, however, is a minor one compared with the oxidation of cysteine to cysteine sulfinate (by an Fe^{2+}-containing dioxygenase). Upon transamination, cysteine sulfinate releases sulfite, which is further oxidized to sulfate and excreted in the urine (Figure 4.14). The catabolism of cysteine represents the **only pathway for the excretion of sulfur as sulfate,** since methionine transfers its sulfur to cysteine (see Figure 4.12).

The major degradative pathway of **glycine** leads to CO_2, NH_3, and methylene tetrahydrofolate in a reversible reaction catalyzed by **glycine synthase.** In a minor pathway, glycine can be oxidatively deaminated by **glycine oxidase** to glyoxylate. Further oxidation of glyoxylate leads to **oxalic acid,** which forms very insoluble Ca salts and may lead to the formation of kidney stones. (The only other sources of oxalate in the human body are ascorbic acid and a diet rich in oxalate.)

Tryptophan belongs only partly to the pyruvate family. Its side chain is cleaved off to form alanine, whereas the degradation of the ring system

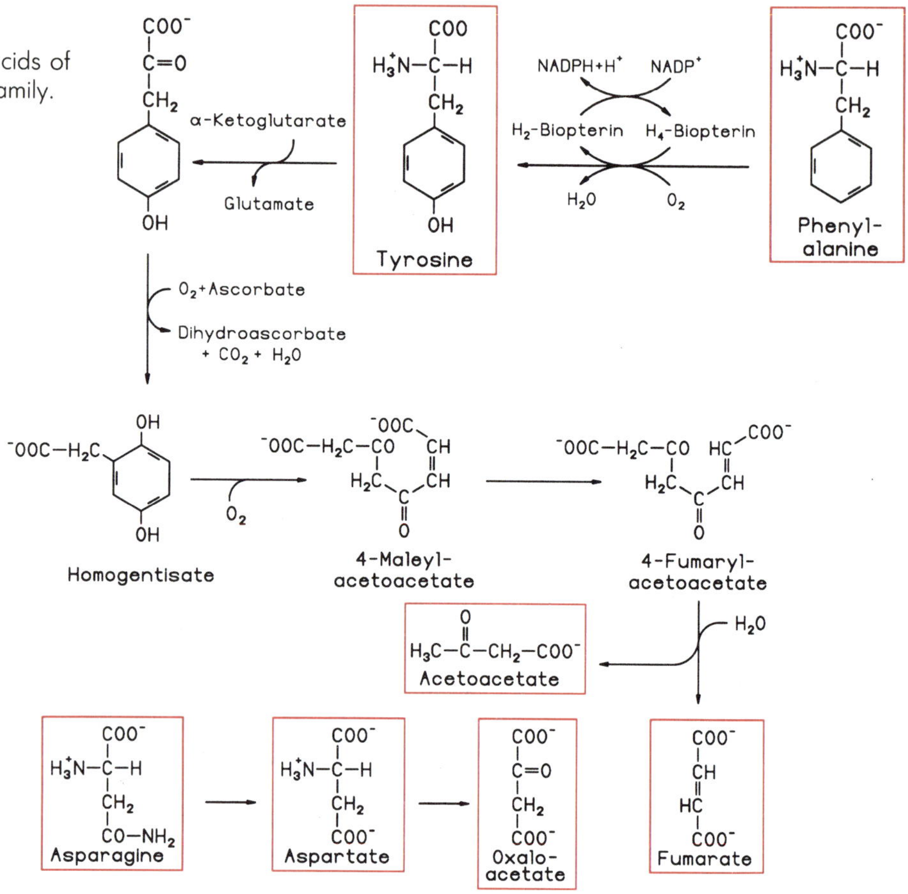

Figure 4.13.
Degradation of the amino acids of the fumarate-oxaloacetate family.

leads to acetoacetyl CoA and a variety of other products (see Figure 4.15).

4.7.5 The Acetoacetyl CoA Family

The amino acids **leucine, lysine,** and **tryptophan,** although structurally unrelated, are all catabolized to acetoacetyl CoA. The degradation

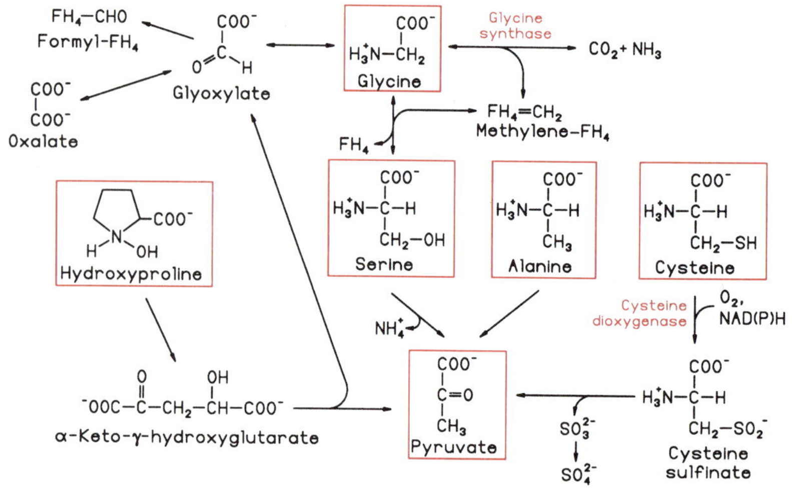

Figure 4.14.
Degradation of the amino acids of the pyruvate family.

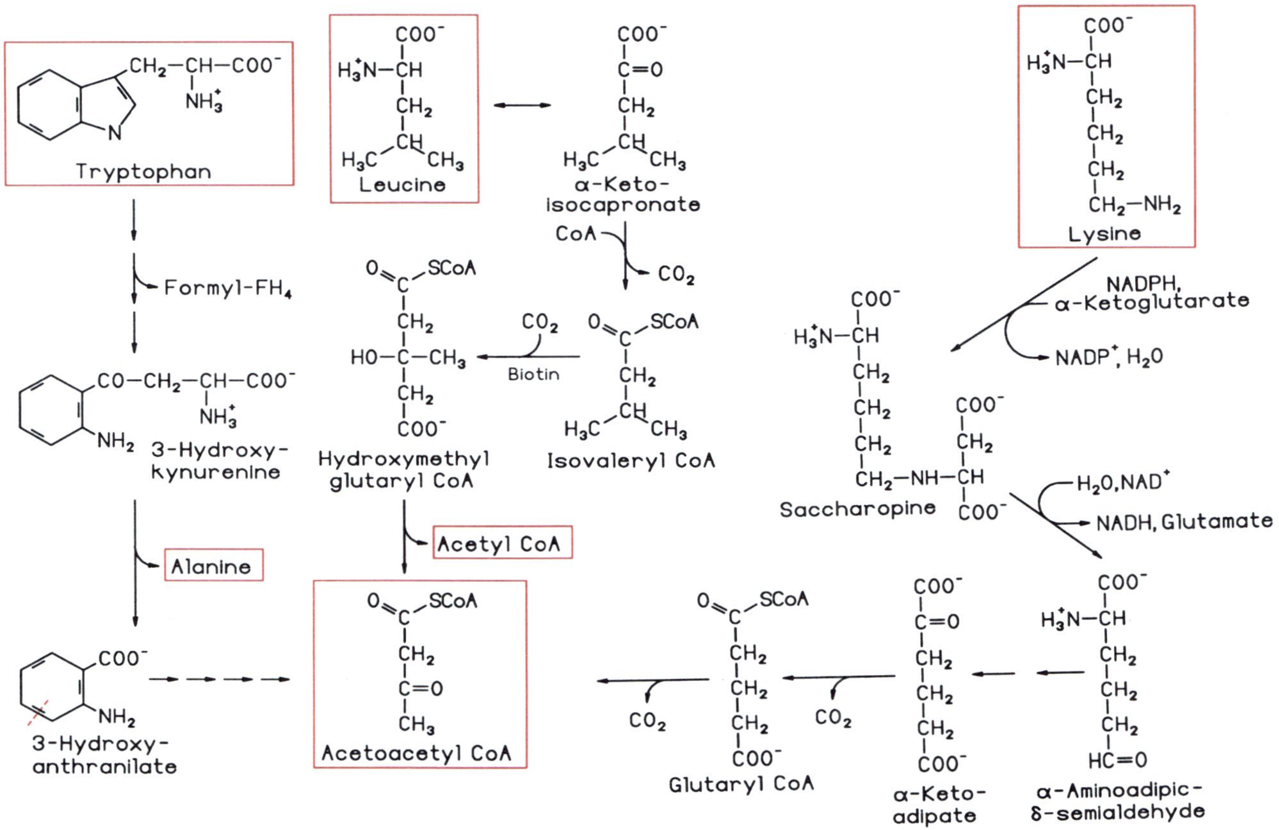

Figure 4.15. Degradation of the amino acids of the acetoacetyl CoA family.

of leucine is similar to that of the other branched-chain amino acids isoleucine and valine. The initial transamination to α-ketocaproate is followed by oxidative decarboxylation and β oxidation. The next step involves a biotin-dependent carboxylation. The product, α-hydroxy-β-methylglutaryl CoA (HMG-CoA), is then cleaved to acetyl CoA and acetoacetate (Figure 4.15).

The degradation of tryptophan begins with the oxidative opening of the pyrrole ring by the dioxygenase **tryptophan pyrrolase.** The resulting *N*-formylkynurenine (see Figure 2.38) then transfers its formyl group to tetrahydrofolate. Kynurenine is hydroxylated to 3-hydroxykynurenine, which loses its alanine moiety to yield 3-hydroxyanthranilate. A complex series of reactions eventually produces acetoacetyl CoA.

Lysine degradation is also complex. Its known degradation pathway leads to acetoacetyl CoA. Nevertheless, lysine is classified as a mixed amino acid, being both ketogenic and glucogenic. This classification is based on an experimental protocol in which diabetic animals are fed amino acids, and the presence of glucose or ketone bodies is subsequently determined in blood and urine. Lysine has been found to elevate the levels of both glucose and ketone bodies.

4.8 ABNORMALITIES IN AMINO ACID CATABOLISM

Genetic deficiencies of many of the enzymes involved in amino acid degradation have been described in the literature. Although most of these diseases are extremely rare, taken as a group they represent a substantial component of genetic diseases. **Phenylketonuria,** one of the more frequent genetic disorders, occurs with a frequency of about 1 in 20,000 births.

For each of the enzymes involved in urea synthesis, partial enzyme deficiencies have been reported in patients. Patients affected with these disorders exhibit markedly elevated blood levels of ammonium ion and show clinical symptoms of ammonia intoxication. The complete absence of any of the early enzymes of the urea cycle results in death of the newborn after its first protein meal (milk).

Table 4.1 lists some of the genetic disorders that affect amino acid metabolism.

4.9 COFACTORS THAT PARTICIPATE IN AMINO ACID METABOLISM

Pyridoxal phosphate, a derivative of vitamin B_6 (pyridoxine), participates in many reactions involving amino acids. All these reactions require the formation of a Schiff base intermediate betweer the aldehyde group of pyridoxal phosphate and the α amino group of the amino acid. Figure 4.16 shows the mechanism of a transamination reaction. Other reactions that employ pyridoxal phosphate include decarboxylation

Table 4.1. Genetic Disorders in Amino Acid Metabolism

Disorder	Affected Enzyme	Symptoms
Albinism	Tyrosinase	No melanin synthesis
Alkaptonuria	Homogentisate oxidase	Relatively benign, black urine
Argininemia	Arginase	Mental retardation
Argininosuccinic aciduria	Argininosuccinase	Mental retardation, seizures, death
Citrullinemia	Argininosuccinate synthetase	Mental retardation
Cystathioninuria	Cystathionase	No consistent biochemical picture
Histideinemia	Histidase	Some mental retardation
Homocystinuria	Cystathionine synthetase	Rather common, minimal abnormalities
β-Hydroxyisovaleric aciduria (cat's urine syndrome)	β-Methylcrotonyl CoA carboxylase	Neurologic problems, urine smells like cat urine
Hyperammonemia I	Carbamoylphosphate synthetase	Very high levels of glutamine in blood, poor prognosis, rare
Hyperammonemia II	Ornithine transcarbamoylase	Increased levels of glycine and glutamine in blood and urine, very rare, fatal
Hyperglycinemia (ketotic)	Propionyl CoA carboxylase	Episodes of ketosis and acidosis
Hyperlysinemia	Lysine-α-ketoglutarate reductase	Rare, severe retardation
Hyperoxaluria type I	α-Ketoglutarate-glyoxylate carboligase	
Hyperoxaluria type II	D-Glycerate dehydrogenase	
Hyperprolinemia I	Proline oxidase	Relatively harmless
Hyperprolinemia II	Δ1-Pyrroline-5-carboxylate dehydrogenase	Possible effect on brain development
Isovaleric acidemia	Isovalerate CoA dehydrogenase	Episodes of acidosis, leucopenia, anemia
Lysine intolerance	Lysine-NAD-oxidoreductase	Episodes of vomiting, convulsions, coma, can be controlled by low-protein diet
Maple syrup disease	Keto acid decarboxylase	Convulsions, coma, dietary control possible
Methylmalonic aciduria	Methylmalonyl CoA carboxymutase	Mental retardation if not treated by dietary restrictions
Pyroglutamic aciduria (oxoprolinuria)	Pyroglutamate hydrolase	Severity depending on completeness of enzyme deficiency, retardation, acidosis
Tyrosinemia	Tyrosine aminotransferase	Mental retardation
Tyrosinosis	*p*-Hydroxyphenylpyruvate oxidase	Excretion of hydroxyphenylpyruvate, renal complications
Valinemia	Valine aminotransferase	

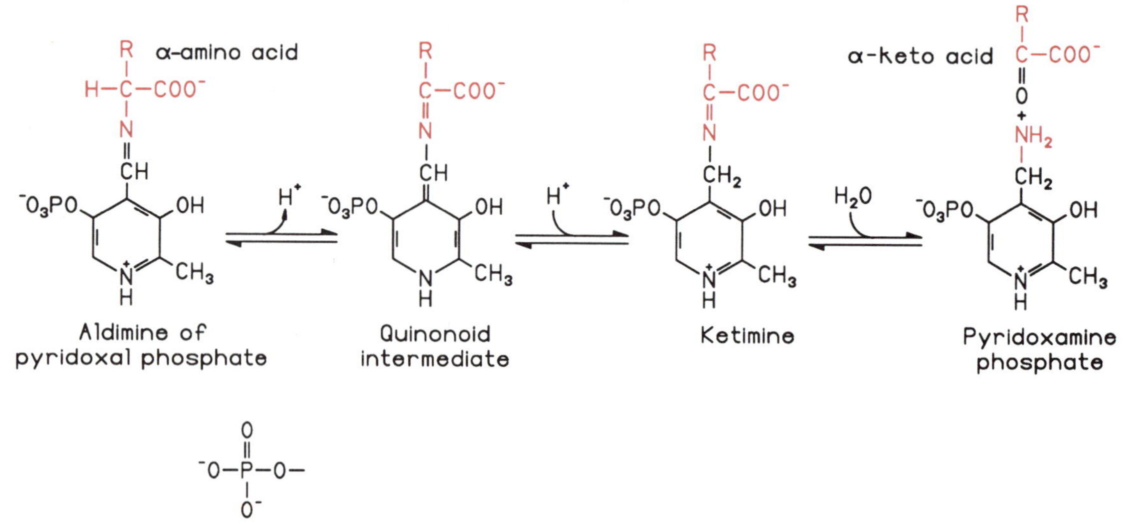

Figure 4.16.
Mechanism of transamination.

(e.g., histidine decarboxylase), dehydration (e.g., serine dehydratase), and deamination.

The involvement of **lipoamide** and **thiamine pyrophosphate** in the oxidative decarboxylation of α-keto acids has been discussed in Section 1.11.

Tetrahydrobiopterin is required as cofactor of monooxygenases for the transfer of hydrogen ions from NADPH to oxygen and the substrate (Figure 4.17). Hydroxylation of phenylalanine, tyrosine, and tryptophan involve this cofactor. The human body can synthesize the pteridine ring system de novo.

Cobalamin coenzymes are derived from vitamin B_{12} (hydroxycobalamin). In the form of **5′-deoxyadenosyl-cobalamin** (Figure 4.18), this cofactor participates in the reaction catalyzed by **methylmalonyl CoA mutase.** The unusual cobalt-carbon bond in this cofactor is unstable and breaks easily to form a free radical. It is this free radical form of the cofactor that catalyzes the molecular rearrangement in the methylmalonyl molecule. In the form of **methylcobalamin,** the cofactor functions in the reaction catalyzed by **homocysteine transmethylase.** It accepts a methyl group from N^5-methyl tetrahydrofolate and transfers it to homocysteine to regenerate methionine (see Figure 4.21).

Pernicious anemia is a vitamin B_{12} deficiency that is caused by the absence of **intrinsic factor,** a stomach glycoprotein required for the absorption of the vitamin from the intestine.

Ascorbic acid, vitamin C, is a strong reducing agent (Figure 4.19). It is involved in tyrosine catabolism, in the formation of hydroxyproline (see Figure 4.23) and hydroxylysine during collagen synthesis, and in the biosynthesis of carnitine (Figure 4.26).

Tetrahydrofolate (FH_4) serves as carrier of one-carbon fragments. These fragments are attached to the N^5 and/or the N^{10} positions of FH_4 (Figure 4.20) and can exist in three different states of oxidation (Table 4.2). Although FH_4 contains the same pteridine ring system as tetrahydrobiopterin (see Figure 4.17), it cannot be synthesized de novo in the human body and, therefore, is considered a vitamin.

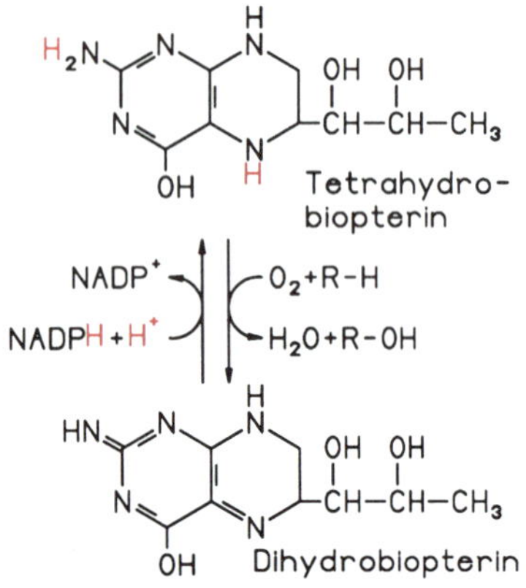

Figure 4.17.
Electron transfer between dihydrobiopterin and tetrahydrobiopterin.

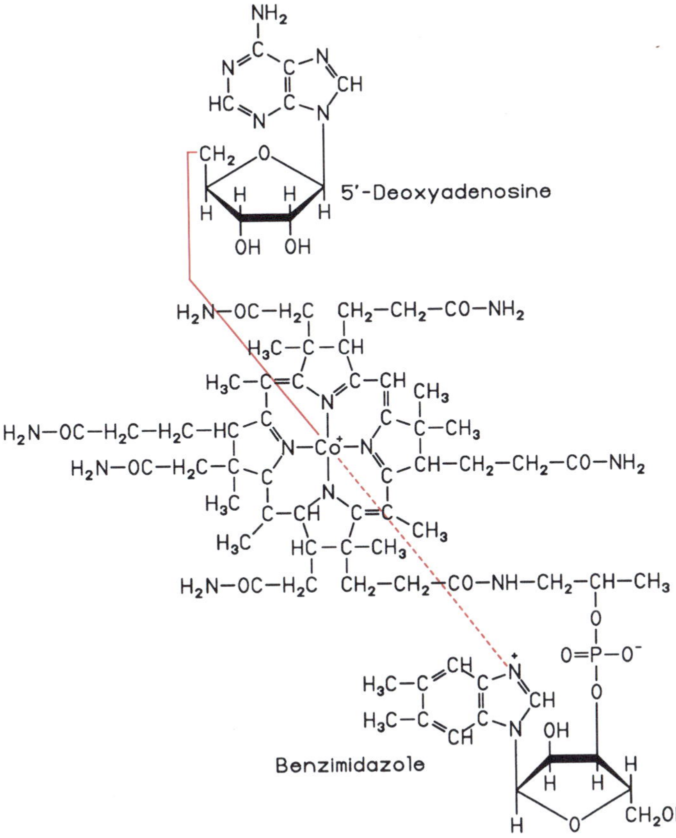

Figure 4.18.
5′-Deoxyadenosyl cobalamine. The labile Co—C bond is indicated in red.

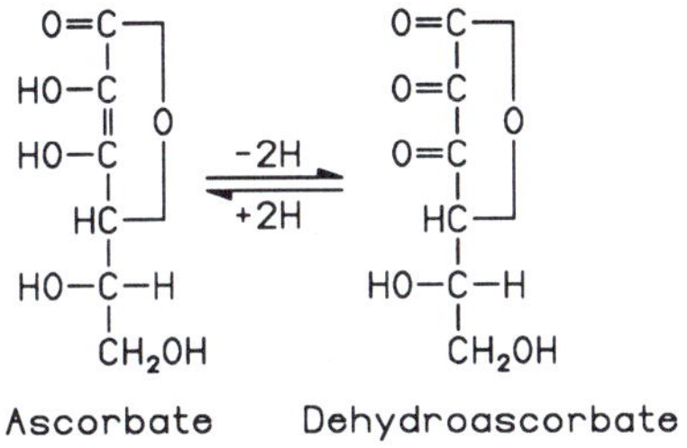

Figure 4.19.
Electron transport between ascorbate and dehydroascorbate.

4.9.1 Metabolism of One-Carbon Fragments

The major source of one-carbon units in the human body is glucose, which, via serine and glycine, yields N^5, N^{10}-methylene FH_4 and N^{10}-formyl FH_4 (see Figures 4.14 and 4.21).

The interconversions of different tetrahydrofolate derivatives are all enzyme catalyzed. Folic acid, a vitamin, is absorbed in the intestine and then reduced to tetrahydrofolate by NADPH in two steps. This reaction is blocked by amethopterin (methotrexate); the use of this drug in cancer therapy is discussed in Section 5.2.

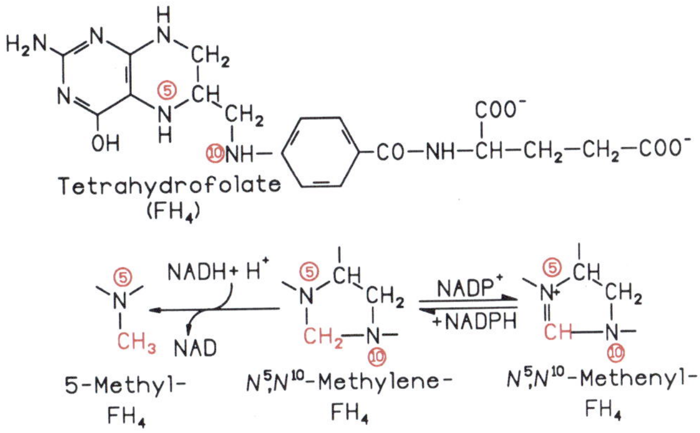

Figure 4.20.
Tetrahydrofolate and derivatives.

Table 4.2. Oxidation States of One-Carbon Fragments Carried by FH_4

State of Oxidation	One-Carbon Unit	Structure	Pathway or Reaction
Equivalent to CH_3OH (methanol)	Methyl (—CH_3)	N^5-Methyl-FH_4	Methionine regeneration
Equivalent to $H_2C{=}O$ (formaldehyde)	Methylene (—CH_2—) Hydroxymethyl (—CH_2OH)	N^5,N^{10}-Methylene-FH_4	Thymidine synthesis, serine hydroxymethylase, glycine cleavage enzyme
Equivalent to HCOOH (formic acid)	Formyl (—CHO)	N^{10}-Formyl-FH_4	Formic acid activation and oxidation, formylmethionyl tRNA synthesis
	Methenyl (—CH=) Formimino (—CH=NH)	N^5,N^{10}-Methenyl = FH_4 N^5-Formimino-FH_4	Purine synthesis Histidine degradation

Note: tRNA = transfer ribonucleic acid.

In addition to the tetrahydrofolate derivatives, **choline,** endogenous or dietary, is an important source of methyl groups for transmethylation reactions. Choline is first oxidized to **betaine** (trimethylglycine), which then transfers one methyl group to homocysteine to form methionine and dimethylglycine. The remaining two methyl groups cannot be transferred directly. Instead, the N—CH_3 bond undergoes oxidative cleavage, with the eventual formation of formic acid. In an ATP-requiring

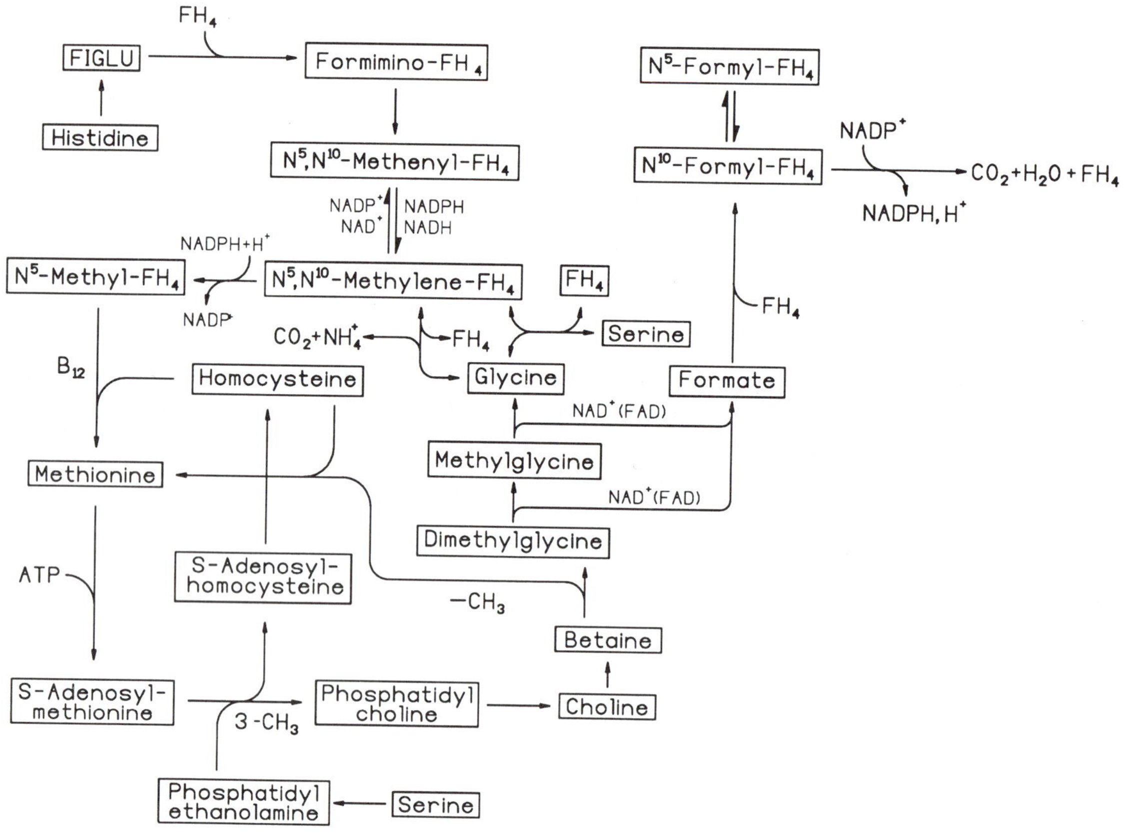

Figure 4.21.
Summary of metabolism of one-carbon fragments.

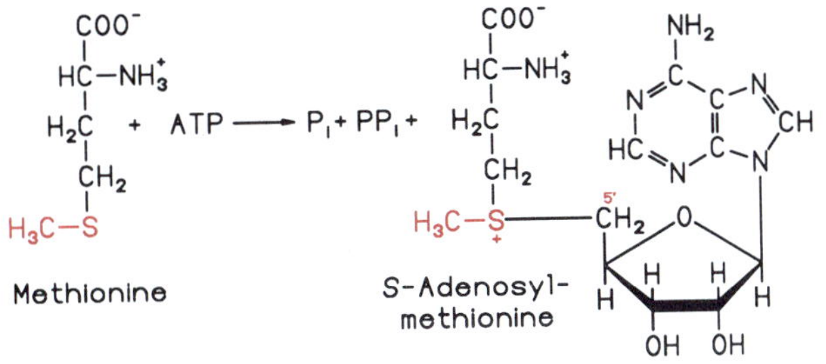

Figure 4.22.
The synthesis of *S*-adenosylmethionine.

reaction, tetrahydrofolate can combine with formate to yield N^{10}-formyl tetrahydrofolate, which can be reduced to the methyl level or oxidized to CO_2. Figure 4.21 illustrates the interconversions of the one-carbon fragments.

The reduction of N^5, N^{10}-methylene FH_4 to N^5-methyl FH_4 by NADH and reduced flavin adenine dinucleotide ($FADH_2$) is irreversible. N^5-methyl FH_4 transfers its methyl group to homocysteine to regenerate methionine in a reaction that requires vitamin B_{12}. In the absence of this vitamin **(pernicious anemia),** most FH_4 eventually is converted to N^5-methyl FH_4. As a consequence, one-carbon metabolism is slowed down, causing defective nucleotide synthesis (see Section 5.2). The resulting defect in DNA replication leads to **megaloblastic anemia.**

The major donor of methyl groups in many biochemical reactions is ***S*-adenosylmethionine.** AdoMet is formed from ATP and methionine in an unusual reaction in which the 5′-triphosphate of ATP is replaced by the methionyl group (Figure 4.22). The sulfur atom in methionine becomes directly linked to the 5′-carbon of adenosine. This makes the methyl group labile and allows its transfer to other substrates. (Compare the structure of 5′-deoxyadenosyl cobalamin and its conversion to methylcobalamin.)

Dietary restriction of choline and amino acids in experimental animals can cause accumulation of triacylglycerides in the liver. This effect can be reversed by feeding methionine. The reason for the lipid accumulation is not completely understood. It is possible that a limited supply of phosphatidyl choline, required for the construction of lipoproteins, slows down the export of triacylglycerides from the liver. Alternatively, the choline restriction may result in a deficiency of methyl groups. This would reduce the synthesis of carnitine, which is required for fatty acid transport into the mitochondria for oxidation, thus leaving the fatty acids in the cytoplasm where they accumulate in the form of triacylglycerides. Although choline can be synthesized in humans, the limited supply appears to require supplementation by dietary choline. This situation is similar to that observed with arginine, which also is in short supply and needs to be supplemented in the diet.

4.10 BIOSYNTHESIS OF NONESSENTIAL AMINO ACIDS

The pathways for the synthesis of essential amino acids in bacteria and plants will not be discussed here. As mentioned before, only the nonessential amino acids can be synthesized in humans. However, this syn-

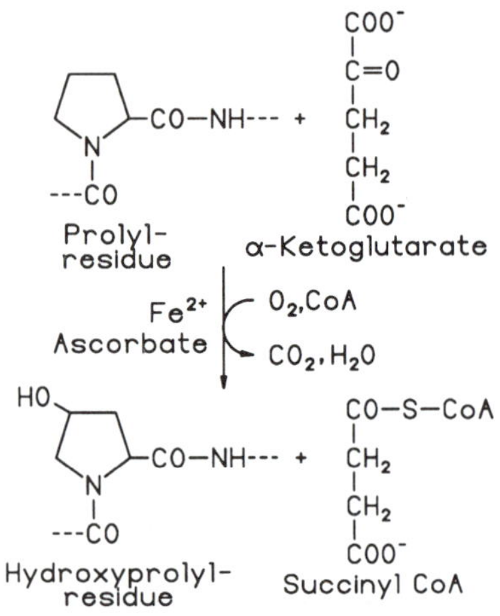

Figure 4.23.
Hydroxylation of prolyl residues in collagen.

thesis does not constitute a complete de novo synthesis, since α amino groups must be provided by other amino acids.

Alanine, aspartate, and **glutamate** are synthesized by direct transamination of pyruvate, oxaloacetate, and α-ketoglutarate, respectively. Glutamate can also be synthesized by glutamate dehydrogenase from α-ketoglutarate and ammonia (see Figure 4.5).

Asparagine and **glutamine** are synthesized from aspartate and glutamate by asparagine synthetase and glutamine synthetase, respectively (see Figure 4.10). In addition, asparagine can be formed from aspartate and glutamine by transamidation in a reaction that requires ATP.

Serine is formed from 3-phosphoglycerate, a glycolysis intermediate, by dehydrogenation with NAD^+ and transamination. These two reactions can occur either directly with 3-phosphoglycerate, leading to phosphoserine, or with free glycerate, yielding serine.

Glycine is formed from serine by **serine hydroxymethyltransferase** (see Figure 4.14) or directly from CO_2, NH_4^+, and N^5, N^{10}-methylene tetrahydrofolate by **glycine synthase.** In addition, glycine can be made from choline by oxidation and transfer of its methyl groups (see Figure 4.21).

Cysteine receives its sulfur atom from methionine, an essential amino acid, and its carbon chain from serine (see Figure 4.12) via cystathionine. **Tyrosine,** although not considered an essential amino acid, can only be made from the essential amino acid phenylalanine by phenylalanine hydroxylase. **Proline** is synthesized by reversal of the reactions that lead to its degradation (see Figure 4.11). However, degradation and biosynthesis of proline are catalyzed by different enzymes. Degradation occurs in the mitochondrion and is essentially irreversible, whereas biosynthesis takes place in the cytoplasm. The generation of glutamic semialdehyde from glutamate requires ATP. By transamination, glutamic semialdehyde can also be converted to **ornithine** and thus enter into the route to **arginine.**

Hydroxyproline is synthesized from proline after it has been incorporated into the nascent procollagen chain. The reaction requires molecular oxygen, ascorbate, Fe^{3+} ions, and α-ketoglutarate. Hydroxylation of proline is coupled to the decarboxylation and oxidation of α-ketoglutarate to succinate. The oxygen (O_2) is incorporated equally into hydroxyproline and H_2O. The oxidative decarboxylation of α-ketoglutarate provides the reducing equivalents required for the formation of water (Figure 4.23).

Hydroxylysine, another constituent of collagen, is formed in a similar reaction. **Arginine** is being constantly formed via the urea cycle, but virtually all of it is degraded in the formation of urea. Arginine may, therefore, become a dietary requirement when demands are high (for instance, during growth).

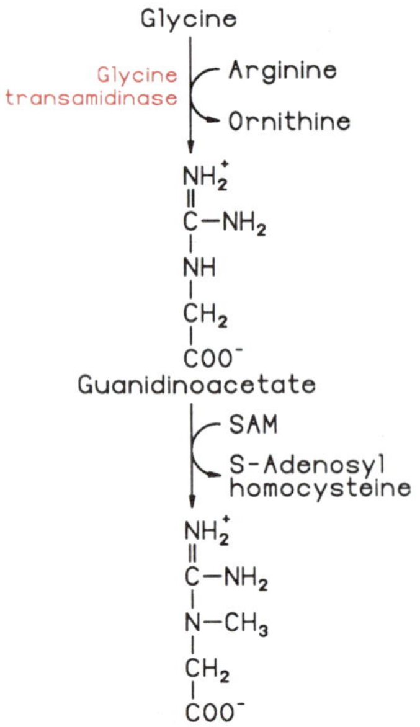

Figure 4.24.
Biosynthesis of creatine.

4.11 NITROGENOUS COMPOUNDS MADE FROM AMINO ACIDS

4.11.1 Creatine

Creatine phosphate serves as a storage compound of chemical energy in muscle and brain. Creatine is synthesized in the liver from glycine

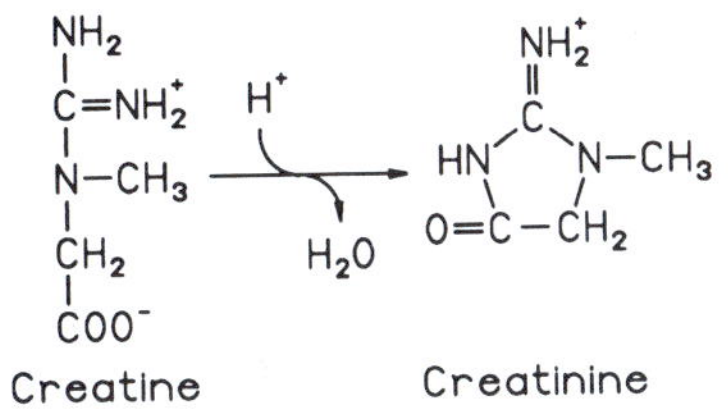

Figure 4.25.
Formation of creatinine from creatine.

and arginine with the participation of methionine, which donates a methyl group via *S*-adenosyl methionine (Figure 4.24).

Creatine is phosphorylated by ATP in a reversible reaction that serves to reduce the intracellular concentration of ATP but regenerates it from ADP when levels fall too low.

$$\text{Creatine} + \text{ATP} \xrightleftharpoons{\textit{Creatine kinase}} \text{Creatine phosphate} + \text{ADP}$$

Creatine phosphate, as well as creatine itself, undergoes a slow nonenzymatic cyclization process to yield **creatinine,** which is excreted in the urine (Figure 4.25). Since the production of creatinine over a 24 h period is relatively constant, its urinary concentration provides a convenient measure of glomerular filtration rate.

Creatine kinase (CK) [old term: creatine phosphokinase (CPK)] is found in the form of different isoenzymes in several tissues. The heart muscle isoenzyme of CK appears in the blood during a **myocardial infarction** and serves as a useful diagnostic indicator.

4.11.2 Carnitine

Carnitine, the fatty acyl carrier across the mitochondrial membrane, is synthesized from lysine. First, the ε-amino group of lysine is methylated by AdoMet. Then, after deamination and loss of carbon atoms 1 and 2, the product γ-butyrobetaine is hydroxylated in a reaction similar to the hydroxylation of proline (Figure 4.26).

Figure 4.26.
Biosynthesis of carnitine.

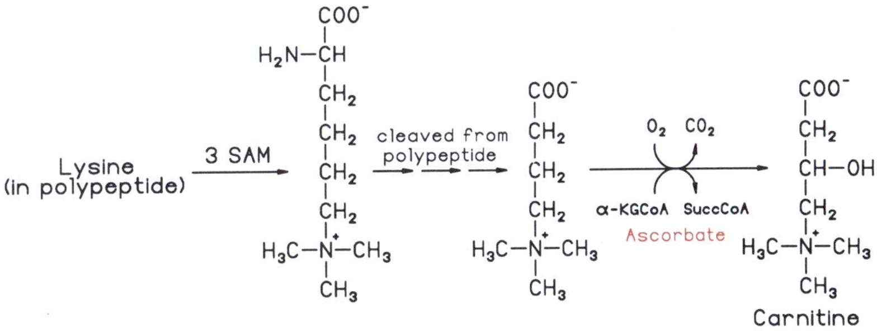

4.11.3 Neurotransmitters

The neurotransmitter **acetylcholine** is synthesized from serine via ethanolamine, choline, and acetylation with acetyl CoA (Figure 4.27). The catecholamines **epinephrine** and **norepinephrine** are derived from tyrosine. The intermediates in this reaction, **L-dopa** and **dopamine,** also function as neurotransmitters (Figure 4.28).

After release from the nerve terminal, norepinephrine is inactivated by the actions of **catecholamine O-methyl transferase (COMT)** and **monoamine oxidase (MAO).** The sequence of actions by these two enzymes depends on where the inactivation takes place. In the receptor cell, norepinephrine is first oxidized by MAO, and the resulting dihydroxymandelaldehyde is transported to the liver where COMT converts it to vanillyl mandelate, which is excreted in the urine. Some of the norepinephrine escapes into the blood. It is inactivated in the liver by methylation and oxidation, in that order (Figure 4.29).

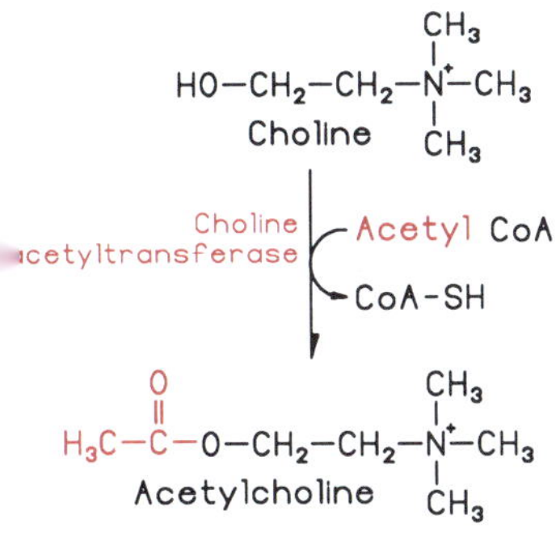

Figure 4.27.
Synthesis of acetylcholine.

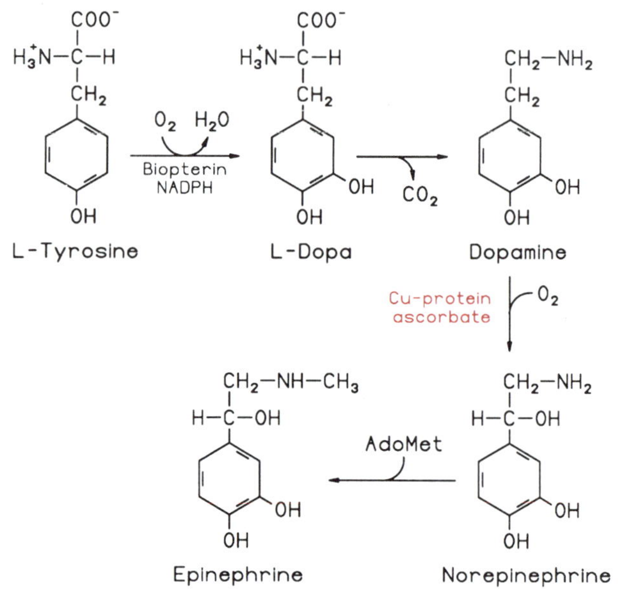

Figure 4.28.
Biosynthesis of catecholamines.

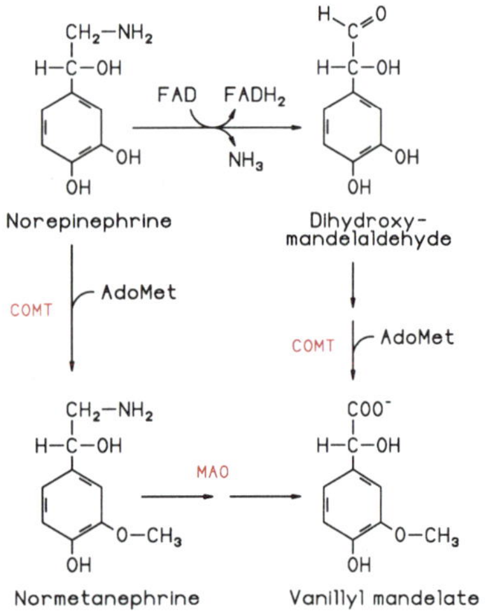

Figure 4.29.
Catabolism of catecholamines.

γ-Aminobutyric acid (GABA) is formed by decarboxylation of glutamate. It serves as an **inhibitory neurotransmitter** in the central nervous system. Most of the released GABA is taken up again by the neuron, where it is catabolized to succinate. Synthesis and catabolism of GABA thus represent a bypass of the α-ketoglutarate dehydrogenase reaction in the citric acid cycle. In contrast to the citric acid cycle reaction, which produces four high-energy phosphate bonds, the **GABA shunt** yields only three ATPs (Figure 4.30).

Serotonin (5-hydroxytryptamine) also functions as a neurotransmitter. It is synthesized from tryptophan by hydroxylation followed by decarboxylation (Figure 4.31).

Histamine, a potent vasodilator, is made in mast cells from histidine by **histidine decarboxylase** and is released during anaphylaxis, allergic reaction, or in response to certain drugs. It may also function as a neurotransmitter.

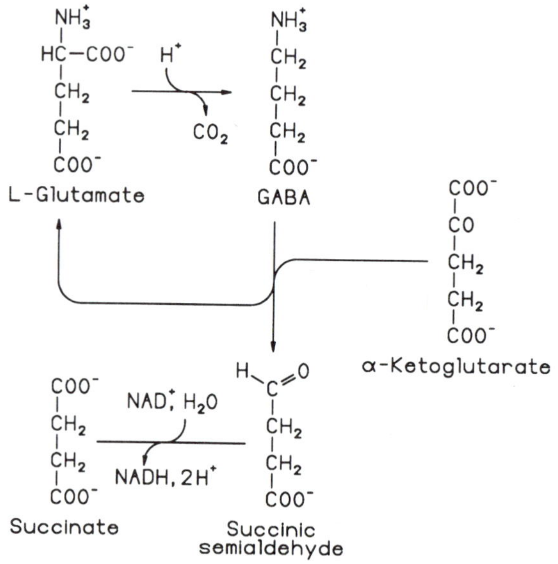

Figure 4.30.
The GABA shunt.

4.11.4 Polyamines

The polyamines **putrescine, spermidine,** and **spermine** are found in association with nucleic acids. Their function is still not precisely known. They are derived from ornithine by decarboxylation and transfer of one or two aminopropyl groups from decarboxylated AdoMet (Figure 4.32).

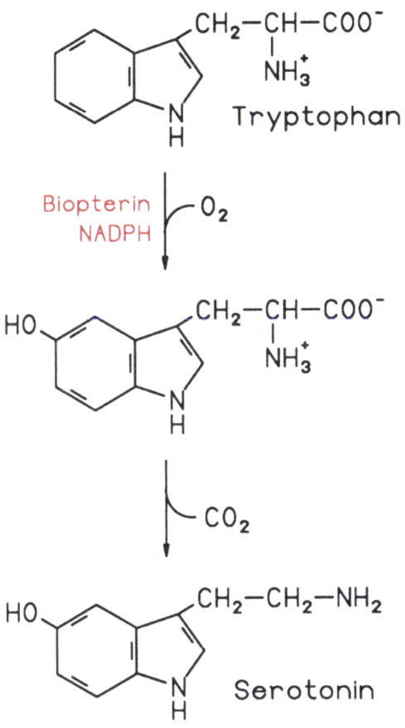

Figure 4.31.
Biosynthesis of serotonin.

4.11.5 Pigments

Melanins are polymeric oxidation products of tyrosine; they constitute the black and brown pigments of the skin, hair, retina, and other structures. Blond and red pigmentation is produced by tyrosine oxidation with the participation of cysteine; the reaction yields a series of **trichochromes.** The genetic absence of **tyrosinase,** the melanosomal enzyme that oxidizes tyrosine to melanin via L-dopa, causes **albinism.** (Note that this is not the same enzyme that participates in epinephrine biosynthesis.)

4.11.6 Heme

Heme is the iron-containing pigment of hemoglobins and cytochromes. Its synthesis begins with the condensation of glycine and succinyl CoA to **δ-aminolevulinate.** Two molecules of δ-levulinate are then condensed to **porphobilinogen.** The enzymes responsible for the two reactions, **δ-aminolevulinate synthetase** and **porphobilinogen synthetase,** are inhibited by the end products heme, hemoglobin, and other heme proteins. Four molecules of porphobilinogen are condensed to **uroporphyrinogen III** (Figure 4.33), which is further converted via **coproporphyrinogen** and **protoporphyrin** to **heme.** δ-Aminolevulinate synthetase is an inducible enzyme with a half-life of only about 1 h. Its synthesis is feedback-inhibited by heme.

If heme biosynthesis is reduced, either by genetic deficiencies in some of the enzymes that convert δ-aminolevulinate to heme or by certain drugs that inhibit these enzymes, the synthesis of δ-aminolevulinate synthetase is no longer controlled. As a result, enzyme levels increase dramatically, leading to the production of excessive amounts of porphyrin precursors, such as δ-aminolevulinate and porphobilinogen. These substances are excreted in the urine, giving it a wine-red color because of photooxidation products.

Disorders of porphyrin metabolism are generally called **porphyrias.** There are two major types of porphyria:

1. **Erythropoietic porphyria** is rare and is inherited as an autosomal recessive trait. Deposition of excess porphyrins in

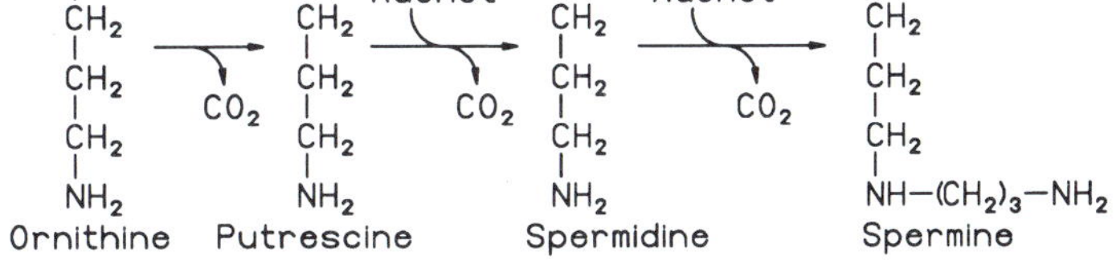

Figure 4.32.
Biosynthesis of polyamines.

Figure 4.33.
Biosynthesis of porphyrins.

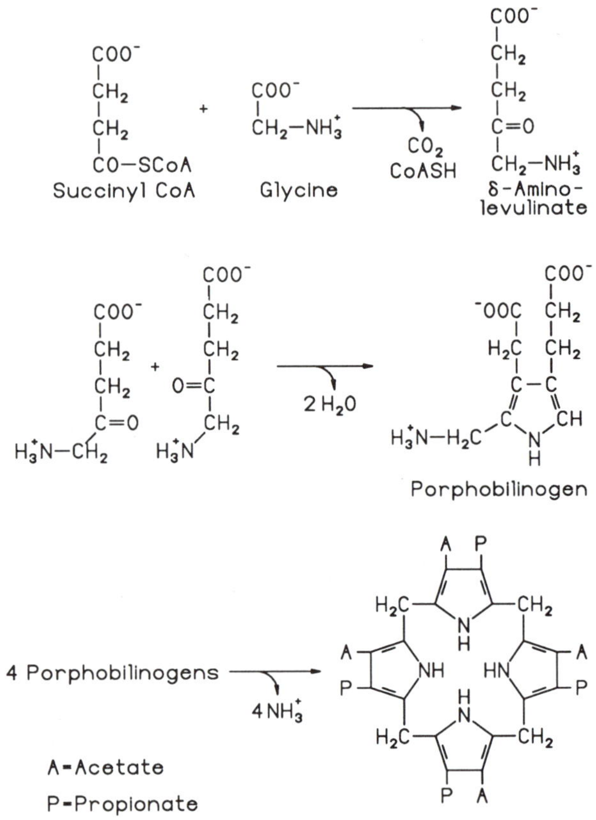

the tissues sensitizes the skin to light and causes dermatitis, scarring, and infections. The disorder usually causes early death.

2. **Hepatic porphyria,** inherited as an autosomal dominant trait, usually remains asymptomatic until an acute attack of porphyria is triggered by the intake of some drug or alcohol. It is due to deficiencies in some enzyme in heme biosynthesis, a situation that is exacerbated further by the action of the inhibiting drug. The overproduction and deposition of porphyrin precursors in the liver produces abdominal pain and colic and may cause a large variety of other symptoms.

For a more extensive discussion of the metabolism of amino acids and related compounds, consult the following textbooks:

Mathews et al. (1990): Chapters 20, 21

Stryer (1988): Chapters 21, 24

Further References

Bessman SP, Carpenter CL. The creatine-creatine phosphate energy shuttle. *Annu Rev Biochem.* 1985; 54:831–862.

Bloom FE. Neurotransmitters: past, present, and future directions. *FASEB J.* 1988; 1:32–41.

Umbarger HE. Amino acid biosynthesis and its regulation. *Annu Rev Biochem.* 1978; 47:533–557.

Wellner D, Meister A. A survey of inborn errors of metabolism and transport in man. *Annu Rev Biochem.* 1981; 50:911–948.

CHAPTER

5 Nucleotide Metabolism

OBJECTIVES

After studying this chapter, the student should be able to

1. Draw the chemical structures of the nucleotides found in DNA and RNA.
2. Point out the principal difference between the biosynthetic pathways of purine and pyrimidine nucleotides.
3. State the origins of the atoms in the purine and pyrimidine rings.
4. Explain the central role of PRPP in the biosynthesis of purine and pyrimidine nucleotides.
5. Describe the control of AMP and GMP synthesis from IMP.
6. Describe the reaction catalyzed by aspartate transcarbamoylase.
7. Describe the reaction steps involved in the conversion of CDP into TMP.
8. Explain the mode of action of the antitumor drug amethopterin.
9. State several possible reasons for the overproduction of uric acid.
10. Explain the mode of action of allopurinol in the treatment of gout.

Nucleotides are the building blocks of nucleic acids. Most cells are capable of synthesizing the nucleotides required for the biosynthesis of deoxyribonucleic acid (DNA) and ribonucleic acid (RNA). Dietary nucleotides are not required. In addition to this capability of **de novo biosynthesis** of nucleotides, many cells possess **salvage pathways** that allow the conversion of degradation products of nucleic acids, such as nucleosides or free bases, to nucleotides. Genetic defects in some of the salvage pathway enzymes have profound pathologic consequences.

5.1 STRUCTURE AND NOMENCLATURE

The monomeric units of nucleic acids, **RNA** and **DNA,** are called **nucleotides.** They consist of a purine or pyrimidine base, the five-carbon sugar **ribose** or **deoxyribose,** and a **phosphate** group attached to the 5′ end of the sugar. Figure 5.1 gives the structures of the purine and pyrimidine bases commonly found in nucleic acids.

When a base is linked through a glycosidic bond to the 1 position of ribose or deoxyribose, the molecule is called a ribonucleoside or deoxyribonucleoside, respectively. If the 5′-hydroxyl group (the prime indicates a carbon atom in the sugar as opposed to carbon atoms in the purine or pyrimidine ring) is esterified to phosphoric acid, the molecule is called a nucleoside 5′-phosphate or simply 5′-nucleotide. Two examples of nucleotides are illustrated in Figure 5.2.

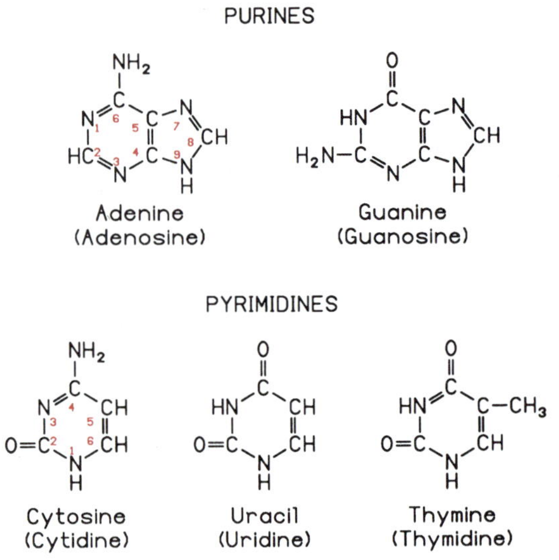

Figure 5.1.
Structures of the common purine and pyrimidine bases found in nucleic acids. The names of the corresponding ribonucleosides are given in parentheses. Deoxyribonucleotides carry the prefix "deoxy" with the exception of thymidine, which is usually used without the prefix.

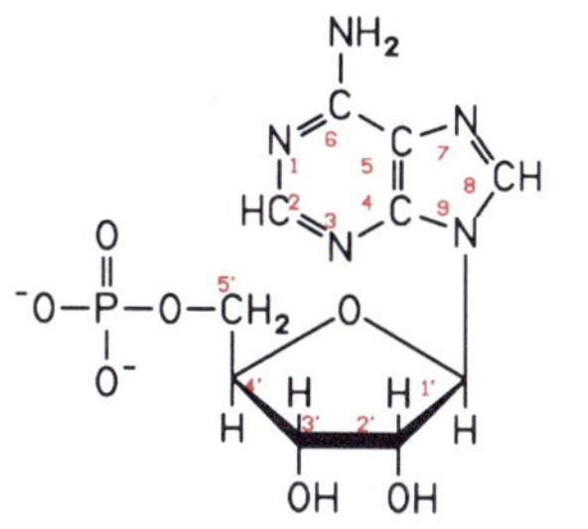

Adenosine 5'-phosphate (AMP)

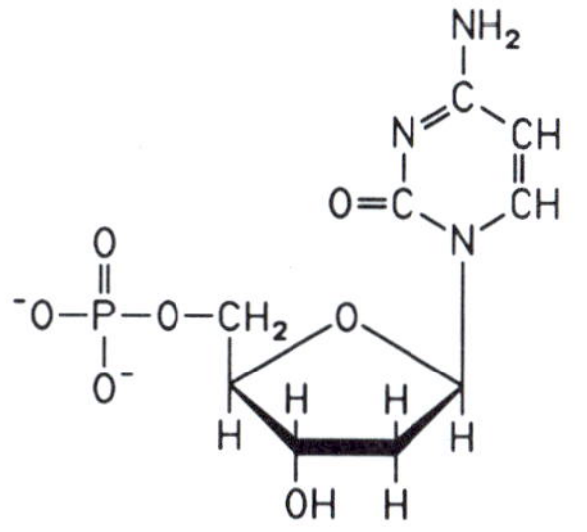

Deoxycytidine 5'-phosphate (dCMP)

Figure 5.2.
Structures of two nucleotides.

5.2 DE NOVO BIOSYNTHESIS OF NUCLEOTIDES

The principal difference in the biosynthesis of purine and pyrimidine nucleotides is the fact that the former are synthesized directly as nucleotides, starting with ribose 5-phosphate, whereas the latter are derived from orotic acid, a free base to which the ribose moiety is added on later.

5.2.1 Purine Nucleotides

The purine skeleton is synthesized from **glycine, aspartate, glutamine, CO_2** and two **one-carbon fragments** provided by tetrahydrofolate derivatives (Figure 5.3.)

The first step in purine biosynthesis is the phosphorylation of ribose 5-phosphate to **5-phosphoribosyl-1-pyrophosphate (PRPP;** Figure 5.4). The next step, the formation of **5-phosphoribosyl-1-amine,** is the committed step in purine biosynthesis (Figure 5.5). The enzyme **amidotransferase** is feedback-inhibited by the end products of the pathway, adenosine triphosphate (ATP) and guanosine triphosphate (GTP). [Adenosine monophosphate (AMP) and guanosine monophosphate (GMP) also inhibit the reaction.]

In a series of nine further reactions (Figure 5.6), the purine nucleotide

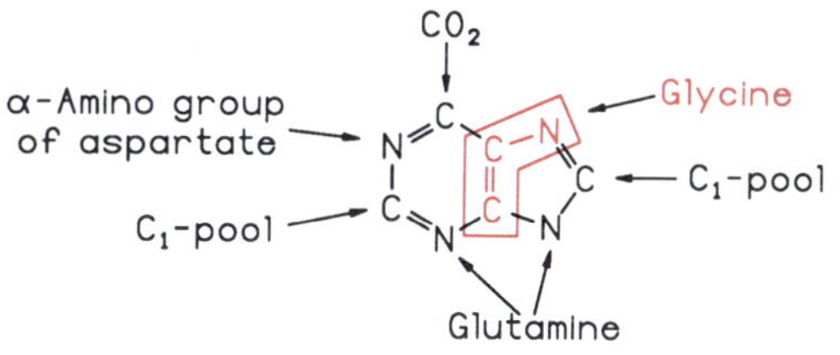

Figure 5.3.
Origin of the atoms in the purine skeleton.

inosine monophosphate (IMP) is formed. (Note that the purine base in IMP is called **hypoxanthine.**)

Inosine monophosphate is converted to guanosine monophosphate via xanthosine monophosphate and to adenosine monophosphate via adenylosuccinate. The initial step in each reaction is inhibited by its end product (GMP or AMP, respectively) and activated by the triphosphate of the other reaction's product. This cross-regulation ensures the balanced synthesis of adenine and guanine nucleotides (Figure 5.7).

5.2.2 Pyrimidine Nucleotides

The pyrimidine skeleton is synthesized from **carbamoyl phosphate** and **aspartate** (Figure 5.8).

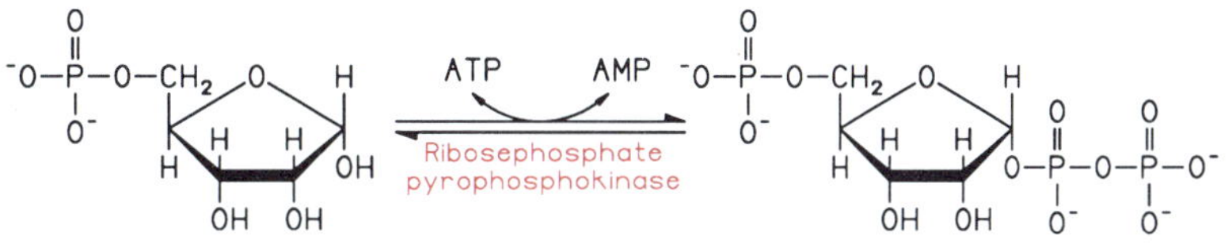

Ribose 5-phosphate

5-Phosphoribosyl-1-pyrophosphate (PRPP)

Figure 5.4.
Synthesis of 5-phosphoribosyl-1-pyrophosphate (PRPP).

Figure 5.5.
Synthesis of 5-phosphoribosyl-1-amine.

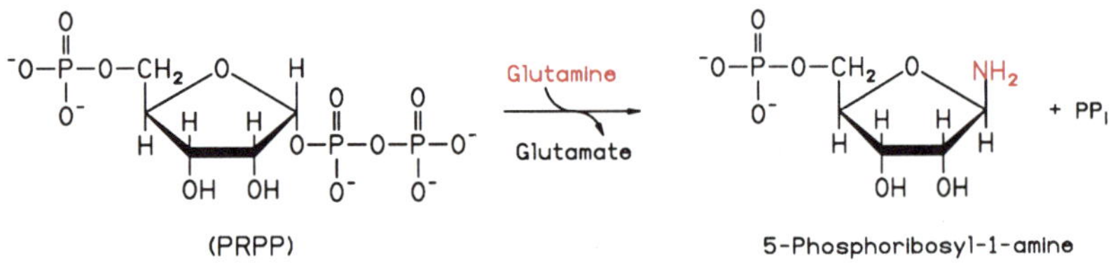

The synthesis of carbamoyl phosphate from glutamine, CO_2, and two ATPs occurs in the cytoplasm of the cell and is inhibited by uridine triphosphate (UTP). (Compare this reaction with the synthesis of urea in the mitochondrion, which employs ammonium ion instead of glutamine.)

The committed step in pyrimidine biosynthesis is the formation of ***N*-carbamoyl aspartate.** The enzyme **aspartate transcarbamoylase** is allosterically inhibited by cytosine triphosphate (CTP) and activated by ATP (see Figure 1.46). The binding of ATP and CTP to the regulatory

Figure 5.6.
The purine nucleotide biosynthetic pathway.

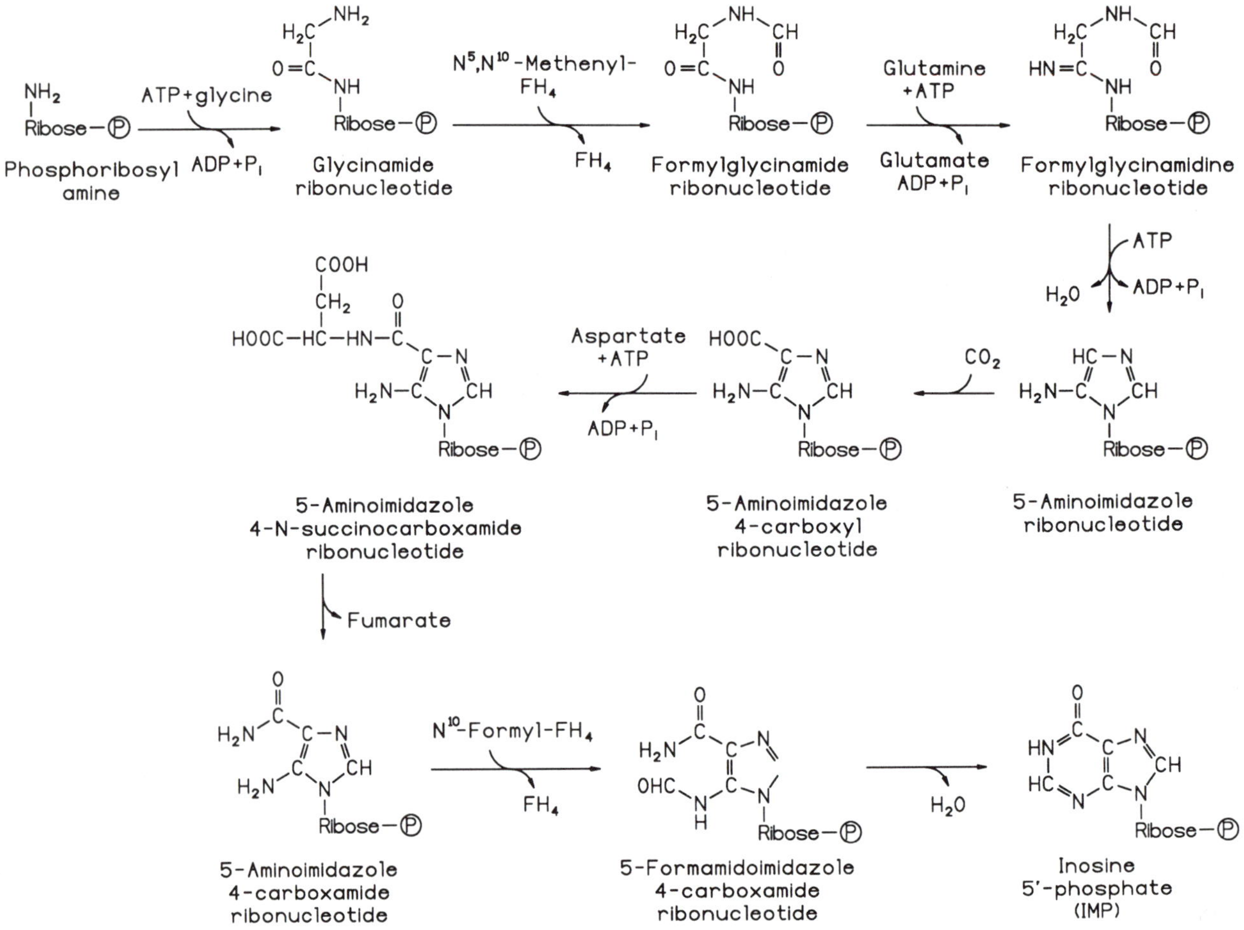

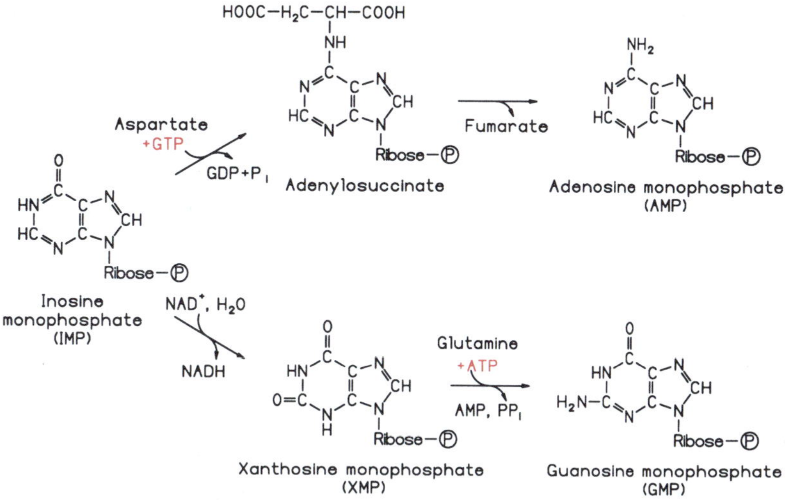

Figure 5.7.
The conversion of IMP into AMP and GMP.

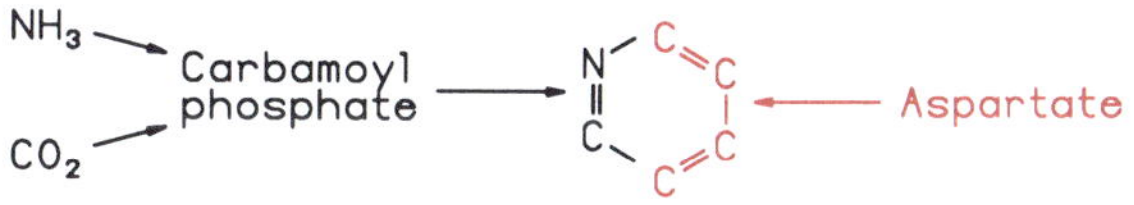

Figure 5.8.
Origin of the atoms in the pyrimidine skeleton.

site of the enzyme is competitive. This mechanism seems to balance the synthesis of purines and pyrimidines.

Carbamoyl aspartate cyclizes to **dihydroorotate,** which is then dehydrogenated to orotate by oxidized nicotinamide adenine dinucleotide (NAD^+). Orotate is converted to **uridine monophosphate (UMP)** by condensation with phosphoribosyl pyrophosphate, followed by decarboxylation (Figure 5.9).

Phosphorylation of nucleoside monophosphates to the triphosphates, the substrates for nucleic acid biosynthesis, occurs in two steps:

$$\text{NMP} + \text{ATP} \rightleftharpoons \text{NDP} + \text{ADP}$$

$$\text{NDP} + \text{XTP} \rightleftharpoons \text{NTP} + \text{XDP}$$

where NMP = nucleoside 5′-monophosphate, NDP = nucleoside 5′-

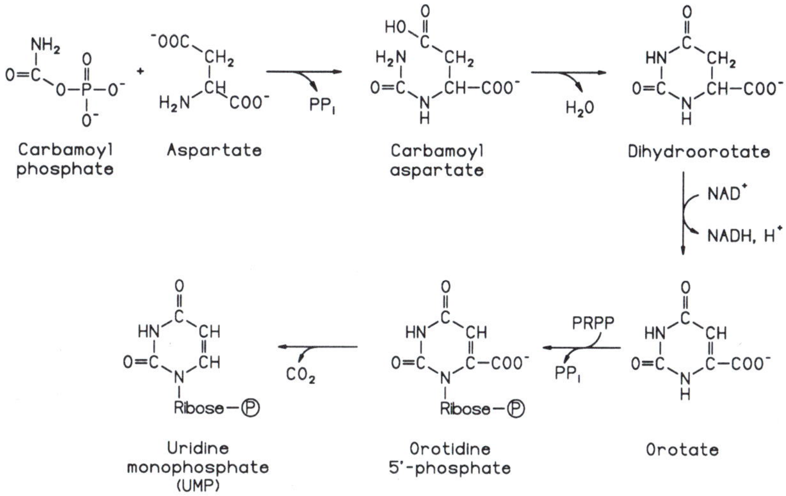

Figure 5.9.
The pyrimidine nucleotide biosynthetic pathway.

phate, NTP = nucleoside 5′-triphosphate, ADP = adenosine diphosphate, and X denotes any of the four bases.

The first reaction is catalyzed by **nucleoside monophosphate kinases,** enzymes specific for each particular nucleotide. ATP is the phosphate donor. In contrast, the phosphorylation of ribonucleoside and deoxyribonucleoside diphosphates is catalyzed by a single enzyme, **nucleoside diphosphate kinase.** Any nucleoside triphosphate can function as the phosphate donor in this reaction.

Cytosine nucleotides are synthesized by the amination of UTP. Glutamine is the amino donor, and one ATP is hydrolyzed in the reaction (Figure 5.10).

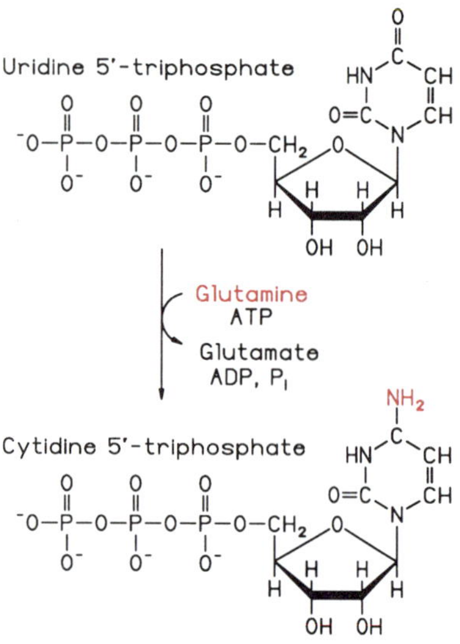

Figure 5.10.
The synthesis of CTP from UTP.

5.2.3 Deoxyribonucleotides

The reduction of ribonucleotides to the deoxyribonucleotides is catalyzed, at the diphosphate level, by the enzyme **nucleotide reductase.** This enzyme is unusual in that it carries a free radical at its active site. By removal of its hydroxyl group, the 2′-carbon of the ribose is transiently converted into a free radical bound enzyme. Then a hydrogen atom is added.

In *Escherichia coli,* the immediate hydrogen donor is one of two proteins: **thioredoxin** or **glutaredoxin.** In the reaction, thioredoxin is converted to its disulfide form (Figure 5.11), which is subsequently reduced again with reduced nicotinamide adenine dinucleotide phosphate (NADPH). Glutaredoxin carries glutathione as a prosthetic group. It functions in a fashion similar to thioredoxin. In mammalian cells, analogous cofactors participate in the reaction.

Ribonucleotide reductase activity is under complex control with many ribonucleotides and deoxyribonucleotides acting as effectors. For example, thymidine triphosphate (dTTP) inhibits the reduction of cytosine diphosphate (CDP) to dCDP. High concentrations of thymidine (which is converted intracellularly to dTTP) will therefore block DNA synthesis through the inhibition of dCTP synthesis. This property of thymidine is often used to synchronize cells in the cell cycle (see Section 6.2).

Figure 5.11.
The reduction of ribonucleotides to deoxyribonucleotides.

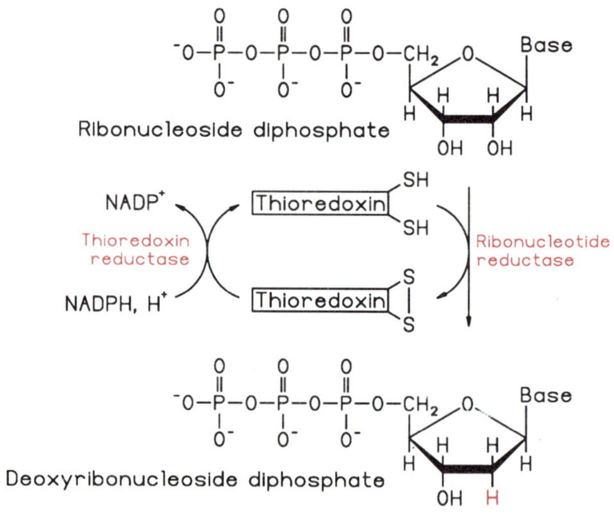

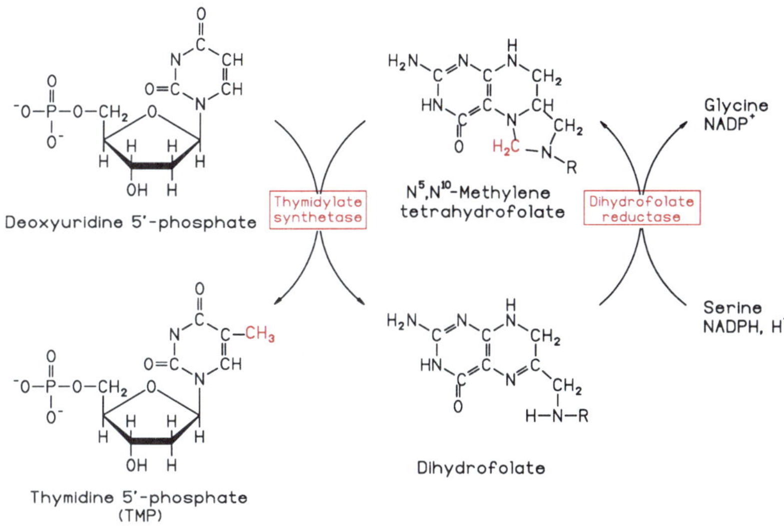

Figure 5.12.
The synthesis of thymidine monophosphate.

Thymidine nucleotides are formed from deoxyuridine monophosphate (dUMP) by the enzyme **thymidylate synthetase.** The dUMP used for this reaction is generated by the deamination of dCMP or, probably more extensively, by the hydrolysis of dUTP. Cells contain a powerful dUTPase that keeps intracellular levels of dUTP at low levels, thus reducing the chance of it being used for DNA synthesis.

The conversion of dUMP to deoxythymidine monophosphate (dTMP) is catalyzed by thymidylate synthetase (Figure 5.12). N^5,N^{10}-methylene tetrahydrofolate serves as both methyl donor and reducing agent (the methylene group is one oxidation level above that of the methyl group introduced into dUMP; see Table 4.2). The dihydrofolate (FH_2) that is generated in this reaction must be reduced again to tetrahydrofolate (FH_4) by the enzyme **dihydrofolate reductase** and NADPH.

5.2.4 Anticancer Drugs that Interfere with Nucleotide Synthesis

Cancer cells divide more rapidly than most normal cells. Thus, by interfering with nucleotide synthesis, one can often preferentially starve cancer cells of vital precursors for their DNA synthesis. Rapidly dividing normal cells, such as those in bone marrow and intestinal epithelium, are also affected by these drugs, thus limiting their use.

One of the most successful anticancer drugs is **amethopterin,** a folic acid analog that inhibits the enzyme dihydrofolate reductase. Another drug, trimethoprim, inhibits preferentially the bacterial enzyme and thus is used primarily as an antibiotic. As discussed before, without dihydrofolate reductase, the cellular FH_4 pool is quickly depleted when thymidylate synthesis occurs such as in rapidly proliferating cancer cells. This leads to cell death. Nondividing cells synthesize little, if any, thymidine nucleotides and therefore are little affected by this drug.

5-Fluorouracil (Figure 5.13) is another often used anticancer drug. In the cell, it is converted into its deoxyribonucleotide monophosphate, 5-fluorodeoxyuridine monophosphate, which directly inhibits thymidylate synthetase.

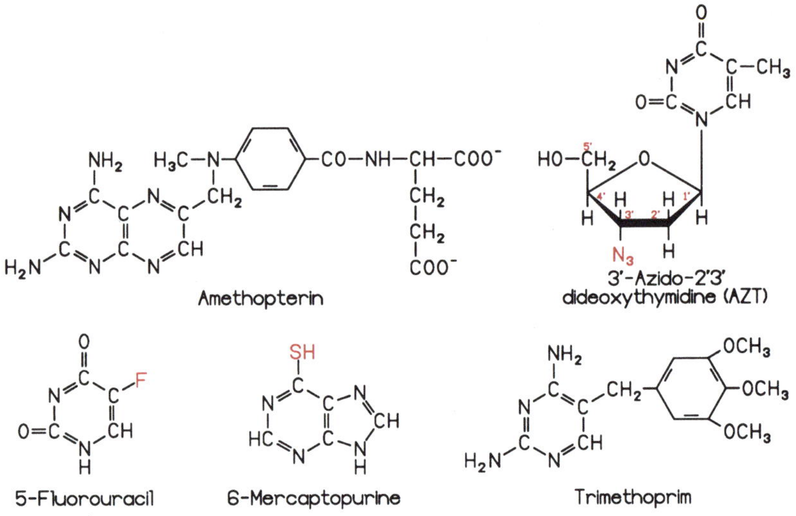

Figure 5.13.
Structures of some anticancer drugs. Zidovudine (also known as AZT) is used in the treatment of acquired immunodeficiency syndrome (AIDS).

6-Mercaptopurine, when converted to its ribonucleoside 5′-phosphate, inhibits several enzymes in purine biosynthesis. It is used in the treatment of leukemia.

5.3 SALVAGE PATHWAYS

The turnover of messenger RNA (mRNA), the repair of DNA, or the degradation of DNA during erythrocyte maturation all produce nucleotides that may be further degraded to nucleosides or free bases. Being uncharged, these molecules can pass through the cell membrane into the bloodstream, where they are carried to other tissues. Reutilization (salvage) of these breakdown products for nucleic acid synthesis is energetically much less expensive than complete de novo synthesis. Most cells possess several salvage pathways for the conversion of nucleosides and free bases back to the nucleotide level. There is also some evidence that purine biosynthesis in the liver provides precursors for DNA synthesis in other cells, such as bone marrow cells. The following enzymes and reactions are involved in salvage synthesis of nucleotides:

1. **Nucleoside phosphorylases**

$$\text{(Deoxy)ribonucleoside} + P_i \rightleftharpoons \text{base} + \text{(deoxy)ribose 1-phosphate}$$

2. **Phosphoribosyl transferases**

$$\text{Base} + \text{PRPP} \rightleftharpoons \text{ribonucleoside monophosphate} + \text{PPi}$$

3. **Nucleoside kinases**

$$\text{(Deoxy)ribonucleoside} + \text{ATP} \rightleftharpoons \text{(deoxy)ribonucleoside monophosphate} + \text{ATP}$$

Decreased levels of nucleoside phosphorylases and phosphoribosyl transferases are found in some patients with gout and in all patients with Lesch-Nyhan disease. The specific salvage pathways for purine nucleotides are included in Figure 5.15.

5.4 CATABOLISM

Excess purines and pyrimidines that cannot be utilized through salvage pathways are catabolized. Pyrimidines are completely degraded to ammonia and carbon dioxide. Purines, on the other hand, are oxidized to **uric acid,** which in humans cannot be degraded further and must be excreted as such.

5.4.1 Pyrimidine Catabolism

Uracil and cytosine are converted to β-alanine, thymine is converted to β-aminoisobutyrate. Both amino acids can be oxidized further to CO_2 and NH_3 (Figure 5.14), but β-aminoisobutyrate is metabolized only slowly and is often found excreted directly. (β-Aminoisobutyrate can be converted to methylmalonate, which is then rearranged, as a CoA derivative, to succinyl CoA in a vitamin B_{12}–dependent reaction.)

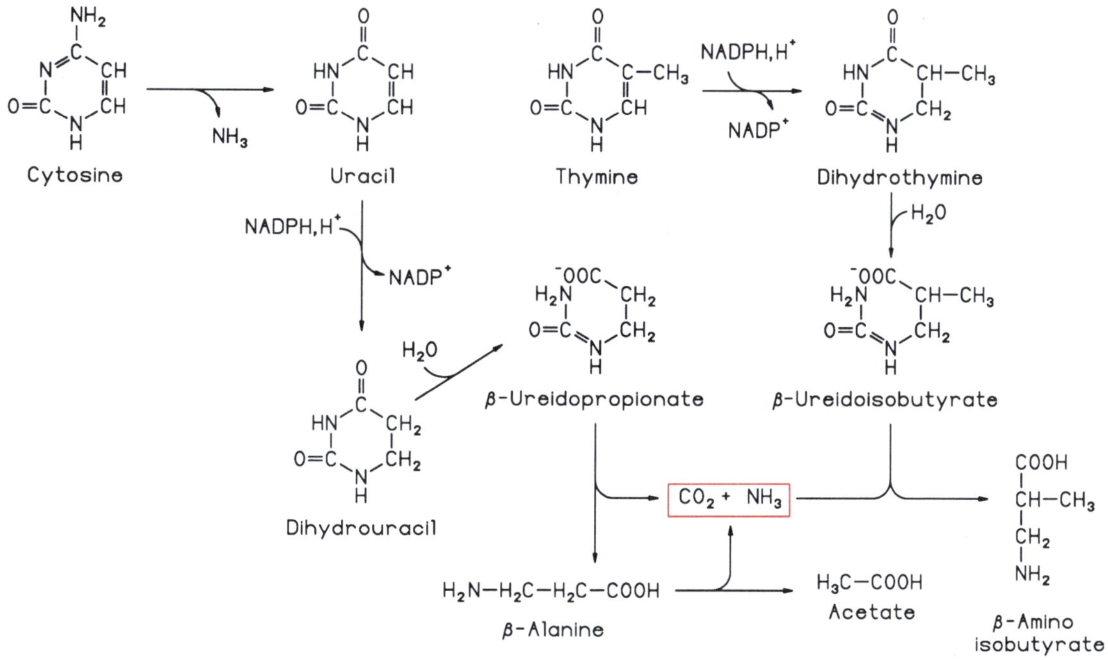

Figure 5.14.
Pyrimidine catabolism.

5.4.2 Purine Catabolism

Uric acid, the final product of purine catabolism in humans, is formed by successive oxidations of hypoxanthine and xanthine, both catalyzed

Figure 5.15.
Overview of purine metabolism.

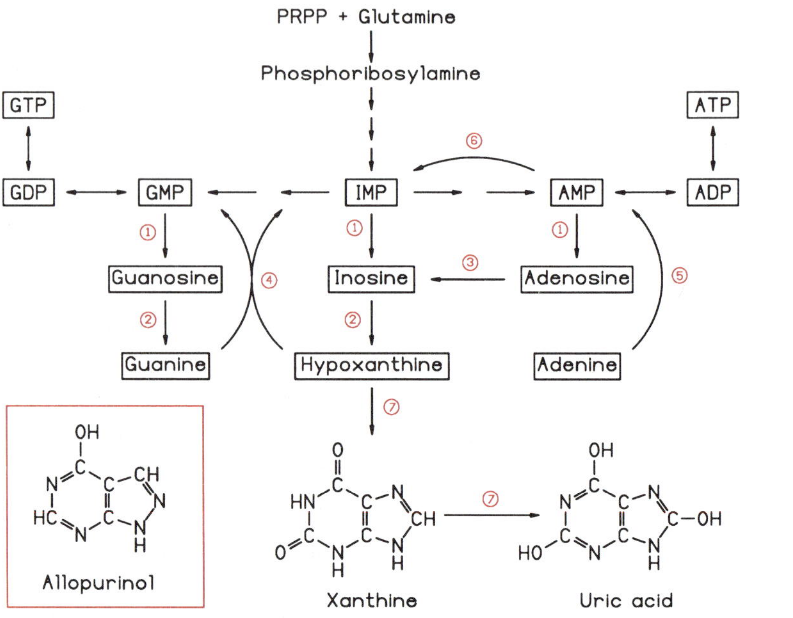

1. 5'-Nucleotidase
2. Nucleoside phosphorylase
3. Adenosine deaminase
4. Hypoxanthine-guanine-phosphoribosyl transferase (HGPRTase)
5. Adenine phosphoribosyl transferase
6. AMP deaminase
7. Xanthine oxidase

by the molybdenum-containing enzyme **xanthine oxidase.** The complete interplay of de novo synthesis, salvage synthesis, and catabolism of purine nucleotides is illustrated in Figure 5.15.

5.5 DISORDERS OF PURINE NUCLEOTIDE METABOLISM

Uric acid is rather poorly soluble in water. Normal adult humans synthesize between 600 and 700 mg of uric acid per day, and dietary purines contribute another 300–600 mg (total miscible pool: 1200 mg). This amount of uric acid must be excreted daily through the kidney (66%) and the gastrointestinal tract (33%). Serum levels below 7 mg/100 ml of uric acid are considered normal. Hyperuricemimia (>7mg/100 ml) may lead to **gout.**

Acute gouty arthritis is caused by a burst of microcrystallization of sodium urate in a joint, which is followed by an inflammatory reaction. Slowly growing urate crystals also form deposits **(tophi),** which may lead to gross deformations. **Uric acid nephrolithiasis** (kidney stones) causes renal damage and is generally considered the greatest threat of hyperuricemia.

Hyperuricemia may be caused by either a reduced rate of renal excretion or an overproduction of uric acid. **Renal excretion** of uric acid proceeds as follows. Uric acid is filtered freely in the glomerulus but is subsequently reabsorbed to at least 98% in the proximal tubules. Active secretion of uric acid in the distal tubules accounts for most of the urate that finally appears in the urine. Organic acids, such as lactate or ketone bodies, interfere with urate excretion. This is why diabetes mellitus, glycogen storage disease I, alcohol intoxication, and starvation are often accompanied by hyperuricemia. Uricosuric drugs, such as **probenecid** or **phenylbutazone,** initially cause urate retention when given in low doses because the drugs inhibit tubular secretion. When given at higher doses, however, reabsorption is also inhibited, thereby allowing all filtered urate to remain in the tubules. In **chronic renal disease,** the capacity of the kidney to secrete uric acid is lost more quickly than the reabsorption capability, thus causing hyperuricemia.

Overproduction of uric acid may be due to an increased turnover of nucleic acids. This condition may be observed in leukemia, polycythemia vera, sickle cell anemia, psoriasis, or in disorders in the regulation of purine metabolism. The rate of the de novo biosynthesis of purines is determined by the availability of PRPP, whose synthesis is regulated by purine ribonucleotides through end product inhibition. In glycogen storage disease type I, the increased levels of glucose 6-phosphate may cause an overproduction of PRPP (via the pentose phosphate pathway), which in turn stimulates purine biosynthesis.

In the **Lesch-Nyhan syndrome,** the enzyme **hypoxanthine-guanine-phosphoribosyl transferase (HPRTase)** is defective. Patients with this disease have elevated levels of uric acid and show an increased rate of de novo synthesis of purines. The genetic disorder is X-linked recessive and occurs only in males, since patients with this disease usually do not reach sexual maturity and so cannot pass on their defective X chromosome. In addition to hyperuricemia, the patients show neurologic disorders such as severe self-mutilation and mental retardation.

In another rare familial abnormality, the enzyme **adenine-phosphoribosyl transferase (APRTase)** is defective. Patients show mental retardation but have normal levels of uric acid.

Several catabolic enzymes, if deficient, interfere with the normal regulation of purine metabolism and consequently affect DNA synthesis. For example, the absence of **adenosine deaminase** or **nucleoside phosphorylase** causes immunodeficiency. A defective **AMP deaminase** may cause muscular cramps during strenuous exercise (see Figure 4.6).

The enzymatic defect in primary gout is often not known. In some cases, however, the level of the PRPP amidotransferase was found to be elevated or the enzyme was resistant to feedback inhibition.

Allopurinol (see Figure 5.15), which is widely used in the treatment of gout, inhibits xanthine oxidase. Xanthine and hypoxanthine are much more soluble than uric acid and are, therefore, excreted more rapidly. In addition, allopurinol is converted into allopurinol monophosphate and therefore competes for PRPP. This reduces the size of the PRPP pool and thus lowers the rate of de novo synthesis of purines. Both IMP and allopurinol ribonucleotide inhibit PRPP-amidotransferase.

Acute gout is treated most effectively with **colchicin,** a drug that inhibits microtubule assembly. Since microtubules function in the ameboid movement of macrophages, these cells can no longer ingest urate crystals by phagocytosis in the presence of colchicin, interrupting the inflammatory process. Uricosuric drugs, such as probenecid, are used to increase renal urate excretion.

5.6 DISORDERS OF PYRIMIDINE NUCLEOTIDE METABOLISM

Orotic aciduria, a very rare hereditary disorder, is a pyrimidine auxotrophism in humans. In one known case, the enzyme orotidylate decarboxylase was found defective; in another case, both orotidylate pyrophosphorylase and decarboxylase were defective. Replacement therapy with uridine is possible.

For a more extensive discussion of nucleotide metabolism, consult the following textbooks:

Mathews et al. (1990): Chapter 22

Stryer (1988): Chapter 25

Further References

Kornberg A. DNA replication. 2nd ed. San Francisco, Ca: W. H. Freeman; 1989. (Chapter 1 covers nucleotide metabolism.)

Reichard P. Interactions between deoxyribonucleotide and DNA synthesis. *Annu Rev Biochem.* 1988; 57:349–375.

Reichard P, Ehrenberg A. Ribonucleotide reductase, a radical enzyme. *Science.* 1983; 221:514–519.

CHAPTER

6 Biochemical Genetics

OBJECTIVES

After studying this chapter, the student should be able to

1. Define the following terms: genome, haploid, diploid, allele, cistron, homozygous, heterozygous, plasmid, genotype, phenotype.
2. Explain the basic features of the DNA double helix, such as base-pairing, $5' \rightarrow 3'$ polarity, base-stacking. Chargaff's rules, hypochromism, denaturation, intercalation.
3. Describe the different structures and chemical properties of RNA and DNA.
4. Name at least three different types of cellular RNA and explain their functions.
5. Explain the following features of the genetic code: triplet codon, degenerate, nonoverlapping, commaless, universal.
6. Describe the principle of the Meselson-Stahl experiment and state its conclusion.
7. Name the three enzymatic activities of DNA polymerase I and describe their probable functions.
8. Describe the basic mechanism of discontinuous DNA replication and name some of the proteins participating in it.
9. State the difference between template and primer in DNA synthesis.
10. State the function of a primosome.
11. Describe the mechanism of excision repair.
12. Explain the significance of heteroduplex intermediates in genetic recombination.
13. Give an example of a point mutation and name at least two mechanisms by which it can arise.
14. State the differences in structure and synthesis of prokaryotic and eukaryotic mRNA.
15. Outline the basic mechanism of RNA splicing.
16. Give at least one reason why eukaryotic mRNA is always monocistronic, whereas prokaryotic mRNA often is polycistronic.
17. Describe the function of the nucleolus and explain the following terms: 45S precursor rRNA, spacer, gene amplification.
18. Explain the functions of the σ factor and the ρ factor in transcription.
19. Name two inhibitors of transcription and describe their mode of action.
20. Describe the subunit structure of a ribosome and name its basic components.
21. Draw an outline of a tRNA molecule, identify its various arms, and describe their functions.
22. Describe the reactions involved in amino acid activation.
23. Describe in detail the four basic steps of the translation mechanism: initiation, elongation, translocation, and termination.
24. Name several inhibitors of translation and describe their mode of action.

25. Explain the principle of tRNA-mediated suppression.
26. Explain how proteins are targeted for excretion or membrane insertion.
27. Describe the molecular mechanisms of enzyme induction and repression.
28. Explain the nature of *cis* dominant and *trans* dominant mutations and give examples.
29. Define the following terms: promoter, operator, operon, repressor, inducer, corepressor, CAP protein, attenuation, polar mutation.
30. Explain the significance of Cot analysis of eukaryotic DNA.
31. Describe the basic features of chromatin structure.
32. Explain Lyon's hypothesis and define the term "Barr body."
33. Give an example of genetic mosaicism.
34. Describe the genotypes and phenotypes of Klinefelter's and Turner's syndromes.
35. Explain the term "alternative splicing."
36. State the significance of homeotic genes.

6.1 THE FLOW OF GENETIC INFORMATION

The information that enables a cell to live and to divide, giving rise to two identical daughter cells, is carried on **genes.** The chemical material of a gene is **deoxyribonucleic acid (DNA)** (in some viruses, it is ribonucleic acid [RNA]). Several observations led to this conclusion. In 1928, Griffith observed that when mice were injected with both live nonpathogenic *Pneumococcus* bacteria of the R type (rough colonies lacking a capsular polysaccharide) and heat-killed pathogenic bacteria of the S type (smooth colonies with capsular polysaccharide), the mice died, even though injection of killed S-type bacteria alone did not kill the animals. Live S-type pneumococci could be isolated from the dead animals. Griffith concluded correctly that the genes for the synthesis of the pathogenic polysaccharide must have been taken up from the coinjected dead bacteria and incorporated into the live bacteria, changing their **genotype** (genetic makeup) to that of the pathogenic bacteria.

In 1944, Avery, McLeod, and McCarty demonstrated that purified DNA isolated from the dead bacteria, when added to live bacteria, could accomplish the same **transformation.** Persuasive confirmation that it was indeed DNA that contained the genetic information came from an ingenious experiment by Hershey and Chase in 1952. Hershey labeled **bacteriophage T2** (a bacterial virus containing DNA packaged in a protein coat) with ^{32}P in its DNA and with ^{35}S in its coat proteins and then infected *Escherichia coli* bacteria with this phage. During the infection process, most of the ^{32}P-labeled DNA entered the bacterial cell, whereas the ^{35}S-labeled proteins remained on the outside of the cell and could be sheared off by stirring the bacterial culture in a blender. The ^{35}S-free but ^{32}P-containing bacteria, separated from the phage by centrifugation, nevertheless produced progeny phage particles, as would be expected if the information for the synthesis of new phage was contained solely in the DNA (Figure 6.1).

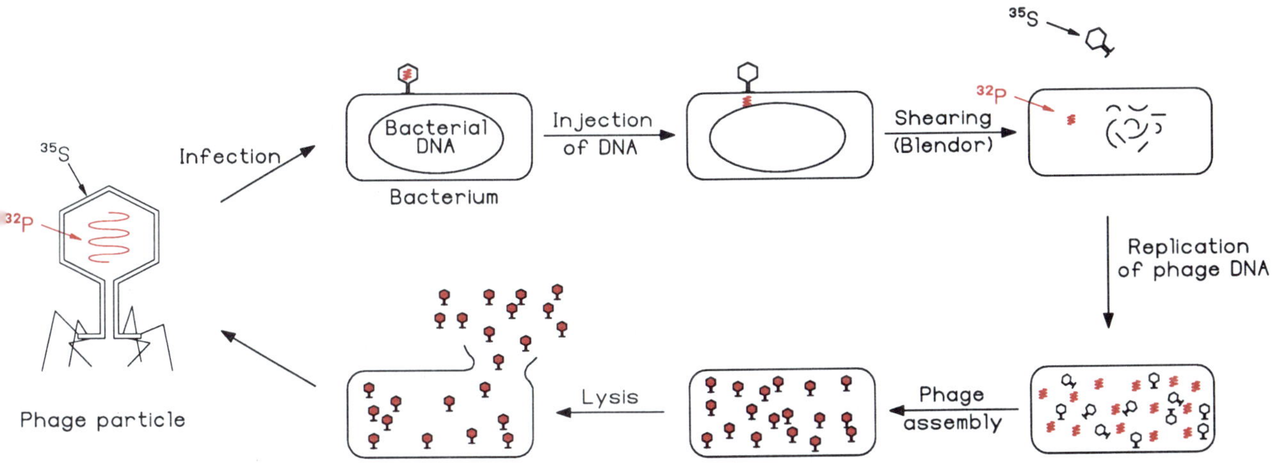

Figure 6.1.
The Hershey-Chase experiment.

6.1.1 Introduction to Genetics

A **gene** contains the information for the synthesis of a polypeptide. It is a segment within a very long strand of DNA called a **chromosome.** Chromosomes contain many different genes. Prokaryotic cells (bacteria and blue-green algae) contain a single chromosome consisting of one very long double-stranded DNA molecule. **Eukaryotic cells** (all other organisms) usually contain several chromosomes located within a nucleus. **Prokaryotic cells** do not have nuclei. The entire genetic content of a cell, whether in a single or several chromosomes, is called its **genome.** Viruses also contain genomes, usually packaged in a protein coat that in some cases is enveloped by a membrane. However, viruses can express their genetic information only in the cell they infect.

The genomes of prokaryotes and viruses are usually **haploid;** that is, there is only one copy of a chromosome per cell. Most eukaryotic cells, on the other hand, are **diploid,** containing two copies of each chromosome. Only the maternal and paternal **gametes** (germ cells) are haploid. Upon fertilization, the two gametes fuse to form a diploid **zygote** from which all cells of a multicellular organism are derived.

In a diploid cell, a particular gene may be present in two identical or two different forms, called **alleles** or mutational states; the cell is said to be either **homozygous** or **heterozygous,** respectively. Often,

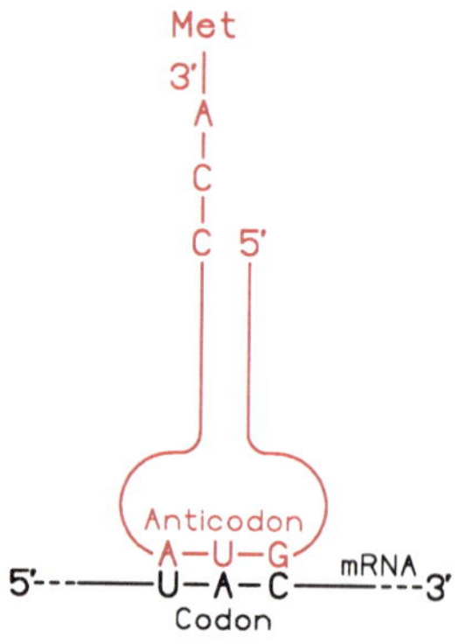

Figure 6.2.
Schematic drawing ot a tRNA molecule.

only one allele in a heterozygote is visibly expressed; that is, its gene product determines the **phenotype** (detectable expression of the **genotype**) of the cell, the other being invisible. The allele that determines the phenotype of the cell is called **dominant,** the other one is **recessive.** Usually the **wild-type** allele (i.e., the normal allele as it occurs in nature) is dominant over a defective **mutant** allele, but sometimes it is reversed.

Some cells contain extrachromosomal DNA in the form of **plasmids** (found in bacteria and some eukaryotes) or in subcellular particles such as mitochondria and chloroplasts.

6.1.2 Information Transfer

The genetic information for the synthesis of a protein is contained in the nucleotide sequence of DNA. There are only four nucleotides that make up the genetic alphabet: adenine, guanine, cytosine, and thymine. Chromosomal DNA usually contains two strands of DNA wound around each other in a **double helix** (see next section). The bases on one strand of the double helix form specific hydrogen bonds to bases on the other strand. A always pairs with T, and C with G. Thus, the genetic information is present in duplicate in the double helix, once directly in the sequence of the **sense strand** and a second time in complementary form in the other strand **(antisense strand).**

In order to be expressed, a DNA sequence must first be copied into a strand of RNA. This **transcription** process uses the antisense strand as template for the synthesis of a **messenger RNA (mRNA)** strand that is complementary to the antisense strand but identical to the information in the sense strand. Each gene gives rise to many identical mRNA molecules. The mRNA is then transported from the nucleus to the cytoplasm (a process not necessary in prokaryotes), where it is translated into the amino acid sequence of a protein. This essentially unidirectional flow of genetic information is sometimes called the **central dogma** of molecular biology.

$$\text{DNA} \xrightleftharpoons{\text{transcription}} \text{RNA} \xrightarrow{\text{translation}} \text{Protein}$$

The information transfer from DNA to RNA is reversed during the multiplication of RNA tumor viruses (see Chapter 7). A specific enzyme, called reverse transcriptase, synthesizes a DNA copy from an RNA template.

The translation of the mRNA sequence into the amino acid sequence of a protein requires the participation of **transfer RNA (tRNA).** This small RNA molecule of only about 80 nucleotides has an amino acid attached to one end and is capable of base-pairing, through its **anticodon** loop, with a specific **trinucleotide codon** sequence on the mRNA (Figure 6.2).

Each tRNA molecule is specific for only one amino acid. Thus, by lining up along the mRNA via its anticodon loop, the tRNA places the amino acid at its proper position in the growing polypeptide chain (Figure 6.3). The translation process takes place on the **ribosome,** a cytoplasmic particle containing both proteins and **ribosomal RNA (rRNA).**

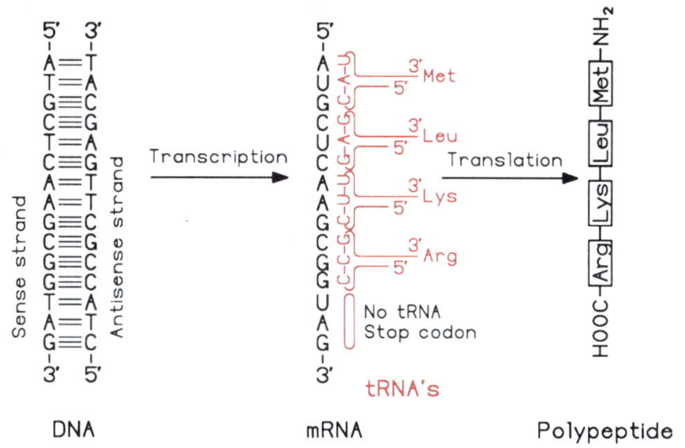

Figure 6.3.
The flow of genetic information.

6.1.3 The Genetic Code

As already mentioned, the genetic code uses triplet codons, that is, a sequence of three nucleotide codes for one amino acid. There are 64 (4^3) possible codons for only 20 different amino acids found in proteins. Sixty-one of these codons code for amino acids, usually several for the same amino acid. Thus, the genetic code is said to be degenerate. Only two amino acids have unique codons, Met and Trp. The three codons not coding for amino acids are used as stop codons. Stop codons are also called **nonsense codons** since mutations in genes that convert a normal codon into a stop codon cause the translation process to stop prematurely, thus producing a defective protein.

The code is **commaless,** that is, one triplet on the mRNA is directly followed by the next one without any additional signal between them, and nonoverlapping. The reading frame is set by the first tRNA molecule that binds to the mRNA and initiates the translation process. All proteins are initiated with methionine. The ribosome then moves along the mRNA in steps of three nucleotides.

Sometimes two or even three genes partially overlap. This occurs if the translation of the mRNAs, derived from the DNA segment coding for these genes, is initiated at sites that produce different reading frames for the translation process. The genetic code is essentially **universal** for all living things on this earth. Only in mitochondria and chloroplasts of some species and in some protozoa have a few altered codons evolved.

The first amino acid codon was deciphered by Nirenberg and Matthaei, who observed that a synthetic RNA containing only uridine monophosphate (UMP) residues (poly-U) directed the synthesis of poly-phenylalanine in a cell-free translation system. They correctly concluded that the codon for phenylalanine must be UUU.

The other codons were elucidated by measuring the in vitro incorporation of amino acids into protein directed by synthetic ribonucleotide monopolymers and heteropolymers and by studying the binding of amino acid–carrying tRNAs to ribosomes in the presence of specific oligoribonucleotides. The complete genetic code is presented in Table 6.1.

Many tRNA molecules can recognize several codons that differ in their third positions. This is explained by Crick's **wobble hypothesis.** Crick proposed that the base at the 5′ end of the anticodon of the tRNA

Table 6.1. The Genetic Code

First Position (5′ end)	Second Position U	C	A	G	Third Position (3′ end)
U	Phe	Ser	Tyr	Cys	U
	Phe	Ser	Tyr	Cys	C
	Leu	Ser	Stop	Stop	A
	Leu	Ser	Stop	Trp	G
C	Leu	Pro	His	Arg	U
	Leu	Pro	His	Arg	C
	Leu	Pro	Gln	Arg	A
	Leu	Pro	Gln	Arg	G
A	Ile	Thr	Asn	Ser	U
	Ile	Thr	Asn	Ser	C
	Ile	Thr	Lys	Arg	A
	Met*	Thr	Lys	Arg	G
G	Val	Ala	Asp	Gly	U
	Val	Ala	Asp	Gly	C
	Val	Ala	Glu	Gly	A
	Val*	Ala	Glu	Gly	G

*The Met codon is used as the initiation codon.
GUG codes for Met when used as the initiation codon.
U = uracil, C = cytosine, A = alanine, G = guanine.

is not as rigidly confined by the mRNA-ribosome complex as the other two bases, thus allowing it to form hydrogen bonds with any of several bases located at the 3′ end of the codon. Table 6.2 summarizes the wobble rules that explain the degeneracy of the code. According to these rules, a total of only 32 tRNA species would be required for the translation of all codons. In reality, cells contain about twice as many different tRNA molecules. It has been speculated that some of the extra tRNAs are involved in the regulation of gene expression.

Table 6.2. Wobble Pairing Rules

Base in Anticodon (at 5′ end)	Base in codon (at 3′ end)
G	U, C
C	G
A	U
U	A, G
I	A, U, C

I = inosine.

6.2 STRUCTURES AND PROPERTIES OF DNA

6.2.1 Chemistry and Structure of DNA

DNA is a linear polymer of four types of deoxyribonucleotides linked together by 3′,5′-phosphodiester bonds (Figure 6.4). The four bases

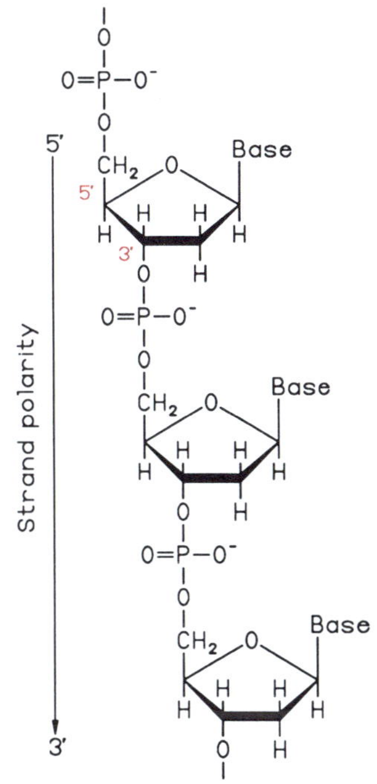

Figure 6.4.
The chemical structure of DNA.

found in DNA are adenine (A), cytosine (C), guanine (G), and thymine (T).

Through model building and analysis of X-ray crystallographic data by Wilkins and Franklin, Watson and Crick were the first to recognize that a DNA molecule usually consists of two strands of DNA wound around each other to form a right-handed double helix. With regard to the direction of their 3′,5′-phosphodiester bonds, the two strands in the DNA helix have opposite polarity (Figure 6.5).

The hydrophobic bases are arranged on the inside of the double helix at right angles to the helix axis and form specific interstrand hydrogen bonds: A pairs with T, and G pairs with C (Figure 6.6). The specific base-pairing explains **Chargaff's** earlier observation that the ratio of purines to pyrimidines in double-stranded DNA always equals unity, and that the ratios A/T and C/G also equal unity. An exception is single-stranded DNA, found in certain viruses, where this **rule** obviously need not hold.

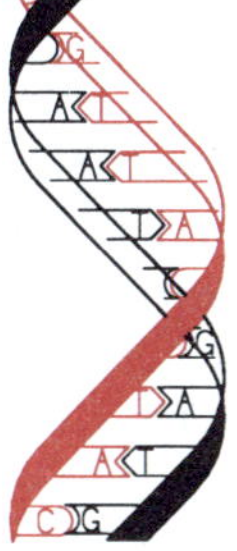

Figure 6.5.
The DNA double helix.

The C + G composition of DNA, on the other hand, may vary widely between different species (e.g., *E. coli,* 0.5; crayfish, 0.35, ribosomal RNA genes, 0.65).

The stability of the double helix is not only determined by the hydrogen bonds between paired bases but also by the longitudinal interaction of neighboring bases, called **base-stacking.** Base-stacking is due to the overlapping of the π electron orbitals of the planar bases. Because of this π electron interaction, double-stranded DNA exhibits a lower light absorbance at 260 nm than single-stranded DNA. This effect is called **hypochromism.**

Chemically, DNA is a rather stable molecule. In contrast to RNA, it is not hydrolyzed by alkali. Acids, on the other hand, cause depurination, that is, the hydrolysis of the glycosidic bond between the deoxyribose and a purine base. If such depurinated DNA is subjected to alkaline pH,

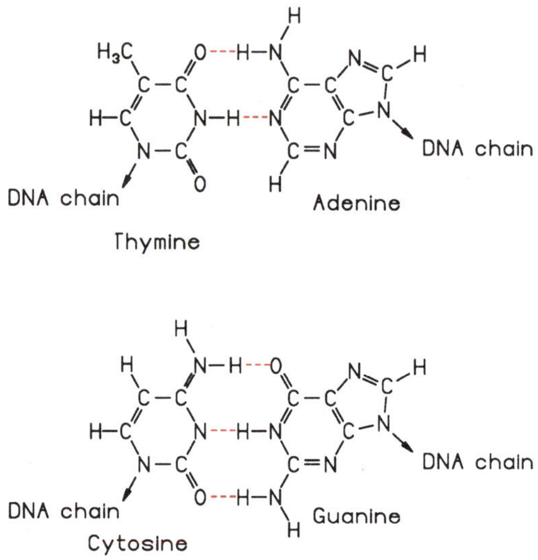

Figure 6.6.
Base-pairing in DNA.

the phosphodiester bond next to a depurinated deoxyribose residue is hydrolyzed, causing a break in the DNA chain.

6.2.2 Types of DNA Helix

The DNA double helix has a wide and a narrow groove, and there are 10 base pairs per turn of the helix. In addition to this most **common B form,** DNA can assume two other conformations. Upon dehydration, the B form is converted into the A form, with approximately 11 base pairs per turn. Pure **A form** probably does not exist in a cell. RNA cannot form B form helices. Double-stranded RNA, found in some viruses, and RNA-DNA hybrids, which are formed over short distances during the transcription process, exhibit A-like configurations.

A left-handed double helix, called **Z-DNA,** with 12 base pairs per turn, forms preferentially in regions with alternating purines and pyrimidines. Its formation is also facilitated by the methylation of cytosine residues in the DNA. Since methylation may affect gene expression, it is tempting to speculate that this occurs through the formation of regions of Z configuration.

The local configuration of a long DNA molecule fluctuates. Depending upon its local sequence, DNA can form kinks or bends. Certain DNA-binding proteins (e.g., histones, restriction enzymes) cause bending of the DNA.

6.2.3 Denaturation of DNA

High temperature or high pH cause the double helix to denature, releasing two single DNA strands (Figure 6.7). The **melting temperature** of a DNA molecule is dependent on its G + C content, since the G—C pair, forming three hydrogen bonds, is stronger and thus requires a higher melting temperature than the C—T pair, which has only two hydrogen bonds.

Denaturation is accompanied by an increased UV absorbance (hyperchromicity), which is caused by the "unstacking" of the bases. The denaturation reaction is reversible. When complementary single strands

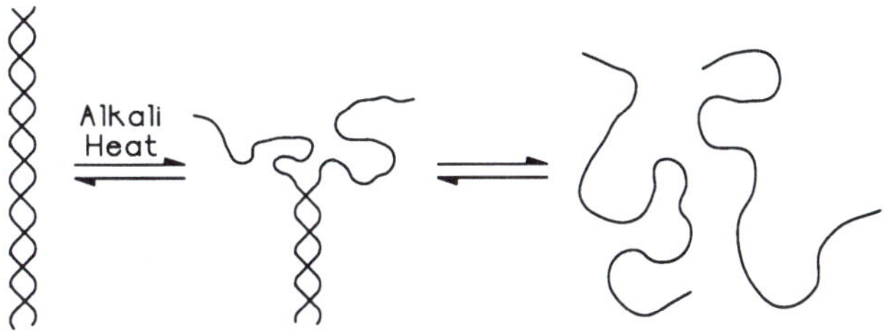

Figure 6.7.
Denaturation of DNA.

of DNA are incubated just below the **melting point,** double helices reform. This process is called **annealing** and is used to study relationships between DNA sequences.

Most DNA is of very high molecular weight and, as a consequence, is easily broken during isolation by mechanical shearing forces. A bacterium, for example, contains 2.5×10^6 kd of DNA all contained in a single 2 mm–long molecule. In comparison, the average human chromosome contains about 10^8 kd of DNA. Stretched out, this molecule would be 80 mm long. (The diploid human cell contains 46 chromosomes.)

6.2.4 Circular DNA

In some bacteria and viruses, as well as in mitochrondria, the DNA molecule is circular. In other species, the DNA is linear but often exhibits terminally repeated sequences (terminal redundancy), which prevent loss of genetic information from the ends during DNA replication and, in some cases, allow circularization.

A special case of terminal redundancy is exhibited by the DNA of bacteriophage λ. In this phage particle, λ DNA is linear, but its ends are single-stranded and complementary to each other. These cohesive ends can anneal to one another, thus creating a double-stranded circular DNA molecule (Figure 6.8). This occurs normally when the phage DNA enters a bacterium. The replication of λ DNA occurs on the circular form.

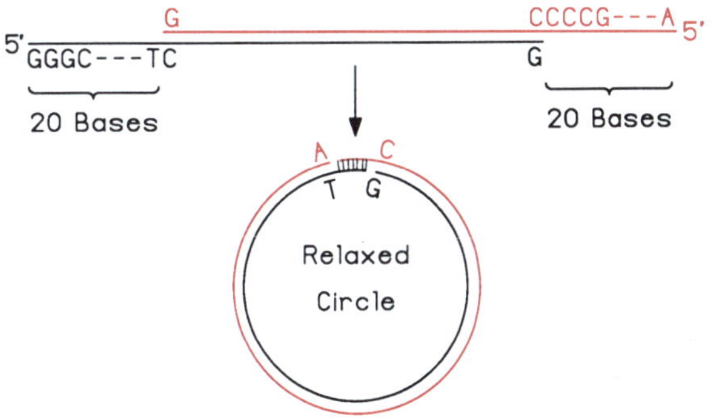

Figure 6.8.
The "cohesive ends" of bacteriophage λ DNA.

6.2.5 Supercoils

When **circular DNA** molecules, in which both strands are covalently closed (the "nicked" circle in Figure 6.9 can be closed in the cell by DNA ligase), are isolated from cells, they are usually found in supercoiled form. In a supercoil, segments of the double helix are twisted around each other. It arises from changing the pitch of the double helix. Supercoils can only occur in circular DNA or when the ends of linear

DNA are attached to some structure. In free linear DNA, the ends can rotate freely, thus releaving any torsional stress.

Supercoiling can be quantitatively described by the **linking number,** a number that indicates the number of times one strand of DNA is wound around the other. Relaxed B DNA has a linking number of one per 10 base pairs. If the number is decreased, the DNA is underwound or negatively supercoiled; if increased, the DNA forms a **positive supercoil** (Figure 6.9). Circular DNA isolated from cells usually is negatively supercoiled.

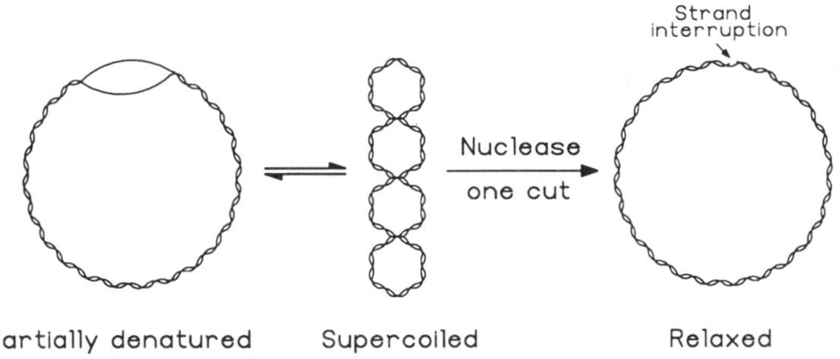

Figure 6.9.
Change in DNA topology by altering its linking number.

A single cut (nick) in supercoiled circular DNA will relax the DNA (Figure 6.10). Similarly, local unwinding of a DNA segment will cause a **negative supercoil** to become relaxed.

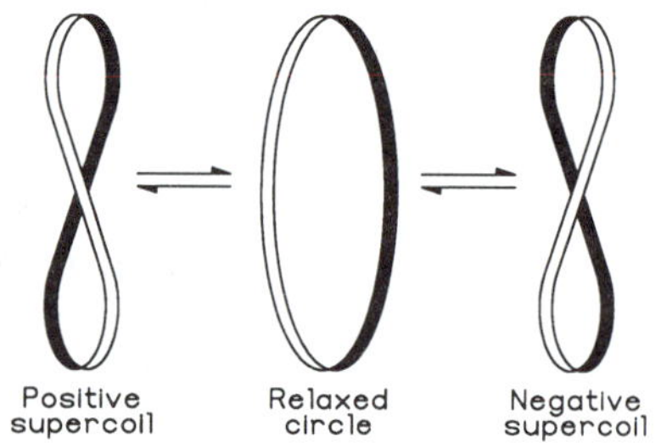

Figure 6.10.
Supercoiled and relaxed DNA.

6.2.6 Topoisomerases

Cells contain enzymes, so-called topoisomerases, that can change the linking number of DNA. **Topoisomerase I** changes the linking number of a circular DNA molecule in single steps, topoisomerase II in steps of two. Usually these enzymes increase the linking number thereby relaxing negatively supercoiled DNA. They never go beyond the point where they would generate positively supercoiled DNA. However, local positive supercoils may be produced at the replication fork (see the next section).

A special type of **topoisomerase II** is **gyrase,** isolated from bacteria. This enzyme can actually introduce negative supercoiling into DNA (decreased linking number). It requires adenosine triphosphate (ATP) for this reaction. Gyrase is required for the transcription of some genes (see Section 6.6). No gyrase has been found in eukaryotes. However, in contrast to bacterial topoisomerase I, eukaryotic topoisomerase I can also relax positive supercoils. Eukaryotic DNA is normally wound around complexes of histone proteins, called nucleosomes (see later), and in this state is relaxed. Only when the histones are removed does the DNA become negatively supercoiled.

DNA topoisomerase I from *E. coli* operates by first cleaving a phosphodiester bond, creating a nick in one of the two strands. In this reaction, the enzyme becomes covalently linked, via a tyrosine residue, to the phosphate group. It then somehow allows the other DNA strand to rotate and then reforms the phosphodiester bond. No energy is required for this reaction. The reaction mechanism for topoisomerase II is more complex. In order to be able to change the linking number in steps of two, a double-stranded cut must be made in the DNA and another segment of the same DNA molecule passed through this gap before it is released. Topoisomerase II also unlinks interlocked circles of DNA (see DNA replication).

6.2.7 Intercalation

Due to its compact structure, supercoiled DNA sediments faster than relaxed DNA of the same size. **Ethidium bromide,** a heterocyclic aromatic compound, intercalates between the base pairs of double-stranded DNA, thereby increasing its winding number (this is an artificial increase, since the ethidium bromide molecules count as bases even though they are not part of the DNA (Figure 6.11). Intercalation of just the right amount of ethidium bromide into a negative supercoil lets the DNA become relaxed without breaking a phosphodiester bond. Further addition of ethidium bromide produces a positive supercoil.

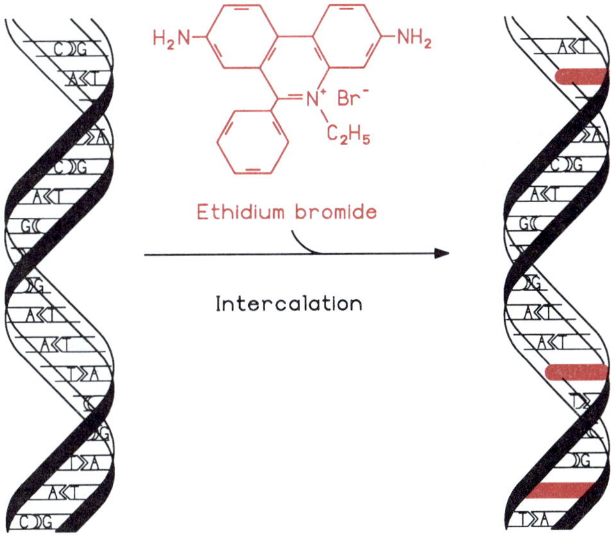

Figure 6.11.
Intercalation of ethidium bromide into DNA.

6.3 DNA REPLICATION

There are two types of DNA synthesis: **DNA replication,** which is responsible for the duplication of the genome before the cell divides, and **repair synthesis,** which corrects damaged or altered DNA regions. In addition to these processes, DNA is subject to changes in nucleotide sequence caused by genetic recombination and by mutagenesis.

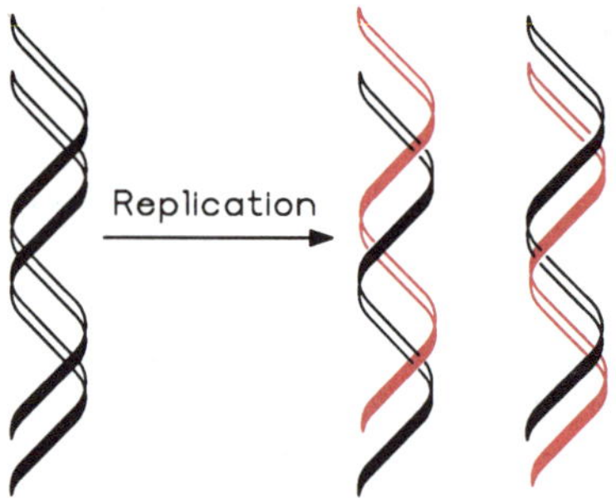

Figure 6.12.
Semiconservative DNA replication.

6.3.1 Semiconservative DNA Replication

During DNA replication, the two strands of the double helix separate, and each parental strand serves as the template for the synthesis of a new daughter strand. Each of the two daughter helices thus consists of one old and one new strand of DNA (Figure 6.12).

This semiconservative mode of replication had been predicted by Watson and Crick and was subsequently supported by Meselson and Stahl in a density-labeling experiment: *E. coli* bacteria were cultured for several generations in a growth medium containing the heavy nitrogen isotope ^{15}N in the form of NH_4Cl. The cells were then transferred to a light ^{14}N-containing medium, and the DNA was isolated at various times after this transfer and analyzed by CsCl density gradient centrifugation. The principle of this analysis is as follows. When a solution of CsCl is centrifuged at high speed, a concentration gradient develops as the consequence of the equilibrium reached between sedimentation and diffusion. The Cs salt of DNA, if mixed into the centrifuge tube, will form a narrow band within a couple of days at the CsCl concentration that corresponds to its buoyant density. ^{14}N- and ^{15}N-containing DNA molecules have different buoyant densities and thus can be distinguished in CsCl density gradients.

Meselson and Stahl observed that, after transfer to the light medium, more and more of the DNA, which originally banded at the heavy ^{15}N position in the gradient, formed a new band at a position of lighter density until, after one generation, essentially all DNA was of a hybrid density in-between that of fully ^{15}N-labeled heavy DNA and that of light ^{14}N-labeled DNA. In the course of a second round of replication (generation) in the light medium, a new band gradually appeared at the position of light DNA whereas the hybrid band remained unchanged in quantity (Figure 6.13).

This result was consistent with the predicted semiconservative replication mechanism. (The final proof of this mode of replication was provided by Baldwin, who demonstrated that hybrid DNA indeed consisted of double-stranded DNA helices and not four-stranded intermediates as had been speculated by some.)

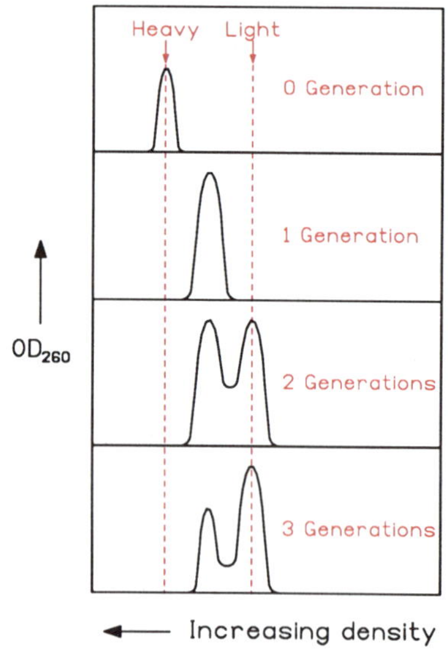

Figure 6.13.
The Meselson-Stahl experiment. Distribution of parental (heavy) and daughter (light) DNA helices in a CsCl density gradient.

6.3.2 The Structures of Replicating DNA

The process of DNA replication at the level of an individual molecule was first visualized by John Cairns. He labeled bacteria with radioactive ^{3}H-thymidine and then carefully lysed the cells to release the DNA in unbroken form and allowed it to attach to a nitrocellulose filter. The contours of the radioactively labeled DNA molecules were then visualized by **autoradiography.** Cairns observed that replicating bacterial DNA had a theta-like structure with two branch points. The branch points represent replication forks. DNA replication is initiated at a specific nucleotide sequence, the origin of replication, and proceeds in two directions until the forks meet at the opposite side of the circular chromosome (Figure 6.14). At completion, the two circular daughter molecules are interlinked and must be separated by topoisomerase II.

In fast-growing bacteria, multiple rounds of replication are initiated before the first one is completed. This leads to multiforked structures. During replication of certain bacteriophages, concatemers, long linear

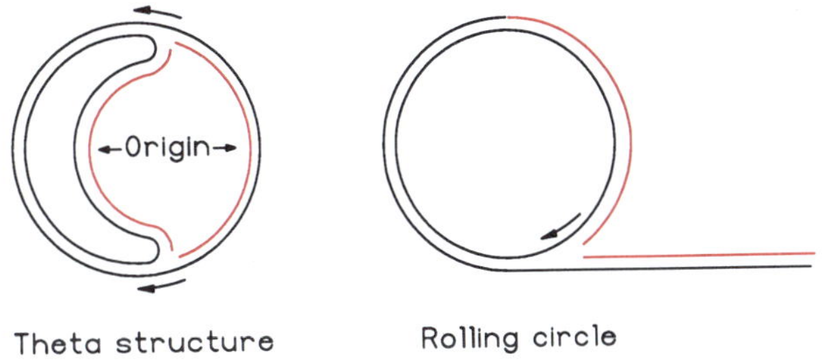

Figure 6.14.
Two modes of DNA replication: the theta structure and the rolling circle.

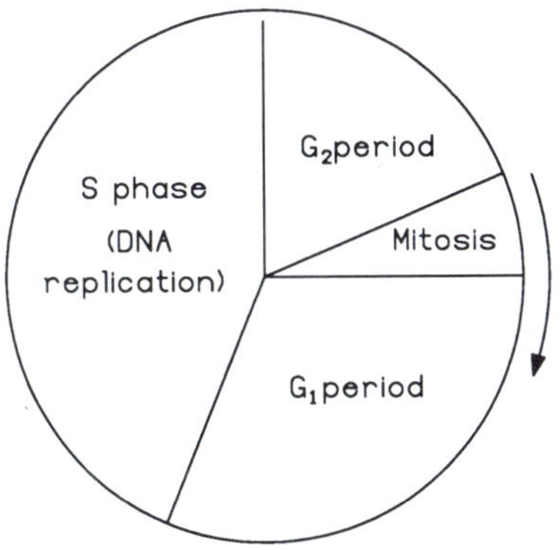

Figure 6.15.
The eukaryotic cell cycle.

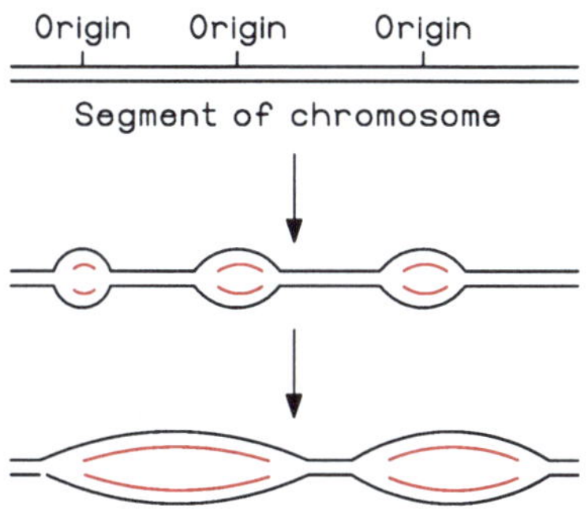

Figure 6.16.
Multiple replicons in eukaryotic DNA.

multiples of the genome, are formed. They arise when one of the forks in the theta structure opens and the other fork then continues unidirectionally around the circular parental strand (rolling circle replication).

The rate of replication (fork movement) in bacteria is about 50,000 nucleotide pairs per minute at 37°C. At this rate, the parental duplex has to be unwound at the rate of 5,000 rpm (since there are 10 base pairs per turn). This simple calculation gives some idea of the mechanical complexity of the replication process.

In contrast to bacteria, in which DNA is being replicated at all times, eukaryotic cells replicate their DNA only during a fraction of the **cell cycle.** The time span between two mitotic divisions **(M),** called **interphase** because the chromosomes are not visible, is divided into three distinct phases: **G1, S, and G2** (Figure 6.15).

DNA replication occurs only during the S phase. The rate of replication, as measured by fork movement, is considerably slower than that observed in prokaryotes. However, considering the large size of most eukaryotic chromosomes, the time required to replicate an entire chromosome is not long. This is because replication can occur simultaneously at many different sites along the chromosome. Each replication unit, or **replicon,** has an **origin** from which two replication forks move in opposite directions until they meet another fork coming from the neighboring replicon (Figure 6.16).

6.3.3 DNA Polymerases

The first enzyme capable of synthesizing DNA from low-molecular-weight precursors was isolated from *E. coli* by **A. Kornberg.** DNA polymerase uses deoxyribonucleoside triphosphates as substrates and requires both a high-molecular-weight DNA strand as a template and a short polynucleotide strand as a primer (Figure 6.17). Elongating the 3′ end of the primer strand DNA polymerase creates a new DNA strand hydrogen-bonded, and thus complementary, to the template strand.

As it turned out, Kornberg's DNA polymerase, now named **DNA polymerase I,** is not the principal replication enzyme, although it has an essential function in the replication process (see later). *E. coli* has two additional DNA polymerases, polymerase II and polymerase III. Polymerase II is probably dispensable; its precise function in the cell is not known. **Polymerase III,** however, is the primary replication enzyme.

All bacteria DNA polymerases possess a **3′→5′ exonuclease** activity in addition to their polymerizing activity. This exonuclease removes nucleotides from the 3′ end of a growing DNA strand if, and only if, they are not properly base-paired with the template strand. This prop-

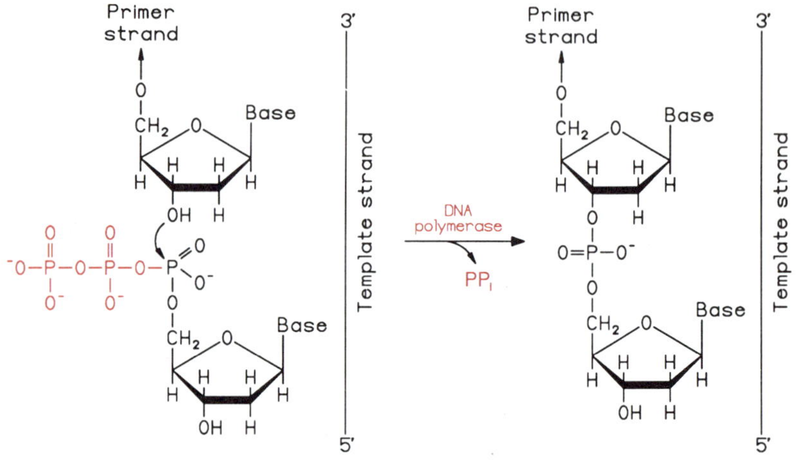

Figure 6.17.
The mechanism of polymerization of DNA.

erty allows the polymerase to proofread its product. If a wrong nucleotide is added to the growing chain, it is immediately removed before the synthesis resumes.

Polymerases I and III also have **5′→3′ exonuclease** activities that may function in the removal of thymine dimers (see next section of DNA repair). This exonuclease activity in polymerase I is essential for the removal of the RNA primers that initiate the synthesis of DNA strands (see below).

Eukaryotic cells also contain several DNA polymerases, named α, β, γ, and δ. Polymerases α and γ are involved in chromosomal DNA replication. Polymerase β is a smaller enzyme with different properties whose precise function is not yet known. Polymerase γ is the mitochondrial replication enzyme.

6.3.4 Discontinuous DNA Replication

The mechanism of DNA replication has turned out to be much more complicated than originally thought. The main difficulty arises from the fact that all DNA polymerases elongate DNA strands only at their 3′ ends. However, each replication fork contains both a 3′- and a 5′-terminated new DNA strand. The problem is solved through a discontinuous replication mechanism.

In the replication fork, only one of the new DNA strands, the **leading strand** with the 3′ terminus, is elongated directly. The other so-called **lagging strand,** carrying the 5′ terminus, is synthesized in the form of short pieces, generally called Okazaki fragments after their discoverer Reiji Okazaki (Figure 6.18), which are later joined together by DNA ligase. (The term "fragments" is really a misnomer, since it implies that they originate from fragmentation of something larger: they should be called **Okazaki pieces** since they do not.)

Since all DNA polymerases require primers, the synthesis of Okazaki pieces is initiated by an RNA polymerase called primase, that does not require primers. The RNA primer is very short, in the range of only three to five nucleotides. After the synthesis of an Okazaki piece, the RNA primer is removed by the 5′→3′ exonuclease function of polymerase I.

Priming with RNA rather than DNA probably has evolved because

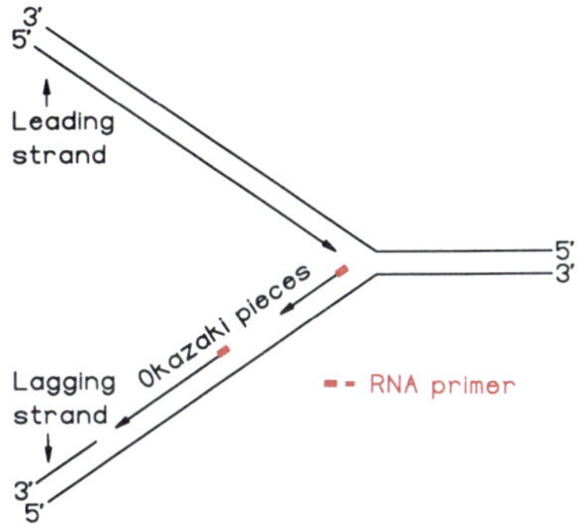

Figure 6.18.
Discontinuous DNA replication.

synthesis of the first few nucleotides complementary to the template is likely to be inaccurate. There are not enough hydrogen bonds to keep the growing strand aligned with the template for the polymerase to use its proofreading function (3′→5′ exonuclease). Since the RNA primer is removed later, any mismatches with the template strand can be repaired when the RNA strand is replaced by DNA through extension of the more recently made Okazaki piece.

6.3.5 The Replication Fork

DNA polymerase III (pol III) is a multi-subunit **holoenzyme** made of two subunits, each consisting of at least seven different polypeptides, called α (*dnaE* protein), ϵ, θ, β, τ, γ, and δ. It is a highly **processive** enzyme; that is, the enzyme remains tightly bound to the growing DNA strand during synthesis. The α subunit carries the polymerizing and the 5′→3′ exonuclease activities. The epsilon subunit is a 3′→5′ exonuclease. The precise functions of the other subunits are not known.

In addition to polymerase III holoenzyme, several other proteins participate directly or indirectly in the replication process; they are listed in Table 6.3.

Figure 6.19 illustrates the location of some of these additional proteins within the replication fork. Ahead of the fork, two **helicases** cooperate in the unwinding of the parental duplex. They require ATP hydrolysis for their movement along the DNA. The dnaB protein forms a complex with dnaC, primase (dnaG), and four other polypeptides. This so-called **primosome** complex moves along the parental DNA strand that serves as a template to the lagging strand and initiates the synthesis of Okazaki pieces. dnaB protein is a DNA helicase and requires ATP hydrolysis for this movement. **Single-strand binding protein (SSB)** binds to the single strands of parental DNA that are generated by the unwinding of the parental double helix and thereby keeps them in extended form. DNA polymerase I removes the RNA primer from Okazaki pieces and fills in the gap by extending the 3′ end of the next Okazaki piece. The topoisomerases facilitate replication by releaving torsional stress in the parental double helix ahead of the replication fork.

DNA ligase connects Okazaki pieces by first activating the terminal

Table 6.3. DNA Replication Proteins in *E. coli*.

Protein	Function
dnaB protein	Major protein of primosome
dnaC protein	Complexes with dnaB
DNA polymerase II	Replicates DNA
Primase (dnaG)	Synthesizes RNA primer
SSB	Single-strand binding protein
DNA polymerase I	Removes RNA primer, fills gap
DNA ligase	Connects Okazaki pieces
Topoisomerase I	Relaxes negative supercoils
Topoisomerase II (gyrase)	Creates negative supercoiling
Rep protein	Helicase on leading strand
Helicases II and III	Helicases for lagging strand

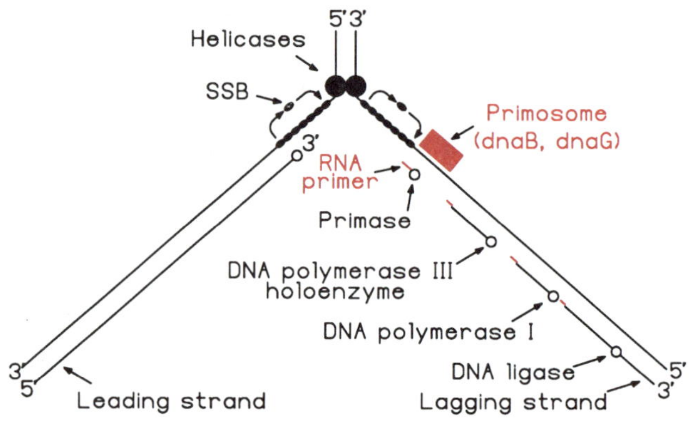

Figure 6.19.
Location and function of replication proteins in the replication fork.

5′-phosphate group of one DNA strand by attachment of adenine monophosphate (AMP). The source of AMP is ATP (in *E. coli,* it is the oxidized form of nicotinamide-adenine dinucleotide [NAD^+]). A phosphodiester bond is subsequently formed by reaction of the AMP-activated 5′-phosphate with the 3′-hydroxyl terminus of the apposing DNA strand, and AMP is released. (Note that the activation of the 5′ end involves the formation of a phosphoanhydride bond.)

A recent model of the replication fork proposes the formation of a loop of the lagging strand. This arrangement would allow the polymerase holoenzyme dimer to synthesize both new strands simultaneously (Figure 6.20), thus ensuring that leading strand synthesis occurs in synchrony with lagging strand synthesis. It is likely that the two halves of the holoenzyme dimer differ slightly in their processivity properties. (The γ and τ subunits have been implicated in this asymmetry because they are homologous, differing only at their carboxyl termini.) The holoenzyme half that synthesizes the lagging strand must dissociate transiently from the template in order to release the loop and begin synthesis of a new Okazaki piece.

In eukaryotic cells, it appears that a similar asymmetry of polymerases exists at the replication fork, with polymerase δ being the leading strand polymerase and polymerase α the lagging strand polymerase.

6.3.6 Postreplicative Modification of DNA

Bacteria and some bacterial viruses have evolved a mechanism that recognizes foreign DNA and degrades it. This **restriction-modification** phenomenon was first discovered in the T-even bacteriophages of *E. coli* that incorporate into their DNA modified bases, such as hydroxymethylcytosine, in order to make their DNA resistant to degra-

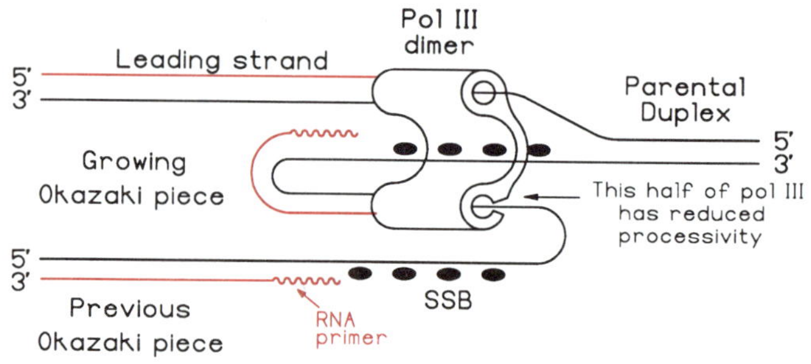

Figure 6.20.
Looping model for polymerase III action.

dation by bacterial nucleases. A more widely found type of **DNA modification** is achieved through **methylation** of specific bases, located at unique short sequences in the DNA. The modification enzyme uses S-adenosylmethionine as methyl donor. The modified sequence is thereby protected from degradation by the **restriction enzyme,** an endonuclease that cleaves the DNA if the specific sequence is not methylated. Each bacterium has its own restriction and modification enzymes, thus ensuring preservation of its own DNA but degradation of foreign DNA. Restriction enzymes have found wide application in the analysis of DNA and in recombinant DNA technology (see next chapter).

In addition to the restriction-modification mechanism, bacteria also possess enzymes that methylate specific adenine and cytosine residues in the DNA. In *E. coli,* the *dam* gene codes for a **deoxyadenosine methylase,** the *dcm* gene for a **deoxycytidine methylase.** These methylases serve to distinguish parental DNA from newly replicated DNA. This is possible because methylation of the new DNA strands occurs several minutes after replication. Thus, the two daughter duplexes emerging from the replication fork each have one methylated and one unmethylated strand. The *dam* methylase, for example, has an important function in mismatch repair (see next section). Other functions of methylated bases may include regulation of transcription and/or replication.

In eukaryotes, there are no restriction enzymes. However, postreplicative methylation of cytosine residues, found in the sequence C*pG, has been shown to affect the expression of certain genes (see later).

6.3.7 Regulation of DNA Replication

DNA replication in *E. coli* is initiated at a **unique origin** (*oriC*). The *oriC* sequence contains three AT-rich 13-mer tandem repeats and four 9-mer repeats, called dnaA boxes. The dnaA protein is required for initiation at this site. From 20 to 40 molecules of *dnaA protein* bind tightly to the four 9-mer repeats. This causes stepwise melting of the three adjacent 13-mer repeats, producing an open complex. The dnaB-dnaC complex then enters this melted region, creating the **prepriming complex.** The **dnaB protein** has two functions. First, it functions as a **helicase** that extends the initiation bubble, and second, it serves as a guide for DNA primase (dnaG protein), which synthesizes the RNA primer for subsequent DNA synthesis.

Replication of a linear chromosome by the discontinuous mechanism would create a problem at the molecule's ends unless special provisions are made to prevent loss of genetic information. The problem is that the 3′ ends of the DNA cannot be duplicated. Even if Okazaki pieces were initiated at the very end of the molecule, the subsequent removal of the RNA primer would leave a single-stranded region at the end. Thus, on subsequent replications, the molecule would become progressively shorter. Many bacteriophages circumvent this problem by carrying terminal repeats in their sequence that allow the formation of concatemers in which two replicated DNA strands are joined end to end. This allows completion of replication by 5′→3′ extension (Figure 6.21).

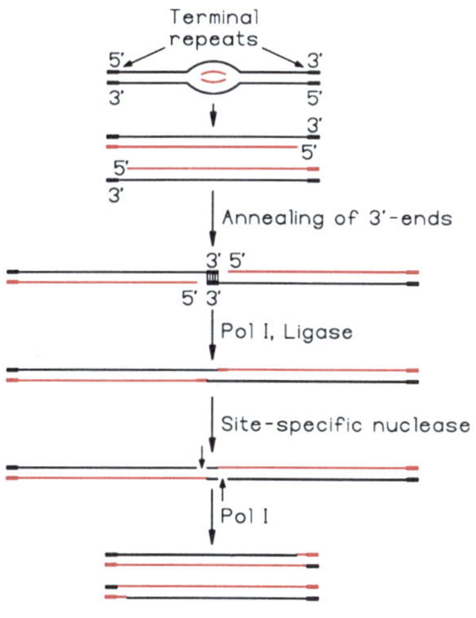

Figure 6.21.
Replication of ends by concatemer formation.

Eukaryotic chromosomes also carry terminal repeats, called **telo-**

meres. Telomeres are tandem repeats (30–50) of an oligonucleotide sequence that are added to the ends of chromosomes by **telomerase,** a new type of DNA polymerase. Telomerase is a ribonucleoprotein whose RNA component contains a sequence complementary to the telomeric tandem repeat ($1\frac{1}{2}$ times). The enzyme is capable of rebuilding the telomeres to their full lengths, thereby preventing the shortening of the telomeric repeats during subsequent replications.

In order to assure that the two copies of the bacterial chromosome are segregated into the two daughter cells, the origin of the bacterial chromosome attaches transiently to proteins in the bacterial membrane. This attachment occurs shortly after DNA replication has been initiated and ceases as soon as the new DNA becomes fully methylated. (The postreplicational methylation of DNA will be discussed in the section on DNA repair synthesis.)

Eukaryotic chromosomes are segregated through their **centromere,** a 250 nucleotide sequence (in yeast) to which proteins are tightly bound. During mitosis, the kinetochore region attaches to the microtubules of the mitotic spindle, which pulls the two daughter chromosomes to opposite poles before the cell divides.

6.4 GENETIC RECOMBINATION

Genetic recombination involves the exchange of DNA segments between related DNA molecules. It is the process that is mainly responsible for genetic variation and for evolution. It also allows the recovery of damaged DNA sequences by genetic exchange with a homologous chromosome. One distinguishes between **homologous (legitimate) recombination,** the most frequently occurring type of recombination, which is based on the genetic exchange between homologous sequences, and **nonhomologous (illegitimate) recombination,** which occurs rarely between unrelated sequences. In addition, there is site-specific recombination catalyzed by enzymes that recognize specific short DNA sequences without regard for extended homology. This type of recombination is found during the integration of some virus genomes into chromosomes, the rearrangement of immunoglobulin genes (see Chapter 9), and the transposition of mobile genetic elements.

In eukaryotes, recombination occurs primarily (but not exclusively) during the diplotene stage of meiosis. The crossing-over (chiasma) between homologous pairs of paternal and maternal chromosomes leads to the formation of recombinant chromosomes of mixed parental origin. Thus, if each chromosome carried a mutation at a different location, recombination between the sites of these two mutations would create both wild-type and double-mutant recombinants (Figure 6.22).

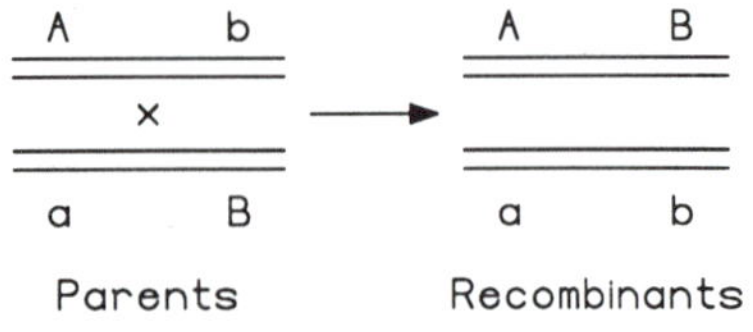

Figure 6.22.
Reciprocal crossing-over leading to the formation of wild-type (A,B) and double-mutant (a,b) recombinants.

6.4.1 The Molecular Mechanism of Recombination

Although the precise mechanism of recombination is still not completely understood, much has been learned during the last few years. Recombination involves the breakage and reunion of the participating DNA molecules. This had been shown by J. Weigle in a density-labeling experiment similar to the Meselson-Stahl experiment. Intermediates in the recombination process have been isolated. They consist of two DNA duplexes that are joined together at a crosslike junction, called the **Holliday structure.** The existence of such a structure was predicted from genetic experiments that revealed **heteroduplex regions** in the recombinant DNA molecules. (A heteroduplex is a DNA molecule in which the two strands are not completely complementary.)

Holliday structures probably arise through invasion of a double helix by an homologous single-stranded DNA tail from another DNA molecule. This is followed by nucleolytic cleavage of the displaced DNA strand and reciprocal base-pairing and ligation of this strand with the other DNA molecule (Figure 6.23). The Holliday intermediate is resolved again into two linear DNA molecules by nucleolytic cleavage. Depending upon the plane of cleavage, either two recombinant or two unchanged parental molecules are formed.

The **crossing-over** point in the recombination intermediate usually migrates before the structure is resolved. This produces heteroduplex recombinant molecules in which the two strands in the DNA duplex are derived from different parents. Replication usually resolves these heteroduplex structures. However, repair synthesis may eliminate one or the other parental allele in the heteroduplex, thus leading to unequal representation of parental alleles in the recombinants. Such an event is known by geneticists as **gene conversion.**

Under rare circumstances, recombination may occur between nonhomologous regions of DNA. This illegitimate recombination, or unequal crossing-over, may lead to the creation of gene duplications and deletions as illustrated in Figure 6.24.

Many enzymes have been shown to contain stretches of similar or identical amino acid sequence. They are thought to have evolved from the same ancestral gene by the process of gene duplication followed by independent mutational changes.

Figure 6.23.
Model of genetic recombination. After Potter H, Dressler D. *Proc Natl Acad Sci USA*. 1977;74:4168.

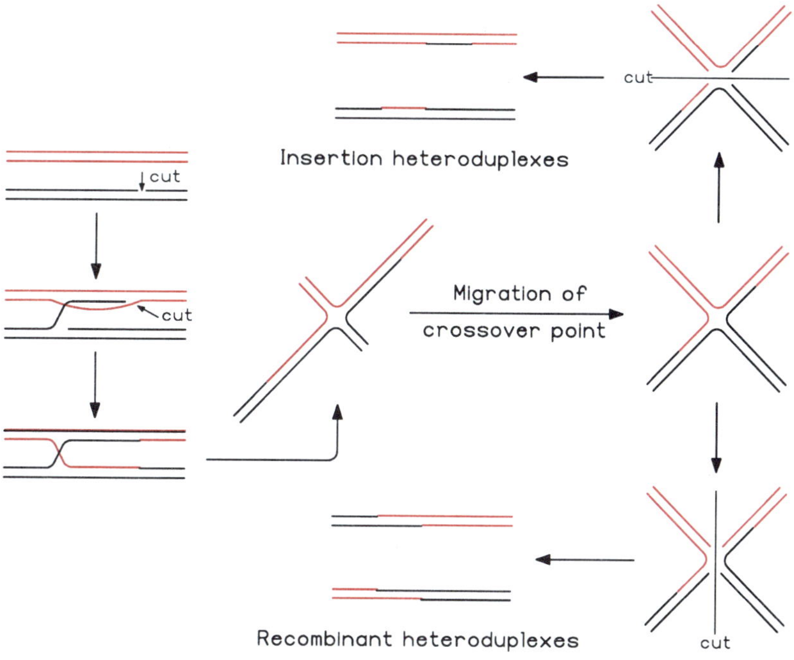

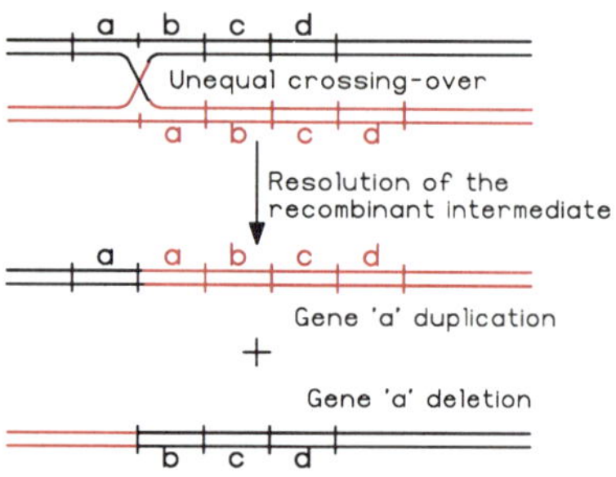

Figure 6.24.
Unequal crossing-over producing gene duplication and deletion.

6.4.2 Enzymes Involved in Genetic Recombination

Many enzymes involved in DNA replication also participate in genetic recombination. This includes DNA polymerase, DNA ligase, single-strand-binding protein, and topoisomerase I. In addition, recombination in *E. coli* requires participation of the **recA protein.** Mutants of the *recA* gene are deficient in genetic recombination. The recA protein binds cooperatively to single-stranded DNA, each protein molecule covering five nucleotides. The resulting DNA–recA-protein filament, in the presence of ATP, then invades an homologous double helix, where the single-stranded DNA anneals to its complementary strand, displacing the resident strand (Figure 6.25). This displacement reaction proceeds in a 5′→3′ direction, with respect to the invading single strand, and is driven by the hydrolysis of ATP. Proteins similar to recA probably catalyze recombination in eukaryotes. A protein with recA-like properties has been isolated from the fungus *Ustilago.*

The requirement of recA protein for single-stranded DNA explains why genetic recombination occurs most frequently at single-stranded gaps in the double helix or at the ends of DNA molecules. Nicks and gaps in double-stranded DNA are the sites where recombinational strand exchanges catalyzed by recA are initiated. Such sites are generated by the **recBCD complex.** This large enzyme complex, consisting of three polypeptides coded by the genes *recB, recC,* and *recD,* has helicase activity and unwinds double-stranded DNA from one end driven by the hydrolysis of ATP. The single DNA strands generated by this action reanneal more slowly than the rate at which the duplex is unwound, thus generating two single-stranded loops. The process comes to a halt at a so-called **Chi site,** where the recBCD enzyme makes a cut, producing a single-stranded tail and a gap in the duplex, thus generating substrates for recA protein. Chi sites have the sequence 5′-GCTGGTGG-3′. The recBCD enzyme recognizes this sequence coming from its 3′ end and makes a cut four to six nucleotides before the site (Figure 6.26).

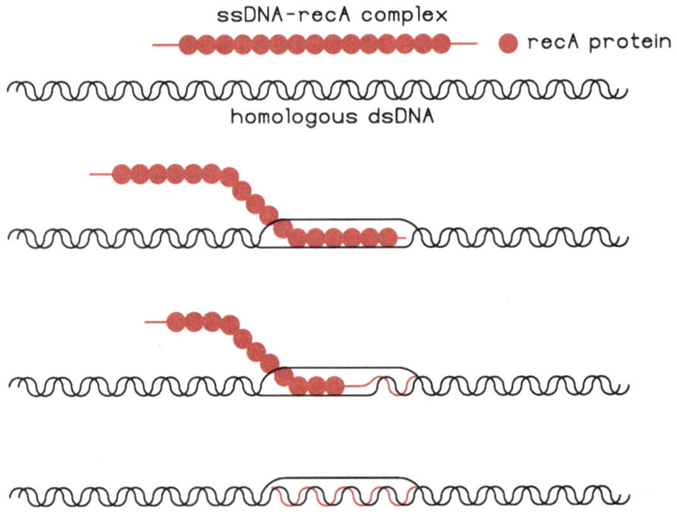

Figure 6.25.
Invasion of a DNA duplex by recA-bound single-stranded DNA.

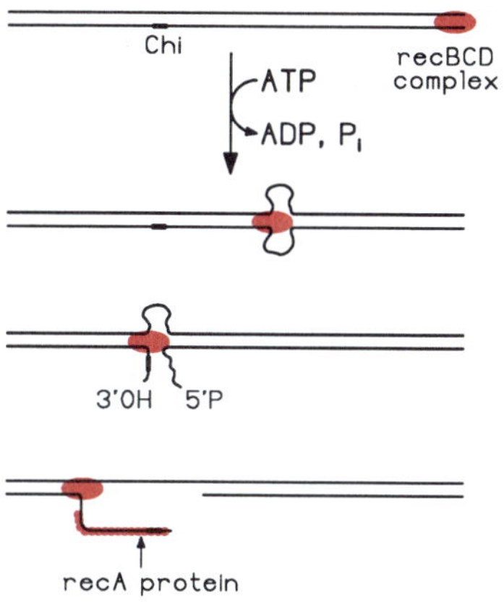

Figure 6.26.
Generation of single-stranded DNA tails by the recBCD enzyme. Genetic recombination can be initiated at such sites by the recA protein.

6.4.3 Site-Specific Recombination in Bacteriophage Lambda

Infection of an *E. coli* cell with bacteriophage λ can have two completely different results: the phage either multiplies, producing many progeny phage particles and killing the bacterium in the process, or the phage genome is integrated, via genetic recombination, into a specific site on the *E. coli* chromosome and is then passed on to progeny cells as a silent **prophage.** In the prophage state, only one gene of the phage λ genome is expressed, the repressor gene *cI.* The repressor protein shuts down all other prophage genes (see the discussion on repressors in Section 6.8). A bacterium carrying such a prophage is called a **lysogen;** it can be induced, by UV irradiation or damage to its DNA, to excise the phage genome and produce progeny phage particles.

As observed in homologous recombination, the crossing-over event between the phage and the bacterial genomes is reciprocal, but the nucleotide sequences involved in the genetic exchange are only partly identical. The sites on the bacterial and phage genomes involved in this site-specific recombination are called *attB* and *attP,* respectively. Integration is catalyzed by a phage-encoded enzyme, **integrase,** together with a bacterial protein IHF (for integration host factor). The reaction can be reproduced in vitro with two smaller circular plasmid DNA molecules that carry the *attB* and *attP* sites. No other proteins are required, indicating that all strand cutting and religation is done by the integrase complex without any requirement for ATP. Figure 6.27 illustrates the integration process and compares the bacterial and phage attachment sites.

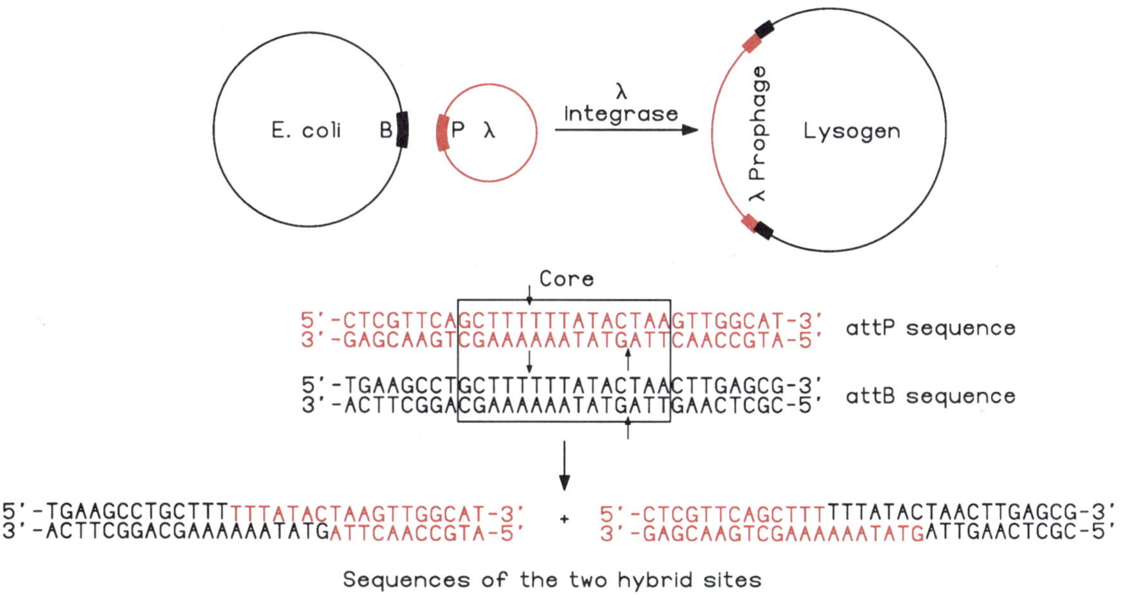

Figure 6.27.
Integration of bacteriophage λ DNA into the *E. coli* chromosome by site-specific recombination. The nucleotide sequences of the attachment sites are shown underneath.

Only a core sequence of 15 nucleotides is shared by the recombining sites. The integrase protein has been shown to bind to sequences more than 100 nucleotides to either side of this core sequence.

6.4.4 Transposons

Transposons are mobile genetic elements that catalyze their own transposition within and between chromosomes. Transposons were first discovered as sequences on **R-factor plasmids** that carry **antibiotic resistance genes.** There are two types of transposons: simple and complex. Simple transposons can be as short as 768 nucleotide pairs. They are also called **insertion sequences (IS elements)** because of their ability to inactivate a gene by insertion into its coding sequence. Such an insertion mutation is often **polar;** that is, it affects the expression of other genes downstream from the insertion site. This effect is caused by the presence of promoter (initiation site) or terminator sequences for transcription within the insertion sequence. Depending on the direction of insertion, the same sequence may express either of these two regulatory functions.

All transposons carry **inverted repeats** at their ends. These are 20–40 nucleotides that are repeated in reverse order at the two ends of the transposon. In addition, when a transposon has inserted into a **target sequence,** several nucleotides of the target sequence are found **directly repeated** on each end of the transposon. The number of repeated nucleotides is specific for each type of transposon (Table 6.4).

While insertion sequences contain just the information that is required for their transposition, **complex transposons** carry several genes, usually one or more **antibiotic resistance genes** and the gene for a **transposase,** an enzyme that catalyzes their transposition. Many complex transposons have insertion sequences at both ends.

The mechanism of transposition is still poorly understood. A possible mechanism for transposition is illustrated in Figure 6.28 on the example of Tn3, a transposon that carries genes for a β-lactamase (degrades penicillin), a transposase and a resolvase, an enzyme that resolves the cointegrate intermediate.

Transposition occurs in the absence of recA protein. It can occur into any sequence, although certain sequences seem to be preferred. Transposons can cause deletions, inversions, and duplications in addition to transpositions. The basic model shown in Figure 6.28 allows all these events.

Transposons are also found in many eukaryotes. They include the **Ty**

Table 6.4. Some Transposons of *E. coli*

Transposon	Size (bp)	Target Sequence Repeat (bp)	Resistance to
Insertion sequences			
IS1	768	9	—
IS2	1327	5	—
IS4	1426	12	—
Complex transposons			
Tn3	4957	5	Penicillin
Tn5	5700	9 (IS50 at end)	Kanamycin
Tn10	9300	9 (IS10 at end)	Chloramphenicol
Tn2571	23000	9 (IS1 at end)	Hg, streptomycin, and others

Figure 6.28.
Model for transposition of Tn3. At first, transposase makes staggered cuts at the target site (this later creates the target site repeats) and at the ends of the transposon. The ends of the transposon are then ligated to the ends created in the target sequence. DNA polymerase copies the two transposon strands beginning from the 3′ ends of the target DNA. This generates a cointegrate intermediate that is resolved by a site-specific recombinational event catalyzed by resolvase.

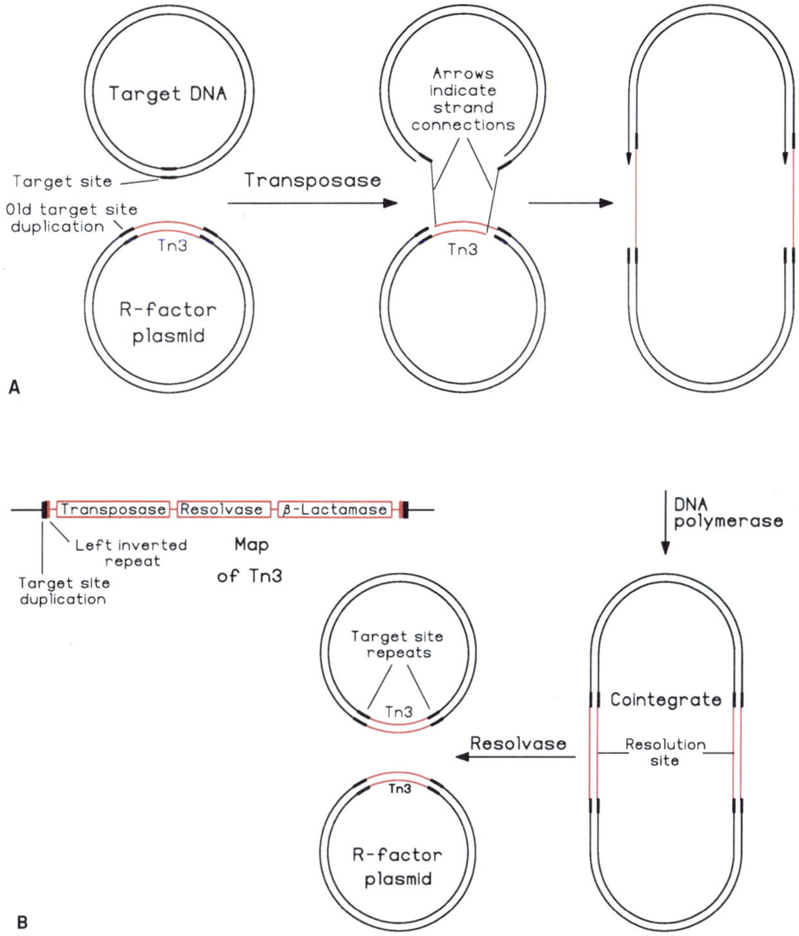

elements in yeast, the **copia** and **P elements** in *Drosophila,* and the widely distributed **retroviruses.** However, the mechanism of transposition is, at least in some eukaryotic transposons, very different. Transposition occurs via transcription of the transposon into RNA, which is subsequently copied by reverse transcriptase into a DNA copy that is inserted into the new site.

6.5 MUTATION, DNA DAMAGE, AND ITS REPAIR

DNA is a remarkably stable molecule. Recently, a piece of DNA from a 2,400-year-old mummy was isolated and sequenced. Nevertheless, the DNA in chromosomes is constantly subjected to damage caused by chemicals, heat, or radiation. Changes in DNA structure may also occur spontaneously. Such alterations are constantly being repaired by a variety of different repair processes. Sometimes the nucleotide sequence in the DNA cannot be restored to its original state, or the repair process produces an altered sequence; then a mutation arises.

6.5.1 Types of Mutation

Several types of mutation can be distinguished. The change of a single base pair is called a **point mutation.** If this change creates a nonsense codon, the mutation is a **nonsense mutation;** if it creates a codon for another amino acid, it is called a **missense mutation.** A missense mutation that destroys the function of a gene completely is called a **null mutation.** However, many missense mutations have more subtle effects upon gene function. For example, they may alter the function of the protein in such a way that the protein is only active at certain temperatures. Such **temperature-sensitive mutations** are valuable tools since they allow the study of the effects of a certain gene product on the organism.

There are two kinds of point mutations. If a purine base is replaced by another purine, or a pyrimidine by another pyrimidine, the mutation is called a **transition mutation.** Replacement of a purine by a pyrimidine, or vice versa, is called a **transversion.** As a general rule, point mutations are reversible; that is, the altered nucleotide pair can revert to the original situation by a mutational event similar to that which caused the mutation in the first place.

Loss of one or more nucleotide pairs is called a **deletion mutation.** Deletions can arise from an unequal crossing-over event (see Figure 6.24). In contrast to point mutations, they are not reversible (for obvious reasons).

The **insertion** or deletion of any number of nucleotide pairs that is not a multiple of three causes a shift in the reading frame of the genetic code. Such mutations are, therefore, called **frameshift mutations.** A frameshift mutation often generates a nonsense codon in the reading frame, thus causing premature termination of protein synthesis.

A frameshift mutation caused by the single insertion of a base pair can be "reverted," or more accurately "suppressed," by either a single base pair deletion or two additional single base pair insertions in the vicinity of the original insertion. This reestablishes the reading frame and may give rise to a functional protein, if the new amino acids, specified by the altered nucleotide sequence between the mutational sites, are compatible with protein function.

6.5.2 Origin of Mutations

Many so-called **spontaneous mutations** arise from errors made during DNA replication. Although DNA replication is an extremely accurate process, about one mistake is made for every 10^9 incorporated base pairs. This high degree of accuracy is achieved partly through the $3' \rightarrow 5'$ exonuclease activity of the DNA polymerase that serves a proofreading function. DNA polymerase holoenzyme with decreased exonuclease activity exhibits an increased rate of misincorporation of nucleotides. The proofreading nuclease activity is carried on the ϵ subunit of polymerase III. Mutations in the gene for ϵ *(mutD)* are mutagenic. A further increase in accuracy is achieved by the mismatch repair system, which screens both daughter duplexes that emerge from the replication fork for mismatched bases (see later).

Runs of identical bases in the DNA may cause the DNA polymerase to slip and make either the template or the growing strand loop out, thereby causing the creation of deletions or additions. This generates frameshift mutations.

Spontaneous mutations also arise from the deamination of cytosine to uracil, which in subsequent replications would be replaced by thymine since it pairs with adenine instead of guanine. In the average human cell, about 100 such **deaminations** occur per day. Another cause for mutations is the **depurination** of bases, which occurs about 5,000 times per day per human cell. Both of these alterations are efficiently repaired (see later).

In addition to this spontaneous mutagenesis, mutations are generated by many **chemical mutagens. Nitrous acid,** for example, a chemical used in the preservation of meat, deaminates adenine to hypoxanthine (which pairs with cytosine) and cytosine to uracil, thus creating mutations. (Guanine is deaminated to xanthine, which still pairs with cytosine. Thus, no mutations result from this deamination.)

Base analogs, such as **5-bromouracil** or **2-aminopurine,** are incorporated into DNA in place of the normal bases. However, the base analog has a tendency to assume a different **tautomeric form** with altered base-pairing properties. In the case of 5-bromouracil, for example, the normal tautomer base-pairs with adenine and, therefore, is incorporated into DNA in the place of thymine. The other tautomer of 5-bromouracil, however, base-pairs with guanine (Figure 6.29). This lactam-lactim tautomerism (often incorrectly called keto-enol tautomerism) is also observed in normal bases. This may be one of the reasons for misincorporation of nucleotides during DNA replication. However, the rare tautomeric form is assumed much more frequently by 5-bromouracil than by thymine. Incorporation of base analogs into DNA causes transition mutations.

Heterocyclic aromatic compounds, such as **ethidium bromide** or **acridine,** induce **frameshift mutations.** These compounds **intercalate** (see Figure 6.11) and thereby distort the double helix, thus causing errors during replication.

The most powerful chemical mutagens are **alkylating reagents,** such as ***N*-methyl-*N*′-nitro-*N*-nitrosoguanidine (MNNG),** which al-

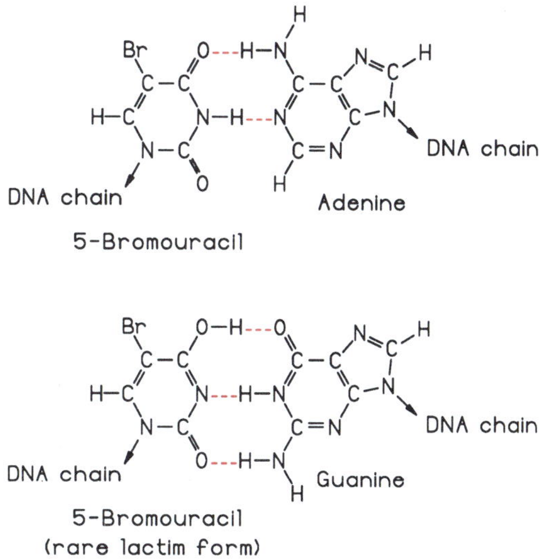

Figure 6.29
Base-pairing of the two tautomeric forms of 5-bromouracil.

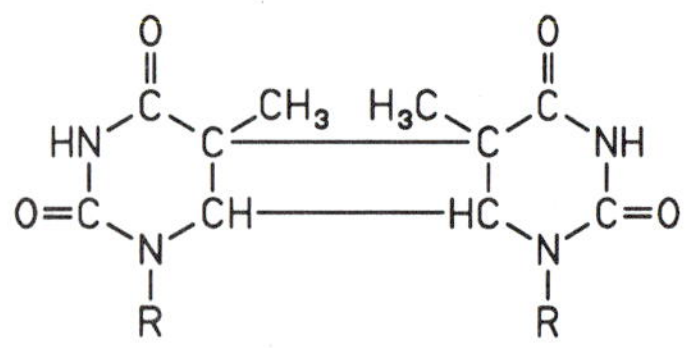

Figure 6.30.
Structure of a thymine dimer.

kylate guanine residues at O^6 or N^7 and adenine residues at N^3. O^6-methylguanine base-pairs with both C and T, thus causing transition mutations. N^3-adenine is a cell-killing lesion if it is not repaired.

Environmental factors also damage DNA and lead to mutations. Heat accelerates deamination of bases. **UV irradiation** leads to the formation of **pyrimidine dimers** from two adjacent pyrimidines in the same DNA strand (Figure 6.30). The dimerized pyrimidines form a cyclobutane ring. Another result of UV irradiation is the 6-4-photoproduct in which two pyrimidines are joined between carbon 6 of the 5′-pyrimidine and carbon 4 of the 3′-pyrimidine. Pyrimidine dimers cause the DNA polymerase to skip over the damaged region, leaving a single-stranded segment in the DNA. Repair of such dimers may generate mutations. Ionizing radiation causes ring opening of bases and single- and double-strand breaks of the DNA.

Crosslinks between the two strands of the double helix are produced by a variety of different agents, including **mitomycin C, formaldehyde, psoralen** (which intercalates and is activated by light), nitrous acid, and others. The repair of such lesions often creates mutations.

Since most carcinogens are mutagens, a simple test has been designed to test the mutagenicity of chemicals used in every-day life. In the **Ames test** (developed by Bruce Ames), a chemical is tested for its ability to revert a *Salmonella* strain, carrying a mutation in one of the genes for histidine biosynthesis, to wild-type. This test can be performed very quickly by plating the bacteria in the presence of the chemical on a growth medium lacking histidine. Only the reverted bacteria will grow. The number of revertant colonies can be counted and compared with the low level of spontaneous revertants occurring without exposure to the chemical. Since many chemicals only act as carcinogens (mutagens) after being modified in the liver (see Chapter 10), incubation of the chemical with a mammalian liver extract usually precedes the Ames test.

6.5.3 DNA Repair Processes

In order to maintain the integrity of their genes, cells have evolved a large number of repair processes. These can be divided into two classes:

1. **Direct repair** processes that simply reverse the chemical damage without breaking the phosphodiester bond.
2. **Excision repair** in which the damaged base or nucleotide is excised and replaced. The second process makes use of the redundant genetic information present in the complementary strand.

One type of direct repair, **photoreactivation,** is expressed in single-cell organisms such as bacteria and yeast. This repair process requires visible light to remove pyrimidine dimers. It is catalyzed by **photolyases,** enzymes that bind to the cyclobutane ring of a pyrimidine dimer, are activated by light, and then simply restore the original pyrimidines.

A second type of direct repair is the removal of O^6-methylguanine from DNA that had been subjected to a methylating chemical mutagen such as MNNG. The methyl group is directly transferred to a cysteine residue of the enzyme, and an unmodified guanine is left in the DNA. The enzyme **O^6-methylguanine-DNA methyltransferase** is irreversibly inactivated in this process. This repair process is inducible; that is, the enzyme levels increase when cells are exposed to low levels of alkylating agents. It occurs in both prokaryotes and eukaryotes.

Base excision repair occurs without breaking the DNA chain and is catalyzed by the DNA glycosylases. **Uracil-DNA glycosylase** removes uracil residues that arise from the deamination of cytosine or that have been mistakenly incorporated into DNA. The reaction leaves a deoxyribose in the DNA chain without a base attached to it that is later excised and repaired (see below). Uracil-DNA glycosylase mutants (*ung*$^-$) exhibit an increased rate of mutation because uracil residues that arise from the deamination of cytosine are not removed efficiently. Thus, transition mutations arise during subsequent DNA replication. (Note that deamination of 5-methylcytosine, which produces thymine, cannot be recognized by this repair system. Thus, regions of the DNA where cytosine is methylated are hot spots for spontaneous mutations.) Other **DNA glycosylases** include hypoxanthine-DNA glycosylase, 3-methyladenine-DNA glycosylase, and a large number of other glycosylases, each specific for a different type of damaged base.

Excision of uracil residues mistakenly incorporated into DNA (about one deoxyuridine monophosphate [dUMP] is incorporated every 2,000 nucleotides) contributes to the number of Okazaki fragments. In deoxyuridine triphosphatase–deficient (**dUTPase**-deficient; *dut*$^-$) mutants, the average size of an Okazaki fragment is decreased, reflecting the increased incorporation of dUMP. This type of Okazaki fragment arises from both strands emerging from the replication fork.

The removal of bases by DNA glycosylases leaves apurinic/apyrimidinic sites **(AP sites)** in the DNA that are recognized by **AP endonucleases.** Several AP endonucleases with different properties exist. The major AP endonuclease cleaves the phosphodiester bond 5′ to the AP site. Most likely, a deoxyribose phosphatase then removes the deoxyribose from the 5′ terminus, and DNA polymerase I replaces a segment of DNA with a newly synthesized piece using the intact strand as template.

In nucleotide excision repair, damaged bases are removed from the DNA in the form of an oligonucleotide that is cut out, and the resulting gap is repaired by polymerase I and DNA ligase. The substrates for this type of repair are pyrimidine dimers, 6-4 photoproducts, as well as base adducts of psoralen, mitomycin C, benzopyrene, and other chemicals. The enzyme responsible for this excision is the **ABC excinuclease complex.** (The genes for the three subunits of this complex are *uvrA, uvrB,* and *uvrC.*) The ABC excinuclease cuts the eighth phosphodiester bond on the 5′ side of the damaged nucleotide and the fourth or fifth bond on the 3′ side, producing a fragment 12 or 13 nucleotides long. There is some evidence that helicase II (*uvrD*) is required for the release of the oligomer. The gap is then repaired by polymerase I.

Xeroderma pigmentosum (XP) is a rare human skin disease characterized clinically by extreme sensitivity to light. It is due to an excision repair deficiency that is transmitted genetically as an autosomal recessive trait. The homozygote is extremely sensitive to UV irradiation

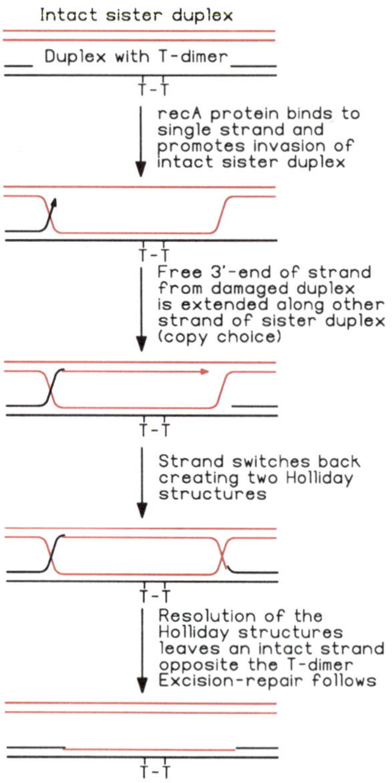

Figure 6.31.
Model for recombinational repair of a thymine dimer.

and easily develops skin cancer. Mutations in at least nine different genes can produce the XP phenotype.

Mismatched bases that have escaped the proofreading activity of the DNA polymerase can be repaired by a **mismatch repair system** that scans newly replicated DNA for unpaired nucleotides. The system appears to recognize both a mismatched base pair and an unmethylated d(GATC) sequence nearby. Since d(GATC) sites are methylated by the ***dam* methylase** only some time after replication, the mismatch repair enzyme system is capable of distinguishing between parental and new DNA and thus will selectively repair any mismatch according to the information in the parental strand. (Note that a thymine derived from the deamination of 5-methylcytosine will be recognized and repaired by this system.)

When a thymine dimer or other nucleotide adduct is encountered during DNA replication, the polymerase stops and reinitiates about 1,000 nucleotides beyond the adduct, leaving a gap in the DNA. This gap cannot be filled by the normal excision repair process because there is no intact template strand left for the DNA polymerase to copy. However, the information is still present in the other daughter duplex, and repair can be achieved by a recombinational event. **Recombinational repair** requires the participation of the recA protein. A possible model for this repair mechanism is depicted in Figure 6.31.

A combination of excision repair and recombinational repair is also capable of repairing crosslinks between the two DNA strands. A likely scenario for this repair is that excinuclease first incises on both sides of the crosslinked base. The excised oligomer is then displaced by recA protein, which facilitates strand invasion by the sister duplex. Finally, excinuclease cuts out the other DNA strand with the crosslinked oligonucleotide still attached to it, and the gap is filled by DNA polymerase I.

When DNA is damaged very severely, excision repair of two neighboring sites on opposite strands may lead to loss of the original genetic information. To cope with this situation, cells have evolved an inducible **error-prone repair** system that does not require a template. Although it creates mutations, it allows a cell to survive when otherwise unrepairable lesions accumulate in its DNA. Together with this repair system, a large number of other genes is induced by exposure of the DNA to damaging agents (e.g., UV). These so-called **SOS genes** are listed in Table 6.5.

Table 6.5. List of Some SOS Genes

Genes	Function of Protein
uvrA,B	Subunits of excinuclease
uvrD	Helicase II
umuC,D	Required for error-prone repair
sulA,B	Inhibit cell division, possibly to allow repair
lexA	Repressor for all SOS genes
recA	Recombination enzyme, accelerates autocatalytic cleavage of lexA protein
recN,Q	Other recombination-promoting proteins
dinA,B,D,F	Damage-inducible genes of unknown function

Single-stranded DNA (probably gaps) activates the recA protein, which binds the lexA repressor and thereby accelerates its autocatalytic cleavage and inactivation (see Section 6.8 on repressor function). This turns on all SOS genes, which include some of the excinuclease genes. Excision of the damaged segment in one of the two DNA strands now leaves a gap that is filled in by the error-prone repair enzymes (Figure 6.32). It is not clear at this time how this is accomplished. Either the umuC,D proteins represent subunits of a new DNA polymerase that may insert a nucleotide opposite a damaged template or they inhibit the proofreading function of the DNA polymerase. SOS repair is the main reason why UV radiation causes mutations.

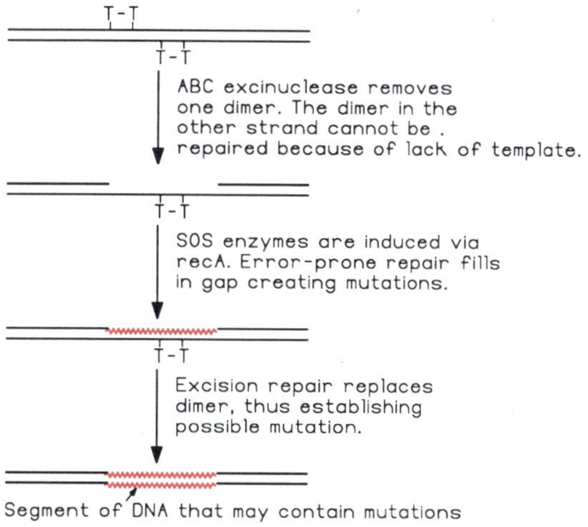

Figure 6.32.
Error-prone repair of two thymine dimers located on opposite strands of the DNA duplex.

6.6 RNA METABOLISM

The genetic information stored in the DNA can only be expressed through transcription into RNA. There are some viruses, however, whose genomes consist of RNA rather than DNA. Their properties will be discussed at the end of this chapter. In many respects, RNA is similar to DNA. However, chemically it is much less stable. In addition, it can fold into three-dimensional structures that are not too different from those of proteins. More and more pure RNA molecules are being discovered that possess enzymatic activities (see ribosomes, later).

6.6.1 Structure and Properties of RNA

Like DNA, RNA is a linear polymer of nucleotides linked together by 3′,5′-phosphodiester bonds. In contrast to DNA, however, the nucleotides in RNA are ribonucleotides, and the base uracil is found instead of thymine (Figure 6.33).

The presence of a 2′-hydroxyl group in the ribose moiety of RNA makes the molecule much more susceptible to **hydrolysis.** RNA is quickly hydrolyzed to 3′-nucleotides by **alkali.**

RNA is usually single-stranded (the exceptions are the genomes of some double-stranded RNA viruses). However, it can form short intramolecular double-helical regions that can give the molecule a three-

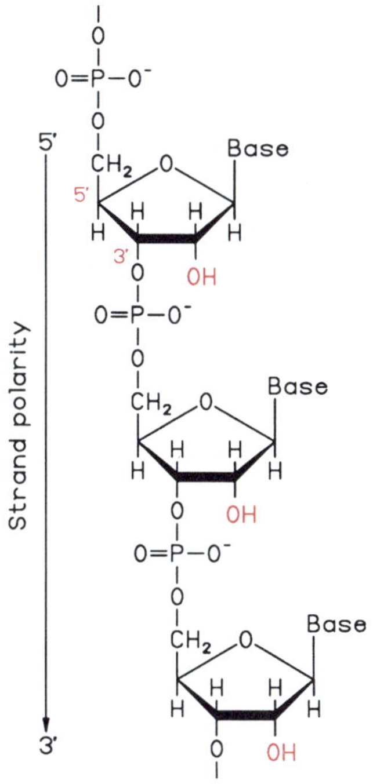

Figure 6.33.
The structure of RNA.

dimensional shape. In addition to AU and GC pairs, weaker GU pairs also form. There are several types of RNA, all of which are transcribed from DNA:

1. **Messenger RNA (mRNA),** as the name indicates, carries the information for the amino acid sequence of a polypeptide chain to the ribosomes where protein synthesis takes place. mRNA is usually metabolically unstable.
2. **Transfer RNA (tRNA)** is a relatively small molecule containing only about 80 nucleotides. It has a three-dimensional structure and is known as the adaptor molecule in protein synthesis (see Figure 6.2).
3. **Ribosomal RNA (rRNA)** is part of the ribosome, a complex ribonucleoprotein particle that functions in the synthesis of polypeptides. Both tRNA and rRNA are metabolically stable.
4. **Small nuclear RNA (snRNA)** and **small cytoplasmic RNA (scRNA)** are found in complexes with proteins that are called small nuclear ribonucleoprotein particles and small cytoplasmic ribonucleoprotein particles (sometimes referred to as "snurps" and "scurps") in the nucleus and cytoplasm, respectively, of eukaryotes. They function in the processing of mRNA and nascent proteins (see later).

A cell contains about 10 times more RNA than DNA, mostly in the form of rRNA.

6.6.2 Messenger RNA Synthesis in Prokaryotes

In prokaryotes, mRNA synthesis accounts for about 50% of total cellular RNA synthesis. mRNA is a short-lived molecule with a half-life of only a few minutes. For this reason it represents only a small percentage of the total RNA content of the cell, the remaining RNA being ribosomal and transfer RNA.

During transcription of a particular gene, only one strand of the DNA double helix, the antisense or template strand, is copied into RNA. Different genes, however, may use different strands of the DNA as their sense strand. The direction of transcription is 5′ to 3′, which means that the template strand is read 3′ to 5′. In contrast to DNA polymerases, RNA polymerases do not require a primer. The first nucleotide in a growing RNA chain retains its triphosphate group, and chain elongation begins at the 3′-hydroxyl group of this nucleotide.

During transcription, RNA polymerase partially unwinds the DNA helix. As the enzyme moves along the DNA, a region of about 17 base pairs is kept unwound. Since nucleotides are incorporated into RNA based on their base-pairing with the template DNA strand, a short 12 base pair–long DNA-RNA hybrid transiently forms directly behind the site of polymerization. Then the RNA strand leaves the template, and the two DNA strands rejoin (Figure 6.34).

E. coli has a single **RNA polymerase** that is responsible for all transcription in the cell. The enzyme consists of five subunits that are listed in Table 6.6. The core enzyme $\beta\beta'\alpha_2$ is capable of copying both strands

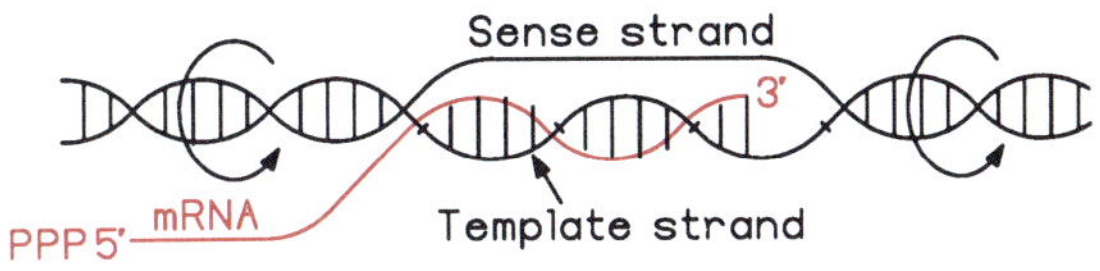

Figure 6.34.
Model of a transcription complex during the elongation phase of transcription.

of the DNA but lacks specificity. Addition of the σ factor makes the RNA polymerase **(holoenzyme)** recognize promoters, specific starting signals in the DNA. After initiation of an RNA chain, the σ factor dissociates again from the polymerase.

The RNA polymerase holoenzyme is capable of sliding along the DNA helix in search of promoters without unwinding it. As soon as a promoter site is encountered, the **closed promoter complex** changes into an open promoter complex in which the DNA duplex is partially denatured (see Figure 6.34). This opening of the double helix is usually facilitated by **negative supercoiling** of the DNA. For the transcription of some genes, however, negative supercoiling is inhibitory. The gene for DNA gyrase is such a gene. Perhaps this feature is used to regulate the degree of overall negative supercoiling of the *E. coli* chromosome.

Comparison of different prokaryotic promoters revealed two common sequences located about 10 and 35 nucleotides upstream of the transcription start site. Although these sequences were not identical for all promoters, they exhibited enough homology that a consensus sequence could be derived:

−10 consensus sequence: TATAAT (TATA box)

−35 consensus sequence: TTGACA

The open complex is thought to form at the **TATA box** facilitated by the weaker hydrogen bonds between As and Ts. Most TATA boxes are flanked by regions of high G + C content.

Only those prokaryotic promoters that are recognized by σ have these sequences. In addition to the regular σ70 factor, several other σ factors can direct RNA polymerase to different sets of genes. For example, in response to heat shock, *E. coli* expresses a σ32 protein, and after infection with bacteriophage T4 a σ55 is synthesized. Thus, σ factors provide a means of regulating gene expression.

In addition to the two promoter consensus sequences, many genes have additional upstream sequences (up to 150 nucleotides 5′ to the start site) that are required for efficient transcription. Some of these sites are recognized by activation factors for RNA polymerase, others might provide attachment sites for topoisomerases.

Table 6.6. Subunits and Transcription Factors of *E. coli* RNA Polymerase

Subunit	Gene	Size (kd)	Number per Holoenzyme	Function
β	*rpoB*	151	1	RNA polymerization
β′	*rpoC*	155	1	Binds to template
α	*rpoA*	37	2	Required for polymerase assembly
σ	*rpoD*	70	1	Binds to promoter

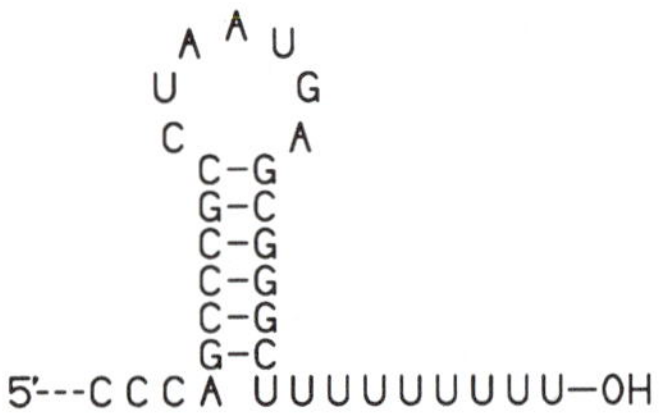

Figure 6.35.
Structure of a ρ-independent termination site at the 3′ end of the attenuated trp mRNA. After Yanofsky. *Nature.* 1981; 289:751.

Most prokaryotic mRNA is initiated with purines as the first nucleotide. After initiation, the σ factor dissociates from the RNA polymerase, and the core enzyme, which is bound more tightly to DNA than the holoenzyme, continues the elongation process of transcription.

Transcription is terminated at **termination sites.** There are two types of transcription termination sites in bacteria. One type consists of a nucleotide sequence that allows the formation of a **base-paired hairpin loop** in the mRNA. The hairpin is followed by a **stretch of U residues** (Figure 6.35). A possible scenario for transcription termination is that the formation of the hairpin loop prematurely pulls out several nucleotides from the transcription complex. The few bases that remain base-paired to the template do not provide enough anchorage, especially if they are held by the weaker A—U hydrogen bonds, so that the transcription complex dissociates.

The other transcription termination site, called *rut* (for ρ utilization), is recognized by a specific termination factor, called ρ. The sites recognized by ρ are different from those described above. The ρ-dependent sites lack the stretch of Us. ρ is a hexameric protein (46 kd subunits) that hydrolyzes ATP while interacting with the nascent mRNA strand. The current model for ρ-mediated transcription termination is that the nascent RNA strand is wound around the ρ hexamer and is thereby pulled away from the DNA template and the RNA polymerase. There is no obvious consensus sequence for *rut* sites except that they are rich in cytidine but poor in guanosine nucleotides.

ρ-independent sites are located exclusively at the ends of structural genes, whereas *rut* sites are also found within genes. It is thought that such sites become transcription termination sites when translation slows down or stops. This would allow the premature termination of unused transcripts. When translation is active, ribosomes follow the RNA polymerase rather closely. This would leave too little time for the ρ factor to recognize and bind to *rut* sites before the ribosome covers them, and transcription would continue.

A specific inhibitor of transcription in both eukaryotes and prokaryotes is **actinomycin D** (Figure 6.36). This antibiotic intercalates into the DNA, which blocks transcription but has little effect upon DNA replication. Another antibiotic, **rifamycin B,** inhibits only initiation of transcription but has no effect upon elongation. Mutants resistant to rifamycin have an altered β subunit.

Figure 6.36.
The structures of two inhibitors of transcription.

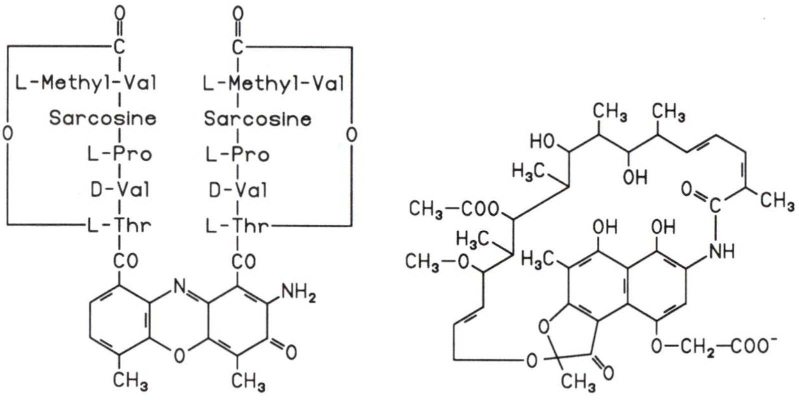

Rifamycin B

Actinomycin D

The synthesis of prokaryotic tRNA and rRNA will be discussed after a description of eukaryotic mRNA synthesis.

6.6.3 Messenger RNA Synthesis in Eukaryotes

In eukaryotes, all RNA is synthesized in the nucleus. mRNA is synthesized in a precursor form, called **heterogeneous nuclear RNA (hnRNA),** of which only about 20% later appears in mRNA. The cell uses three different types of RNA polymerase for the synthesis of the different types of RNA. **RNA polymerase I** transcribes most ribosomal RNA genes, **polymerase II** synthesizes precursors of mRNA and some snRNA and scRNA species, and **polymerase III** synthesizes tRNA, 5S rRNA, and the remainder of snRNAs and scRNAs.

The polymerases can be distinguished by their sensitivity to the mushroom poison α-amanitin. Polymerase I is resistant to this poison, polymerase II is inhibited at very low concentrations, and polymerase III is inhibited only at high concentrations.

Eukaryotic RNA polymerase II is considerably more complex than *E. coli* RNA polymerase. The yeast enzyme, which is best known, consists of 10 polypeptides with molecular weights ranging from 10 to 220 kd. The three largest subunits, RPB1, RPB2, and RPB3, are essential for yeast viability and are related to the prokaryotic core RNA polymerase subunits β′, β, and α, respectively. Parts of RPB4 resemble the prokaryotic σ subunit. Subunits RPB5, RPB6, and RPB8 are essential components in all three eukaryotic RNA polymerases.

The largest subunit, RPB1, has a long tandem repeat of the sequence Ser-Pro-Thr-Ser-Pro-Ser-Tyr at its carboxyl terminus. Yeast RNA polymerase carries 26 repeats, whereas mammalian RNA polymerases carry 52 repeats. The functional role of this heptapeptide repeat is still unclear. It has been suggested that the tyrosine residues in the repeats intercalate into the DNA.

Common eukaryotic promoter elements include a **TATA box** at position −25 and a **CAAT box** or **GC box,** usually present in multiple copies, between positions −40 and −110 (Table 6.7). GC boxes are often found in constitutively expressed genes coding for cellular "house-keeping" proteins. Some genes lack a TATA box completely; in these cases, transcription is initiated at multiple sites.

RNA polymerase II requires the participation of various transcription activation factors for initiating transcription at promoters. One such factor is **SP1,** which recognizes and binds to GC boxes. There are also factors that bind to TATA boxes and CAAT boxes. In addition, **upstream**

Table 6.7. Eukaryotic Promoter Elements

Name	Location	Consensus Sequence
TATA box	−40	TATA(A,T)T
CAAT box	−40 to −110	GG(C,T)CAA(A,T)CT
GC box	−40 to −110	GGGCGG or CCGCCC

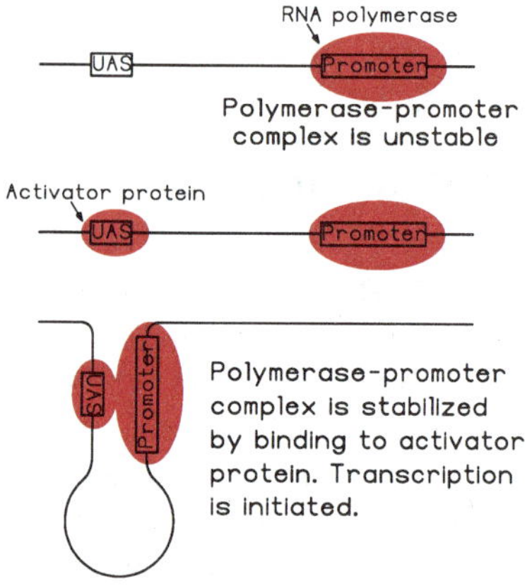

Figure 6.37.
Model for the activation of eukaryotic promoters.

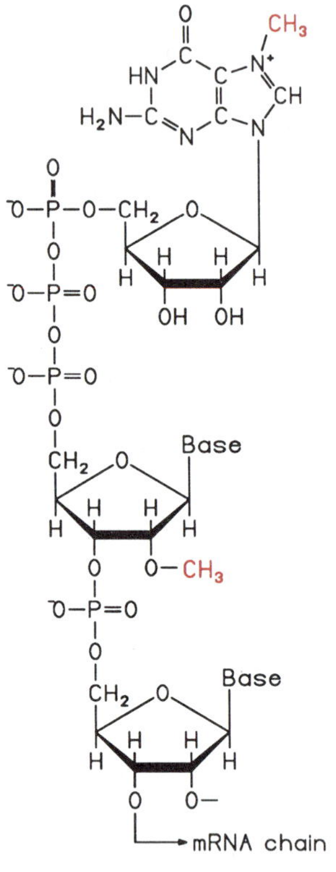

Figure 6.38.
The 5′ cap structure of eukaryotic mRNA.

activating sequences (UAS) are often required. For example, in yeast the transcription of the genes responsible for galactose metabolism requires the binding of the *gal4* protein factor to such a UAS. A special example of UAS are the **enhancer** sequences, which can be located several thousand base pairs upstream or downstream of the promoter. As with UAS, specific protein factors bind to the enhancer and thereby activate the promoter. The most likely model explaining this activation over such long distances is the **looping model.** It proposes that the DNA between the enhancer (UAS) and the promoter loops out, allowing the proteins that are attached to the two sequences to come in close contact for interaction. Transcription initiation would then depend on this interaction (Figure 6.37). Enhancers are tissue-specific, because the factors that bind to them are only expressed in certain cells.

As in prokaryotes, transcription is usually initiated with a purine nucleotide. However, the triphosphate group of the 5′-terminal nucleotide is immediately converted into a **cap structure,** which consists of a methylated GTP connected to the 5′-terminal nucleotide via a 5′-5′-triphosphate linkage (Figure 6.38). There are two such cap structures. In cap 1, the 2′-hydroxyl of the terminal nucleotide is methylated, in cap 2 the 2′-hydroxyls of the two terminal nucleotides are methylated. The cap structure is required for the subsequent translation of the mRNA in the cytoplasm.

The mechanism of transcription termination still waits to be elucidated. However, the nascent transcript is cleaved before termination by a specific nuclease, and a **poly(A) tail** of about 200 AMP residues is added to the 3′ end. The mechanism of 3′-polyadenylation and its function will be discussed later.

Most eukaryotic pre-mRNA molecules contain segments within their sequences that are not found in the final mRNA. This is because many eukaryotic genes contain **intervening sequences** or **introns** that interrupt the coding sequence. The segments of the gene represented in mRNA are called **exons.** The introns must be spliced out of the pre-mRNA in order to restore its coding sequence continuity (Figure 6.39). The mechanism of this **mRNA splicing** will be discussed together with the splicing of tRNA and ribosomal RNA in the section on posttranscriptional processing.

What is the function of introns? Why do most prokaryotic genes lack introns whereas eukaryotic genes contain them? A clear-cut answer is not possible. However, most introns are located at boundaries between different domains of a protein. It is conceivable that introns are remnants of evolution. Complex proteins arose by recombination between genes that encode simpler proteins. In order to preserve the coding segments of these simpler genes, the recombinational crossing-over event had to occur outside of them, thus creating noncoding interven-

Figure 6.39.
The processing of pre-mRNA.

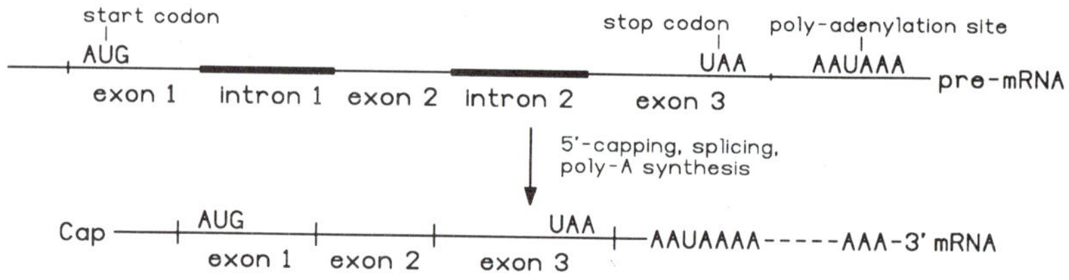

ing sequences that have to be spliced out of the composite genes in order to allow proper expression.

In eukaryotes, the splicing-out of introns occurs at the level of the mRNA. In prokaryotes, most introns probably have been eliminated by recombination. These simpler organisms have much shorter generation cycles than eukaryotic cells, thus increasing the chance for recombination. In addition, noncoding DNA sequences would slow down the generation time of a bacterium considerably more than that of a eukaryotic cell. Thus, there is a strong selective pressure against the maintenance of introns in prokaryotes.

After the completion of processing, the mRNA has to be transported from the nucleus into the cytoplasm where it is translated. Little is known about the mechanism of **transport of mRNA.**

6.6.4 The Synthesis of Ribosomal and Transfer RNA

In both prokaryotes and eukaryotes, **ribosomal RNA** is synthesized in the form of a large **precursor** molecule that is later processed and cut to proper length. Some of the bases in prokaryotic pre-rRNA are methylated. Eukaryotic precursor rRNA is in addition methylated at some of its 2′-hydroxyl groups. The methylation seems to control the processing of precursor rRNA.

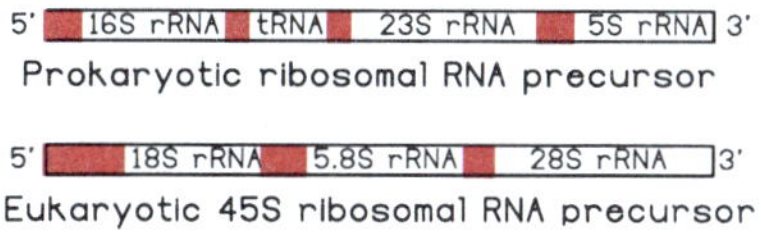

Figure 6.40.
The primary transcripts of ribosomal RNA. Regions shown in red are degraded.

Ribosomal RNA genes occur in multiple copies in the genome. The tandem repeats of the genes are separated by untranscribed **spacer** sequences (Figure 6.40). The primary transcript of *E. coli* rRNA contains the sequences of 16S, 23S, and 5S rRNA as well as the sequence of a tRNA. The processing of this precursor molecule is done by ribonuclease III, an enzyme that has itself an RNA molecule in its active center.

The **45S primary transcript** of eukaryotic rRNA is synthesized by RNA polymerase I in the nucleolus, a particlelike area in the nucleus adjacent to the tandemly repeated rRNA genes (nucleolar organizer). This transcript is then cleaved into 28S, 18S, and 5.8S rRNA (Figure 6.40).

Oocytes (immature egg cells) accumulate a very large number of ribosomes for subsequent distribution, after fertilization, among daughter cells. This requires a greatly increased rate of rRNA synthesis, which is accomplished by an **amplification of the ribosomal genes** from the initial 200–500 copies to a thousand times that number. The process of ribosomal gene amplification produces extrachromosomal circles of rRNA genes, each circle containing about 10 45S rRNA genes.

The 5S rRNA, also found in the ribosome, is synthesized by RNA polymerase III. In contrast to the other rRNAs, this molecule is not processed. It retains a triphosphate group at its 5′-terminus and has at its 3′ end the same sequence that terminates 5S rRNA molecules synthesized in vitro. This suggests that 5S rRNA is an unprocessed primary transcript.

The promoter for a **5S rRNA** gene is unusual in that it does not precede the gene but instead lies in the middle of the gene. A specific **transcription factor, TFIIIA,** must bind to a 50-nucleotide internal control region of the gene to allow polymerase III to initiate transcription at the start of the gene. TFIIIA has an interesting structural feature that may also be present in other transcription factors. It forms nine

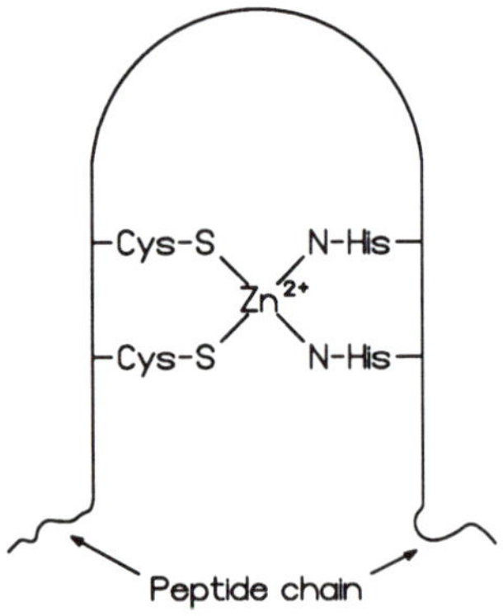

Figure 6.41.
Structure of a zinc finger motif in TFIIIA.

zinc fingers, short identical domains, each containing two cysteines and two histidines that are complexed with a Zn^{2+} ion. It is thought that the fingers fit into neighboring DNA grooves, forming hydrogen bonds with the base pairs (Figure 6.41).

Transfer RNAs are also synthesized in the form of larger precursor molecules that must be processed. One of the *E. coli* tRNA precursors is a 950-nucleotide–long primary transcript that is cleaved by **ribonuclease P** into seven different tRNA molecules. Ribonuclease P is itself a ribonucleoprotein containing a 377-nucleotide RNA molecule that is the actual enzyme. This was the first example of a so-called **ribozyme,** an enzyme that consists of RNA rather than polypeptide chains. Another example of ribozymes is the self-splicing mitochondrial rRNA of *Tetrahymena* (see below).

Many of the bases in pre-tRNA are modified extensively before the molecule matures into tRNA (see later).

Ribosomal RNA precursors in *Tetrahymena,* a ciliated protozoon, and some tRNA precursors in *E. coli* and yeast contain introns that must be spliced out in addition to the normal processing.

6.6.5 Posttranscriptional Processing of RNA

Posttranscriptional processing of primary transcripts includes the splicing out of introns, polyadenylation, and RNA editing. These processes will be discussed in this order.

The Mechanism of RNA Splicing

Among closely related species, both the size and sequence of an intron in a structural gene may vary widely. This is understandable since most introns do not code for anything and, therefore, are not subject to selective constraints. However, there are two highly conserved sequences in all introns of vertebrate mRNAs. These are the sequences straddling the **intron-exon boundaries.** All introns begin with the sequence GU and end with AG. The **consensus sequence** at the 5′ junction of vertebrate introns is AG:GUAAGU, at the 3′ junction it is YYYYYYYYYYNCAG:G (the colon marks the splice site, Y is a pyrimidine, and N is any base) (Figure 6.42). There is also an internal conserved site located about 20 nucleotides upstream of the 3′ splice site. This so-called **branch site** varies in vertebrates but in yeast has the sequence UACUAAC.

Figure 6.42.
Consensus sequences of vertebrate intron-exon boundaries (Y = pyrimidine, N = any nucleotide).

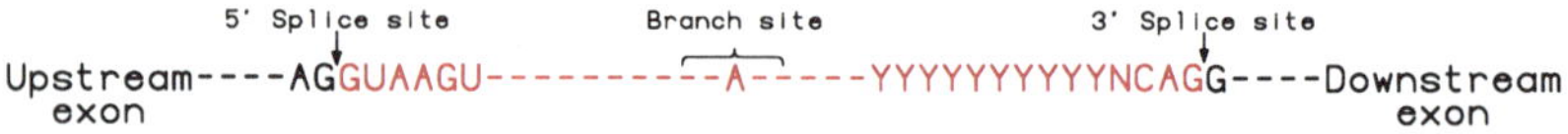

The splicing process must be highly accurate in order to preserve the correct reading frame for translation of the mRNA. This accuracy is accomplished with the help of several **small nuclear ribonucleoprotein particles (snRNPs).** The sequences of the snRNAs in these particles are partially complementary to the consensus sequences of the splice sites. The splicing process occurs in a large (60S) ribonucleoprotein complex called a **spliceosome** that contains several snRNPs (see Table 6.8).

Table 6.8. Functions of Some Small Nuclear Ribonucleoprotein Particles in mRNA Splicing

snRNP	Size of snRNA	Function
U1	165	Recognizes 5′ splice site
U2	185	Recognizes branch site
U5	116	Recognizes 3′ splice site
U4 + U6	145(U4) 106(U6)	Required for assembly of spliceosome

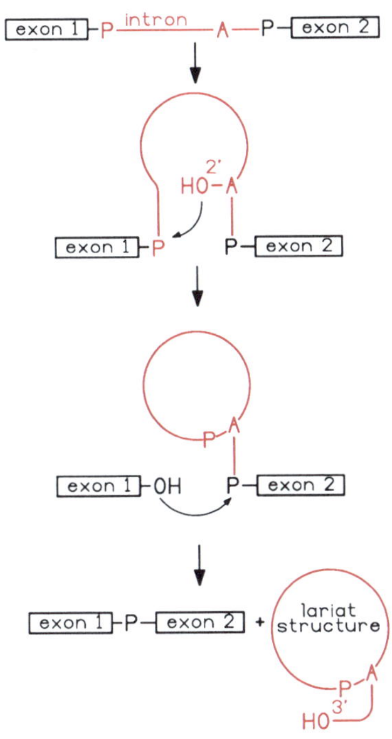

Figure 6.43.
Splicing mechanism for eukaryotic mRNA.

The sequence of reactions that occur during the excision of an intron and the splicing of two exons in eukaryotic mRNA are illustrated in Figure 6.43. First, the phosphodiester bond at the 5′ end of the intron is cleaved, leaving a free 3′-hydroxyl group at the upstream exon and connecting the 5′-phosphate of the intron to the 2′-hydroxyl group of an adenosine at the branch site of the intron. The newly created 3′-hydroxyl group of the upstream exon is then transferred to the 5′ end of the downstream exon whereas the phosphodiester bond at the 3′ end of the intron is cleaved. The intron is released in the form of a **lariat** that is quickly degraded.

The analysis of this process was made possible by the availability of antibodies against snRNPs. Such antibodies are found in patients with **systemic lupus erythematosus,** an autoimmune disease. Many details of the splicing mechanism still need to be worked out. For example, there does not appear to be a strict order in which multiple introns are excised from precursor mRNA. Furthermore, how is splicing restricted to one molecule? In other words, what mechanism prevents the connection of two exons from different RNA molecules? This so-called ***trans*** **splicing** has been shown to occur in **trypanosomes,** the organism causing sleeping sickness, and other biological systems. It is not known what mechanism allows this *trans* splicing. Another still open question is whether splicing is required for subsequent transport of mRNA out of the nucleus. There is evidence both for and against such a requirement.

The excision of introns from other types of RNA occurs by different splicing mechanisms. Many eukaryotic tRNA precursors contain a short intron just 3′ to the anticodon sequence. The mechanism of splicing tRNA involves the intermediate formation of a 2′-3′ cyclic phosphate at the 3′ end of the upstream exon. The reaction is catalyzed by two enzymes, an endonuclease and an **RNA ligase** with a number of other activities (Figure 6.44).

Figure 6.44.
The mechanism of tRNA splicing (yeast tyrosine tRNA).

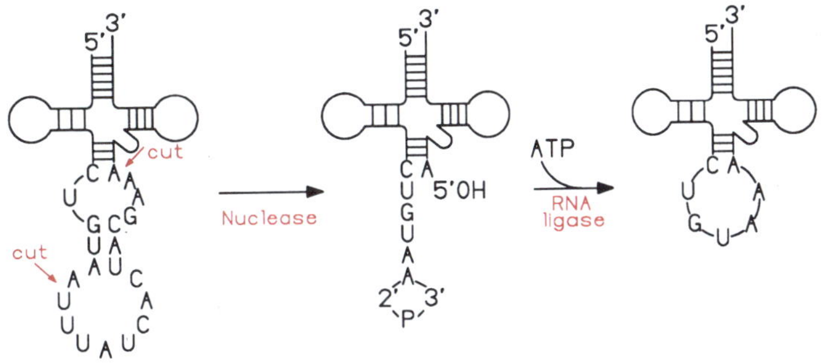

Introns in mitochondrial genes are divided into two groups, both of which exhibit **self-splicing** properties. This self-splicing capability of RNA was first observed by T. Cech with an intron in the pre-rRNA of *Tetrahymena*. In the splicing of **group I introns,** a guanosine nucleotide (monophosphate, diphosphate, or triphosphate) participates as a cofactor in the reaction and becomes attached to the 5′ end of the intron. The intron is released as a linear molecule. In **group II introns,** which also includes chloroplast pre-mRNA, an internal adenosine residue of the intron assumes the function of the cofactor, similar to the branch site in eukaryotic pre-mRNA introns. The intron is released in the form of a lariat.

Although rare, introns have also been found in some prokaryotic genes. For example, five introns have been discovered in bacteriophage T4. Four of these belong to the group I introns; that is, they are self-splicing. The fifth is an unspliced intron of 50 nucleotides in the T4 gene for topoisomerase II that is precisely bypassed by the translation machinery (see "translational" frameshifting later).

The Mechanism of Polyadenylation

In mammalian cells, polyadenylation of the pre-mRNA begins in the nucleus with the cleavage of the primary transcript within 30 nucleotides downstream of the **sequence AAUAAA** and just upstream of a GU-rich sequence. This cleavage is quickly followed by the addition of about 250 A residues to the newly created 3′ terminus by the enzyme **poly(A) polymerase.** After the splicing out of introns, the mRNA is transported to the cytoplasm where the poly(A) tail undergoes a slight shortening while maintaining a narrow size distribution. Thereafter, during translation, the poly(A) tails gradually shorten, but the rate of shortening differs for different mRNA species. There is evidence that some cytoplasmic addition of A residues occurs in all cells, partly counteracting the shortening process.

In early oogenesis, some mRNA species receive only short segments of poly(A) tails. These mRNA species are translationally inactive. Later during development, the poly(A) tails are elongated to 400 or more residues, at which time the mRNA becomes actively translated. The involvement of the poly(A) tails in translational regulation will be discussed later in this chapter.

RNA Editing

Bases in some mRNA species are edited before the mRNA functions as a template for protein synthesis. This editing may change the coding properties of the mRNA. For example, there are two forms of apolipoprotein B (apo B) in human plasma: apo B-48 and apo B-100. The 512 kd protein apo-B100 is synthesized in the liver, whereas the shorter apo B-48 (242 kd), consisting of the amino-terminal half of apo B-100, is made in intestinal cells. The pre-mRNA for both proteins is transcribed from a single apo B gene. However, codon 2153, CAA (for Gln), is changed in the intestinal mRNA to UAA, which is a stop codon. Thus, translation of the intestinal mRNA produces a shortened apo-B protein.

5′···ATA—CAA—UUU···3′

Ile Gln Phe

RNA ↓ editing

5′···ATA—UAA—UUU···3′

Ile Stop Phe

The mechanism of this posttranscriptional editing of RNA is still obscure. Examples of RNA editing have also been reported in other systems. They include the posttranscriptional addition or deletion of other nucleotides.

6.6.6 Replication of RNA Viruses

The genomes of several viruses consist of RNA instead of DNA. Since this RNA must be replicated during the intracellular multiplication of the virus, and since the host cell does not possess an RNA replicase, the virus genome must contain the information for the synthesis of such an enzyme. However, this information can only be expressed in the host cell if the virus RNA can be used directly as mRNA. In many viruses, this is not the case, and infection is only possible if the virus brings its own replicase into the host cell. (Here, Hershey's blender experiment would have given misleading results.)

There are four different types of RNA-containing viruses:

1. Viruses containing single-stranded RNA **(+RNA)** that functions as mRNA in the infected host cell. Examples are R17 bacteriophage and **poliovirus.**
2. Viruses containing anti–messenger RNA **(−RNA)** that has to be copied into complementary mRNA before translation can begin. Examples are **rabies virus** and **vesicular stomatitis virus.**
3. Viruses with double-stranded RNA **(±RNA)** that also requires transcription into mRNA. The genomes of these viruses are segmented. An example is **reovirus.**
4. **Retroviruses,** containing single-stranded RNA **(+RNA)** that must first be transcribed into DNA. This DNA subsequently functions, in double-stranded form, as a template for mRNA synthesis. Examples are **Rous sarcoma virus, leukemia viruses,** and **HIV.**

It is obvious that viruses of types 2, 3, and 4 must carry their own replicases or transcriptases into the cell they infect. Only type 1 viruses can use the host cell's translation machinery directly for the synthesis of their replicases from the virus genome.

The synthesis of complementary RNA from a single-stranded RNA template is accomplished by RNA replicase. Parent and daughter strands do not form a double helix but separate immediately after synthesis of the daughter strand. RNA-dependent DNA polymerase, or **reverse tran-**

scriptase, is found packaged in the virions of retroviruses. This enzyme is used extensively in in vitro recombinant DNA technology (see Chapter 7).

6.7 PROTEIN METABOLISM

The translation of the nucleotide sequence of mRNA into the amino acid sequence of a polypeptide chain takes place on the ribosome, a complex cytoplasmic ribonucleoprotein particle. This process requires the cooperation of over 100 different macromolecules and represents one of the most complex biochemical systems. The discussion of the translation process begins with the description of two of its participants: the ribosomes and transfer RNA.

6.7.1 The Ribosome

Ribosomes consist of **two subunits:** a large one and a small one. Complete ribosomes only exist in a complex with mRNA. The synthesis of a new polypeptide chain is initiated when the small subunit binds to the mRNA. The large subunit then joins to form a functional ribosome that is ready to translate the mRNA molecule. Usually, several ribosomes follow one another on a single mRNA molecule. The resulting structure is called a **polysome.** As soon as the polypeptide chain is finished, ribosomes leave the mRNA and dissociate again into their subunits.

Prokaryotic and eukaryotic ribosomes differ in size and properties. Prokaryotic ribosomes have a molecular weight of about 2.8×10^3 kd and sediment at 70S, whereas the larger eukaryotic ribosomes have a particle size of 4.5×10^3 kd and sediment at 80S. Mitochondria and chloroplasts (in plants) have ribosomes that resemble prokaryotic ribosomes both in size and properties.

Each ribosomal subunit consists of one copy each of one or more specific rRNA molecules and between 20 and 50 different proteins. Subunits and rRNA molecules are usually identified by their sedimentation values. The protein and rRNA composition of ribosomal subunits is tabulated in Table 6.9.

Bacterial ribosomes can be disassembled and reconstituted from their protein and rRNA components. It was originally thought that the re-

Table 6.9. Composition of Ribosomes

Prokaryotes 70S Particle		Eukaryotes 80S Particle	
50S Subunit	30S Subunit	60S Subunit	40S Subunit
23S rRNA 5S rRNA	16S rRNA	28S rRNA 5.8S rRNA 5S rRNA	18S rRNA
34 Proteins	21 Proteins*	~50 Proteins	~30 Proteins

* There are only 20 different proteins, one is present in four copies. Some of these copies are acetylated, giving rise to additional protein species.

constitution of partial assemblies would lead to an understanding of the function of individual ribosomal proteins. For example, streptomycin, at low concentrations, causes **misreading** of mRNA. Streptomycin-resistant mutants show the same degree of translational fidelity as wild-type bacteria in the absence of the drug. By reconstituting hybrid ribosomes and using proteins isolated from normal and streptomycin-resistant cells, it was possible to identify the protein S12 of the 30S subunit as being responsible for the streptomycin sensitivity. (The letter S designates proteins of the small subunit, L those of the large subunit. The protein with the slowest electrophoretic mobility is given the number 1.) This result suggested that protein S12 somehow is concerned with the proper codon-anticodon recognition between mRNA and tRNA.

This approach, however, did not lead much further. Newer methods for studying the relationship of the ribosomal proteins and the rRNA have employed antibodies against individual proteins for surface mapping (by electron microscopy) or cross-linking reagents to determine the spacial relationships between proteins and rRNA within the ribosome. Neutron diffraction studies of concentrated solutions of ribosomes have led to the determination of the locations of all proteins within the 30S subunit of the ribosome.

Ribosomal RNA exhibits extensive secondary structure with many double-helical base-paired regions. In view of the recent discovery of ribozymes, it appears likely that ribosomal RNA participates in the catalytic functions of the ribosome as a multienzyme complex.

6.7.2 Transfer RNA

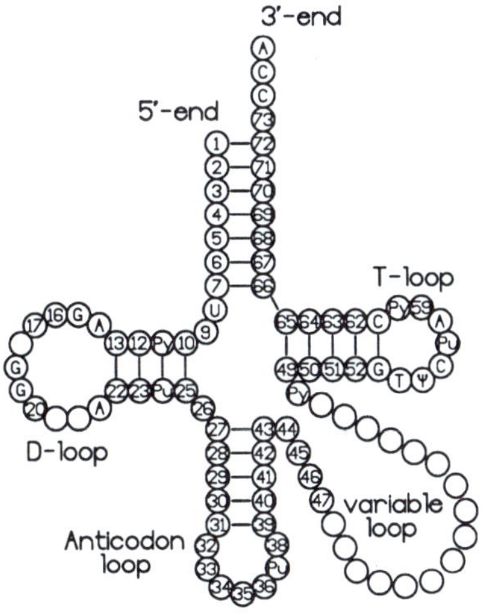

Figure 6.45.
Structure of yeast tRNA^{Phe}. The indicated bases are found in most tRNA molecules. At three places, after bases 17, 20, and 47, additional nucleotides are found in different tRNA species. They are indicated by the numbers 20:1, 20:2, etc. This numbering system was proposed at a 1979 Cold Spring Harbor meeting on tRNA structure.

After being predicted as the adaptor molecule in protein synthesis by Francis Crick in 1958, transfer RNA was identified shortly afterward by Paul Zamecnik. All tRNA molecules, prokaryotic as well as eukaryotic, have the same basic structural features:

1. The molecule can be drawn as a **cloverleaf structure** in which three Watson-Crick–type helical regions separate four unpaired loops (Figure 6.45). The base-paired regions are called stems or arms. The loops are the dihydrouracil loop **(D loop),** the **anticodon loop,** the **variable loop,** and the thymine-pseudouracil loop **(T loop).**
2. Every tRNA molecule has the same **C-C-A-3′-OH terminus** to which the amino acid is attached via an ester linkage.
3. tRNA contains many **unusual bases** such as hypoxanthine, thymine, dihydrouracil, dimethylguanine, pseudouracil, and others. The significance of these altered bases is not known. They may serve to give the tRNA molecule specificity for recognition by its aminoacyl-tRNA synthetase (see below) and the ribosomal proteins.
4. The **three-dimensional structure of tRNA is L shaped.** It arises from the interaction of nucleotides in the D loop with nucleotides in the T loop (Figure 6.46).

Most tRNA molecules are derived from large precursor molecules, some of which contain the sequences of several different tRNAs, by

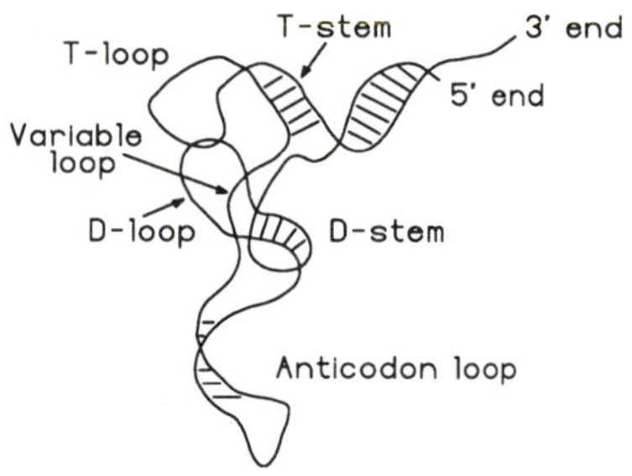

Figure 6.46.
Three-dimensional structure of tRNA. After Kim SH. Transfer RNA. Cold Spring Harbor, NY: Cold Spring Harbor Monographs; 1979.

ribonuclease P, an enzyme that has itself an RNA at its catalytic site. The 3′-terminal CCA sequence often is not encoded by the tRNA gene. There are enzymes that can remove and resynthesize this sequence.

Each tRNA molecule contains two specific sites: (1) the anticodon site, which recognizes and base-pairs with a triplet codon in the mRNA, and (2) a site that is recognized by aminoacyl-tRNA synthetase, the enzyme that attaches the amino acid to the 3′ end of its cognate tRNA (activation). This second site cannot be localized precisely. Instead, it appears to represent the three-dimensional structure of the molecule that is recognized by the activating enzyme.

6.7.3 Amino Acid Activation

Amino acids are attached to their cognate tRNAs by specific **aminoacyl-tRNA synthetases.** There are at least 20 different tRNA synthetases, each one highly specific for only one amino acid.

At first, the enzyme catalyzes the formation of a mixed anhydride between the amino acid (aa) and AMP. The **activated amino acid** is then transferred to the 3′-terminal adenosine residue of one of its cognate tRNAs, forming an ester linkage in the process.

$$\text{aa} + \text{ATP} + \text{synthetase} \rightarrow \text{aa-AMP-synthetase} + \text{PPi}$$
$$\text{aa-AMP-synthetase} + \text{tRNA} \rightarrow \text{aa-tRNA} + \text{AMP} + \text{synthetase}$$

The same enzyme catalyzes both reactions. For the **fidelity of translation,** the enzyme's specificity in recognizing its tRNA and amino acid is as important as the specificity of the codon-anticodon interaction on the ribosome. Once attached to tRNA, the placement of the amino acid in the growing polypeptide chain is determined solely by the anticodon sequence of the tRNA.

For example, if the cysteine residue in Cys-tRNA^{Cys} (tRNA^{Cys} designates a cysteine-specific tRNA) is chemically reduced to alanine, the resulting Ala-tRNA^{Cys} is still recognized by the *Cys* codon. As a consequence, alanine is incorporated into the polypeptide chain at a position normally occupied by cysteine.

Sometimes the tRNA synthetase makes a mistake and binds to the wrong amino acid. Such a mistake might occur, for example, when the isoleucine-specific tRNA synthetase binds to valine rather than isoleucine. The side chains of these two amino acids are similar. When this happens, the **amino acid–AMP intermediate** is hydrolyzed as soon as tRNA^{Ile} enters the enzyme complex.

$$\text{Val} + \text{ATP} + \text{synthetase}^{Ile} \rightarrow \text{Val-AMP–synthetase}^{Ile} + \text{PPi}$$
$$\text{Val-AMP–synthetase}^{Ile} + \text{tRNA}^{Ile} \rightarrow \text{Val} + \text{AMP} + \text{synthetase}^{Ile}$$

Because of this "checking" reaction, the overall fidelity of amino acid activation is very high.

6.7.4 Mechanism of Translation

In prokaryotes, the synthesis of every polypeptide chain starts with formyl-methionine (fMet). There are two methionine-specific tRNAs,

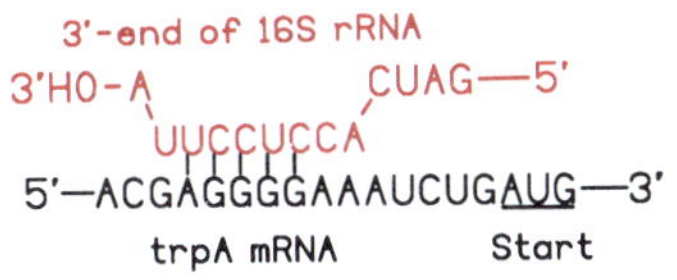

Figure 6.47.
Sequence complementarity between the Shine-Dalgarno sequence at translation start sites in mRNA and the 3′ end of 16S rRNA in the 30S ribosomal subunit.

tRNA^{fMet} and tRNA^{Met}, both of which are first charged with methionine. The amino group of methionine in Met-tRNA^{fMet} is subsequently formylated to fMet-tRNA^{fMet} by a transformylase that uses N^{10}-formyl tetrahydrofolate as formyl donor. Met-tRNA^{Met} is not formylated and is used for the incorporation of methionine at internal positions in the polypeptide chain.

Specific deformylases later remove the formyl group from the amino terminus of the finished polypeptide chain. Met-specific amino-peptidases cleave the N-terminal methionine residue from some of the polypeptide chains.

The codon for methionine is **AUG** (for a few proteins, GUG is also used as the **initiation codon**). In bacteria, the distinction between initiation and internal Met is based upon the nucleotide sequence immediately preceding the AUG codon. At **translation start sites,** a short purine-rich sequence, called the **Shine-Dalgarno sequence,** partially complementary to a stretch of nucleotides near the 3′ end of the 16S rRNA in the small ribosomal subunit, allows the attachment of the 30S ribosomal subunit (Figure 6.47).

In eukaryotic cells, polypeptide chain initiation also uses methionine. As in bacteria, there are two methionine-specific tRNAs, only one of which functions in chain initiation. In contrast to bacteria, however, no formylation takes place.

The first step in polypeptide synthesis is the formation of an **initiation complex** comprising fMet-tRNA^{fMet}, the 30S ribosomal subunit, mRNA, and GTP. The reaction requires Mg^{2+} ions and the participation of three **initiation factors:** IF-1, IF-2, and IF-3. The three factors become attached to the 30S ribosomal subunit as soon as the 70S ribosome dissociates into its two subunits after termination of a round of translation. IF-3 specifically keeps the small subunit from associating with its larger companion. After GTP binds to IF-2, fMet-tRNA^{fMet} and mRNA join the complex while IF-3 is released. In this complex, fMet-tRNA^{fMet} is positioned so that its anticodon forms hydrogen bonds with the initiation codon AUG on the mRNA. This interaction establishes the reading frame for the translation of the mRNA.

The release of IF-3 allows binding of the 50S ribosomal subunit to create a fully functional 70S ribosome. This is accompanied by hydrolysis of GTP to GDP and P_i, which is quickly followed by the release of IF-1 and IF-2. GTP is not used for the phosphorylation of a ribosomal protein or factor. Instead, its hydrolysis causes a conformational change in the translation factors.

In eukaryotes, the initial binding of the smaller ribosomal subunit to the mRNA does not seem to involve sequence homology between mRNA and rRNA. The 40S subunit attaches to the "capped" 5′ end of the mRNA (ATP is required here instead of GTP) and then migrates down the mRNA until it encounters an AUG codon. It is not always the first AUG at which translation is initiated. A consensus sequence for optimal initiation has been derived from comparing many translation start sites. The 60S subunit is then added (GTP is required), and translation is initiated. Here the multiplicity of initiation factors is considerably greater than in prokaryotes.

The absence of a Shine-Dalgarno sequence in eukaryotic mRNA explains why eukaryotic mRNA can only code for a single polypeptide chain. Eukaryotic mRNA is said to be **monocistronic** ("cistron" is a synonym for gene). In contrast, bacterial mRNA often is **polycistronic:** it can code for several proteins because internal Shine-Dalgarno sequences allow initiation of translation at multiple sites on the mRNA.

The next stage of translation is called **elongation** (Figure 6.48). The ribosome has two distinct sites for the attachment of tRNA, the **aminoacyl site (A site)** and the **peptidyl site (P site).** The existence of these sites was deduced from the inhibitory action of the antibiotic puromycin on translation (see next section). fMet-tRNA^{fMet} is initially bound to the P site, whereas the A site is free to receive the next aminoacyl-tRNA.

The next amino acid is placed into the A site in the form of a complex containing aminoacyl-tRNA (aa-tRNA), **elongation factor Tu (EF-Tu),** and GTP. If the anticodon is complementary to the codon, GTP is hydrolyzed and EF-Tu is released complexed with GDP. If the codon and anticodon do not match, the aa-tRNA is released. This can happen both before and after the **hydrolysis of GTP,** thus providing two chances for correcting a misplaced amino acid before it becomes incorporated.

The released EF-Tu–GDP complex combines with another **elongation factor, EF-Ts.** This causes the release of GDP, which in turn allows the binding of a new GTP molecule, then EF-Ts associates again. As during initiation, GTP hydrolysis causes a conformational change of a translation factor, in this case EF-Tu.

When both A and P sites on the ribosome are occupied, the enzyme peptidyl transferase, an intrinsic part of the 50S ribosomal subunit, catalyzes the transfer of the fMet-residue (or any peptidyl chain at a later stage of translation) from the P-site to the α-amino group of the aa-tRNA at the A-site, forming a new peptide bond and leaving a free tRNA^{fMet} at the P site. The resulting dipeptidyl-tRNA now occupies the A site.

The formation of the peptide bond is followed by the **translocation** of the peptidyl-tRNA from the A site to the P site. The uncharged tRNA at the P site is ejected during translocation. The peptidyl-tRNA remains

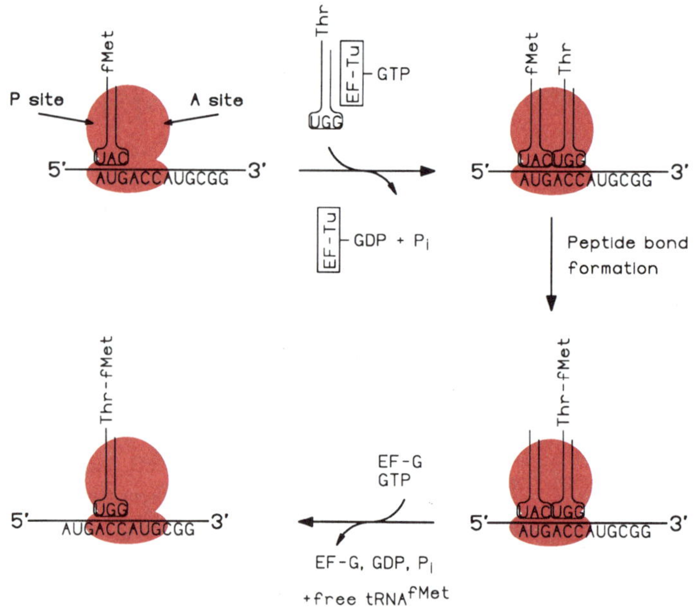

Figure 6.48.
Schematic representation of the elongation phase of translation.

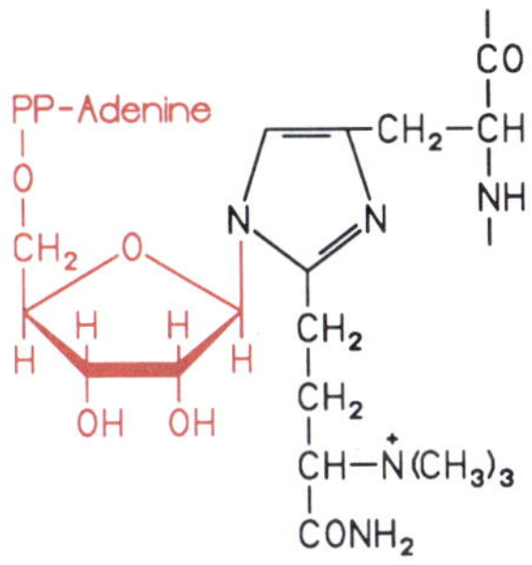

Figure 6.49.
Inactivation of the eukaryotic elongation factor EF-2 by diphtheria toxin. An ADP-ribose residue is attached to the diphthamide residue of EF-2.

hydrogen-bonded to its codon on the mRNA during translocation, causing the ribosome to move along the mRNA by exactly three nucleotides. Translocation is catalyzed by **elongation factor G (EF-G)** (translocase). The factor is bound to a GTP molecule. Hydrolysis of GTP allows EF-G to dissociate again from the ribosome.

In eukaryotes, the elongation factors EF-1α and EF-1βγ perform the roles of EF-Tu and EF-Ts, respectively. The eukaryotic counterpart of EF-G is called EF-2. This translocation factor is specifically inactivated by the **Diphtheria toxin,** formed by *Corynebacterium diphtheriae.* (The gene for the toxin is actually located on a prophage present in *C. diphtheriae.*) EF-2 contains a modified histidine residue, called diphthamide. The diphtheria toxin catalyzes the transfer of an ADP-ribose residue from NAD+ to **diphthamide** (Figure 6.49).

The sequence of elongation and translocation (Figure 6.48) is repeated until a termination codon on the mRNA is reached. Then **release factors** hydrolyze the peptidyl chain from the tRNA, thereby releasing both the polypeptide chain and the tRNA from the ribosome. In *E. coli,* there are two release factors, RF1 and RF2, recognizing the sequences AUU or UAG and UAA or UGA, respectively. In eukaryotes, a single release factor recognizes all three stop codons. GTP is hydrolyzed during this reaction.

Translation proceeds from amino-terminal to carboxy-terminal amino acid. In the process, the ribosome moves along the mRNA molecule in the 5′ → 3′ direction. This is the same direction in which the mRNA is being synthesized. It is, therefore, possible for bacterial ribosomes to attach to the 5′ end of mRNA and begin translating it before the synthesis of the mRNA molecule is completed. This is, of course, not possible in eukaryotes, where transcription and translation occur in different cellular compartments.

6.7.5 Inhibition of Translation

Many **antibiotics** are inhibitors of specific stages of translation. **Puromycin,** for example, inhibits elongation. Because of its structural similarity with the 3′ end of aminoacyl-tRNA (Figure 6.50), this inhibitor binds to the A site of the ribosome and is a substrate for peptidyl transferase. However, further translocation is not possible, and the polypeptide chain is prematurely released covalently bound to puromycin.

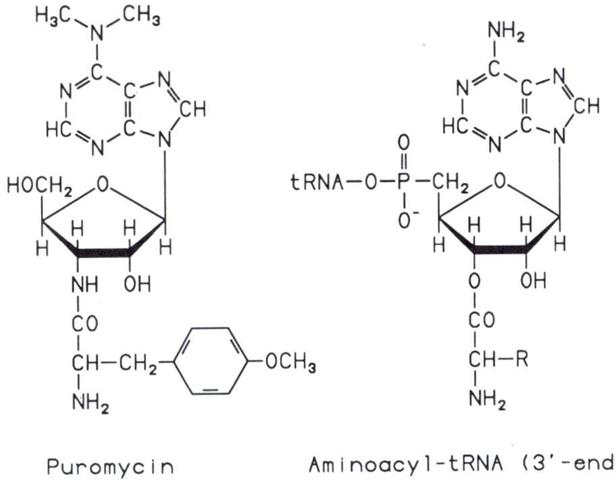

Figure 6.50.
Comparison of the structures of puromycin and the 3′ terminus of aminoacyl-tRNA.

Puromycin was used to define the two tRNA-binding sites on the ribosome. A peptidyl- or aminoacyl-tRNA bound to the A site prevents binding of puromycin; the peptide chain cannot be released under these conditions. Only after the peptidyl-tRNA is transferred to the P site can puromycin affect the ribosome.

Tetracyclines (Figure 6.51) inhibit the binding of aminoacyl-tRNA to the A site of bacterial ribosomes.

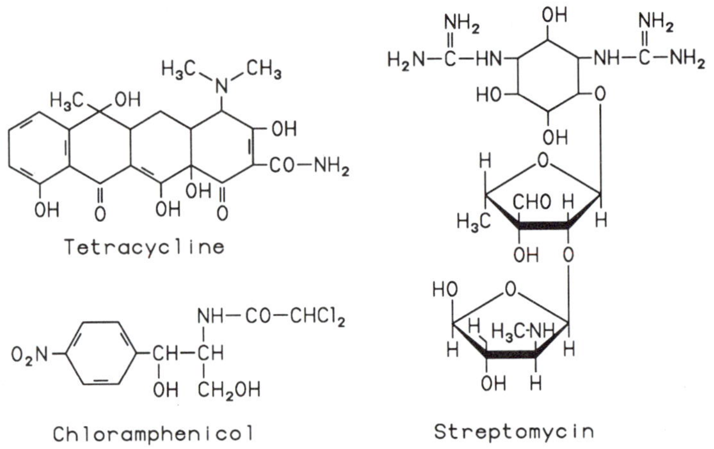

Figure 6.51.
Structures of some antibiotics that inhibit translation.

Chloramphenicol, lincomycin, and **sparsomycin** inhibit peptidyl transferase in bacteria. Chloramphenicol (Figure 6.51), although not effective in eukaryotic translation, affects mitochondrial protein synthesis. In humans, chloramphenicol may cause leucopenia and pancytopenia (lethal). Monitoring of white blood cells during chloramphenicol therapy is therefore mandatory. (This effect of chloramphenicol does not seem to be related to its inhibition of mitochondrial translation.)

Cycloheximide (Figure 6.51) has the same effect as chloramphenicol except that it is specific for eukaryotic cells.

Erythromycin specifically inhibits bacterial translocation.

Streptomycin (Figure 6.51), at low concentrations, causes translational misreading. At higher concentration it inhibits the formation of the initiation complex. Streptomycin can cause damage to the eighth nerve. If taken during pregnancy, it may cause deafness in the child. The biochemical basis for this effect is not known.

6.7.6 Suppression

Both nonsense and missense mutations can be suppressed by a second mutation, called **suppressor mutation,** which is located outside the affected gene. This intergenic suppression results in the synthesis of the protein, albeit at a somewhat reduced rate. The molecular basis of suppression is as follows.

A **suppressor mutation for a nonsense mutant** changes the anticodon sequence of a tRNA in such a way that it now base-pairs with a nonsense codon. As a result, polypeptide synthesis does not come to a halt at the nonsense codon in the mRNA, but a full length protein is synthesized that contains, at the site of the original mutation, the amino acid for which the altered tRNA is specific. If this amino acid differs from the original amino acid at that position, the enzyme may still be active if the amino acids are similar or if the substitution occurred in a region of the protein not required for enzymatic function. Such suppressor tRNAs usually originate from one of a set of redundant tRNA genes coding for tRNAs specific for the same amino acid. Suppressor tRNAs may interfere with normal peptide chain termination. Many structural genes, however, are terminated by two stop codons so that normal termination still is possible.

In addition to nonsense suppressors, there are also **missense suppressors.** Here an altered amino acid, incorporated into the polypeptide chain because of a missense mutation, is replaced by a different amino acid, one that allows the protein to function. Missense suppression, however, is necessarily very inefficient because, if it occurred at higher efficiency, it would change too many correct amino acids in other cellular proteins, causing cell death. Missense-suppressing tRNAs sometimes have a changed specificity for the aminoacyl-tRNA synthetase but retain their original anticodon sequence.

An unusual case of tRNA-mediated suppression is the **suppression of frameshift mutations.** If the frameshift mutation is due to a single base-pair insertion, the suppressing tRNA has in its anticodon four bases instead of the usual three. As a result, the ribosome, at the site of the insertion, moves along the mRNA by four nucleotides instead of three. The existence of such mutants sheds light on the mechanism by which the ribosome moves along the mRNA. Apparently, as the peptidyl-tRNA is translocated from the A site to the P site, it remains bound to its triplet (or quadruplet in this case) codon and thus moves the mRNA along with it.

6.7.7 Intracellular Targeting of Proteins

Although most eukaryotic proteins are synthesized in the cytoplasm (the exception are a few mitochondrial and chloroplast proteins), some proteins are exported from the cell, and others are inserted into the plasma membrane or are transported into the nucleus, the mitochondria, or other intracellular compartments. This targeting of newly synthesized proteins is accomplished through signals in the amino acid sequence of nascent proteins. These sequence signals are recognized by other proteins that act as specific receptors or as modifying enzymes, thereby allowing sorting of the nascent proteins.

All translation is initiated on cytoplasmic ribosomes. If the protein is to remain in the cytoplasm, its synthesis is completed on free cytoplasmic ribosomes. However, if the protein is destined for export or for insertion into the plasma membrane, its amino terminus has a hydrophobic **signal peptide sequence.** As this sequence emerges from the ribosome, it is recognized by a cytoplasmic particle, the **signal recognition particle (SRP)**, which binds to it and temporarily stops the translation process. The SRP consists of six different protein subunits and a 7S RNA molecule. The nascent protein, with the SRP bound to it, then diffuses to the membrane of the **endoplasmic reticulum (ER),** where it binds to a docking protein or **SRP receptor.** This allows the ribosome on the nascent protein chain to bind to two integral ER membrane proteins, **ribophorin I** and **II,** whereupon the SRP is released.

The translation process then resumes but the polypeptide chain is now released through a pore in the ER membrane into the lumen of the ER. This **translocation process** requires the hydrolysis of ATP. In the ER, a **signal peptidase** cleaves off the signal sequence, and if the protein requires glycosylation, the addition of sugar residues is started in the ER. **Shuttle vesicles** then carry the partially glycosylated, or nonglycosylated, protein to the **Golgi apparatus,** where further posttranslational modification occurs. Finally, the protein is carried via shuttle vesicles to the plasma membrane, where the vesicle fuses with the membrane and releases its contents to the outside of the cell (Figure 6.52).

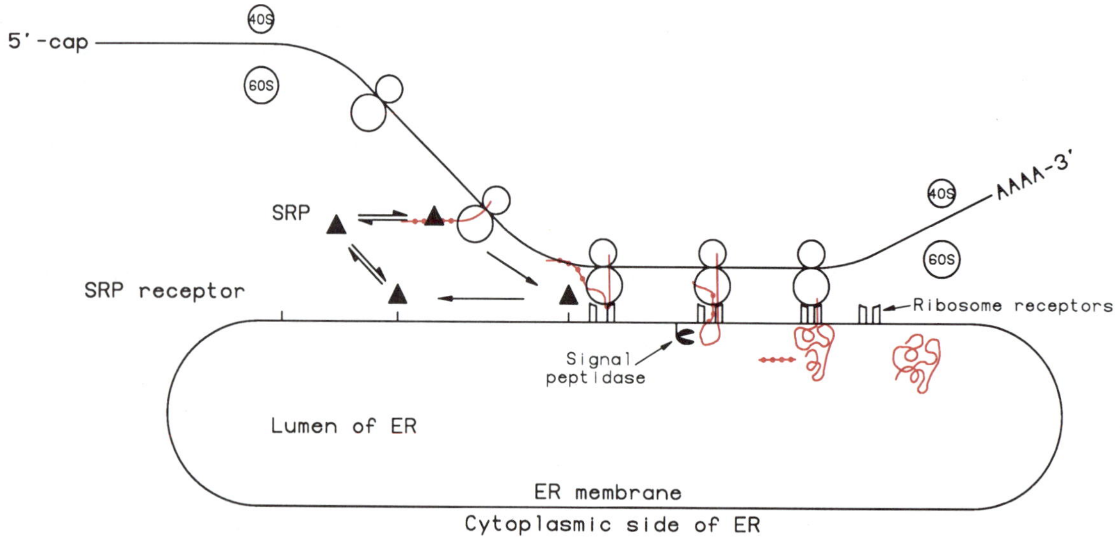

Figure 6.52.
Schematic illustration of synthesis, posttranslational modification, and transport of a secreted protein.

Signal peptide sequences do not exhibit a clear consensus sequence. Their lengths vary from 13 to 40 amino acids, with a central stretch of hydrophobic residues. The site that is later cleaved by the signal peptidase is located near the carboxyl end of the sequence and is preceded by a short segment of polar residues. The amino-terminal end of the sequence contains one or more positively charged amino acids.

If the protein is destined to become an integral plasma membrane protein, the translocation of the nascent polypeptide into the ER is arrested when a **stop-transfer sequence** enters the ER membrane. The remainder of the polypeptide chain (carboxyl terminus) remains on the outside of the ER. The protein is now anchored in the membrane of the ER. The **posttranslational modification** of this protein will occur only on the part of the molecule that has entered the lumen of the ER. When the protein finally arrives, via shuttle vesicles, at the plasma membrane, it becomes an integral membrane protein, with its carbohydrate portion exposed to the outside of the cell. All glycosylated integral membrane proteins have their carbohydrate protein exposed on the outside of the cell.

Some membrane proteins become inserted into the membrane with opposite polarity, that is, their amino terminus faces the cytoplasm; others cross the plasma membrane several times. This appears to be accomplished by a combination of multiple internal signal sequences, absence of signal peptidase recognition sequences, and multiple stop-transfer sequences. There is evidence that, in addition to hydrophobic stop-transfer sequences, positively charged amino acids in the hydrophilic cytoplasmic domains of an integral membrane protein are required for proper membrane insertion. Figure 6.53 illustrates the insertion of such proteins in the ER membrane. Remember that the part of the protein that extends into the lumen of the ER finally becomes extracellular as the shuttle vesicle, derived from the ER membrane, fuses with the plasma membrane.

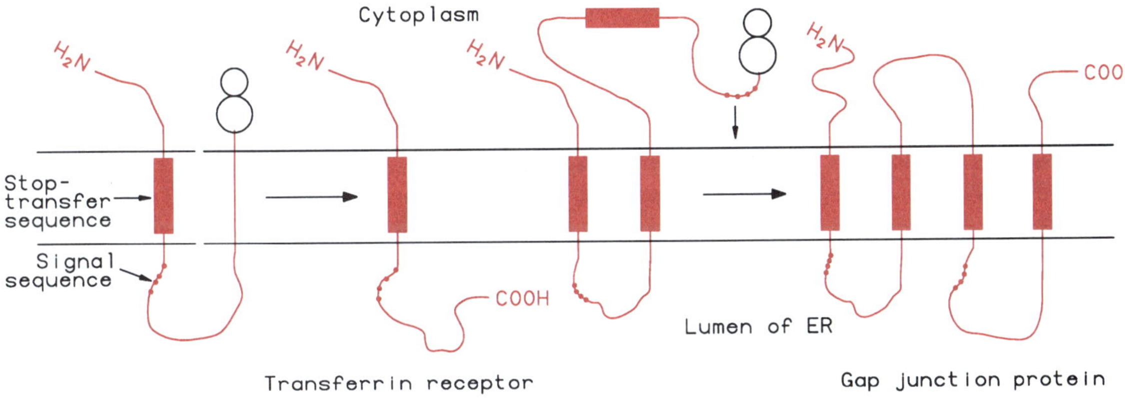

Figure 6.53.
A model for the insertion of two integral membrane proteins, asialoglycoprotein receptor (a) and connexin-32 (b), into the membrane of the endoplasmic reticulum. Modified from Wickner WT, Lodish HF. *Science.* 1985; 230:400.

Other membrane proteins are synthesized in their entirety on free cytoplasmic ribosomes. One example is cytochrome *b*5, which is inserted into the ER membrane only after its synthesis is completed. Insertion is accomplished through an insertion sequence that does not require the participation of SRP.

Targeting of proteins for specific intracellular compartments can also occur by specific tagging with oligosaccharides. **Lysosomal enzymes,** for example, use **mannose 6-phosphate** as their **tag.** Proteins destined to remain in the ER share the carboxy-terminal sequence Lys—Asp—Glu—Leu. Deletion of this sequence results in the secretion of the protein. The **sorting of proteins** in the ER and the Golgi apparatus for delivery to the various subcellular compartments is complicated and far from being completely understood. Bacteria use the same basic mechanism for delivery of proteins to their plasma membrane, their periplasmic space, or to the outside.

Some proteins appear to lack any obvious membrane-binding sequences and are synthesized entirely within the cytoplasm. For these proteins to become targeted to the membrane, specific amino acids are covalently linked to long-chain fatty acids or isoprenoid groups that provide the hydrophobicity for membrane anchorage. The hydrophobic groups include myristate (linked to amino-terminal glycine residues), palmitate (linked to internal cysteines via a thioester linkage), farnesyl alcohol (linked via a thioether bond to a carboxyterminal cysteine that occurs in the sequence CAAX, where A is an aliphatic and X any amino acid, AAX being removed before farnesylation), and higher isoprenoid groups such as geranylgeranyl.

6.7.8 Intracellular Degradation of Proteins

Cellular proteins are constantly synthesized and degraded. The average lifetime of a protein can vary from a few minutes to hundreds of hours. Usually proteins that serve important regulatory functions have the shortest half-lives, whereas housekeeping proteins have the longest. It was recently shown that one factor that determines the half-life of a protein is the nature of its amino-terminal amino acid. The relationship between amino-terminal amino acid and average half-life of cytoplasmic proteins is summarized in Table 6.10. This relationship is found in eukaryotes as well as prokaryotes. It appears that the nature of the amino-terminal amino acid determines the probability of a protein to become tagged by ubiquitin for degradation (see below).

In addition to this routine degradation of proteins, chemically damaged or denatured proteins are also tagged for degradation. How is this tagging accomplished? There are two known cellular systems that function in the breakdown of proteins in eukaryotic cells: the lysosomal system and the ATP-dependent system.

The lysosome is a cytoplasmic vesicle full of degradative enzymes that work optimally in the acid environment of the lysosome (pH 4.8). The low pH in the lysosomes is generated by proton pumps in the lysosomal membrane. **Lysosomal protein degradation** can be inhibited by the weakly basic antimalaria drug **chloroquine,** which is taken up by lysosomes and raises their luminal pH. Substrates enter the lysosome encapsulated in vesicles that fuse with it. The lysosome is responsible for the degradation of membrane proteins, extracellular proteins taken up by endocytosis, and some cytoplasmic proteins that are

Table 6.10. Relationship between Half-lives and Amino-Terminal Residues of Cytoplasmic Proteins

Amino-Terminal Amino Acid	Half-life (approximate)
Stable proteins	
Methionine	>20 h
Alanine	
Serine	
Threonine	
Valine	
Unstable proteins:	
Isoleucine	30 min
Glutamate	
Tyrosine	10 min
Glutamine	
Proline	7 min
Labile proteins:	
Leucine	3 min
Phenylalanine	
Aspartate	
Lysine	
Arginine	2 min

Source: Bachmair et al. *Science*. 1986; 234:179.

taken up in the form of vesicles derived from the ER. For example, during starving conditions, the lysosomes enlarge because they have taken up junks of cytoplasm. The proteins are degraded and released as amino acids that can be reutilized for the synthesis of vital proteins.

The **ATP-dependent system** uses the protein **ubiquitin** to tag proteins for degradation. Ubiquitin becomes covalently attached via an isopeptide bond to the ϵ-amino group of one or more lysine residues. Ubiquitin is first "activated" by ATP at its carboxyl end. The ubiquitin-AMP intermediate then reacts with the lysine residues. Ubiquitin is a heat-shock protein, i.e., its synthesis is increased at high temperature or when the cell contains many abnormal proteins. Ubiquitinated proteins are degraded by a very large protease complex that only attacks such tagged proteins. It requires the hydrolysis of ATP for activity.

Bacteria do not employ ubiquitin for tagging proteins for degradation. However, they have a similar ATP-dependent protein degradation system as eukaryotic cells. The **La protease,** coded by the *lon* gene, responsible for this activity has two sites: a proteolytic site and a regulatory site. When a denatured segment of a protein binds to this regulatory site, ATP binds to the enzyme, thereby activating its proteolytic site. During proteolysis, ATP is hydrolyzed to ADP, which inhibits the protease activity. Only upon binding of another damaged or denatured polypeptide segment to its regulatory site does the protease regain its activity. For every cleaved peptide bond, two ATPs are hydrolyzed. The resulting oligopeptides are then degraded by other peptidases. Like ubiquitin, the La protease is a heat-shock protein.

Proteins that are incorrectly folded, perhaps because of cellular stress conditions, are not directly tagged for degradation. Such proteins are recognized by so-called stress proteins, which facilitate the refolding of the protein. One of these stress proteins is the **heat shock protein hsp70.** It has been found to form complexes with incorrectly folded proteins. It has been suggested that hsp70 helps to solubilize and refold the abnormal protein in a process that is coupled to the hydrolysis of ATP.

Another type of stress proteins that may serve the same function are the **glucose-regulated proteins (GRPs)** found in the ER. The synthesis of these proteins is specifically induced by incorrectly folded proteins.

6.8 CONTROL OF GENE EXPRESSION IN PROKARYOTES

Many enzymes that serve housekeeping functions, such as those involved in oxidative phosphorylation, citric acid cycle reactions, or glycolysis, are required at all times. In order to maintain a constant cellular level of these enzymes, they must be synthesized continuously since all proteins have finite lifetimes and are being degraded at all times. These are the **constitutive enzymes.** On the other hand, there are enzymes that are required only in special conditions. These enzymes are normally synthesized at very low levels, but their synthesis can be induced to proceed at much higher rates. For example, when lactose is added to *E. coli* bacteria growing on glycerol as their sole carbon source, the bacteria begin synthesizing β-galactosidase, an enzyme that converts the undigestible disaccharide lactose into the metabolizable monosaccharides glucose and galactose. β-galactosidase is an **inducible** enzyme. The molecular mechanism of enzyme induction has been worked out by **Jacob and Monod** and is known as the **operon hypothesis.** Its mechanism of operation is now well established.

6.8.1 The Lactose Operon

i: Gene for repressor
p: Promoter
o: Operator
z: Gene for β-galactosidase
y: Gene for galactoside permease
a: Gene for thiogalactoside transacetylase

Figure 6.54.
The lactose operon and its regulator gene.

When **lactose** is added to a culture of *E. coli* growing in the presence of glycerol, rapid synthesis of **β-galactosidase** begins within about 2 min. Together with β-galactosidase, two other enzymes, galactoside permease and thiogalactoside transacetylase, are **induced.** (The permease is required for uptake of lactose; the transacetylase's function in the cell is not known.) The genes for these three enzymes belong to a single **operon,** their structural genes lie next to each other on the chromosome and are transcribed into a single mRNA molecule. The expression of this operon is regulated by the interaction of a regulatory factor, the **repressor,** with the **operator,** a segment of DNA preceding the **structural genes** that encode the enzymes (Figure 6.54).

The mechanism of regulation of the lactose operon is as follows: The **regulator gene *i*** (which need not be located next to the operon it controls) codes for the repressor protein. The repressor has a very high affinity for the **operator sequence *o*.** RNA polymerase attaches to the **promoter *p*,** located just upstream of, and partially overlapping, the operator, but is prevented from binding if a repressor occupies the operator. Thus in the presence of an active repressor, transcription of the entire operon is shut down. However, in the presence of the **inducer** lactose (the intracellular inducer actually is **allolactose,** or β-1,6-galactosyl glucose, which is made from lactose by β-galactosidase), the repressor forms a complex with the inducer. This results in a conformational change of the repressor, which in turn drastically reduces its affinity for the operator. As a result, RNA polymerase can now bind to the promoter and begin the synthesis of the multicistronic mRNA coding for the three proteins of the operon.

The lactose repressor is a tetrameric protein with a dyad symmetry. It shows a very high affinity for the lactose operator sequence ($K_{diss} = 2 \times 10^{-12}\ M$). In the presence of inducer, the operator-repressor complex is about 2×10^4 times weaker. In wild-type cells, the concentration of the repressor is about 1,000 times the dissociation constant, which would keep the operator closed 99.9% of the time. After induction, on the other hand, the operator would be closed only 5% of the time.

Some of the genetic data that led to the operon model need to be discussed. Operator mutations, o^c (*c* for **constitutive**), exist in which the operator no longer binds to the repressor. β-Galactosidase is then produced constitutively. Partially diploid bacteria can be constructed that carry two copies of the lactose operon: one on the chromosome and another on a plasmid. These **merodiploids** make it possible to study the functional interaction of two different alleles of the lactose operon present in the same cell. For example, o^+z^-/o^cz^+ merodiploids (one operon has a normal operator but a mutation in the *z* gene, whereas the other operon has a normal *z* gene connected to an o^c operator) were found to produce β-galactosidase constitutively. The reciprocal merodiploid $o^c z^- / o^+ z^+$, however, synthesized β-galactosidase only upon induction. These results indicate that the *oc* mutation can confer constitutivity only on the adjacent structural genes; it has no effect on another lactose operon (see behavior of the reciprocal merodiploid). The mutant o^c operator, therefore, is said to be *cis* dominant over the wild-type operator o^+.

A mutation in the regulator gene *i* can render the repressor either unable to bind to the operator (i^-), or unable to bind to the inducer (i^s the *s* stands for superrepressor). Although i^+ (wild-type) is *trans* dominant (i.e., dominant irrespective of its location with respect to the *z* gene) over i^-, *it* is *trans* dominant over both i^- and i^+. This is because the **superrepressor** cannot bind inducer and thus covers all lactose operators in a cell whether or not inducer is present. Phenotypically, the i^s-mutant cell is lac^-, i.e., defective in β-galactosidase synthesis. i^- mutants are constitutive β-galactosidase producers.

As a rule, ***trans*-acting mutations** affect genes that produce a product, in this case the repressor, which can diffuse through the cytoplasm of the cell and interact with any DNA sequence. ***Cis*-acting mutations,** on the other hand, occur in DNA regions that do not code for any product, such as operators or promoters. They can only affect the expression of genes that are located right next to them.

The mechanism of enzyme induction is an example of **negative gene control.** The repressor is a negative controlling element that has to be removed from the DNA in order to allow transcription to occur. There are also cases of **positive gene control.** An example is the σ factor, a controlling element that activates RNA polymerase to allow transcription initiation.

6.8.2 Catabolite Repression

The induction of β-galactosidase synthesis by lactose does not occur if the growth medium contains glucose. This phenomenon was named **"catabolite repression"** because it had originally been thought to be caused by a catabolite of glucose. We now know that the intracellular level of **cyclic AMP (cAMP)** is lowered from about 10^{-4} *M* to 10^{-7} *M* in the presence of glucose. cAMP stimulates the expression of several genes that are subject to catabolite repression. cAMP binds to a **catabolite gene activator protein (CAP),** thereby enabling it to bind to the promoters of genes that are subject to catabolite repression. The cAMP-activated CAP functions as a **transcription factor** for RNA polymerase similar to the transcription factors that bind to upstream activating sequences. Catabolite repression is another example of positive control.

The binding site for CAP (−87 to −49) is located immediately upstream of that of RNA polymerase (−48 to +5). Similar to the site where the repressor binds, the CAP-binding site has a dyad symmetry (Figure 6.55). Such **palindromic sequences** are often found at sites where regulatory proteins bind. They allow the protein to slide along the DNA and recognize its binding sequence coming from either direction. (A palindrome is a sentence that can be read forward and backward, giving the same meaning; for example, "DNA HELD IN AN IDLE HAND".)

The CAP protein and some other DNA-binding proteins have been

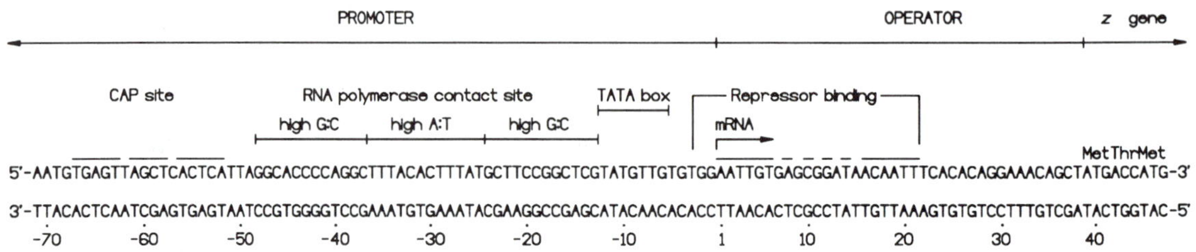

Figure 6.55.
The nucleotide sequence of the lactose promoter. Palindromic sequences are indicated by lines drawn over the base letters.

analyzed by x-ray crystallographic studies. They all have an exposed α helix that appears to fit into the major groove of the DNA helix where the side chains of amino acids form hydrogen bonds with bases in the DNA. The CAP protein is a dimer in which the two α helices are related to each other by a twofold axis of symmetry and are 34 Å apart. This distance corresponds to the distance between two neighboring wide grooves on the double helix. In vitro mutagenesis experiments (see next chapter) on a phage repressor have shown that when specific amino acids in this DNA-binding α helix are altered, the specificity of the repressor can be changed to that of a related phage.

6.8.3 The Tryptophan Operon

E. coli cells growing on a synthetic medium containing ammonium ions as the only nitrogen source express the enzymes required for the synthesis of all 20 amino acids. However, if one of the amino acids, for example tryptophan, is added to the culture medium, the synthesis of the enzymes required for tryptophan biosynthesis is rapidly reduced to very low levels. This phenomenon is called **enzyme repression.** The mechanism of repression is very similar to that of enzyme induction. In repressible operons, the repressor cannot bind to the operator unless it is first activated by binding to a **corepressor.** In the case of the tryptophan operon, a transcriptional unit specifying the structures of five enzymes required for tryptophan biosynthesis, the corepressor is tryptophan itself. In the absence of tryptophan, the operon is said to be **"de-repressed"** (in contrast to **"induced"**); that is, it is fully active.

The tryptophan operon contains a second control site, the **attenuation** region, located about 140 nucleotides downstream from the mRNA initiation site. mRNA synthesis is aborted prematurely at this site if the level of tryptophan rises above a certain level. The mechanism of attenuation is illustrated in Figure 6.56. The leader transcript codes for a 14-amino-acid oligopeptide that includes two adjacent Trp residues. It also contains four regions (1, 2, 3, and 4) capable of forming three **stem** and **loop structures,** SL-1 : 2, SL-2 : 3, and SL-3 : 4. SL-3 : 4 is a transcription termination site; it is immediately followed by a stretch of Us. SL-3 : 4 can only form if SL-2 : 3 is prevented from forming.

In the absence of tryptophan (i.e., when full expression of the operon is required), ribosomes pile up at the two Trp codons located in region 1. As a consequence, SL-1 : 2 cannot form, and SL-2 : 3 forms instead. This prevents the formation of SL-3 : 4, and mRNA synthesis is not aborted since SL-2 : 3 is not recognized as a transcription termination signal.

In the presence of tryptophan, however, ribosomes move into region 2 of the leader transcript before it can base-pair with region 3, thus preventing the formation of SL-2 : 3. This allows the formation of SL-

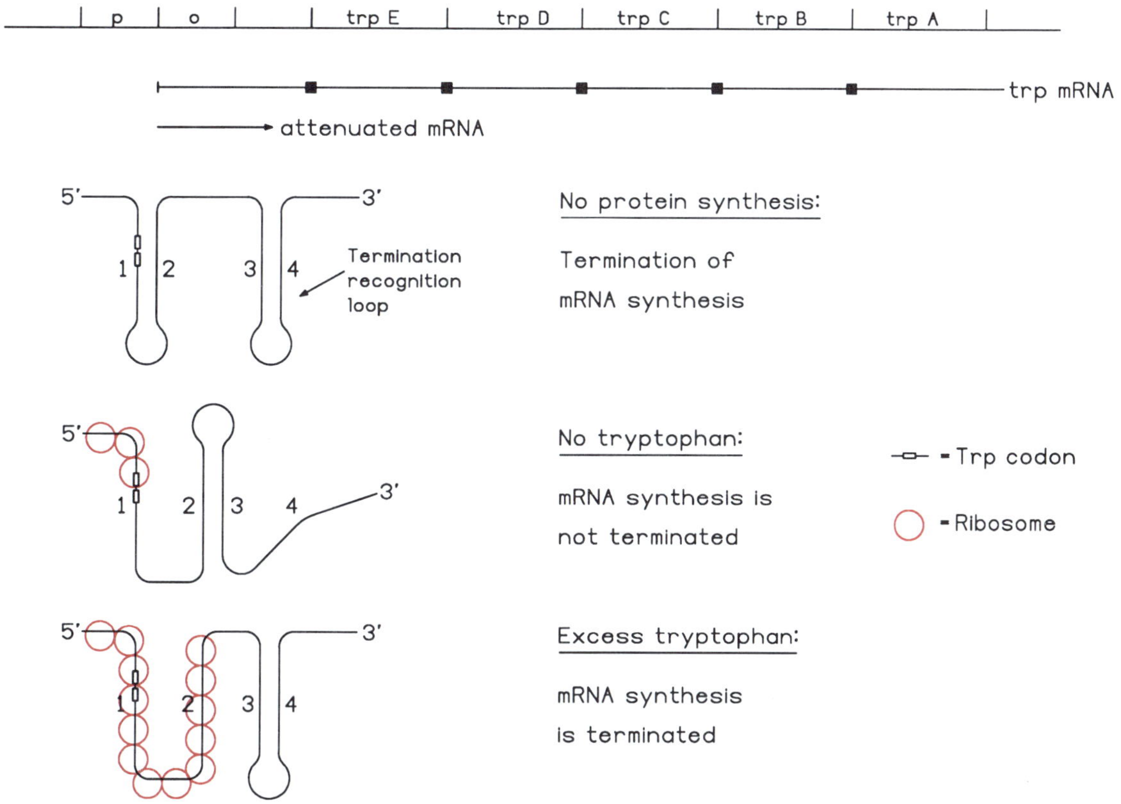

Figure 6.56.
Mechanism of attenuation of Trp mRNA synthesis. From Oxender DL, et al. *Proc Natl Acad Sci USA.* 1979; 76:5524.

3:4, which results in premature transcription termination. Attenuated transcription also occurs when no translation takes place. Under those conditions both SL-1:2 and SL-3:4 form.

It should be noted that transcription attenuation appears to be the main mode of control for several other operons involved in the biosynthesis of amino acids. The leader sequences of the mRNAs transcribed from the histidine operon, the threonine operon, and the phenylalanine operon all contain seven or eight codons for the respective amino acid. This allows for very sensitive attenuation of transcription depending on the cytoplasmic levels of the amino acids.

6.8.4 Regulation of rRNA and tRNA Synthesis

An *E. coli* cell contains about 10^4 ribosomes. In order to be able to synthesize so many ribosomes within only 20 minutes, the generation time of *E. coli,* the chromosome contains seven copies of the rRNA precursor gene. The rate of rRNA and tRNA synthesis is closely regulated in relation to the growth rate of the cell. When the rate of protein synthesis is reduced, the rate of stable RNA synthesis also decreases. The mechanism of this so-called **stringent response** is as follows. A **guanosine tetraphosphate, ppGpp,** is generated from GTP on the ribosome when an uncharged tRNA occupies the A site. Studies with cell-free systems suggest that ppGpp binds to RNA polymerase and reduces its affinity for rRNA and tRNA promoters.

6.8.5 Translational Regulation

As described in the previous sections, most prokaryotic gene regulation occurs at the transcriptional level. From an energetic point of view this makes sense. It is clear, however, that translational control exists as well. Often the enzymes of a single operon are synthesized at strikingly different rates. This can be explained by variations in the Shine-Dalgarno sequence making some sites more favorable for attachment of ribosomes.

An interesting kind of translational control is the regulation of expression of the translation release factor RF2 in *E. coli.* It is accomplished by **translational frameshifting.** The mRNA for RF2 contains a UGA stop codon after amino acid 25 (Leu). Thus, in the presence of enough RF2 protein, the translation of RF2 mRNA comes to a premature stop. However, if there is not enough RF2 in the cell, a translational frameshift occurs in which the U residue in the UGA codon is skipped and the next codon is read as GAC (Asp).

UGA = stop

5′ . . . GGG U AU CUU U GAC U AC GAC . . . 3′

Gly23 Tyr24 Leu25 Asp26 Tyr27 Asp28

This frameshifting establishes an open reading frame for the complete translation of the RF2 protein. Translational frameshifting is catalyzed by certain types of tRNA. It commonly occurs during the translation of some retroviral mRNAs (see Chapter 7).

As discussed earlier, the genes for **ribosomal proteins** are contained in several different operons, some of which also include the genes for subunits of RNA polymerase, elongation factors, as well as DNA primase. The translation of these polycistronic mRNAs is regulated by free ribosomal proteins acting as **translational repressors** on their own mRNA. For example, the mRNA transcribed from the L11 operon, which codes for L1 and L11, is specifically inhibited from being translated when free L1 protein binds to the 5′ end of the mRNA. Although L1 has a higher affinity for 23S rRNA, it will bind to its own mRNA when no 23S rRNA is available. This type of mechanism also works with the mRNAs of the other ribosomal proteins, thereby balancing their synthesis with rRNA synthesis. The inclusion of the RNA polymerase subunits in these regulatory units suggests that the synthesis of RNA polymerase is balanced with overall protein synthesis as well.

6.9 EUKARYOTIC GENE ORGANIZATION AND EXPRESSION

The human genome contains about 1,000 times as much DNA as a bacterium. This DNA is present in the form of 23 chromosomes (haploid number), each containing one continuous double strand of DNA condensed into a tight package, called **chromatin.**

The selective expression of different sets of genes in different somatic cells underlies the process of cell differentiation. The mechanisms of eukaryotic gene regulation are still only partly understood. In many respects, regulatory mechanisms are similar to those observed in prokaryotes. However, because of the organization of eukaryotic DNA into chromatin, additional control mechanisms for gene expression have evolved. Furthermore, negative control elements, so predominant in prokaryotic gene regulation, appear to be used much less frequently. However, before we can discuss these mechanisms we have to describe the organization of eukaryotic genomes and the structure of chromatin.

6.9.1 DNA Sequence Organization

Although there is 1,000 times more DNA in a human cell than in a bacterium, the human cell certainly does not have 1,000 times as many genes. Even if we consider the presence of introns in eukaryotic genes and multiply the amount of DNA needed to code for an average protein by a factor of 10, the total amount of DNA in higher eukaryotic cells still exceeds by far the coding capacity required to encode all cellular proteins.

Analysis of the **renaturation kinetics** of eukaryotic DNA revealed that much of this extra DNA is contained in repetitive nucleotide sequences. When eukaryotic DNA is denatured and then allowed to renature, the observed reassociation kinetics reveals the presence of three types of sequences:

1. Unique sequences (1 copy per genome)
2. Intermediate repetitive sequences ($10^2 - 10^5$ copies)
3. Highly repetitive sequences ($> 10^5$ copies)

The relative amounts of the three classes of DNA sequences can be determined by measuring the rate of **renaturation** of single-stranded DNA (Figure 6.57).

Renaturation is a bimolecular reaction in which the reaction rate is directly proportional to the product of the concentrations c of the two homologous strands. Knowing that any denatured DNA sample contains as many "Watson" strands as "Crick" strands, the rate of renaturation can be expressed as

$$\frac{dc}{dt} = k_2 c^2 \tag{6.1}$$

where k_2 is a simple rate constant whose value is inversely proportional to the sequence complexity of the DNA. Integration of equation (6.1) gives

$$\frac{c}{c_0} = \frac{1}{(1 + k_2 c_0 t)} \tag{6.2}$$

where c is the concentration of single-stranded DNA (expressed in moles of nucleotides per liter) at time t (in minutes), and c_0 is the concentration of DNA at time zero.

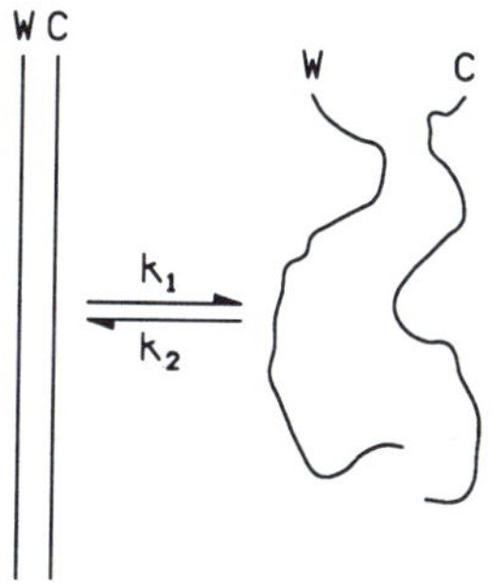

Figure 6.57.
Renaturation kinetics of DNA.

By convention, one usually measures the time that is required for the renaturation of one half of the DNA. Expressed in mathematical terms,

$$\text{when } \frac{c}{c_0} = \frac{1}{2}, \qquad \text{then} \qquad k_2 = \frac{1}{c_0 t_{1/2}}. \qquad (6.3)$$

In other words, the product of the initial concentration of single-stranded DNA, c_0, and the time required to renature 50% of the DNA, $t_{1/2}$, is inversely proportional to the rate constant k_2 of the renaturation reaction. This $c_0 t_{1/2}$ value, usually simply called the **Cot value,** is directly proportional to the complexity of the genome (defined as the number of unique sequences of the genome). Figure 6.58 illustrates the **Cot curves** for DNAs of different complexity.

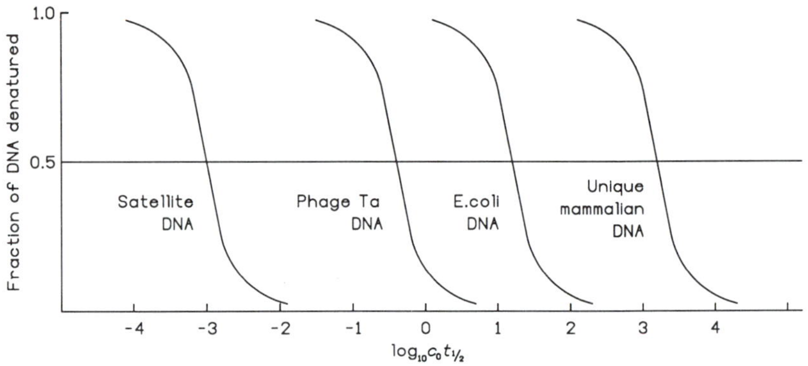

Figure 6.58.
Comparison of Cot curves for DNA of different sequence complexity.

When human DNA is sheared to fragments of about 500–1,000 nucleotides, denatured, and then allowed to renature, the formation of double-stranded DNA occurs in three phases. At first, the highly repetitive sequences (10%; low complexity) renature with Cot values of about 10^{-3}. Intermediate repetitive sequences (20%) follow with Cot values between 10^{-2} and 1. Finally, unique sequences (70%; high complexity) renature with a Cot value of about 10^3 (Figure 6.59).

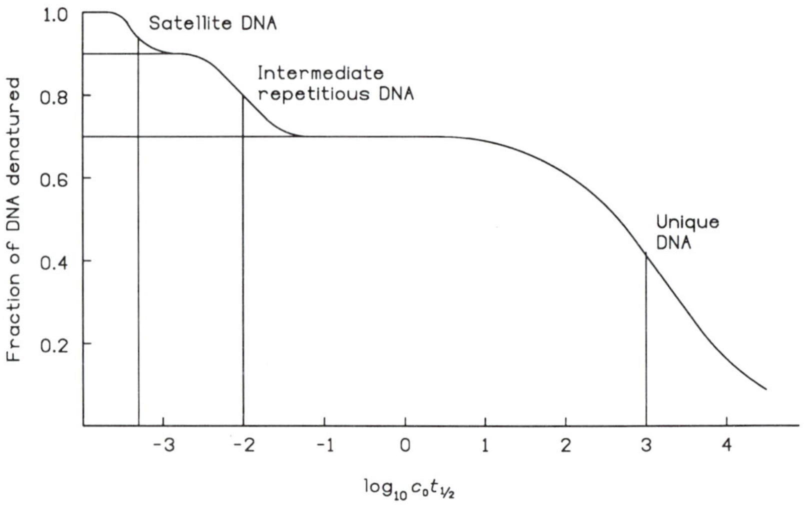

Figure 6.59.
Cot analysis of human DNA.

Highly repetitive DNA usually consists of tandemly arranged sequences, often only a few nucleotides in length, that are not transcribed and are frequently located in centromeres of chromosomes. It is likely that these sequences function in the recognition and alignment of chromosomes during meiosis and mitosis. If the base composition of such a highly reiterated sequence is sufficiently different from the overall base composition of the chromosomal DNA, the repetitive DNA can be seen as a satellite peak in CsCl density gradients and isolated as **satellite DNA** (Figure 6.60).

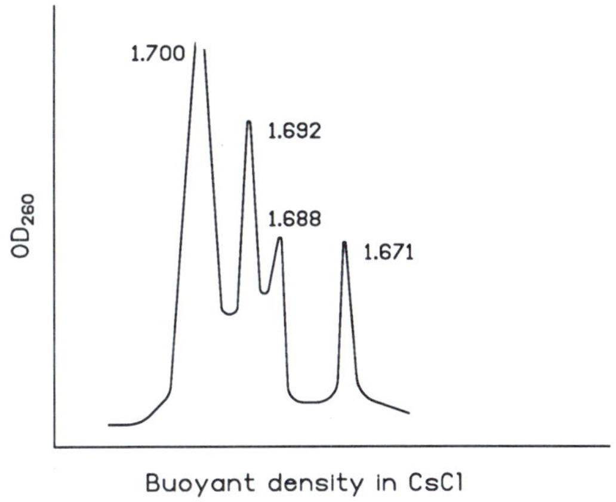

Figure 6.60.
Satellite DNAs in *Drosophila virilis*. Sheared *Drosophila* DNA was centrifuged to equilibrium in CsCl. About 25% of the genome is satellite DNA.

Other repetitive sequences may be dispersed throughout the genome. In the human genome there are almost a million copies of a 300 nucleotide sequence, called the ***Alu* sequence,** because it contains a single *AluI* restriction endonuclease site. The dispersion of the *Alu* sequences throughout the human genome is thought to have occurred by reverse transcription of an RNA template. *Alu* sequences have poly(A) tails and are often flanked by direct repeats. This suggests some kind of transposition event involving reverse transcriptase. It is interesting that the 7SL RNA found in SRP (see Figure 6.61) exhibits strong homology to the Alu sequence at both ends. In addition to the *Alu* sequences, the human genome contains other **interspersed repeated sequences** that are repeated about 104 times.

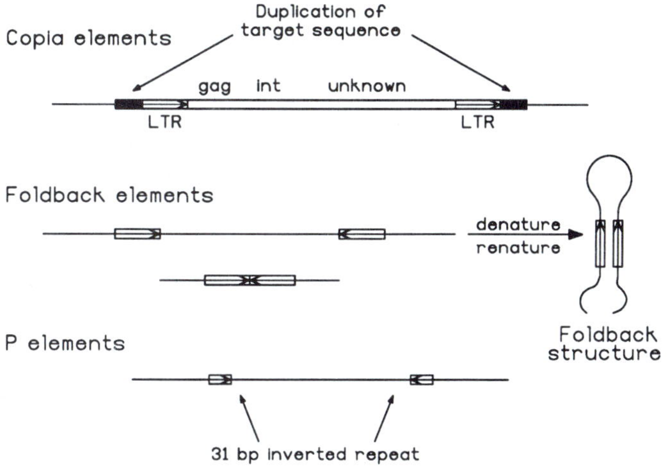

Figure 6.61.
The transposable elements found in *Drosophila*.

Eukaryotic genomes also contain many copies of **transposable elements** (similar to the prokaryotic transposons) capable of moving around in the genome and causing various kinds of rearrangements. In *Drosophila*, for example, three types of such mobile elements are known (Figure 6.61).

The **copia element** is about 5,000 base pairs long and characterized by long terminal repeats of a few hundred base pairs. At both ends of the element are short inverted repeats and direct repeats of the target sequence into which they had inserted themselves, similar to those found in prokaryotic transposons. The copia element is actively transcribed and appears to code for a reverse transcriptase. In many respects it resembles retroviruses that also insert themselves into the genome by reverse transcription. (A similar type of element, called the **Ty-1 element,** is found in yeast.)

Another type of transposable element in *Drosophila* is the foldback (FB) element. It is characterized by terminal inverted repeats that form stem and loop structures upon denaturation. Their mechanism of transposition may be very similar to prokaryotic transposons.

Finally, the **P elements** are between 500 to 1,400 nucleotides long and have perfect 31 base pair inverted repeats at their ends. They are distributed throughout the *Drosophila* genome. P elements are actually defective P factors containing a large internal deletion. Complete P factors are found in so-called P strains of *Drosophila*. The P factor is about 3,000 base pairs long and appears to code for a transposase. P factors are used as vectors for the insertion of DNA into the *Drosophila* chromosome.

As the example with *Drosophila* shows, mobile genetic elements are found in many eukaryotic organisms. They are certainly responsible for part of the repetitive sequences found in all higher eukaryotic genomes.

Another source of repetitive DNA are **gene families.** This includes the genes for ribosomal RNAs and the histone genes. In both cases, the genes are tandemly repeated with spacer regions in between. In humans, there are several hundred 45S rRNA genes and 30–40 histone genes. The existence of these large gene families raises an interesting question: What mechanism keeps the sequences in these families homogeneous? The normal rate of mutation and recombination would tend to introduce alterations in the basic sequence. Nevertheless, little variation exists. Since even the large spacer regions that separate the individual genes show little sequence divergence, evolutionary pressure cannot be responsible for sequence maintenance. There must be some kind of rectifying process. This could occur by one of three mechanisms:

1. Gene amplification (as observed with rRNA genes in amphibian oocytes)
2. Recombination
3. Gene conversion, the repair of a gene according to the sequence of a master gene (at this time we have no idea about how this sequence alignment is accomplished)

A eukaryotic cell has the capability to amplify certain sequences when selective pressure is applied. For example, when cancer patients are treated with methotrexate, an inhibitor of dihydrofolate reductase, cells often arise that have become resistant to the drug. When such cells were analyzed, it was discovered that they contained hundreds of **tandem repeats** of the gene for dihydrofolate reductase. (The repeating unit was considerably larger than the length of the gene as would be expected if duplication occurred by a recombinational event.) This allowed the cell to overproduce the enzyme and thereby compensate for the inhibitory effect of methotrexate. Such **gene amplifications** probably occur frequently but are not noticed because of the absence of selective pressure.

Some cancer cells that have become resistant to methotrexate showed no amplification of the dihydrofolate reductase gene. These cells had become resistant not only to methotrexate but also to a large number of other drugs; they had developed **multidrug resistance (MDR).** Multidrug resistance is due to the overexpression of a **P-glycoprotein,** a normal cellular glycoprotein that is involved in the export of foreign substances from the cell. It is a membrane transport protein that requires ATP hydrolysis for active drug transport.

Most genes that code for proteins are **single-copy genes.** The total number of genes in the human chromosome has been estimated to about 100,000. If we assume that the average gene contains about 1,000 nucleotides (excluding any intron sequences), then 100×10^6 nucleotides will suffice to code for all human proteins. This represents only about 2.5% of the entire human genome. We do not know the function of the other 97.5% of the DNA. It has been speculated that most of that DNA is **"selfish DNA,"** that is, it has the ability to replicate as part of the human genome and to expand by transposition but contributes little, if anything, to the cell, with the exception perhaps that it provides a pool of sequences for further evolution.

Many single-copy genes have actually evolved into families of very similar genes. The human hemoglobin genes, for example, have evolved from a single ancestral gene. There are now two families, the α-type genes and the β-type genes. Each subfamily consists of a cluster of very closely related genes located on two different chromosomes. The individual genes are expressed at different times during development (see Chapter 9). In addition to functional hemoglobin genes, the gene clusters also contain several **pseudogenes.** These are genes that contain mutations and, therefore, are not expressed. They appear to be relics of evolution. A special type of pseudogene is the **processed pseudogene.** It contains no introns and often has a poly(A) tail at its 3′ end. Processed genes also lack promoter sequences and are flanked by short direct repeats. This suggests that they have arisen by reverse transcription of an mRNA molecule (compare retroviruses, Chapter 7).

6.9.2 Chromatin Structure

Eukaryotic DNA is highly condensed by being complexed with protein. This complex is called chromatin. The degree of condensation becomes apparent when one compares the total length of the DNA contained in all human chromosomes with the combined length of the chromosomes at metaphase. The contour length of 8 billion base pairs of human DNA is about 2.7 m, while the combined length of the metaphase chromosomes is only 200 μm. How is this 10,000-fold condensation accomplished?

The first level of condensation is the formation of a complex between the DNA double helix and **histone proteins.** There are five different histones, H1, H2a, H2b, H3, and H4. Under the electron microscope, **chromatin** has a **beads-on-a-string structure,** each bead representing a **nucleosome,** a complex of 200 base pairs of DNA with nine histone molecules. There are 140 base pairs of the DNA wound around the nucleosome core, forming 1.75 turns of a left-handed helix. The nucleosome core is a disk-like particle made from two copies each of histones H2a, H2b, H3, and H4. The addition of a single copy of the fifth histone, H1, completes the nucleosome. It is bound to the linker DNA connecting two nucleosome cores.

When DNA complexed with nucleosomes is treated with low concentrations of **micrococcal DNase,** a ladder of fragments is generated that are all multiples of 200 bp. This indicates that the linker DNA connecting two nucleosome core particles is more accessible than the DNA wound around the cores. Digestion with high concentrations of DNase I cleaves even the DNA wound around the nucleosome cores. The single-stranded fragments produced under these conditions differ in size by only 10 nucleotides, reflecting the turns of the double helix. Apparently, the phosphodiester bonds located on the outside of the loops that are wound around the nucleosome are most susceptible to cleavage.

The interaction between H1 molecules on neighboring nucleosomes forces the DNA into a higher order helical superstructure. The exact configuration of this structure is not known. However, it forms a **30 nm chromatin fiber** that can be seen in interphase nuclei.

An even higher order of condensation is achieved by attachment of the 30 nm chromatin fiber to a nuclear protein scaffold, called the **metaphase scaffold.** The loops between attachment points, also called **looped domains,** contain between 35 and 100 kd of DNA. Two proteins have been isolated from the histone-depleted scaffold structure. One of these proteins is a topoisomerase II. Figure 6.62 illustrates the various levels of chromosome condensation.

Figure 6.62.
Model of different levels of chromatin condensation. Adapted from Alberts et al. *Molecular Biology of the Cell,* 2nd ed. New York, NY: Garland Publishers.

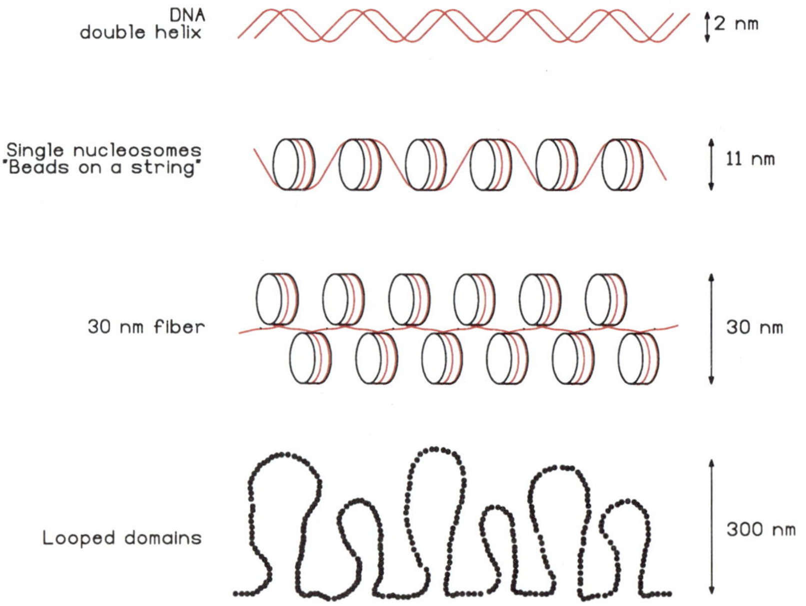

Histone H1 may be involved in the condensation of chromatin that occurs before mitosis. Many of its serine residues are found to be phosphorylated in mitotic chromosomes. Shortly after mitosis, the level of phosphorylation decreases again.

Looped domains can be visualized as bands in the **polytene chromosomes** found in the salivary glands of certain insect larvae. A polytene chromosome consists of about 1,000 copies of the chromosome arranged in a thick bundle in which individual looped domains are in register. They arise by repeated cycles of DNA replication of most genes. Only the centromeric regions are not duplicated, and the ribosomal RNA genes are duplicated less frequently. Upon staining, about 5,000 individual bands can be observed under the light microscope. It is now known that each band contains up to about 10 genes. Thus, there are probably fewer than $10 \times 5{,}000$ genes in the *Drosophila* genome.

The highest degree of condensation of chromatin is found in sperm heads. In order to achieve this compaction of DNA, histones are replaced by the smaller **protamines,** a highly basic class of proteins that appear to form α helices upon binding to DNA.

Histones are synthesized during the S phase of the cell cycle. As the two daughter DNA duplexes emerge from the replication fork, the old nucleosomes stay with the strand that serves as template to the leading strand. The daughter duplex with the lagging strand acquires new nucleosomes.

6.9.3 Activation of Chromatin

Regulation of gene expression in eukaryotes probably occurs at several levels. The first level is the overall accessibility of the chromatin to transcription. For example, some parts of the chromatin remain tightly packed even during interphase. This type of chromatin is called **heterochromatin** and usually includes the centromeric region as well as other nontranscribed regions. Chromatin that is being transcribed is called **euchromatin.** Sometimes chromatin that is active during early development becomes heterochromatic in adult tissue. If a gene is inserted into a heterochromatic region, it becomes inactivated. The same gene can be expressed if inserted into a euchromatic region of the chromosome. This suggests that chromatin structure is important in regulating large segments of chromosomes.

In mammalian females (genotype XX), one of the two **X chromosomes** remains highly condensed during interphase while only the other X chromosome is genetically active. The condensed, and thus largely inactivated, X chromosome is called the **Barr body.** This inactivation of one of the two X chromosomes may be necessary in females in order to maintain a single dose of X-linked genes and thus the same kind of genetic balance in relation to genes located on other chromosomes that is found in males (genotype XY). The Barr body is missing in women with **Turner's syndrome** (genotype X0) and is present in men with **Klinefelter's syndrome** (genotype XXY).

Females who are heterozygous for **glucose 6-phosphate dehydrogenase deficiency** (the gene is located on the X chromosome) have two populations of cells, one with normal levels of enzyme, and the other without any enzyme at all. This **mosaicism** of genotypes in one organism is explained by **Lyon's hypothesis.** Mary Lyon postulated that the decision as to which X chromosome is condensed, and thereby largely inactivated, is made randomly at every cell division during early embryogenesis. At some time during development, however, the decision becomes fixed, and the same X chromosome remains heterochromatic in all descendants of a cell. (Only a few genes are expressed on the heterochromatic X chromosome.) The mechanism for this specific chromosome inactivation is not known.

Active chromatin can be recognized on polytene chromosomes in the form of enlargements or puffs. **Puff formation** can be induced by treatment with the insect molting hormone ecdyson or by heat shock (see later). Because active chromatin has a more open structure, it is also more susceptible to nuclease. Typically, treatment with DNase I produces a ladder of fragments that differ by 200 bp. The activity of a specific gene can be determined by hybridizing these fragments to a labeled probe for this gene. The nuclease-sensitive region usually extends over the entire gene. In addition to this general sensitivity of active chromatin to nuclease, there are so-called **hypersensitive sites** within the 5′-flanking region of active genes.

The **sensitivity of active chromatin to nucleases** is, at least in part, caused by an altered nucleosome structure. We have already discussed the **phosphorylation of histone H1** that occurs just before metaphase. Nucleosomes isolated from active chromatin are deficient in H1 and instead contain two **nonhistone proteins: HMG 14** and **HMG 17 (high-mobility-group proteins).** In addition, some of the other histones are acetylated, and the protein ubiquitin is found covalently attached to some of the H2A histones. The result of these alterations may be an opening up of the 30 nm chromatin fiber structure. Experiments with *Physarum polycephalum,* a slime mold, have shown that histone H3 in active chromatin reacts with SH-specific reagents, whereas H3 in inactive chromatin does not. Thus it appears that, although the histones remain associated with DNA during transcription, histone-histone interaction is reduced.

Activation of chromatin also appears to be correlated with the level of **methylation of the DNA.** On average, about 5% of all C residues are methylated at their 5′ position. This methylation occurs when C appears in the sequence CpG. DNA isolated from genetically active regions was found to be undermethylated. The mechanism by which methylation regulates chromatin activity is not known. Perhaps methylation interferes with the binding of transcription factors. Furthermore, it favors the formation of **Z-DNA.** Whatever the mechanism, it is clear that methylation is only one of many mechanisms of chromatin activation, since *Drosophila* lacks DNA methylation.

6.9.4 Gene Activation at the DNA Level

The activation of the chromatin is necessary but not sufficient for gene transcription. For RNA polymerase to be able to initiate transcription, **upstream activator sequences** and/or **enhancers** are required in addition to a functional promoter. Enhancer elements are sequence elements that can activate promoters over considerable distances when specific proteins bind to them. They function in either orientation and can be located upstream or downstream of the promoter. As discussed in the section on transcription, a possible model for their mechanism of action is the looping model (see Figure 6.63). Alternatively, they might activate a promoter by changing its chromatin structure.

The **glucocorticoid receptor** is an example of a protein binding to an enhancer element. The receptor is first activated by the hormone, and the hormone-receptor complex then binds to the enhancer. In the mouse mammary tumor virus (MMTV) genome there are multiple glucocorticoid receptor binding sites, each site containing a sequence closely related to the consensus sequence AGA(A/T)CAG(A/T) (Figure 6.63).

In response to the binding of the receptor to the enhancer elements, the virus chromatin becomes sensitive to DNase I.

Another example of specific induction of a set of genes are the so-called heat shock genes that occur in all organisms. When the temperature of a cell is suddenly raised, most transcription in the cell stops and a new set of proteins, the heat shock proteins, are synthesized. The activation of the **heat shock genes** is due to the activation of a heat shock transcription factor that binds to several heat shock elements located between 50 and 250 nucleotides upstream of the transcription start sites.

In addition to positively acting regulatory elements, recent evidence has uncovered the existence of **negatively regulating elements** analogous to the repressors in prokaryotes. Specific repressor-type proteins bind to regulatory elements in a promoter and thereby shut down the transcription of the adjacent gene. In analogy to the enhancers discussed above, there are also negatively controlling *cis*-acting elements that can be located at a distance from the transcription start site. They are called **silencers.**

Many of the proteins that recognize specific sequences in the DNA belong to one of three classes of common structural motifs: the **helix-turn-helix,** the **zinc finger,** and the **leucine zipper.**

Examples of DNA-binding proteins with the **helix-turn-helix motif** are the phage λ proteins cI and cro, as well as the catabolite activator protein of *E. coli.* Two α helices, connected by a turn, are arranged so that one helix, the recognition helix, lies in the major groove of the DNA while the other helix, situated at a right angle above the first helix, helps to lock the recognition helix in place.

Zinc finger motif proteins include the 5S ribosomal transcription factor (see Figure 6.41). Two polypeptide segments containing two Cys residues are arranged in antiparallel β sheets. Together with another α-helical segment containing two His residues, the segments form a fingerlike structure, stabilized by a Zn atom complexed between the

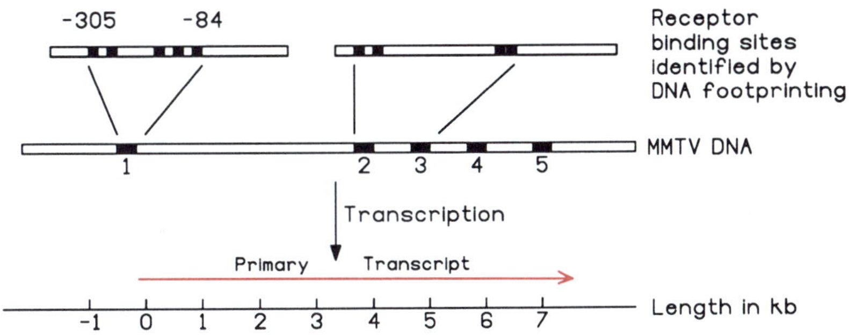

Figure 6.63.
The location of glucocorticoid receptor binding sites in the mouse mammary tumor virus genome.

His and Cys residues, that is thought to extend into the groove of the DNA double helix.

The third motif, the **leucine zipper,** allows two proteins to bind together, through a zipperlike hydrophobic interaction between leucine residues located on one side each of two α helices. Proteins with the leucine zipper motif include several nuclear oncogenes (see next chapter).

The **differentiation** of early embryonic cells into the specialized cells of adult tissue and the development of the body structure are regulated by turning specific genes on and off at the appropriate times. This has been studied most extensively in the fruitfly *Drosophila.* Shortly after fertilization, the nucleus of the *Drosophila* egg undergoes about eight rapid divisions. The 256 nuclei arising from this division then migrate to the plasma membrane where the division continues until about 6,000 nuclei are formed. Then, the formation of new cell membranes creates a monolayer of cells at the periphery of the original egg cell surrounding the yolk. This structure is called the **blastoderm.** The cells in the blastoderm are divided into **14 segments,** Md, Mx, Lb, T1–T3, and A1–A8. Each segment consists of a subgroup of cells that are already predetermined to develop into specific structures of the animal. Md, Mx and Lb form the head, the T segments form the thorax, and the A segments form the abdomen.

During **gastrulation** the cells continue to divide and migrate until the embryo reaches the larval stage. At this point, about 80% of the cells are functional while 20% remain functionally dormant. These groups of cells, called the **imaginal disks,** are arrested in further differentiation until pupation, when, under the influence of the hormone ecdyson, each disk develops into a different adult structure. Mutational analysis indicates that there are three types of genes that affect the segmentation pattern of the developing fly:

1. Maternal effect genes
2. Segmentation genes
3. Homeotic genes

Mutations in **maternal effect genes** affect the development of the embryo along its two main axes, anterior-posterior and dorsal-ventral. For example, the **bicaudal mutation** causes the formation of two posterior segments with mirror symmetry, the dorsal mutation results in an embryo with two dorsal parts, again in mirror symmetry, without any ventral part. Maternal effect mutations can be phenotypically suppressed by the injection of mRNA from an unfertilized egg into the mutant fertilized egg. This indicates that at an early stage, development is controlled solely by proteins arising from the maternal genome.

The products of the **segmentation genes** are responsible for the proper segmentation of the embryo. A widely studied mutation in this group is *fushi tarazu* (*ftz*); Japanese for "not enough segments") which leads to the formation of only one half of the normal number of segments.

Mutations in **homeotic genes** cause transformations of one type of segment into another. For example, the **Antennapedia mutation** transforms the antennae on the head of the fly into legs. Several different homeotic genes have been cloned (see next chapter) and analyzed.

They all code for proteins that share a 60-amino-acid highly conserved region called the **homeobox.** The high content of basic amino acids in this region suggests that the protein binds to DNA. This is supported by the observation that homeotic proteins are found in the nucleus. The homeobox sequence has been found with very few amino acid changes in many other species ranging from *Xenopus* to humans. It appears likely that homeotic genes code for transcription factors important in early development.

6.9.5 Posttranscriptional Regulation in Eukaryotes

The expression of a eukaryotic gene is not only regulated at the transcriptional level. Once the primary transcript is made, additional regulation occurs at the level of RNA splicing, RNA degradation, and translation of the mRNA into protein.

Often the same gene can give rise to two different proteins when expressed in different tissues. This is accomplished by **alternative splicing** of pre-mRNA. An example is the gene for the hormone **calcitonin.** In thyroid cells, it is transcribed into calcitonin mRNA, but in neurons, the same gene is transcribed into CGRP (calcitonin gene-related peptide) mRNA. Figure 6.64 illustrates how this is accomplished. There is some evidence that different tissues express different subsets of snRNPs recognizing alternate splice sites.

The half-lives of mRNAs for different proteins can differ widely. Usually, but not always, shortening of the 3′ poly(A) tail makes mRNA more susceptible to degradation. Some shortlived mRNA species lose their poly(A) tails very rapidly. How this shortening of the poly(A) tail is controlled is still not understood. In addition to the length of the poly(A) tail, the 3′-untranslated (3′ UTR) region of mRNA may play a role in the stability of mRNA. For example, β-actin mRNA has a very short poly(A) region but is relatively long-lived. It is possible that this mRNA has structural features in its 3′ UTR that make it particularly resistant to degradation. On the other hand, mRNAs with high turnover often contain multiple copies of an AU-rich sequence (consensus sequence UAUUUAU) in their 3′ UTRs. When this sequence is transferred to the 3′ UTR of hemoglobin mRNA, a normally long-lived mRNA, the stability of hemoglobin mRNA is reduced dramatically.

All eukaryotic cells contain a poly(A) binding protein (PABP), a 70 kd protein that, at least in yeast, appears to be essential for viability. It is thought that the complex between PABP and poly(A) stabilizes mRNA, although there are exceptions to this rule. There is also evidence that the poly(A)-PABP complex has a role in translation initiation. This would suggest an interaction between the 3′ UTR and the 5′ UTR.

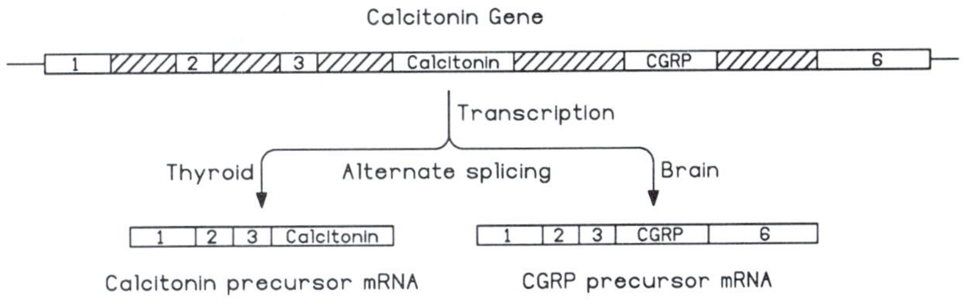

Figure 6.64.
Alternative splicing of calcitonin mRNA.

In addition to the 3′ UTR, there are probably other regions in the mRNA, both 5′ UTRs and internal sequences, that affect the rate of its translation either by forming secondary structures or by binding to specific regulatory proteins. Examples are the rapid degradation of tubulin mRNA that is induced by unpolymerized tubulin subunits and the stabilization of transferrin receptor mRNA by a protein that binds to its 3′ UTR (see Chapter 8).

Not only the stability of an mRNA species seems to be controlled by its nontranslated segments but also its rate of translation. For example, the efficiency of translation of pro-opiomelanocortin mRNA (see Chapter 13) is affected dramatically by changes in its 5′ UTR sequence as well as in its coding sequence. It appears that stable secondary structures form that decrease the rate of translation. Another example of translational regulation is the differential translation of two somatostatin (a growth inhibiting hormone [GIH]) mRNAs found in the pancreatic islet cells of the anglerfish. One mRNA species, GIH-2, contains three copies of an eight-nucleotide repeat (CCAGCAGA) in its 5′ UTR that reduces its translational efficiency. If this sequence is deleted, GIH-2 mRNA is translated with the same efficiency as GIH-1 mRNA that does not contain these regulatory sequences. This regulation of GIH-2 translation can be reproduced in widely divergent cells, suggesting that the octanucleotide sequence is recognized by common *trans*-acting factors.

For a more extensive discussion of biochemical genetics, consult the following textbooks:

Mathews et al. (1990): Chapters 4, 24–28
Stryer (1988): Chapters 4–6, 27, 33

Further References

Argetsinger-Steitz J. "Snurps". *Sci Am.* 1988; 258 (6):56–63.

Beardsly T. Smart genes. *Sci Am.* 1991; 265: 86–95. (A nice introduction to developmental genes.)

Cech TR. Self-splicing of group I introns. *Annu Rev Biochem.* 1990; 59:543–568.

de Robertis EM, Oliver G, Wright CVE. Homeobox genes and the vertebrate body plan. *Sci Am.* 1990; 263:46–52.

Echosl H, Goodman MF. Fidelity mechanisms in DNA replication. *Annu Rev Biochem.* 1991; 60:477–512.

Friedberg EC. *DNA Repair.* New York, NY: WH Freeman; 1985.

Jackson RJ, Standart N. Do the poly(A) tail and 3′ untranslated region control mRNA translation? *Cell.* 1990; 62:15–24.

Johnson PF, McKnight SL. Eukaryotic transcriptional regulatory proteins. *Annu Rev Biochem.* 1989; 58:799–839.

Harrison SC, Aggarwal AK. DNA recognition by proteins with the helix-coil-helix motif. *Annu Rev Biochem.* 1990; 59:933–970.

Hershey WB. Translational control in mammalian cells. *Annu Rev Biochem.* 1991; 60:717–756.

Kornberg A. *DNA replication,* 2nd ed. San Francisco, CA: WH Freeman; 1989.

Landy A. Dynamic, structural, and regulatory aspects of λ site-specific recombination. *Annu Rev Biochem.* 1989; 58:913–950.

Matson SW, Kaiser-Rogers KA. DNA helicases. *Annu Rev Biochem.* 1990; 59:289–330.

McKnight SL. Molecular zippers in gene regulation. *Sci Am.* 1991; 264:54–64.

Moyzis RK. The human telomere. *Sci Am.* 1991; 265:48–55.

Murray AW, Kirschner MW. What controls the cell cycle. *Sci Am.* 1991; 264:56–63.

Noller HF. Ribosomal RNA and translation. *Annu Rev Biochem.* 1991; 60:191–228.

Ptashne M, Gann AAF. Activators and targets. *Nature.* 1990; 346:329–331. (A discussion of the looping model.)

Sancar A, Sancar GB. DNA repair enzymes. *Annu Rev Biochem.* 1988; 57:29–68.

Sapienza C. Parental imprinting of genes. *Sci Am.* 1990; 263 (4):52–60.

Sawadogo M, Sentenac A. RNA polymerase B (II) and general transcription factors. *Annu Rev Biochem.* 1990; 59:711–754.

Schimke RT. Methotrexate resistance and gene amplification: Mechanisms and implications. *Cancer* 1986; 57:1912–1917.

Walker GC. Inducible DNA repair systems. *Annu Rev Biochem.* 1985; 54:425–457.

CHAPTER

7 Recombinant DNA Technology and its Application in Medical Science

OBJECTIVES

After studying this chapter, the student should be able to

1. Explain the significance of restriction endonucleases in recombinant DNA technology.
2. Describe the use of vectors for cloning eukaryotic DNA fragments.
3. Outline the principle of nucleotide sequence analysis of DNA.
4. Explain the term restriction fragment length polymorphism (RFLP) and describe the techniques used in the prenatal diagnosis of sickle cell anemia.
5. Describe the difference between genomic and cDNA libraries and know how to construct such libraries.
6. Describe the methods used in screening a library for specific genes.
7. Know the difference between insertion vectors and replacement vectors.
8. Explain how transcription start sites are mapped.
9. State the conditions that must be considered when a eukaryotic gene has to be expressed in a prokaryotic expression system.
10. Describe the *Xenopus* oocyte expression system.
11. Explain the principles of in vitro site-directed mutagenesis.
12. Explain the terms "primary culture" and "cell line."
13. Give a definition of the terms "oncogene" and "protooncogene" and give at least three different examples of cellular protooncogenes.
14. Draw the basic structure of a retrovirus genome.
15. Explain how DNA tumor viruses may transform normal cells into tumor cells.
16. Explain how retroviruses may transform cells.
17. Describe the life cycle of a retrovirus such as HIV.
18. Explain the mechanism of action of interferon.

Analysis of the complex organization of the eukaryotic genome has only become feasible through the application of in vitro recombinant DNA techniques. This new technology has allowed us to probe the molecular mechanisms underlying the problem of cancer, the functioning of the immune system, and the structure and organization of the acquired immunodeficiency syndrome (AIDS) virus genome.

7.1 PRINCIPLES OF MOLECULAR CLONING

The essence of molecular cloning is the propagation of a piece of eukaryotic (or prokaryotic) DNA in another cell. Usually, bacterial cells are used for this purpose because of their rapid rate of growth. However, eukaryotic cells, such as yeast, can serve the same purpose. The progeny of a single cell is called a **clone** since all cells derived from the original cell, into which the foreign DNA was introduced, contain exactly the same genetic information, including the piece of foreign DNA. This cloning technique thus allows the isolation and propagation of large quantities of a unique piece of eukaryotic DNA that would have been impossible to isolate and purify by other means from the original cell in which it occurred.

For the propagation in bacteria, the eukaryotic DNA fragment is spliced into a **plasmid or bacteriophage vector**. Plasmids are extrachromosomal circular DNA molecules that can replicate independently of the bacterial chromosome. The eukaryotic DNA fragment is inserted in vitro into the plasmid DNA, and the resulting **recombinant plasmid** is then reintroduced into the bacterium, where it replicates and is passed on to the daughter cells when the bacterium divides. Alternatively, the eukaryotic DNA may be inserted into the genome of a bacteriophage that is then used to infect a bacterial cell and produce large quantities of progeny phage, all of which harbor the same piece of foreign DNA.

7.1.1 Restriction Endonucleases

The most important tools for cutting specific DNA fragments from chromosomal DNA are the **restriction endonucleases**. These enzymes are found only in bacteria and are responsible for the restriction and modification of bacterial DNA as discussed in Section 6.2. There are three classes of restriction endonucleases. **Class I enzymes** are multienzyme complexes containing both restriction endonuclease and DNA methylase activities (the methylase always methylates the 6-amino group of adenine). They require *S*-adenosylmethionine and adenosine triphosphate (ATP) as cofactors and recognize specific oligonucleotide sequences in the DNA but cleave at random sites some distance from the recognition sequence. For this reason, they are not useful in recombinant DNA research. The **class II enzymes**, on the other hand, do not require cofactors and cleave the DNA at specific oligonucleotide sequences. (These sequences are modified by class II modifying enzymes, some of which methylate the 6-amino group of adenine, others the 5 carbon of cytosine.) Different bacteria produce restriction endonucleases with different recognition sequences. There is now a very large number of different restriction endonucleases known. Enzymes from different bacteria that recognize the same oligonucleotide sequence are called **isoschizomers**. The third class of restriction enzymes cuts 24–26 bases from the recognition site.

The oligonucleotide sequences recognized by restriction endonucleases usually consist of short **palindromes**. Cleavage occurs on both strands, leaving 5′-phosphoryl and 3′-hydroxyl termini. Often the cleavage is staggered, producing 2-nucleotide-long to 5-nucleotide-long cohesive ends; other enzymes cut both strands at the same position, producing blunt ends. For example, the enzyme **EcoRI** produces **protruding single strands** with 5′ termini, whereas **PstI** produces **protruding 3′-termini**. Table 7.1 lists a few examples of restriction endonucleases with their cleavage sites.

7.1.2 In Vitro Recombination of DNA Molecules

If the vector DNA, plasmid or bacteriophage, is cleaved with the same restriction enzyme that was used to generate the **DNA fragments**, the two molecules can be **annealed** because of their **complementary terminal oligonucleotide sequences**. Figure 7.1 illustrates how an EcoRI-generated *Drosophila* DNA fragment can be annealed to the termini of an EcoRI-cleaved plasmid that has a single EcoRI site. After annealing, the single-stranded nicks in the DNA are covalently linked with DNA ligase.

Another method for joining DNA molecules uses **homopolymer tails** added to the DNA termini by the enzyme **terminal nucleotidyl**

Table 7.1. Sequence Specificities of Some Restriction Endonucleases

Enzyme	Microorganism	Cleavage Site
EcoRI	*Escherichia coli*	G′AATTC CTTAA′G
HindIII	*Hemophilus influenzae*	A′AGCTT TTCGA′A
PstI	*Providencia stuartii*	CTGCA′G G′ACGTC
SmaI	*Serratia marcescens*	CCC′GGG GGG′CCC

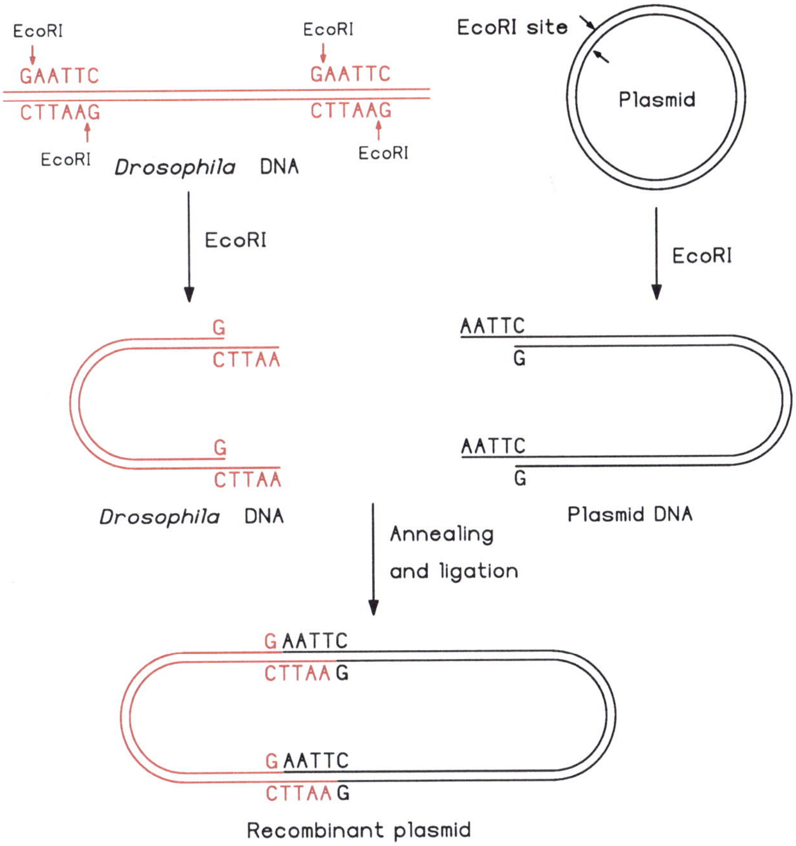

Figure 7.1.
Insertion of an EcoRI restriction fragment of *Drosophila* DNA into a plasmid.

transferase. This enzyme adds deoxyribonucleotides to the 3′-hydroxyl ends of DNA without requiring a template strand. For example, if the termini of the cleaved plasmid DNA are tailed with dG-homopolymers (about 10–20 nucleotides long) and the *Drosophila* DNA is tailed with dC-homopolymers, the ends of the two DNA molecules anneal because they are complementary to each other (Figure 7.2). One advantage of this method is that neither one of the two DNA molecules can cyclize or form long concatemers (linear end-to-end polymers).

7.1.3 Plasmid Vectors

One of the most widely used plasmids for cloning DNA fragments has been **pBR322**, although it is now being replaced by other more specialized vectors (see below). pBR322 was constructed specifically for the purpose of cloning DNA. Unlike some R plasmids (antibiotic resistance transfer factors), which are capable of catalyzing their transmission from one cell to another by the mechanism of conjugation, pBR322 is **nonconjugative**. Furthermore, pBR322 is under relaxed replicative control, which means that its replication is regulated independently from that of the bacterial chromosome. Normally, there are about 20 copies of pBR322 in a cell. This number, however, can be increased to several thousand if chloramphenicol is added to the bacterial culture. Under these conditions, chromosomal DNA replication stops because initiation of bacterial DNA replication requires protein synthesis,

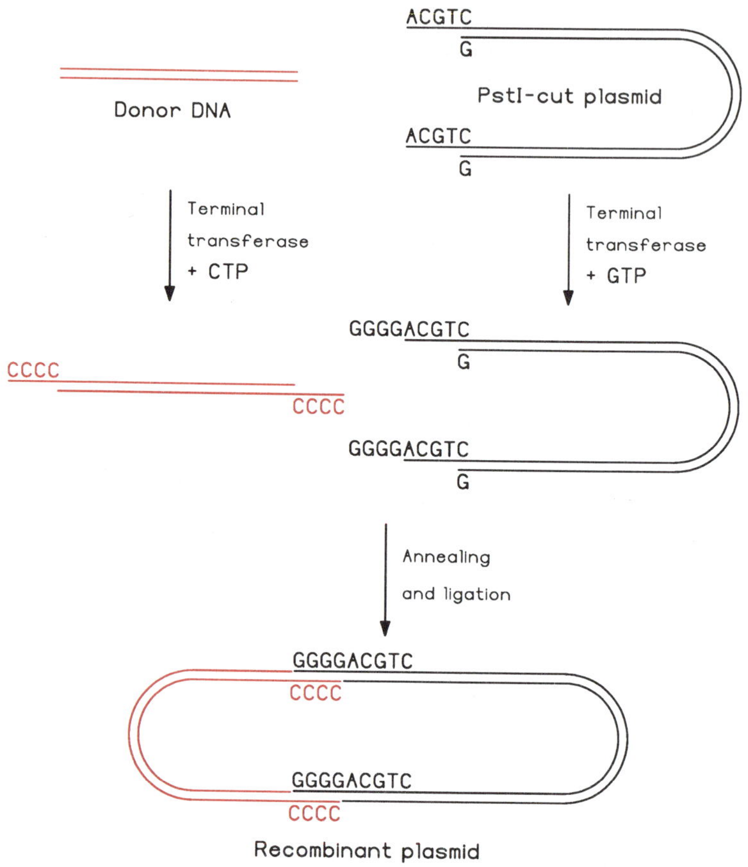

Figure 7.2.
Insertion of a DNA fragment into plasmid DNA by the homopolymer tailing method.

whereas plasmid replication continues until about 50% of the total cellular DNA is plasmid DNA.

The plasmid pBR322 has a molecular weight of 2.6 million; it confers to the host bacterium **resistance to** the antibiotics **ampicillin** and **tetracycline**. Its genetic map shows that the ampicillin resistance gene contains a unique PstI restriction site, and the tetracycline resistance gene contains a unique BamHI site. Insertion of a DNA fragment into either one of these sites will thus inactivate the corresponding gene (Figure 7.3).

After in vitro construction of the recombinant plasmid, the DNA is reintroduced into **bacterial cells** that have been made **competent** for DNA uptake by treatment with $CaCl_2$. (An alternative method for getting DNA, or any nucleic acid, into cells is **electroporation**. In this procedure the cells, bacteria or eukaryotic cells are subjected to a very brief pulse of a high-voltage electrical field.) Since this process of **transformation** (it is called transfection if done with eukaryotic cells) is rather inefficient, cells that have taken up the plasmid DNA must be **selected** for by their ability to grow in the presence of ampicillin or tetracycline. For example, if the foreign DNA is inserted in the BamHI site, the plasmid still confers ampicillin resistance to the bacterium. Any colony growing up on an ampicillin-containing medium must, therefore, have taken up a plasmid. This colony can then be tested for sensitivity to tetracycline, which would indicate the presence of foreign DNA in-

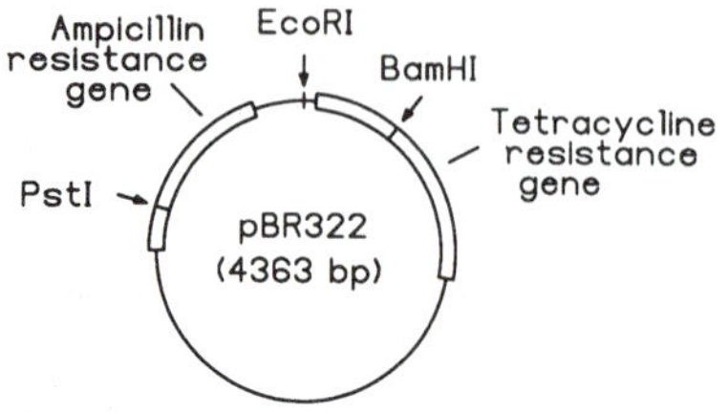

Figure 7.3.
Map of a cloning vector, the plasmid pBR322.

serted into the BamHI site. (Insertion at this site would inactivate the tetracycline resistance gene.)

7.1.4 Bacteriophage Lambda Vectors

Bacteriophage λ has about a 50,000 bp genome. Specific λ strains were created for cloning purposes that contain only one or two EcoRI restriction sites into which foreign DNA could be inserted. Normally, when λ DNA is packaged into phage particles, only about 6 kb (kilobase pairs) of additional DNA can be fitted into a virus particle. Thus, about 6 kb of foreign DNA can be **spliced** into a single EcoRI site located in a region of the λ genome not required for phage multiplication (**insertion vector**). If longer DNA fragments are to be cloned in phage λ, a segment of the λ genome must first be taken out. These specially constructed **replacement vectors** contain two EcoRI sites flanking a nonessential segment of the genome (the stuffer fragment). The vector DNA is cleaved with EcoRI, the two arms are separated from the stuffer, and a piece of foreign DNA, up to 24 kb in size, is inserted between the two EcoRI sites to replace the λ stuffer fragment.

The **recombinant phage DNA**, created in vitro, is then **packaged** by phage proteins to form viable phage particles. This is accomplished with two lysates prepared from mutant phage λ–infected cells, one containing empty phage heads without tails, the other containing only tails. The two lysates complement each other and package recombinant λ DNA to form infective particles. They can then be propagated like normal phage in *E. coli* bacteria. In the case of insertion vectors, both recombinant and nonrecombinant λ DNA is packaged into infectious phage particles. Replacement vector DNA cannot be packaged unless a piece of DNA is inserted between the two λ arms that corresponds in size to the stuffer segment that had been removed before ligation.

Although visible only under the electron microscope, bacteriophage can be counted by the **plaque assay**. When a bacterial culture, containing a few bacteria that were infected with phage λ, is plated on a nutrient agar plate, most of the bacteria grow into a turbid lawn. The few phage-infected bacteria, however, lyse and release about 200 progeny phage particles each. These phage particles immediately infect the surrounding bacteria, causing them to lyse as well, thus creating a clear area, a plaque, on the otherwise turbid bacterial lawn. Each plaque originates from a single phage-infected cell. The cloning of foreign DNA in phage λ is illustrated in Figure 7.4.

7.1.5 Recombinant DNA Libraries

In most situations, the piece of DNA to be cloned is not available in pure form but is part of a mixture of many thousands or even millions of different DNA fragments. When such a mixture of DNA fragments is inserted into vector DNA, a **library of DNA sequences** is created. There are several methods available for identifying the clone of interest in such a library. They will be discussed in Section 7.2.

One distinguishes two types of DNA sequence libraries: the **genomic library** and the **complementary DNA (cDNA) library**. A genomic library contains all sequences of the genome. It is generated by partial digestion of total cellular DNA with a restriction enzyme, usually EcoRI.

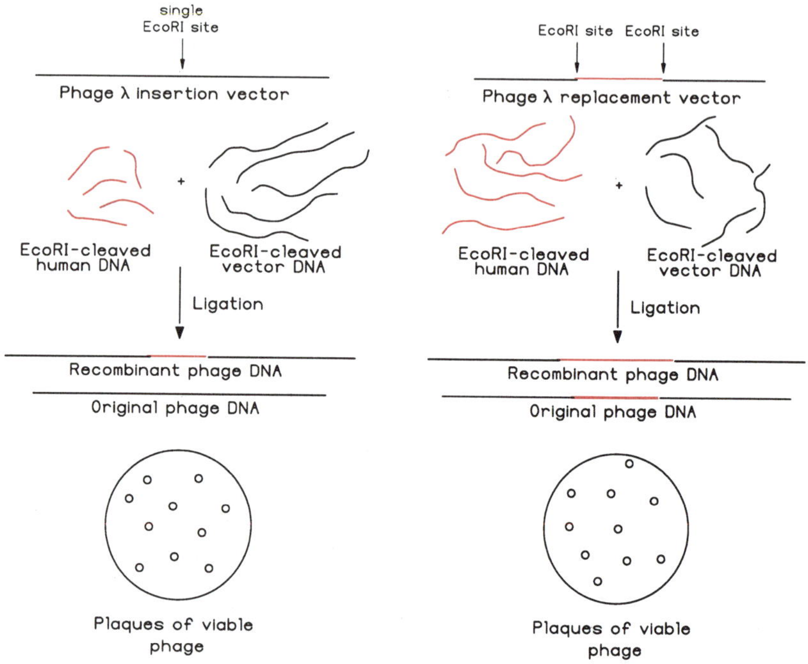

Figure 7.4.
The use of bacteriophage λ as cloning vector. Insertion vectors are shown on the left, replacement vectors on the right.

This creates a mixture of large fragments (about 20 kb in size) that are subsequently inserted into replacement phage vectors.

Larger genomic fragments of up to 45 kb in length can be cloned into **cosmid vectors**. Cosmids are small plasmids, about 5 kb in length, containing the cohesive ends, or *cos* sites, of phage λ DNA that allow the cosmid DNA to be packaged into phage λ particles. Only after the insertion of a foreign piece of DNA into the cosmid vector does the cosmid DNA reach a size suitable for packaging into a phage λ particle.

In recent years it has become clear that some genes, such as the dystrophin gene (1800 kb in length, see later), cannot be cloned as a single fragment in even the largest vectors. Furthermore, a major effort, the **human genome project**, is currently underway with the goal to map and sequence the entire human genome. Gene walking (explained later) within a cosmid library would be too time-consuming and inefficient. This need for a new cloning system led to the development of **yeast artificial chromosomes (YAC)**. YACs are circular DNA molecules with a plasmid replication origin, a multiple cloning site, a centromere region, and two yeast telomere sequences separated by a restriction site. DNA fragments up to 10 million bp (10,000 kb) can be ligated into this vector. After separating the two telomeric regions, the DNA is introduced into yeast cells, where the vector replicates autonomously as a new yeast chromosome. After multiplication, the artificial yeast chromosome can be reisolated for genetic analysis.

A genomic library contains not only DNA that codes for proteins but also **introns** and **regulatory sequences.** A **cDNA** library, on the other hand is made from messenger ribonucleic acid **(mRNA)** that is **reverse-transcribed** into cDNA It includes only the sequences found in mRNA. Since different cell types express different types of proteins a cDNA library is tissue-specific. It contains no introns or regulatory sequences.

cDNA libraries are often constructed with phage λ*gt*10, a phage with a unique EcoRI restriction site located within the λ repressor gene. By plating the recombinant phage on an *E. coli* host that favors lysogenization of the phage, one selects for recombinant phage since these cannot lysogenize because of the disruption of the repressor gene.

cDNA is synthesized from mRNA by reverse transcriptase, an enzyme isolated from retroviruses (see RNA tumor viruses). mRNA can be separated from total cellular DNA by **chromatography on oligo-dT cellulose** columns. Because of their **poly-A tails**, mRNA molecules selectively bind to the column and can be eluted with distilled water. The synthesis of cDNA is initiated on an oligo-dT prime that anneals to the 3′ poly-A tail of the mRNA. After completion of the reaction, the mRNA template is degraded by alkali, and synthesis of the second DNA strand is accomplished by DNA polymerase. Usually the 3′ end of the first cDNA strand serves as the primer for the second strand synthesis, thus creating a **hairpin loop** that must be cleaved before the double-stranded DNA can be inserted into a vector. This is achieved with **S1 nuclease**, an enzyme specific for single-stranded DNA.

Digestion with S1 nuclease removes several nucleotides from the 5′ end of the mRNA, and as a consequence the double-stranded cDNA derived from it also lacks these upstream sequences. This is the reason why most second strand cDNA synthesis today is done by replacement synthesis. After the synthesis of the first strand of cDNA, the DNA-RNA hybrid is treated with low concentrations of ribonuclease H, an enzyme that produces nicks and gaps in the mRNA strand of the hybrid, thereby creating a series of RNA primers that can be used by DNA polymerase for the synthesis of the second strand of cDNA. Both methods are illustrated in Figure 7.5.

Linkers must now be added to the blunt ends of the **double-stranded cDNA** in order to allow ligation into a vector DNA. This can be accomplished either by **homopolymer tailing** (see Figure 7.2) or by ligating **synthetic linker DNA** or **adapters** to the cDNA ends that can be cleaved by restriction enzymes, thereby generating cohesive ends (Figure 7.6). In order to prevent recircularization of the plasmid vector without an insert, the 5′-phosphate groups of the linearized plasmid are

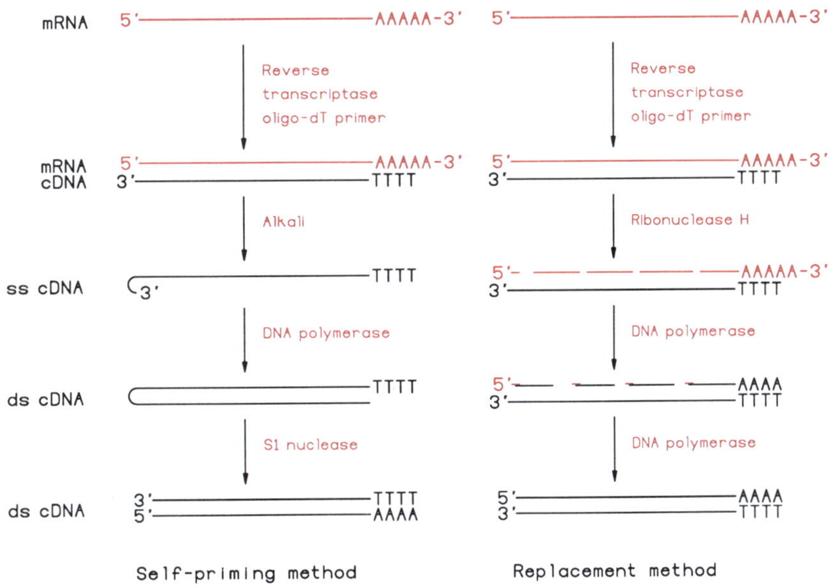

Figure 7.5.
Synthesis of double-stranded cDNA.

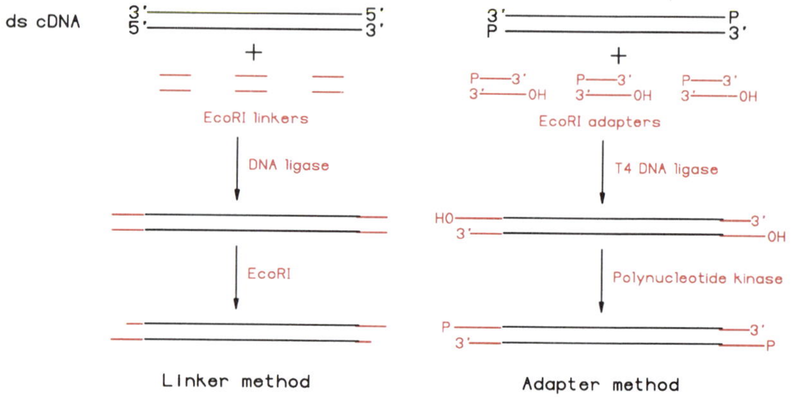

Figure 7.6.
Cloning of double-stranded cDNA with the help of linkers and adapters.

converted to hydroxyl groups by treatment with phosphatase. Ligation to the cDNA still occurs because the cDNA carries phosphate groups on both ends.

Adapters also must be unphosphorylated on their short strands in order to prevent formation of concatemers. Only adapter dimers can form as by-products by ligation at their blunt ends. The blunt ends of the adapters are phosphorylated to allow ligation to the ends of the cDNA. After separation from the adapter dimers, the adapter-terminated cDNA is rephosphorylated at its 5′ ends to allow ligation to a phosphatase-treated vector DNA.

7.2 ISOLATION AND ANALYSIS OF CLONED DNA

7.2.1 Hybridization Probes

In order to identify the clone of interest in a recombinant DNA library, a radioactively labeled **hybridization probe**, such as mRNA or cDNA, representing the sequence of interest must be available. Often the probe itself is a piece of cloned DNA. It is radioactively labeled in a process called **nick translation**. In this procedure, a few single-stranded breaks are introduced into the DNA by an endonuclease. The DNA polymerase I from *E. coli* and radioactive nucleoside triphosphates are added. The 5′→3′ exonuclease activity of the polymerase enzyme removes unlabeled nucleotides, which are immediately replaced by radioactive nucleotides incorporated by the polymerase activity (Figure 7.7).

Often the partial amino acid sequence of the protein whose gene or cDNA is being sought is known. In that case a short **oligonucleotide probe** can be synthesized chemically. The sequence of the probe is deduced from the amino acid sequence. Because of the degeneracy of the genetic code, ambiguities exist at the third position of most codons. In order to avoid mispairing, **inosine** residues are placed into ambiguous positions since inosine can pair with most other nucleotides and thus does not disrupt Watson-Crick base-pairing. Synthetic oligonucleotides are usually labeled at their 5′ ends with ^{32}P-ATP and polynucleotide kinase.

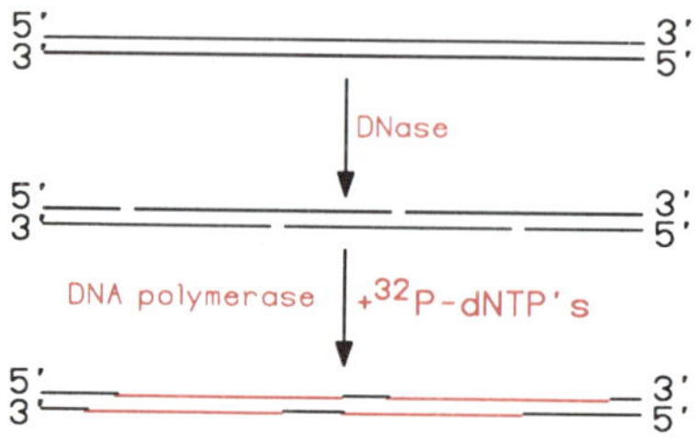

Figure 7.7.
Radioactive labeling of DNA by nick translation.

DNA libraries in plasmids are probed by the **colony hybridization** assay. In this assay, a nitrocellulose filter is placed on the agar plate with the bacterial colonies. In the process, some of the bacteria are transferred to the filter. The transferred bacteria are lysed on the filter by alkali, and the denatured DNA becomes bound to the nitrocellulose. The filter is then incubated with the radioactive probe, which will anneal only to complementary sequences. Thus, colonies that contain sequences homologous to the probe are easily visualized by autoradiography, and they can be picked from the original agar plate for large scale growth and isolation of the recombinant DNA.

Phage libraries are screened with the **plaque hybridization** assay, which is analogous to the colony hybridization assay except that instead of colonies, plaques containing phage particles are transferred onto nitrocellulose.

The **minimum size of a genomic library** is determined by the average size of the cloned fragments and the size of the genome. For example, a mammalian genome contains about 3×10^9 bp. Thus, the minimum number of clones, assuming 20 kb–size inserts, that would include the entire genome is about 150,000. Since individual fragments are cloned on a random basis, there is only about a 50% chance that a particular sequence is included in a library of this size. A good genomic library should be 5–10 times larger than this minimum size, and between 1 and 2 million clones should be screened in order to find a particular gene.

A typical mammalian cell contains about 10,000 different species of mRNA. Most of these mRNA species are represented in only a few copies per cell, and cDNA clones of rare mRNA species may be represented in a cDNA library by as few as one clone in 100,000.

7.2.2 Identification of a Cloned cDNA

Once a clone has been selected on the basis of its hybridization properties, it has to be identified. If an **antibody** to the protein is available, it can be used for this identification. First, total mRNA from the cell where the protein is expressed is translated in vitro, and the mixture of proteins that are synthesized is subjected to polyacrylamide gel electrophoresis. The separated proteins are then transferred onto a sheet of nitrocellulose by application of an electrical field. Thus the nitrocellulose represents a replica of the positions of the various proteins on the polyacrylamide gel. This process of blotting proteins from a gel onto nitrocellulose is called **Western blotting** in analogy to Southern blotting, the transfer of DNA fragments from a gel onto nitrocellulose (see below).

The nitrocellulose sheet is then incubated with antiserum against the protein in question. The antibodies that bind to the protein band can be visualized by reaction with radioactively labeled secondary antibodies. Alternatively, a secondary antibody that is coupled to an enzyme such as horseradish peroxidase or alkaline phosphatase can be used to convert a colorless substrate into a visible product, thus identifying the protein band. This second method is called **ELISA**, which stands for enzyme linked immunosorbant antibody technique.

If the in vitro translated protein is identifiable by these techniques, there are two ways to identify a DNA clone as the sequence that codes for the protein in question. The first method is called hybrid selection. The cloned DNA is bound in denatured form to a nitrocellulose filter. The filter is then incubated with total cellular mRNA under conditions where DNA-RNA hybrids form. If the clone represents a sequence that codes for the protein in question, its mRNA will anneal to the DNA on the filter. After washing off the other mRNA species, the filter-bound mRNA can then be eluted from the filter by a brief incubation at 100°C in distilled water. This enriched mRNA can then be tested in the in vitro translation assay discussed above.

The second method used to identify a clone is called **hybrid arrest**. In this technique, the cloned cDNA is first denatured and then annealed to total cellular mRNA. If the clone represents a sequence that corresponds to the mRNA that codes for the protein in question, it will anneal to the mRNA and prevent its translation in an in vitro system. Thus, if the cloned DNA interferes with the translation of the mRNA, its sequence must be identical to it.

The final identification of a cDNA clone is its direct expression into protein in an **expression vector**. This method will be discussed in section 7.3.

7.2.3 Restriction Enzyme Mapping and Sequence Analysis of DNA

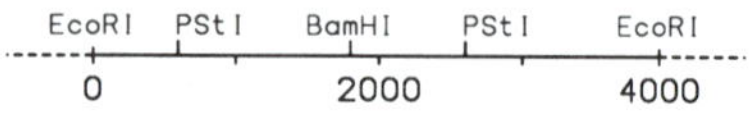

Figure 7.8.
Restriction map of the hypothetical cDNA clone constructed according to the data contained in Table 7.2.

Once a clone harboring a segment of a particular gene or cDNA has been isolated and identified, the positions of restriction endonuclease sites on this DNA are determined. Figure 7.8 shows an example of such a restriction analysis. Suppose a cDNA of 4,000 bp is cloned into phage λ*gt*10. The sizes of the cDNA fragments generated by digestion of the clone with combinations of EcoRI, PstI, and BamHI are listed in Table 7.2. They were obtained by agarose gel electrophoresis of the digested DNA, transfer of the DNA from the gel onto nitrocellulose (see Southern blotting, below), and identification of the cDNA sequences by hybridization with a cDNA-specific probe. This is necessary because the large phage λ DNA contains multiple restriction sites that would interfere with the analysis. Alternatively, the cDNA insert can be cleaved from λ*gt*10 DNA with EcoRI and **subcloned** in a small plasmid vector that contains none of the restriction sites used for this analysis.

Once the positions of several restriction enzyme sites are known, it is possible to cut out from the vector short segments of the cDNA (or genomic fragment) and subclone them for sequence analysis. Often this subcloning is done into a **phagemid vector**. Phagemids are small plasmids that contain the single-stranded DNA replication origin for **phage M13**. M13 is a circular single-stranded DNA phage that first replicates in the bacterium in a double-stranded replicative form but then switches

Table 7.2. Restriction Fragment Sizes Generated from a Hypothetical cDNA Insert in Phage λ*gt*10

Restriction Enzymes	Sizes of Fragments (bp)
EcoRI	4000
EcoRI + PstI	600, 1400, 2000
EcoRI + BamHI	1800, 2200
EcoRI + PstI + BamHI	600, 800, 1200, 1400

to single-stranded DNA synthesis late in infection to produce the single-stranded DNA that is packaged into phage particles (Figure 7.9).

The DNA fragment is inserted into the double-stranded plasmid into a **multiple cloning site**. This is a region of the plasmid DNA that contains multiple unique restriction sites allowing the insertion of DNA fragments created with a variety of different restriction enzymes. At the ends of the cloning site the **phagemid carries two promoters (T7 and T3)** that are recognized by phage-specific RNA polymerases. (The use of these promoters for expression of the inserted DNA will be discussed later.) The recombinant phagemid is then used to transform a bacterial strain that allows the multiplication of single-stranded phage. When a culture of bacteria harboring the recombinant phagemid is infected with an **M13 helper phage**, the phage induces the synthesis of single-stranded circular DNA not only from its own genome but also from the phagemid. This single-stranded recombinant DNA is then packaged into M13 phage particles that can be isolated in large quantities. The M13 helper phage used for this infection is a special mutant that has a 50-fold lower efficiency for packaging its own DNA into phage particles than the phagemid DNA. Thus, most M13 particles isolated from these phagemid-containing bacteria contain phagemid DNA rather than M13 DNA.

The single-stranded phagemid DNA can then be used directly for **sequence analysis** by the **Sanger technique** A synthetic oligonucleotide complementary to one of the two promoter sites flanking the recombinant DNA is annealed to the phagemid DNA, and DNA synthesis from this primer is initiated with radioactive nucleoside triphosphates. DNA synthesis is carried out in the presence of a low concentration of a **chain-terminating** nucleotide. For example, the addition of **2′, 3′-dideoxyadenosine triphosphate** will cause the elongation of a DNA

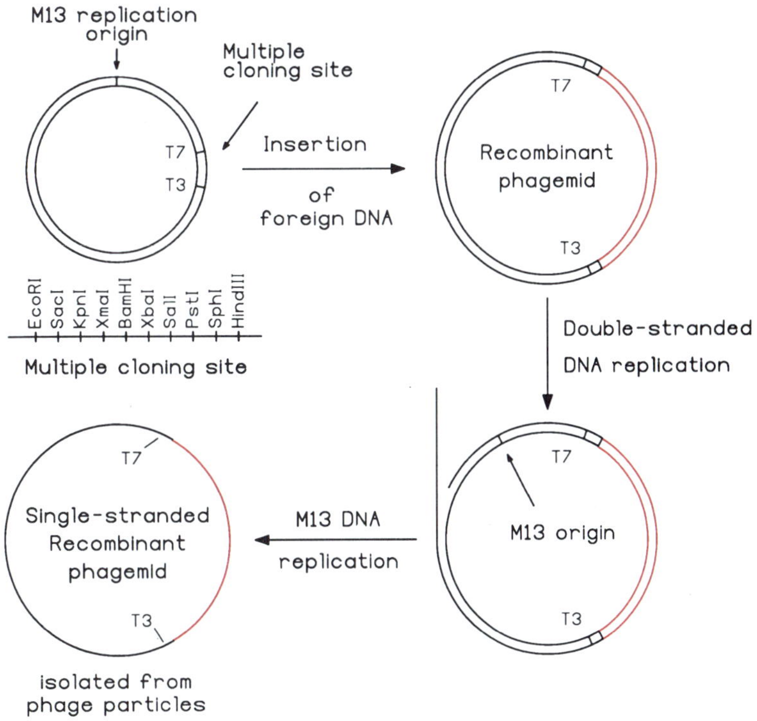

Figure 7.9.
Cloning of DNA into phagemids for sequence analysis.

chain to stop when this nucleotide is incorporated. Of course, only a very small concentration of dideoxynucleotide is added to the mixture, so that chain termination occurs at different A residues in different chains. Thus, a nested set of DNA chains is synthesized, each terminating with an A residue. Analysis of this reaction mixture by **polyacrylamide gel electrophoresis** will reveal the lengths of the different fragments and thus the relative positions of A residues in the DNA. The same reaction is carried out with the other three chain-terminating dideoxynucleotides generating different sets of DNA chains. During electrophoretic analysis, each reaction will thus produce a series of bands whose relative positions on the polyacrylamide gel corresponds to the locations, within the DNA segment that has been synthesized, of the bases at which the reaction was terminated by the incorporation of the dideoxynucleotide. Nucleotide sequences up to 300 bases in length can be read directly from a single gel (Figure 7.10).

In **automated DNA sequencers**, the dideoxynucleotide causing the chain termination is tagged with a fluorescent dye. By using fluorescent-tagged nucleotides with different absorption maxima, the products of all four reactions can be analyzed in a single lane on a polyacrylamide gel, since the nature of the chain-terminating nucleotide is easily recognized. More than 500 bases can be sequenced in one run by the automated sequencers.

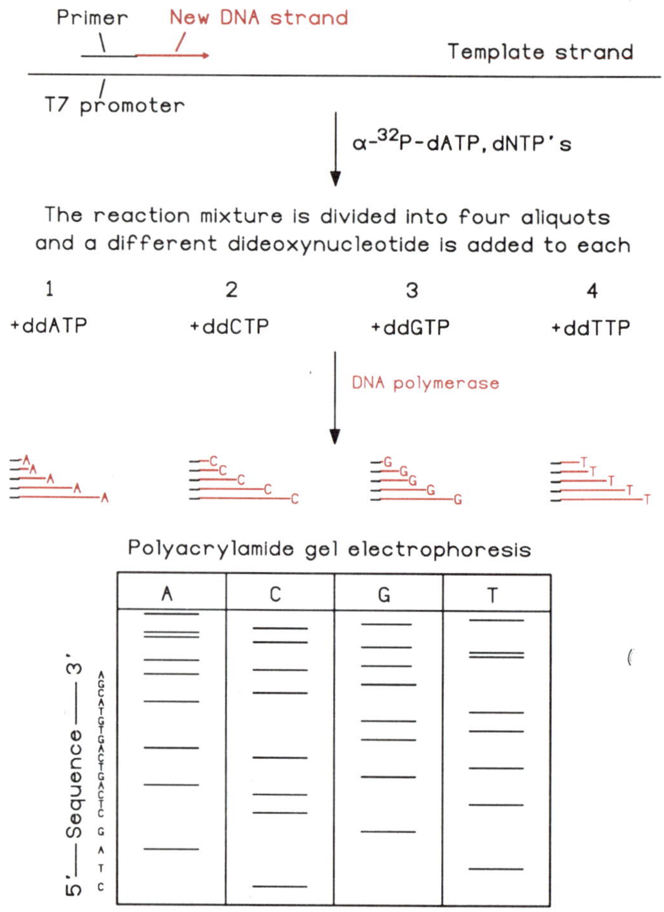

Figure 7.10.
DNA sequence analysis by the Sanger technique.

7.2.4 Mapping of Transcription Start Sites and Introns

Since genomic clones contain all the nontranslated 5′ regulatory sequences in addition to the coding sequences, they are useful for studying gene regulation. They can also be used for mapping the precise transcription start site of a gene. There are two methods available for doing this: **S1 nuclease mapping** and **primer extension** (Figure 7.11). In the first method, a fragment of the genomic clone straddling the presumed transcription start site is radioactively labeled at its 5′ end and annealed to an excess of mRNA from the cell where this gene is expressed. The DNA-RNA hybrid is then treated with S1 nuclease, which degrades all single-stranded nucleic acids, and the size of the S1-resistant fragment is determined by polyacrylamide electrophoresis.

In the primer extension method, a 5′-labeled single-stranded fragment from the genomic clone is used as a primer for reverse transcription of the mRNA. The final size of the extended primer allows the determination of the 5′ end of the mRNA.

The positions of introns in a gene can be mapped in an analogous way. A DNA fragment containing the intron-exon boundary is annealed to mRNA, and the resulting hybrid is subjected to S1 nuclease digestion. Determination of the size of the S1-resistant fragment allows precise localization of the boundary. Another way to locate the positions of introns is to anneal the mRNA to its gene. Under appropriate annealing conditions, the mRNA displaces one DNA strand in exonic segments,

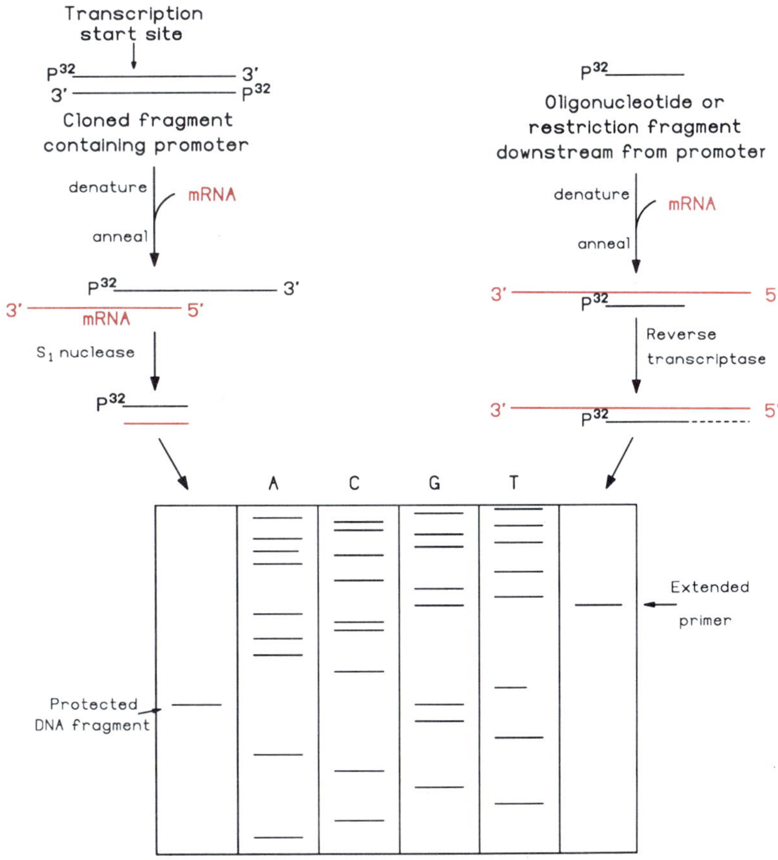

Figure 7.11. Mapping of transcription start sites by S1 nuclease protection (a) and primer extension (b).

forming an RNA-DNA hybrid, while intronic segments of the DNA remain double-stranded. After exposure of S1 nuclease, only the exons will remain intact, allowing their visualization in agarose gels. Under the electron microscope, the exonic segments of the DNA can be seen as **R loops**, so called because the displaced DNA strands loop out. (Figure 7.12).

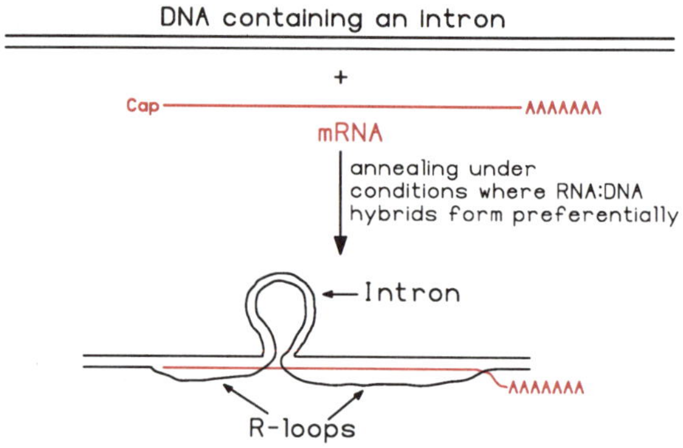

Figure 7.12.
Mapping of introns by R loop formation.

7.3 EXPRESSION OF CLONED GENES

7.3.1 Expression Vectors

Often it is desirable to express a eukaryotic gene in bacteria. For example, a cDNA library can be screened for clones that express a protein for which antibodies are available. Because of the presence of introns in most eukaryotic genes, usually only cDNA clones can be expressed in bacteria. In order to make the gene transcribable and translatable in a prokaryotic system, both a prokaryotic promoter and a Shine-Dalgarno sequence must be placed upstream of the coding sequence. This is easily accomplished by inserting the cDNA sequence into the coding sequence of a prokaryotic gene such as β-galactosidase.

A phage λ vector commonly used for this purpose is λ*gt*11. It carries a β-galactosidase gene with a unique EcoRI site located 17 amino acids from its carboxyl terminus. A cDNA inserted into this site is under the control of the β-galactosidase promoter; that is, transcription can be induced by IPTG (isopropyl thiogalactoside, a synthetic inducer of the lactose operon). The mRNA is recognized by ribosomes because of its Shine-Dalgarno sequence. If the inserted cDNA sequence is in same reading frame as the β-galactosidase, the fusion protein that is synthesized contains most of the amino acids from β-galactosidase at its amino terminus, which usually do not affect the activity of the eukaryotic protein, with the cDNA-coded protein sequence at the carboxyl terminus.

cDNA libraries in λ*gt*11 can be screened with antibodies against the protein whose clone is to be isolated. The proteins contained in the plaques are transferred onto nitrocellulose and can be identified by antibodies in an ELISA assay. However, a larger number of plaques have to be screened than in the hybridization assay, since only one of six cDNA molecules will be in the right reading frame (three forward and three backward; Figure 7.13).

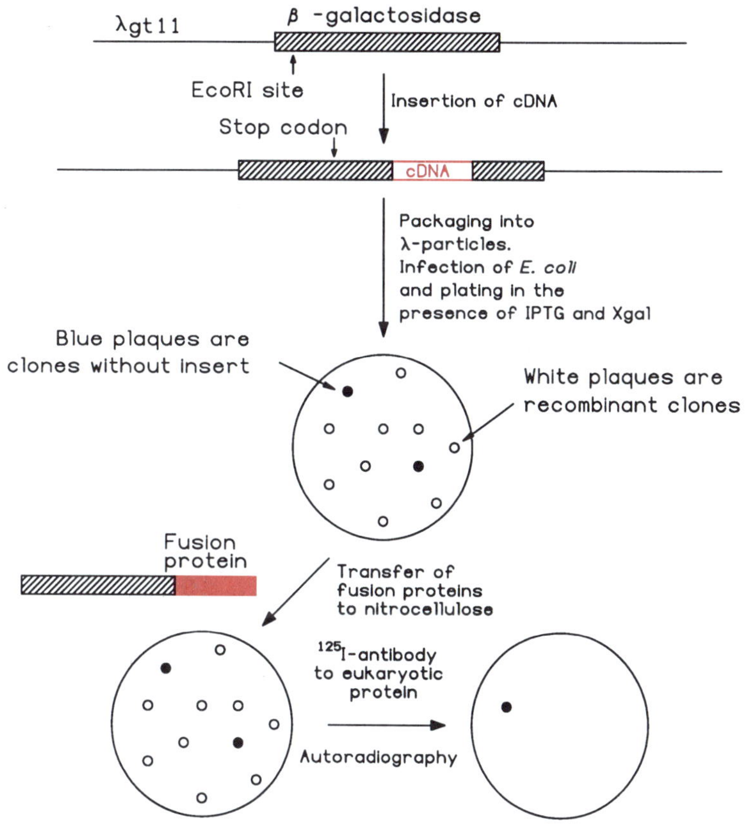

Figure 7.13.
Screening of a cDNA library in the expression vector λ*gt*11 with antibodies.

The insertion of the cDNA into the β-galactosidase gene has an additional advantage. Plaques that contain no cDNA insert will have a blue color when plated on a medium containing **X-gal**, a synthetic β-galactoside that is converted into a blue compound when the β-galactosidic bond is cleaved. Plaques of phages with cDNA inserts are clear, since no functional β-galactosidase is produced.

Prokaryotic expression vectors are also used for the large scale commercial production of a eukaryotic gene product such as insulin or human growth hormone or for the synthesis of in vitro mutagenized proteins for structure-function analysis. Unfortunately, the synthesis of large amounts of a foreign protein often interferes with the life cycle of the bacterium, and the foreign protein may be subject to degradation. In order to avoid such problems, special expression plasmids have been created that express the foreign protein only when induced. One such vector is pET3a, which contains a unique NdeI restriction site located downstream of a **phage T7 promoter** just after the methionine triplet where translation starts. The host bacterium, BL21 (DE3), that is transformed with this plasmid contains the phage T7 RNA polymerase gene under control of the lactose operator. Thus, when IPTG is added to a culture of this strain, the T7 RNA polymerase is produced which, in turn, transcribes the foreign cDNA that is inserted into the plasmid. Since the cell contains many copies of this plasmid, the foreign protein is overproduced until it accounts for more than 50% of the total cellular protein (Figure 7.14).

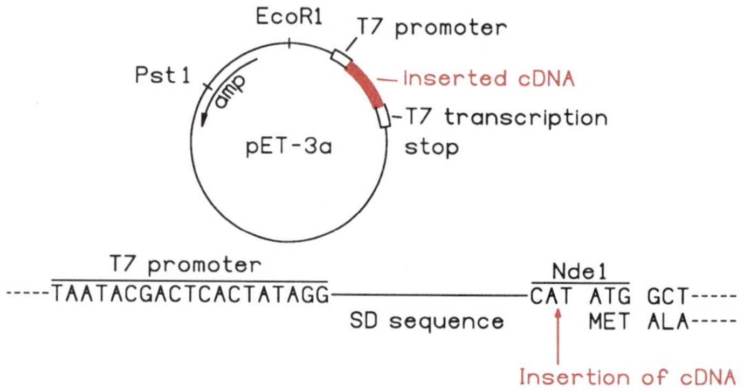

Figure 7.14. Expression of eukaryotic cDNAs in the expression vector pET-3a (plasmid for expression by T7 RNA polymerase). The unique NdeI site located within the initiation codon AUG of a phage T7 gene allows integration of cDNA sequences that are then transcribed by T7 RNA polymerase. This resulting mRNA is efficiently translated in *E. coli*.

7.3.2 Shuttle Vectors

Sometimes it is necessary to have the modified gene expressed in a eukaryotic cell. Special vectors have been constructed for this purpose that can be shuttled between a bacterial cell and a eukaryotic cell. Such a vector usually consists of a prokaryotic plasmid segment containing a replication origin and an antibiotic resistance gene. In addition, the vector must contain a eukaryotic replication origin and a functional eukaryotic promoter located upstream of the inserted cDNA. The plasmid **pSV2-neo** contains the replication origin of the tumor virus SV40 and a gene for the bacterial **neomycin phosphotransferase** that inactivates the antibiotic **G418** (geneticin) to which many eukaryotic cell lines are sensitive (Figure 7.15). This allows the selection of mammalian cells that have taken up the plasmid and as a consequence have become resistant to neomycin. Cells that have become transformed by a pSV2-neo contain the plasmid DNA integrated into their genomes. They are stable and remain resistant to neomycin even under nonselective conditions.

Gene transfer vectors, or **shuttle vectors**, are used for the introduction of new or modified genes into eukaryotic cells, gene expression studies, gene therapy, transformation in plants, and many other applications. A large number of eukaryotic viral vectors, derived from polyoma virus, papilloma virus, herpes simplex virus, or retroviruses, have been constructed for this purpose. These vectors also include powerful eukaryotic promoters, such as the **metallothionein promoter**, which is activated by heavy metal ions (As, Cd, Zn, Cu), or the **heat shock promoter**, which can be activated by raising the temperature.

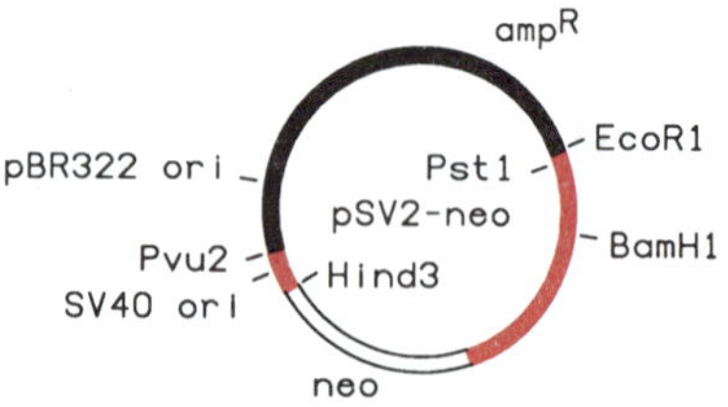

Figure 7.15. The structure of the shuttle vector pSV2-neo. pBR322-derived sequences are represented in black, SV40 sequences in red. cDNA sequences are inserted into the BamH1 site.

There are several methods for the **introduction of a new or modified gene into mammalian cells**. If the recombinant vector can be packaged into a virus to which the cells are susceptible, the cells can be directly infected with the recombinant virus vector. Often this is not possible, and the recombinant DNA must be introduced into the cell by other means. **Microinjection** is one possibility that is highly efficient, but the procedure is tedious and limited regarding both the type and total number of cells that can be injected. The direct introduction of DNA into cells is called **DNA-mediated gene transfer (DMGT)**. In this process, cells are made **competent** by treatment with a variety of agents ($CaCl_2$, DEAE dextran, polyethylene glycol), and the DNA is added in the form of a calcium precipitate. Other methods include **electroporation**, in which the cell is subjected to a high-voltage electric field, or **lipofection**, in which the negatively charged DNA is bound to the surface of cationic liposomes. The efficiency of **transfection** may be as high as 10^{-3}.

7.3.3 The Oocyte Expression System

Structure function analysis of a protein usually requires that a series of mutated proteins be isolated from the expression system and analyzed. Often this is not possible, either because the mutated protein cannot be expressed efficiently in bacteria or mammalian cells or the protein cannot be isolated in an active form. This is often seen with **membrane proteins**. Examples are the acetylcholine receptor, the sodium channel, or the gap junction protein. Mutational analysis of such proteins is possible in the *Xenopus* oocyte expression system.

mRNA for the membrane protein is synthesized in vitro from the cloned cDNA. The phage promoters T3 and T7 that flank the cloning site in the phagemid vector (see Figure 7.16) can be used for this purpose. The mRNA is then microinjected into *Xenopus* oocytes where it is translated into proteins. Any posttranslational modification and transport of the protein to its site of function occurs in the oocyte just as it would occur in another eukaryotic cell. The function of the protein often can be directly assayed in the oocyte. For example, when mRNA synthesized from the cloned cDNA of the **gap junction protein** is injected into oocytes, a large number of precursors of cell-cell channels are assembled in the oocyte membrane. When two mRNA-injected oocytes are brought into close contact, functional **cell-cell channels** are

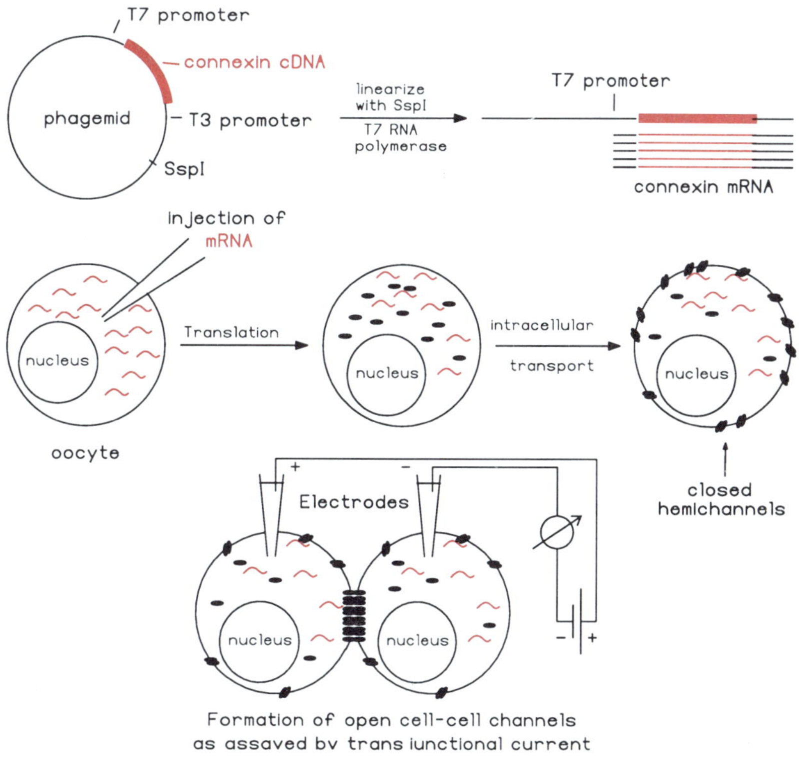

Figure 7.16.
Formation of cell-cell channels between paired *Xenopus* oocytes in response to the injection of connexin-43 mRNA.

formed from the precursors. The properties of these channels can then be studied electrophysiologically by inserting microelectrodes into the two oocytes and measuring transjunctional current in response to transjunctional voltage difference and by injecting potential mediators of channel activity (Figure 7.15).

The oocyte expression system is also useful in the isolation of cDNA clones of membrane receptors, such as peptide or neurotransmitter receptors. In this method, DNA from pools of clones from an expression library are transcribed in vitro, and the mRNA is injected into the oocytes. If a cDNA clone with the entire open reading frame for the receptor protein of interest is in the pool, its expression on the oocyte surface can be observed. Successive subdivision of the positive pool is then performed to obtain a single clone. This method requires no sequence information nor antibody, only the ability to assay the function in a *Xenopus* oocyte.

7.4 IN VITRO MUTAGENESIS

Analysis of mutants has been a very powerful tool in the study of gene expression and protein structure. Until recently, the availability of mutants was restricted to naturally occurring mutations or to mutations induced by mutagens. The isolation and characterization of such mutants often was a time-consuming process, and the location of the mutation was restricted to sequences that affected gene function. With the advent of recombinant DNA technology, however, any conceivable mutation can be generated by site-directed mutagenesis. Of course, this in vitro mutagenesis is only possible on a cloned DNA sequence inserted into an appropriate plasmid vector. For analysis of the mutation, the mutated gene will have to be reintroduced into a cell.

Deletion mutations are most easily obtained by removal of the DNA located between two restriction enzyme sites. After a brief digestion with the specific restriction endonuclease, the linearized plasmid is religated. Shorter deletions can be obtained by first cutting the plasmid at a unique restriction site and then treating the linear DNA with exonucleases, such as Bal31 nuclease, to remove short segments of DNA. The nuclease-treated DNA is then converted back into circular form either directly by T4 ligase (which ligates DNA molecules with blunt ends) or after the addition of small oligonucleotide linkers to the ends. Alternatively, if appropriate linkers are ligated to the ends of a recombinant plasmid cut with a restriction endonuclease, and the plasmid is then recircularized, short insertion mutations of known length are created.

Precisely placed single-base-pair changes are created by oligonucleotide-directed mutagenesis. A 20-base-long oligonucleotide chain containing the desired mutation is synthesized on an **automatic DNA synthesizer** and annealed to the single-stranded phagemid DNA harboring the gene. Despite the slight mismatch of one, or even a few, base pairs, the oligonucleotide can be used as primer for the in vitro synthesis of double-stranded phagemid DNA. This DNA is then used to transfect *E. coli* where the mismatched DNA is repaired and replicated. Normally, the heteroduplex region in this DNA is quickly repaired by *E. coli* enzymes, resulting in at least a 50% reduction of the desired mutation. However, a genetic trick can be used to fool the repair system so that it uses the mutated DNA as template and thus "repair" the non-mutant strand. If the phagemid is grown in a *dut*$^-$, *ung*$^-$ cell, its DNA contains many uracil residues instead of thymine (see Section 6.4). The new DNA strand synthesized from the synthetic primer, however, contains only thymine residues. Thus, after transformation into a *dut*$^+$, *ung*$^+$ cell, the wild-type strand is repaired according to the information contained in the mutated strand, yielding a high percentage of mutants (Figure 7.17).

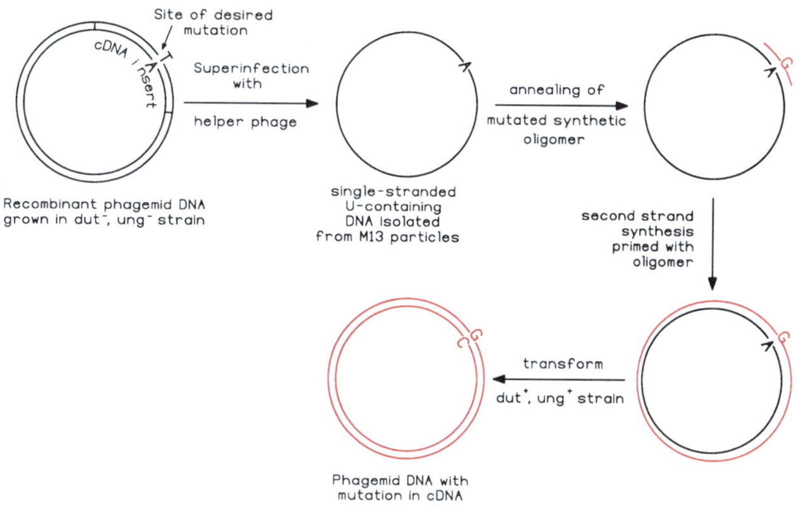

Figure 7.17.
Site-directed mutagenesis using mutated oligonucleotide primers.

When many different mutations must be generated within a gene, **cassette mutagenesis** is the method of choice. Here a segment of the gene located between two restriction sites is replaced by a synthetic piece containing the desired mutation. By using the degeneracy of the code, additional unique restriction endonuclease sites can often be placed into a gene without changing the amino acid sequence. This is accomplished by replacing a wild-type restriction fragment with a synthetic piece of DNA containing the additional sites. The new construct then allows the replacement of relatively short specific segments of the gene sequence by synthetic sequences carrying the desired mutations.

Chemical DNA synthesis is now routinely performed in many laboratories. The first nucleotide is covalently attached to a silica matrix, and subsequent nucleotides are added in the form of **phosphoramidite** precursors. In order to prevent side reactions, reactive groups on the nucleotides must be protected. The synthesis begins at the 3′ end and proceeds toward the 5′ end, just the opposite direction from enzymatic DNA synthesis. Before the addition of a new nucleotide, the **dimethyltrityl** protection group on the 5′ end of the growing DNA molecule must be removed. The quantitation of the released dimethyltrityl group is used for the calculation of the coupling efficiency of each step. The sequence of reactions is illustrated in Figure 7.18.

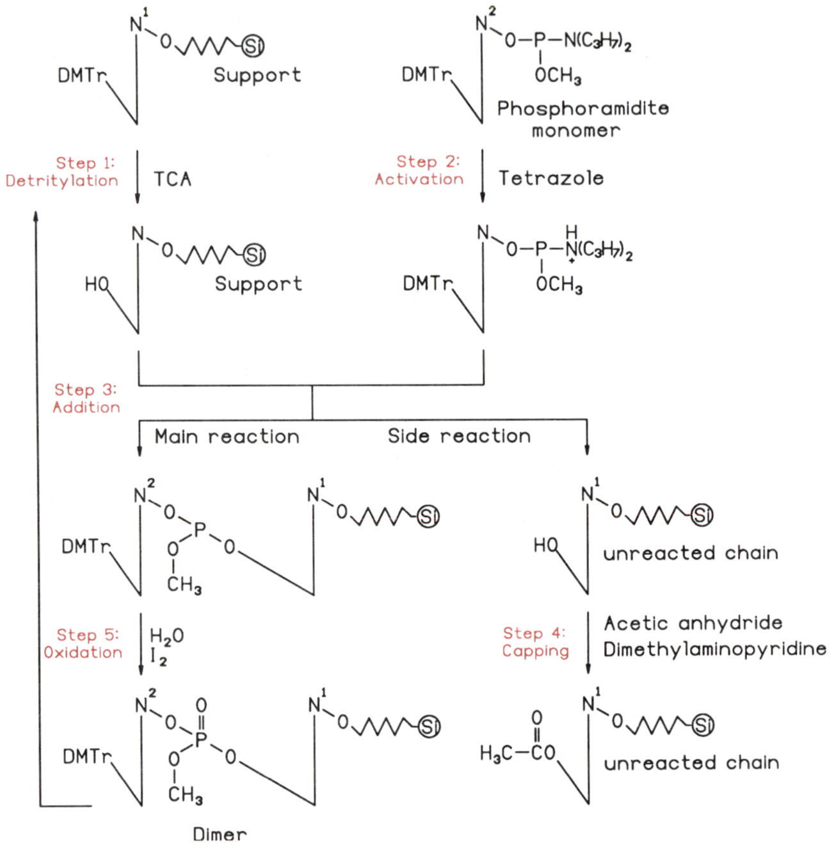

Figure 7.18.
Chemical synthesis of DNA. N = nucleotide, DMTr = Dimethyltrityl. Adapted from *A Procedure for the Manual Synthesis of Deoxyribonucleotides*. Applied Biosystems.

7.5 MOLECULAR ANALYSIS OF INHERITED DISORDERS

The advent of recombinant DNA technology has opened many inherited diseases to study and attack. Not only is it now possible to test the genotype of a fetus by analyzing a few cells obtained through **amniocentesis**, but in several cases analysis of the mutated gene that is giving rise to the disease phenotype has actually led to the characterization of the gene product. In many cases this would not have been possible by using conventional biochemical methods, since the protein involved is present in quantities too small for analysis.

7.5.1 Restriction Fragment Length Polymorphism

Many **inherited disorders** can be **diagnosed** if the sequence of the gene involved in the disorder is known. This is done by using a segment of the gene, usually cloned in a plasmid, as a **hybridization probe** for the analysis of the mutated gene found in fetal cells obtained by amniocentesis. A complete restriction endonuclease digest of chromosomal DNA from such cells is subjected to agarose gel electrophoresis, and the position of a particular genomic fragment, for example the β-globin gene, is then located by a technique called **Southern blotting**.

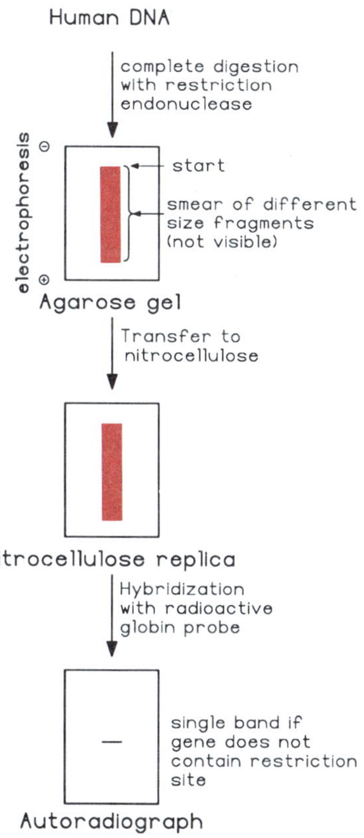

Figure 7.19.
Southern blotting analysis of genomic DNA with a β-globin cDNA probe. Multiple bands would indicate that the restriction enzyme either cuts inside the gene or that the genome contains more than one copy of the gene.

In this technique, the gel is placed into alkali, and the denatured DNA fragments are transferred by blotting onto a sheet of nitrocellulose, where they are immobilized by heat treatment. The nitrocellulose sheet, representing a replica of the pattern of restriction fragments in the agarose gel, is then incubated with a solution containing the radioactively labeled DNA probe, in this example ^{32}P-labeled β-globin cDNA. Specific base-pairing between the cDNA and any fragments containing globin-specific sequences occurs, and after removal of the excess cDNA, these sequences can be visualized by autoradiography (Figure 7.19).

This simple technique can be used for the **prenatal diagnosis** of a genetic disease such as **sickle cell anemia**. In this genetic defect, the amino acid valine replaces glutamate at position 6 in the β-globin protein. The mutation involves a single base pair change: a GAG codon (in mRNA) is replaced by GUG. The normal β-globin probe hybridizes to both normal and mutant DNA fragments, since a single unpaired base pair is not enough to weaken the overall hydrogen bonding between probe and DNA significantly. However, the sickle globin gene can often be recognized by its surrounding sequences. Most people who carry the sickle mutation originate from Northern Africa and the Mediterranean countries and have a genetic background that is different from noncarriers. Thus, the DNA sequences surrounding the mutant globin gene are distinct from those found near the normal globin gene. This phenomenon is called **genetic polymorphism** and can be utilized to identify a sickle globin carrier. The DNA isolated from fetal cells by amniocentesis is subjected to digestion by the restriction endonuclease HpaI, and the mixture of fragments is analyzed by **Southern blotting** using the normal β-globin cDNA as a hybridization probe. The normal β-globin gene is found in a fragment of 7 or 7.6 kb in length, whereas the sickle globin gene is found in a 13 kb fragment (Figure 7.20). The test is sensitive enough to allow identification of a heterozygous carrier who has both a normal and a sickle globin gene, yielding both 7 kb and 13 kb fragments on Southern blots.

Analysis of DNA based on **restriction fragment length polymorphism (RFLP)** can also be used to identify and thus distinguish DNA samples obtained from different persons. Usually the DNA is digested with different restriction enzymes and, after Southern blotting, probed with a variety of probes representing different segments of the genome. This **DNA fingerprinting** is useful in paternity tests and is being applied increasingly in criminal investigations.

Often there is not enough DNA available for RFLP analysis. In such a case, the **polymerase chain reaction (PCR)** is used. PCR allows the amplification of a specific sequence contained in a very small quantity

Figure 7.20.
The generation of β-globin–specific HpaI fragments from normal and sickle cell hemoglobin DNA. The extent of the β-globin gene is indicated in red.

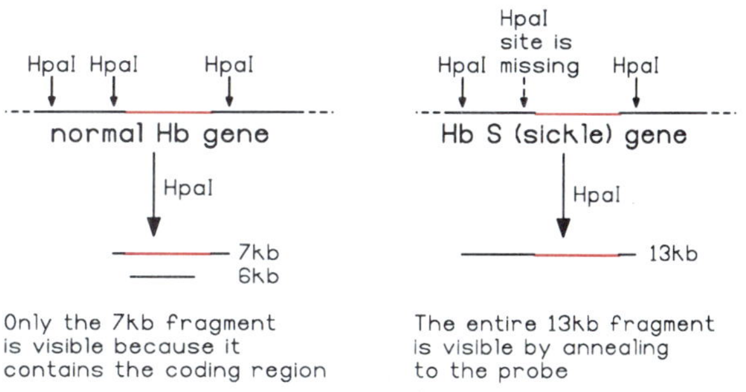

of DNA. The reaction requires two oligonucleotide primers, flanking the sequence of interest that is to be amplified, and the heat-stable **Taq DNA polymerase** from *Thermophilus aquaticus*. A mixture of DNA containing the desired sequence, a large excess of primers, Taq polymerase, and deoxyribonucleoside triphosphates is subjected to the following sequence of steps: (1) denaturation of DNA at 94°C, (2) annealing of primers at 50°C, and (3) DNA synthesis at 72°C initiated from the primers (Figure 7.21). The cycle is then repeated by simply changing the temperature of the reaction mixture between these values. Since the newly synthesized DNA serves as additional template for DNA synthesis, each subsequent round of synthesis will have twice as much DNA template for primer annealing and extension, resulting in the exponential amplification of the original DNA sequence. From 25 to 35 cycles can amplify a DNA sequence more than a million times. Other sequences, contained in the original DNA sample but not bounded by the primers, are not amplified. The PCR product can then be subjected to agarose electrophoresis for identification and isolation of the amplified sequence.

Since its introduction a few years ago, the polymerase chain reaction has found many other applications in molecular biology. They include

1. **Amplification of genomic DNA.** Only two short sequences need to be known, to which primers are synthesized and used

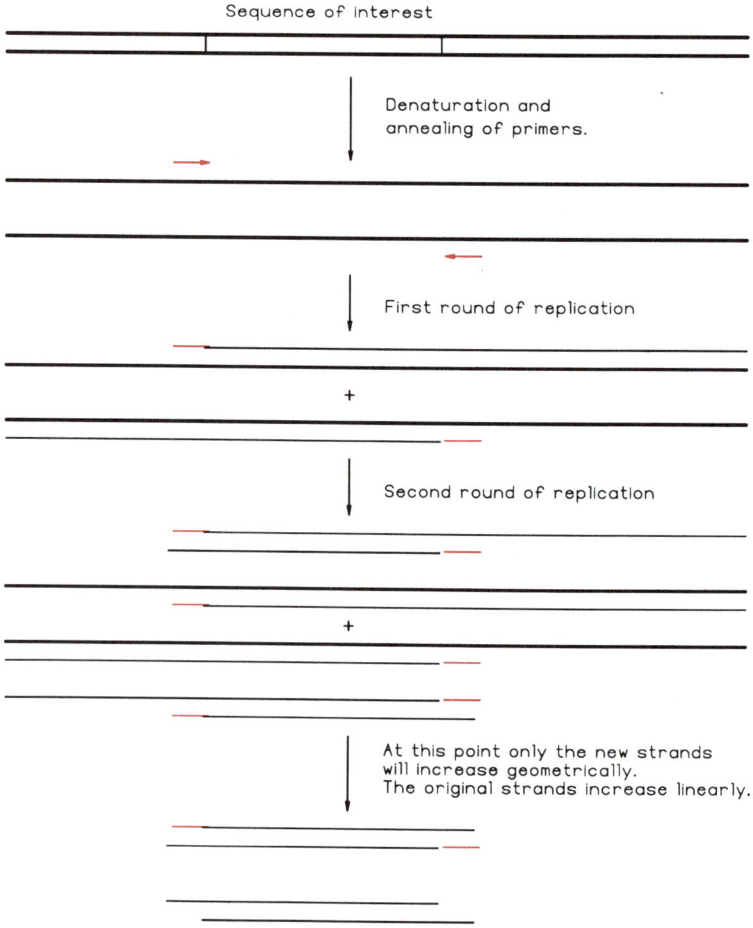

Figure 7.21.
The polymerase chain reaction with the thermostable Taq DNA polymerase. The reaction mixture is simply cycled between 50°C (annealing of primer), 73°C (DNA synthesis), and 90°C (denaturation), which can be done automatically.

in the PCR reaction. (A modification to this technique, called reverse PCR, requires only a single primer.)

2. **cDNA cloning with degenerate primers.** After reverse transcription of total cellular RNA with oligo-dT as primer, the single-stranded cDNA is amplified with a specific primer, corresponding to a sequence near the amino terminus of the protein, and oligo-dT as the second primer.
3. Quantitation of cellular mRNA.
4. Site-directed mutagenesis and generation of recombinant clones containing segments from different parental genes.
5. Identification of latent virus genomes in cells.
6. Determination of familial relationships.
7. Amplification of small samples of DNA in forensic medicine.
8. Identification of genetic defects by amniocentesis.
9. Asymmetric amplification of DNA to produce single-stranded DNA for sequence analysis.
10. Direct sequencing of double-stranded DNA.

7.5.2 Identification of the Gene Involved in Duchenne Muscular Dystrophy

Usually, the most difficult part of initiating the molecular analysis of a genetic disorder is the identification of the gene(s) involved. There are several ways to search for a gene whose product is at least partially known. For example, the partial amino acid sequence of the protein known to be affected in the disease can be used to synthesize an oligonucleotide probe, which in turn can be used to screen a cDNA library obtained from the tissue where the protein is being expressed. Alternatively, antibodies against the protein can be generated and used for screening a λ*gt*11 cDNA library. Once a positive clone is found, its nucleotide sequence must be determined and compared with the sequence of the same gene obtained from the cell of an individual not afflicted by the disease. An altered nucleotide sequence found in the DNA from the diseased cell can be considered strong evidence that it is this gene that is involved in the disease phenotype.

In some genetic disorders, however, the gene product is not known; thus, the approaches described above can not be used. **Duchenne muscular dystrophy (DMD)**, an X-linked degenerative muscle disorder that affects about 1 in 3,500 males born, is such a disease. Nevertheless, the gene for the protein most likely involved in this disorder has recently been isolated and characterized. This feat was made possible by the cytogenetic discovery of a small **deletion in** the short arm of the **X chromosome** near the suspected **DMD locus**. The method uses differential reassociation of DNA molecules and illustrates one of several approaches that can be used to isolate an unknown gene. The entire procedure is illustrated in Figure 7.22.

DNA isolated from a human lymphoid cell line containing four X chromosomes but a normal DMD gene was digested with the restriction enzyme Mbo1 and mixed with a 200-fold excess of DNA from the patient with the DMD deletion that had been sheared to 1,000 bp fragments by sonication. After denaturation, the DNA was allowed to renature.

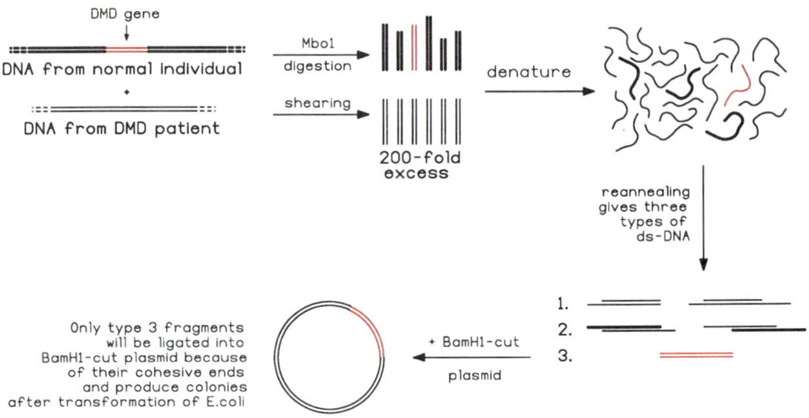

Figure 7.22.
Isolation of part of the gene involved in Duchenne muscular dystrophy by differential DNA reassociation.

Three types of double-stranded DNA fragments are expected to form:

1. The majority of fragments will carry sheared DNA ends.
2. A rare class of fragments will have Mbo1 ends in one strand and sheared ends in the other strand.
3. A very small fraction of fragments will carry Mbo1 ends on both strands.

Only this last class of fragments can be ligated into a BamH1-cleaved plasmid. (BamH1 generates the same cohesive ends as Mbo1 but recognizes six bases instead of four and consequently cuts less frequently.) Transformation of *E. coli* with the ligated plasmid DNA yielded a library enriched for DNA sequences corresponding to the segment of the X chromosome that was deleted in the DMD patient. Clones from this library were tested for their ability to hybridize to DNA from normal individuals but not to DNA from the DMD patient carrying the deletion. One such clone was then used as the starting point for a **chromosome walk**, which allowed the isolation from a genomic library of additional segments of the DMD gene and thus the restriction enzyme mapping of the entire DMD locus of the human X chromosome.

The principle of chromosome walking is as follows. Once a DNA segment from a chromosome is cloned, it can be used as a hybridization probe to identify neighboring, partially overlapping, segments. The second segment thus isolated is used in turn to identify a third segment, and so on. Chromosome walking is most efficiently done with **cosmid libraries** because of the large size of their chromosomal inserts. Each step in a cosmid library genomic walk can cover 35–45 kb of DNA.

A cloned genomic fragment from the DMD locus that was found to be highly conserved among different mammalian species was used as hybridization probe to screen a human skeletal muscle cDNA library. This led to the isolation of several overlapping clones that covered an open reading frame (a nucleotide sequence beginning with an AUG codon that does not contain any stop codons) of 14 kb representing the entire transcript of the DMD gene.

The presence of a 14 kb mRNA species could be detected in normal muscle tissue as well as in some other tissues by **Northern blot hybridization** using the DMD cDNA clone as a hybridization probe. In this technique, total mRNA from a cell is analyzed by agarose gel electrophoresis under denaturing conditions. (The RNA is denatured with formamide, and renaturation is prevented by formaldehyde.) After electrophoresis, the RNA is transferred, by blotting, onto a nitrocellulose sheet, where the RNA is immobilized by heat treatment. The location of the DMD-specific mRNA can then be visualized by incubating the nitrocellulose sheet with the radioactively labeled denatured cDNA probe. After washing the sheet to remove unhybridized cDNA, the location, and therefore the size, of the DMD mRNA can be determined by autoradiography.

In order to identify the gene product of the DMD gene, antibodies were raised against peptides synthesized according to the sequence of the cDNA. These antibodies were found to bind to a 427 kd protein present in muscle extracts from normal individuals but not from DMD patients. The protein, now commonly termed dystrophin, appears to be a minor component (0.002% of the total cellular protein) of the muscle cell membrane. Its function is still unknown.

The identification of the DMD gene and its product, **dystrophin**, is an excellent example of the power of molecular cloning techniques.

7.6 THE MOLECULAR BIOLOGY OF CANCER

Although long suspected, it is now well established that most cancers have genetic origins and arise either as a direct result of a mutation in a growth-regulating gene or from the overproduction (in some cases underproduction) of a gene product involved in cellular growth control. This conclusion is based on the molecular analysis of such mutated genes, called **oncogenes**, or their promoters and will be discussed in this section. However, before we can discuss the possible defects of cellular growth control that must occur in cancer cells, we have to describe what is known about normal growth control.

7.6.1 The Control of the Cell Cycle

The cell cycle has two main phases, **interphase** and **mitosis**. During interphase, which includes G1, S, and G2 (see Figure 6.15), the cell accumulates components, grows in size, and replicates its DNA. In mitosis, first the nuclear envelope breaks down, then the duplicated chromosomes become attached to the **mitotic spindle**, which segregates them into two identical sets of chromosomes, and finally the nuclear envelopes reform around each set of chromosomes, allowing the cell to divide into two identical daughter cells. The entire process must be controlled precisely so that cells that have not completed their DNA replication cannot enter mitosis. This is important because the lack of a chromosome can kill the cell or cause it to become a cancer cell.

Mitosis is initiated by the **maturation promoting factor (MPF)**, a protein factor consisting of at least two subunits, the **cdc2** subunit and **cyclin**. MPF is a protein kinase that, among other functions, phosphorylates the proteins of the nuclear envelope, called **lamins**, thereby initiating the breakdown of the nuclear envelope. During most of the cell cycle, the cdc2 protein is inactive. For activation it must combine with cyclin which accumulates during interphase. The resulting pre-MPF is then converted to active MPF by several other enzymes, one of them being **cdc25**. Active MPF then initiates mitosis, which includes the formation of the spindle apparatus. In addition, it activates enzymes that degrade cyclin. This degradation of cyclin is necessary for mitosis to proceed to completion. A deletion mutant of cyclin, capable of cdc2 activation but resistant to degradation, has been shown to cause the cell to remain arrested in mitosis. As MPF becomes inactive, phosphatases inactivate the enzymes that MPF had turned on by phosphorylating them. This inactivation also includes the enzymes involved in cyclin degradation so that cyclin can begin to accumulate again to initiate a new cell cycle (Figure 7.23).

Figure 7.23.
Control of the cell cycle by MPF.

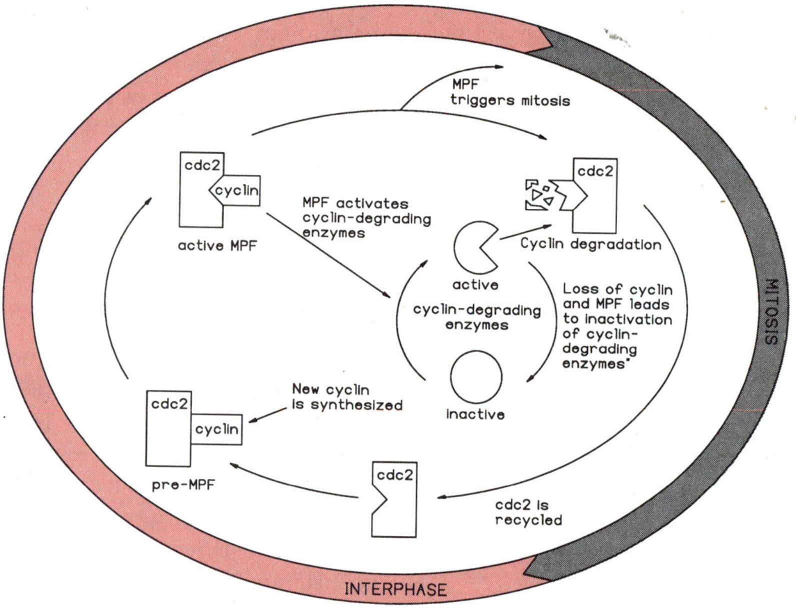

The cell cycle has a second control point, called START, at the entry of S phase. This control also involves the cdc2 protein, which complexes with a second type of cyclin that accumulates in G1. The accumulation of this second cyclin is influenced by nutrients, hormones, and growth factors. Cells that lack nutrients cannot pass through START and begin DNA replication.

7.6.2 The Incidence of Human Cancer

Two types of tumor can be distinguished: the **benign tumor**, which grows into a large mass of cells but stays within its normal tissue boundaries, and the **malignant tumor**, whose cells can invade other tissues and form secondary tumors, called metastases, at sites removed from the site of the original tumor. Although human cancers can occur in almost any tissue, they are generally divided into three types:

1. **Carcinomas** are cancers derived from epithelial tissues: skin, gut, lung, and glands.

2. **Sarcomas** arise from cells of bone, cartilage, muscle, and fibroblasts.
3. **Leukemias** and **lymphomas** originate from blood cells.

Over 90% of all human cancers are carcinomas. Most likely this is because epithelial cells, whether in the skin, the gut, or the lungs, are exposed to environmental influences such as carcinogens or radiation, agents that cause mutations (see Section 6.5). Epidemiological studies suggest that some forms of cancer are clearly caused by **environmental agents**, such as diet, life style, or pollution of the air, although the nature of the carcinogenic agent is often not yet known. For example, colon and breast cancers are most prevalent in the United States and occur much less frequently in Japan. On the other hand, cancer of the stomach is very common in Japan but much less frequent in the United States. The evidence that cigarette smoking causes most lung cancers is now overwhelming, and the role of exposure to excessive sunlight is known to increase the frequency of skin cancer.

In addition to **carcinogens**, certain **viruses** can cause cancer as well. **Hepatitis B** virus, for example, is believed to be at the root of many **liver cancers**. The **DNA viruses polyoma and SV40** produce cancers in hamsters. **Retroviruses** have been shown to cause cancers in many different species. Some of these viruses that cause cancer in humans will be discussed in this chapter.

Most human tumors consist of a **clone of cells** that are all derived from a single mutated cell. This has been concluded from the fact that in females who are heterozygous for the X-linked enzyme glucose 6-phosphate dehydrogenase, all tumor cells express the same isozyme, whereas in the rest of the body different cells have different X-chromosomes inactivated and thus express different isozymes (see Section 6.8).

While the **initial event** in the formation of a cancer cell usually is a **mutation**, the mutated cell often remains in a **precancerous state** for long periods of time. It requires the action of a **tumor promoter** for the expression of the cancer phenotype. Tumor promoters are typically agents that stimulate quiescent cells to divide, a necessary step in carcinogenesis. For example, when the skin of a mouse is treated with **benzopyrene**, a potent carcinogen, tumors rarely develop. However, if the skin is later painted with a **phorbol ester**, such as 12-*O*-tetradecanoylphorbol-13-acetate (TPA, Figure 7.24), tumors will develop. **TPA** by itself is not a carcinogen; instead it is an **activator of protein kinase C** (to be discussed in Chapter 12). The action of protein kinase C causes the mutagenized cell to proliferate and thereby establish its mutated phenotype.

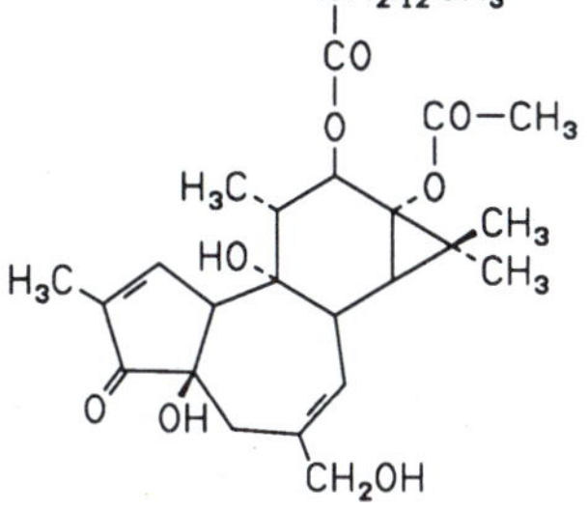

Figure 7.24.
The structure of 12-*O*-tetradecanoylphorbol-13-acetate, a tumor promoter and activator of protein kinase C.

Usually, a single mutation is not sufficient to convert a normal cell into a cancer cell. In some cancers, up to seven individual mutations are required, in others only two suffice. Thus, the incidence of cancer increases as the *n*th power of age, *n* being the number of mutations required for expression of the cancer phenotype.

7.6.3 Isolation of a Human Cancer-Causing Gene

When normal mammalian cells are grown in **tissue culture**, the cells divide until they form a **monolayer**, a single sheet of cells that covers the culture dish. At that point cell growth ceases. The culture is said to be **confluent**. In order to make the cells grow again, they must be dislodged from their support surface and from each other with the proteolytic enzyme trypsin and replated at a lower concentration in a new culture dish. For human cells, this process of propagation can be repeated until the cells have divided about 60 times; then all the cells die. Presumably, a biological clock ticks inside each cell that makes the cell age and eventually die. Only germ cells are immune from aging. The nature of this **biological clock** is not known.

Sometimes, however, a cell in a **primary culture** acquires a mutation that stops the biological clock and thus makes the cell immortal. Cells derived from such a mutant are called a **cell line**; they can be propagated indefinitely but still form a monolayer in culture that must be trypsinized for growth to resume. Tumor cells behave very differently; they are **immortal** like cell lines, but instead of forming monolayers, they continue to grow, causing the cells to pile up on top of each other.

One cell line that has been most useful in the isolation of human cancer-causing genes is the mouse fibroblast **3T3** line. Normally, 3T3 cells grow in a monolayer. However, when DNA isolated from a human tumor is sheared to fragments of about 50 kb and added in the form of a calcium phosphate precipitate to the 3T3 monolayer, some 3T3 cells incorporate the gene that was responsible for the cancer phenotype of the tumor cells and, as a consequence, assume the properties of the tumor cell. Such transformed 3T3 cells form **foci**, clusters of cells piled up on top of one another, that are readily recognized within the 3T3 monolayer (Figure 7.25). DNA isolated from these primary 3T3 transfectants can be used to transfect new 3T3 cultures, producing foci that are called secondary transfectants.

The human gene that was incorporated into the mouse genome of 3T3 cells can be identified by the presence of the highly reiterated human **Alu sequence**, which is present on average about once in every 5 kb segment of the human genome. (Mouse DNA does not contain this sequence.) This is done by making a genomic library from a secondary 3T3 transfectant and screening it with a radioactive Alu probe. The first **human oncogene** isolated this way was the **H-*ras*-1** gene found in a **bladder carcinoma**. It was soon discovered that this gene was identical to the H-*ras* transforming gene of the **Harvey sarcoma virus**, a retrovirus causing tumors in rodents.

By using the Harvey sarcoma virus **H-*ras*** gene as a hybridization probe, an H-*ras* homolog was also found in normal human cells. When this H-*ras*-1 gene was cloned and used to transfect 3T3 cells, no foci formed. Sequence analysis soon revealed that the H-*ras*-1 oncogene isolated from the bladder carcinoma contained a single base pair change as compared with the normal **cellular H-*ras*-1** gene. Apparently, the product of the H-*ras*-1 gene is somehow involved in cellular growth regulation, and the mutation that occurred in this gene caused the cells to proliferate, producing a bladder carcinoma. The H-*ras* gene of the Harvey sarcoma virus was found to have a mutation at the same site in addition to other mutations.

If the development of cancer requires more than one mutation, why then is a single mutated H-*ras*-1 gene capable of transforming 3T3 cells? The answer probably is that 3T3 cells are not really normal cells, even though they do not cause tumors when injected into animals. It is pos-

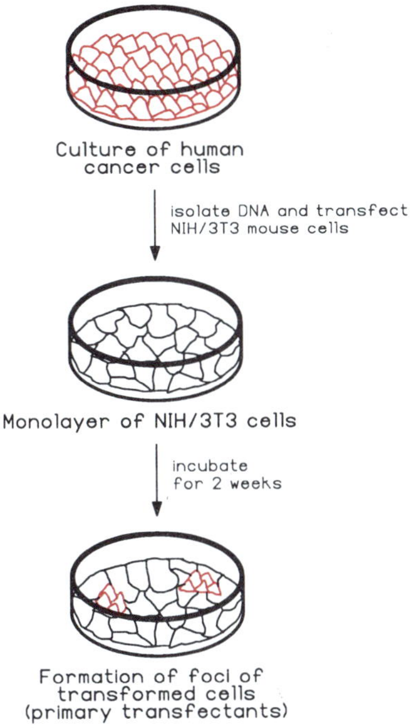

Figure 7.25.
The formation of foci in mouse 3T3 cultures by transfection with oncogenes.

sible that the immortalizing mutation, responsible for making mouse fibroblasts a cell line, together with the mutated H-*ras*-1 gene is sufficient for transformation. When primary mouse cells are transfected with the mutated H-*ras*-1 gene, they do not become transformed.

7.6.4 Oncogenes

As the example with the H-*ras* gene showed, cancer-causing genes, oncogenes, arise by mutation from normal cellular genes, called proto-oncogenes. Some RNA tumor viruses also carry oncogenes that are derivatives of normal cellular **proto-oncogenes**. Since the genome of tumor viruses integrates into its host's chromosomal DNA during part of the virus's life cycle, it is likely that a cellular proto-oncogene was picked up when the provirus genome was transcribed into viral RNA for incorporation into virus particles. There the proto-oncogene mutated into an oncogene, and when the virus now infects a normal cell, it expresses the oncogene and thereby transforms the cell into a tumor cell.

Many different viral oncogenes, called **v-onc**, have been isolated and characterized from retroviruses. Corresponding cellular proto-oncogenes, called **c-onc**, have been found in every case. One might ask what the normal cellular function of these genes is. In several cases, the properties of their gene product or their cellular function is known. In other cases, only the subcellular location and time of expression of the proto-oncogene product are known. Table 7.3 lists some of the oncogenes that can be grouped into five distinct classes.

Most oncogenes discovered by the **3T3 focus assay** were found to be of the *ras* type. This may be because 3T3 cells are especially sensitive to transformation by *ras*. The human genome contains five *ras* proto-oncogenes, two of which (H-*ras*-2 and K-*ras*-1) are pseudogenes that are not expressed. H-*ras*-1 and K-*ras*-2 have been found mutated, and thus activated, in bladder, lung, and colon carcinomas. The proteins of these genes (p21) are highly homologous and resemble the guanine nucleotide–binding elongation factor EF-Tu. The third *ras* proto-oncogene, N-*ras*, found in mutated form in human neuroblastoma, is less homologous. As seen in other cellular proto-oncogenes, the c-*ras* genes contain several introns that are absent from the v-*ras* oncogenes carried by retroviruses.

The normal *ras* p21 protein contains a Gly residue at position 12 that is altered in the oncogenic form. Activated *ras* protein has a decreased GTPase activity. It is possible that *ras* proteins function as **G proteins** that are involved in the signal transduction of growth factor or hormone receptors across the cell membrane. Upon binding of the growth factor to its receptor, the G proteins bind GTP and then activate a **protein kinase**. This activation ceases when the GTP is cleaved to GDP by the intrinsic GTPase activity of the G protein. If the GTPase activity is reduced, the activation lasts longer. Thus, permanently activated *ras* proteins might stimulate the protein kinases continuously without need of growth factor. (For a discussion of G factors, see Chapter 13).

The first oncogene discovered was the v-*src* gene found in **Rous sarcoma virus**. It codes for a 60 kd protein kinase that phosphorylates tyrosine residues in various proteins including itself. The **pp60c-*src*** protein, found in normal cells, also has **tyrosine protein kinase** ac-

Table 7.3. Properties of Some Oncogenes

Oncogene	Retrovirus or Tumor	Location	Function
Class 1: Protein kinases			
src	Rous sarcoma virus (RSV)	Cell membrane	Protein kinase (tyr-specific)
yes	Avian sarcoma virus (AVS)	Cell membrane	Protein kinase (tyr-specific)
abl	Abelson murine leukemia virus (MLV)	Cell membrane	Protein kinase (tyr-specific)
fes	Feline sarcoma virus (FeSV)	Cell membrane	Protein kinase (tyr-specific)
erbB	Avian erythroblastosis virus (AEV)	Cell membrane	EGF receptor
fms	Feline sarcoma virus	Cell membrane	CSF-1 receptor
mos	Moloney sarcoma virus	Cytoplasm	Protein kinase (ser/thr specific)
Class 2: GTP-binding proteins			
H-*ras*	Harvey mouse sarcoma virus (Ha-MSV)	Cell membrane	G protein with GTPase activity
K-*ras*	Kirsten MSV	Cell membrane	G protein with GTPase activity
N-*ras*	Human neuroblastoma	Cell membrane	G protein with GTPase activity
Class 3: Cellular growth factors			
sis	Simian sarcoma virus (SSV)	Secreted	PDGF
Class 4: Hormone and growth factor receptors			
erbA	Avian erythroblastosis virus (AEV)	Cell membrane	Thyroid hormone receptor
Class 5: Nuclear proteins			
myc	Burkitt's lymphoma	Cell nucleus	Unknown
fos	Murine osteosarcoma virus	Cell nucleus	Unknown

EGF = epidermal growth factor, CSF-1 = colony-stimulating factor 1, PDGF = platelet-derived growth factor.

tivity but apparently much less than pp60v-*src*. The level of tyrosine-phosphorylated proteins in RSV-transformed cells is about 10 times that found in normal cells. In part, this is due to the high level of expression of v-*src* under the regulation of the powerful **retroviral long terminal repeats (LTR) promoter**. However, pp60c-*src* and pp60v-*src* also differ in a few amino acids. The role of pp60*src* in growth control is not clear. Its location in the plasma membrane suggests a function in signal transmission.

Both the *ras* and *src* proto-oncogene products need to be anchored in the plasma membrane to be functional. This is accomplished through covalent linkage to hydrophobic long-chain fatty acids or isoprenoid groups. pp60c-*src* is esterified with its amino-terminal glycine residue to myristic acid (C_{14}), whereas *ras* p21 is linked to palmitic acid (C_{18}) via a thioester bond to an internal cysteine residue. At the same time, *ras* p21 is also linked to farnesyl alcohol via a thioether bond to the carboxyterminal cysteine residue.

Two other cellular proto-oncogenes that also possess tyrosine protein kinase activity have been identified as genes coding for growth factor receptors. The c-*erbB* gene codes for the receptor for **epidermal**

growth factor (EGF). The v-*erbB* oncogene has a large deletion at its amino terminus in addition to other mutations. The mutated protein has lost its ability to bind to EGF and, at the same time, is capable of continuously stimulating the cell to divide. The receptor for colony-stimulating factor 1 (CSF-1), a peptide growth factor specific for macrophages, is encoded by the c-*fms* gene. The retroviral v-*fms* gene has mutations near the carboxyl terminus removing a tyrosine residue that is normally phosphorylated in the c-*fms* protein. Apparently this loss of phosphorylation causes the receptor to give continuous stimulatory signals to the cell.

One of the oncogenes codes itself for a cellular growth factor. The v-*sis* oncogene represents the gene for the **platelet-derived growth factor (PDGF)**. It is carried as an additional gene by a feline sarcoma virus. When this virus infects and transforms a cell, PDGF is expressed at high levels because of the strong viral LTR promoter. Only cells that express the receptor for PDGF can be transformed by this oncogene. This is an example of tumorigenesis by autocrine growth stimulation without mutation of the oncogene product.

7.6.5 Transcriptional Activation of Proto-Oncogenes

Karyotype analysis of human tumor cells revealed that some forms of cancer are always associated with chromosomal rearrangements. In some cases a segment from one chromosome was found translocated onto the end of another chromosome; in other cases **mitotic nondisjunction** led to uneven distribution of chromosomes among daughter cells: One cell received both copies, the other none, of a particular chromosome, giving rise to **trisomy and monosomy**, respectively. Two examples of cancers involving **chromosomal rearrangements** are **Burkitt's lymphoma** and **chronic myelogenous leukemia**.

In **Burkitt's lymphoma**, a cancer of B lymphocytes found commonly in Africa, the c-*myc* proto-oncogene is translocated from the long arm of chromosome 8 to the long arm of chromosome 14, where the genes for the heavy immunoglobulin chains are located (Figure 7.26). This translocation brings the *myc* gene under the influence of a powerful enhancer, located near the genes for the immunoglobulin heavy chain constant region (Ig-Cu), resulting in **transcriptional activation of the *myc* gene**. In a few cases of Burkitt's lymphoma, the c-*myc* gene re-

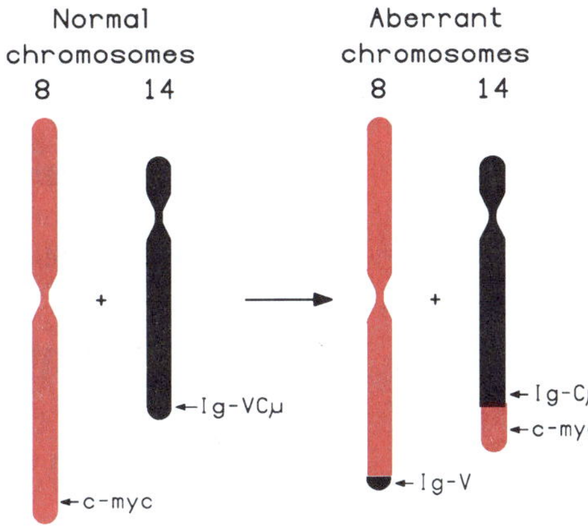

Figure 7.26. The chromosomal translocation found in most Burkitt's lymphomas.

mained on chromosome 8, but segments from chromosomes 2 (λ light chain locus) or 22 (κ light chain locus) were found translocated next to the *myc* locus. This activation of the c-*myc* gene involves no change in the protein sequence. Tumor formation appears to be solely the result of overexpression of c-*myc*. In normal cells, c-*myc* is induced in response to mitogenic stimuli; it is not expressed in quiescent cells. Thus, it is tempting to speculate that c-*myc* is either a transcription factor or a protein involved in DNA replication.

Chronic myelogenous leukemia (CML) is a cancer of myeloid stem cells. It is caused by a **reciprocal crossing-over event** between the long arms of chromosomes 9 and 22. The altered chromosome 22 is called the Philadelphia chromosome. This translocation inserts the c-*abl* gene, normally located on chromosome 9, into the *bcr* gene (named after the point of breakage: "**break point cluster region**") of chromosome 9. As a result, the first 25 amino acids of the c-*abl* protein are replaced by 600–700 amino acids of the *bcr* protein. Apparently, the resulting fusion protein between *bcr* and c-*abl* has tyrosine protein kinase activity, whereas the normal c-*abl* protein does not. (In the Abelson murine leukemia virus, the v-*abl* protein also is expressed as a fusion protein with a viral protein.)

Proto-oncogenes can also be activated by insertion of a strong viral promoter. This explains why some retroviruses are capable of transforming cells even though they do not carry an oncogene. Typically it takes longer for these viruses to produce a tumor in the organism they infect. As will be discussed in the next section, all retroviruses carry strong enhancers within their LTR termini. Thus, if the virus is inserted near a cellular proto-oncogene, this gene is transcribed at much higher levels, which leads to a cancerous phenotype. For example, B cell leukemia cells in chicken, caused by infection with avian leukosis virus (ALV), have the virus genome, often in partially deleted form, integrated into the first exon (nontranslated) of the c-*myc* gene, The c-*myc* gene is now under transcriptional control of the strong promoter located at the 3^1 end of the integrated ALV genome.

Different **leukemia-producing retroviruses** are found to have different preferences for the cell type they transform. Some prefer B cells, other T cells, and still others, erythroid cells, thereby producing different types of cancer. The reason for this target cell specificity lies in the virus's LTR and its transcriptional enhancer, which must itself be activated by tissue-specific protein factors.

Some tumors become progressively more malignant as they grow. This is because an activated oncogene is amplified. This was first observed in a human neuroblastoma, where the N-*myc* gene, a relative of the *myc* gene, was amplified more than 1000-fold. The tandemly duplicated N-*myc* genes can be seen on metaphase chromosomes as homogeneously staining regions (HSRs) or as double minutes (DMs), small extrachromosomal chromatin bodies containing the amplified genes. It is likely that HSRs and DMs are interconvertible by recombination.

7.6.6 Tumor Suppressor Genes

The discovery of cellular oncogenes and their mutational or transcriptional activation in tumor cells led to the hypothesis that the activation of one or more cellular oncogenes in a cell was sufficient to transform the cell into a malignant cell. This theory was further supported by the finding that many proto-oncogenes code for growth factors, growth factor receptors, signal transducers, protein kinases, and transcriptional activators; the abnormal expression of such proteins would be expected to interfere with normal cellular growth control. Thus, it was generally assumed that the activation of an oncogene is dominant in nature.

When tumor cells were fused with normal cells of the same type, however, the hybrid cells were nontumorigenic; in other words, the original mutation responsible for the tumor phenotype was recessive. This phenomenon was called **tumor suppression**. It indicated that normal cells contain a gene product that can suppress the expression of an activated oncogene. Soon after this discovery, it was shown that different tumors could be suppressed by the introduction into the cell of specific chromosomes from normal cells.

The observation that many tumor cells contain chromosomal rearrangements, including deletions, was the basis for the isolation of the first tumor suppressor gene, the **retinoblastoma suppressor gene Rb-1**. This tumor arises from cells of the embryonal neural retina and occurs in young children. Cytogenetic studies of cells from retinoblastoma tumors often were found to have deletions on the long arm of chromosome 13. Chromosome-walking from a nearby known locus led to the isolation of a clone containing the Rb-1 locus. This clone hybridized to a 4.7 kb mRNA found in normal cells but not in retinoblastoma cells. The precise function of the 100 kd Rb-1 protein encoded by this mRNA is still unknown. It appears to be a DNA-binding protein, probably a transcription factor, that is found in most cells. The protein is unphosphorylated during the G0 and G1 phases but becomes multiply phosphorylated during the S and G2/M phases of the cell cycle. It is also found associated with transforming proteins of DNA tumor viruses (see below). There is some evidence that the retinoblastoma protein may play a role in the induction of differentiation. Such differentiation would stop the cell from uncontrolled proliferation.

Another example of a tumor suppressor gene is the gene that codes for the **p53 protein**, a 53 kd cellular phosphoprotein first identified because of its ability to bind to SV40 T antigen (see section on DNA tumor viruses). Normal p53 protein is thought to be involved in controlling cell entry into S phase. When normal cells are cotransfected with the activated *ras* oncogene and a mutated p53 gene, the cells become transformed. Cotransfection with a wild-type p53 gene, however, suppressed the transformed phenotype. Deleted or mutated p53 genes are frequently found in human tumors. In fact, mutations in p53 are the most common genetic change found in human cancers.

Mutations in, or deletion of, other tumor suppressor genes appear to be responsible for Wilms' tumor (chromosome 11), bladder carcinoma (chromosomes 9, 11, and 13), and melanoma (chromosomes 1 and 6) to name a few. In all cases, introduction of a normal chromosome into the tumor cell suppressed the tumorigenic phenotype. Mutations in tumor suppressor genes appear to be recessive; that is, both alleles have to be mutated or deleted to produce the cancer phenotype. It is likely that many of these genes code for transcription factors, regulating gene expression in both positive and negative fashion.

7.7 TUMOR VIRUSES

As discussed in the previous section, many animal viruses are capable of transforming a cell into a tumor cell. The first such virus was the Rous sarcoma virus, isolated from the tumor of a chicken by Peyton Rous in 1909. It is a retrovirus containing RNA as its genome. However, both DNA and RNA viruses are known to cause cancer. We will discuss the DNA viruses first.

7.7.1 DNA Tumor Viruses

SV40 and polyoma are two small viruses with circular double-stranded DNA that induce tumors in newborn hamsters. Certain "permissive" cells allow replication of the virus and are killed in the process. Other cells, such as newborn hamster cells, are nonpermissive for viral replication, and a small fraction of such cells become permanently transformed by the virus. In these transformed cells, the virus DNA has become integrated at random locations into the host genome. SV40 was originally discovered in the rhesus monkey cell lines used in the preparation of the Salk vaccine against polio. It has never been demonstrated to cause tumors in humans.

Although different in their nucleotide sequence, the two viruses exhibit strong homologies in their amino acid sequence. Figure 7.27 shows the map of the **SV40** genome. Early in the replication cycle, two proteins, called little and large **T antigen**, are synthesized. They arise from the same pre-mRNA by differential splicing. The large T antigen is a regulatory protein that increases the expression of several host cell enzymes involved in DNA replication causing the cell to divide. Nonpermissive cells that do not become transformed divide only once and then return to their quiescent state. The three late genes, coding for virus capsid proteins V1, V2, and V3, are only expressed in permissive cells. Their expression requires the presence of T antigen as well.

Large T antigen is required for the transformation of cells. In SV40-transformed cells, the large T antigen is found associated with a cellular protein called p53. In normal cells, p53 is expressed before cells enter S phase and disappears thereafter. Apparently, binding to T antigen makes p53 more stable, so that it continues to stimulate cells to replicate and divide. The role of little T antigen is not fully understood yet.

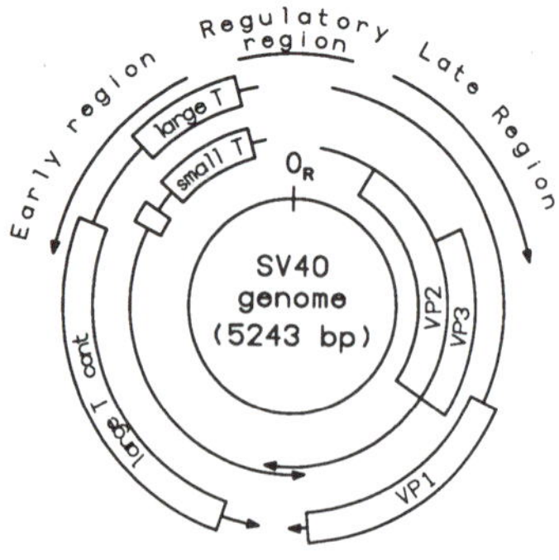

Figure 7.27.
The SV40 genome and its transcription and translation products.

Polyoma virus makes **three types of T antigen**: small T, middle T, and large T. Large T antigen is not capable of transforming cells but is sufficient to make primary cells immortal. Middle T antigen is required to transform immortalized cells. It is found associated with the cellular proto-oncogene p60c-*src*. Thus we may conclude that the p53 protein, which associates with the SV40 large T antigen, serves a function similar to p60c-*src*.

The genome of the **adenovirus** is a linear DNA of 36 kb. It is considerably more complex than polyoma or SV40 and codes for many different proteins. Like SV40 or polyoma, its genome is integrated at random positions into the transformed cell's genome. Only 8% of the adenovirus genome is required for maintaining the transformed state of the cell. The products of two genes, **E1A** and **E1B**, are required for transformation. The E1A gene product resembles polyoma's large T antigen in that it can immortalize cells but by itself is unable to transform. E1B protein will then transform these cells. It associates with the same cellular p53 protein as SV40 large T antigen.

7.7.2 RNA Tumor Viruses

RNA tumor viruses are **retroviruses**; that is, their life cycle involves reverse transcription of the RNA genome into complementary DNA that is then integrated into the host genome. Since cells normally do not contain a reverse transcriptase, the virus must carry this enzyme packaged together with the RNA inside the virion. Most retroviruses have three genes. The ***gag*** gene (for group-specific antigen) codes for a precursor protein from which all capsid proteins are made. The product of the ***pol*** gene is cleaved into reverse transcriptase and an enzyme required for the integration of the virus into the host genome. The ***env*** gene codes for the envelope glycoprotein that covers the surface of the virus particle. *Gag* and *pol* are transcribed into a single mRNA that contains a stop codon at the end of *gag*. In a few percent of the mRNAs, the stop codon is ignored, and a *gap-pol* polyprotein is made that is subsequently cleaved into *gag* and *pol* proteins. This translational frameshifting accounts for the smaller number of reverse transcriptase molecules made as compared with the number of *gag* proteins.

Each retrovirus particle contains two identical copies of linear RNA. The sequence at the ends of the RNA molecules consists of short direct repeats and longer nontranslated segments, U5 and U3. In addition, a ***transfer RNA*** (**tRNA**) molecule is annealed to a sequence (primer-binding site, or PB) near the 5′ end of the RNA molecule. It serves as a primer for the synthesis of the complementary DNA after infection of a cell. The replication of the retroviral RNA is complex and leads to the generation of two identical **long terminal repeats** at the ends of the provirus before it is inserted into the host genome. The LTR regions contain powerful **enhancer and promoter** sequences. The process of reverse transcription is depicted in detail in Figure 7.28.

The mechanism of **integration** of the retro-provirus into the host genome is not completely understood. It may involve cyclization and reciprocal recombination with a suitable site on the host genome. Once integrated, the provirus is never excised. Infected cells usually continue to divide while, at the same time, making new virus particles.

Most retroviruses do not contain oncogenes. As discussed in the previous section, some of these retroviruses can nevertheless cause tumors in animals, but only after very long incubation periods. In the tumor cell, the retrovirus LTR is usually found inserted near a cellular proto-oncogene that becomes activated by the viral LTR.

Only retroviruses carrying oncogenes can transform cells in culture. They are called **acute transforming viruses**. The Rous sarcoma virus is unique among the retroviruses in that it carries the v-*src* gene as an additional fourth gene. All other known acute transforming viruses are replication defective because the oncogene replaces part of a viral gene. Thus, these viruses can only replicate in cells that are simultaneously infected with a nondefective helper virus.

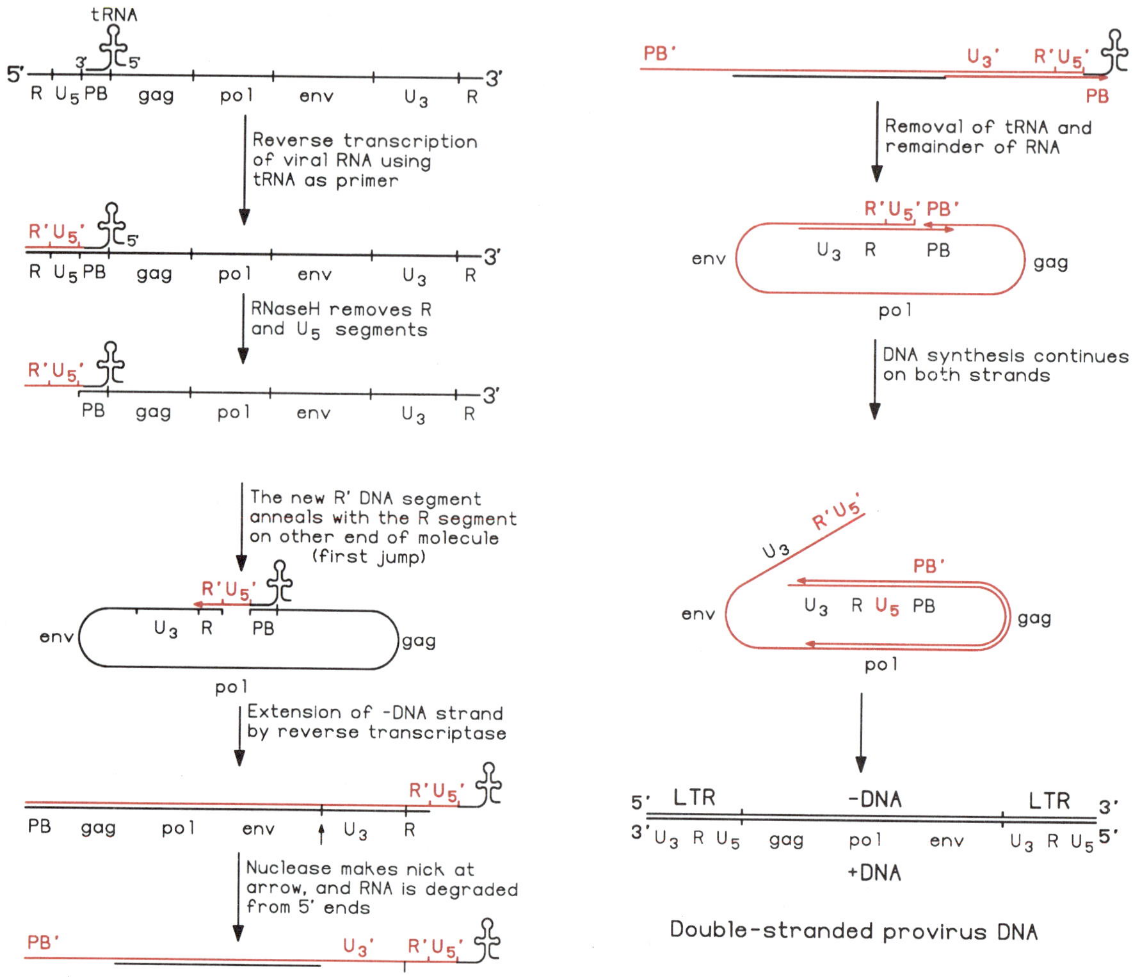

Figure 7.28.
The reverse transcription of a retrovirus.

7.7.3 Human Immunodeficiency Virus

The **human immunodeficiency virus (HIV)** causes the **acquired immune deficiency syndrome (AIDS).** After infecting its host, the virus usually replicates rapidly, and free virus particles can be found in the bloodstream and the cerebrospinal fluid. Later, the virus becomes dormant until several years later, when the multiplication of HIV again flares up, leading to the eventual death of the infected person. During the dormant stage of AIDS, HIV still can be demonstrated not only in **T4 lymphocytes** (helper T cells), the primary target cell for virus replication, but also in other cells of the immune system, in neurons, and in bone marrow cells. In T cells, the virus can stay dormant indefinitely, whereas in macrophages and their precursor cells, the monocytes, HIV continues to replicate slowly without killing the cell.

HIV infects a cell by binding, through its envelope glycoprotein gp120, to the **CD4 receptor** on the lymphocyte surface. Only cells that express this receptor are susceptible to HIV infection. The viral lipid membrane then fuses with the membrane of the infected cell, and the virus core containing the two genomic RNA copies is released into the cell's cytoplasm.

The viral gp120 envelope protein, together with the cellular CD4 receptor, also appears to be responsible for the **cell fusion** observed frequently between HIV-infected cells. Such cell-cell fusions between HIV-infected cells and uninfected cells produce large **syncytia** that cause the death of lymphocytes that are not themselves infected by the virus. Because of the destruction of the helper T cells, the patient can no longer mount an immune response to invading microorganisms or malignant cells arising in his body.

The HIV genome contains, in addition to the usual retroviral genes *pol, gag,* and *env*, six more genes, *tat, rev, vif, vpr, vpu,* and *nef* (Figure 7.29). The products of these additional genes are all regulatory proteins and are responsible for the complex life cycle of HIV.

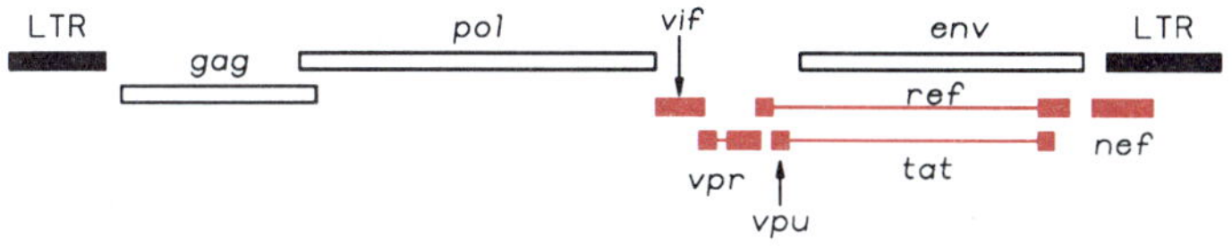

Figure 7.29.
The structure of the HIV genome. Only three genes, *gag, pol,* and *env,* code for the protein components of the virus particle. The other six genes code for regulatory proteins. The mRNA for some of these must be spliced from noncontiguous segments of the genome.

The product of the *tat* gene is a general activator protein affecting all HIV genes, including itself and the other regulatory genes; the *nef* gene product represses all HIV genes (including itself); and the *ref* protein represses only regulatory genes but activates virion genes, thus stimulating virus production. The functions of the other regulatory proteins are less well known. The regulation by the different HIV regulatory proteins may occur at the transcriptional level, at the level of splicing and transport of mRNA, and at the translational level.

Like all RNA viruses, HIV is susceptible to frequent mutations because both the reverse transcriptase, which makes a DNA copy of the viral genome, and the cellular RNA polymerase, which is responsible for new virus RNA synthesis, lack proofreading capabilities. Thus, during the course of an infection with HIV, many different mutants usually arise in a single patient.

Several avenues are being explored to interfere with the proliferation of HIV. Only a few are listed here:

1. **Nucleotide analogs**: 2′, 3′-dideoxynucleotides, when incorporated into the growing chain, block further addition of nucleotides. During normal DNA replication, such blocking nucleotides can be removed by the DNA polymerase's exonuclease activity. The reverse transcriptase, however, lacks this enzyme so virus replication is blocked. The most commonly used drug in AIDS therapy is **3′-azido-2′,3′-dideoxythymidine (AZT)** or **zidovudine**; see Figure 5.13.

2. **CD4 proteins:** injection of soluble forms of CD4, genetically engineered, would bind to the viral gp120 envelope proteins and thereby interfere with the ability of the virus to attach to cellular CD4 sites.

3. **Protease inhibitors**: these would prevent the virus-encoded protease from processing the viral precursor protein into enzymes and proteins required for virus replication.

7.8 INTERFERON

When cells in an animal are infected with RNA or DNA viruses, an antiviral state is usually established that prevents further infection of still uninfected cells. This **antiviral state** is due to the presence of interferon, which is synthesized in response to the virus infection. At this time, we know three types of interferon. α interferon is produced by lymphocytes, and β interferon is produced by fibroblasts; the two proteins are related. In contrast, γ interferon is a rather different protein. Its synthesis is induced in lymphocytes by mitogens, substances that trigger cell division, but not by viral infection.

How is the antiviral state achieved? Interferon blocks viral replication in two ways. First, interferon induces the synthesis of a **protein kinase**, which phosphorylates, and thus inactivates, the α subunit of the translational initiation factor eIF-2. The reaction requires the participation of double-stranded RNA (e.g., viral replication intermediates). Second, interferon induces the synthesis of the enzyme **2′,5′-oligo A synthetase**, which polymerizes ATP into an RNA oligomer containing 2′-5′-phosphodiester bonds instead of the usual 3′-5′-bonds. This 2′,5′-oligo A binds to, and thereby activates, **RNase L**, a latent ribonuclease that cleaves single-stranded RNA.

For a more extensive discussion of recombinant DNA technology and its application in medical science, consult the following textbooks:

Mathews et al. (1990): Chapters 24, 25

Stryer (1988): Chapters 6, 34

Further References

Belten HJ, Gehring WJ. Position effects on eurkaryotic gene expression. 1990; 6:679–714.

Innis MA, Gelfand DH, Sninsky JJ, White TJ. *PCR Protocols: A guide to Methods and Applications.* Academic Press, New York, 1990.

Landel CP, Chen S, Evans GA. Reverse genetics using transgenic mice. *Ann Rev Physiol.* 1990; 52:823–851.

Melton DA, Whitman M. Growth factors in early embryogenesis. *Ann Rev Cell Biol.* 1989; 5:93–118.

Mullis KB. The unusual origin of the polymerase chain reaction. *Sci Am.* 1990; 262:56–65.

Old LJ. Tumor necrosis factor. *Sci Am.* 1988; 258 (5):59–75.

Ransone LJ, Verma IM. Nuclear proto-oncogenes *fos* and *jun. Ann Rev Cell Biol.* 1990; 6:539–558.

Rosenberg SA. Adoptive immunotherapy for cancer. *Sci Am.* 1990; 262:62–69.

Scientific American September issue. What science knows about AIDS: A single topic issue. *Sci Am.* 1988; 259(3).

Singer SJ. The structure and insertion of integral proteins in membranes. *Ann Rev Cell Biol.* 1990; 6:247–296.

Stanbridge EJ. Human tumor suppressor genes. *Ann Rev Genet.* 1990; 24:615–657.

Vaishnav Y, Wong–Stahl F. The biochemistry of AIDS. *Ann Rev Biochem.* 1991; 60:577–630.

Verma IM. Gene therapy. *Sci Am.* 1990; 263:68–85.

Westphal H, Gruss P. Molecular genetics of development studied in the transgenic mouse. *Ann Rev Cell Biol.* 1989; 5:181–196.

White JM. Viral and cellular membrane fusion proteins. *Ann Rev Physiol.* 1990; 52:675–697.

Young JDE, Cohn ZA. How killer cells kill. *Sci Am.* 1988; 258(1):38–44.

CHAPTER

8 Nutrition

OBJECTIVES

After studying this chapter, the student should be able to

1. Name the digestive enzymes released as proenzymes from the pancreas and explain how they are activated.
2. Outline the probable mechanism of HCl secretion in the stomach and describe its regulation.
3. Explain why bomb calorimetry of proteins gives caloric values that exceed their caloric energy yield during metabolism.
4. Define the term "respiratory quotient" and give RQ values observed during the metabolism of carbohydrates, lipids, and proteins.
5. Explain the term "basal metabolic rate."
6. State the average specific dynamic action observed with a mixed diet and explain why different types of food exhibit different SDAs.
7. Explain why, in addition to the minimum daily requirement of 13 g of essential amino acids, a 70 kg man needs to take in nonessential amino acids in order to maintain nitrogen balance.
8. Name the essential amino acids in the human diet.
9. Name the amino acids that exhibit a sparing action on the requirements of Met and Phe, respectively, and understand the process.
10. Explain why steatorrhea (fatty stool) might lead to blood-clotting problems.
11. Explain the participation of vitamin A in the light response of the human photoreceptor and describe how this initial response is converted into a neural signal.
12. State which of the four cations, Ca^{2+}, Mg^{2+}, and K^+, and Na^+, are predominantly intracellular, which are extracellular, and which are regulated in pairs by similar mechanisms.
13. Explain in what respect iron homeostasis is different from the regulation of other cations.
14. Outline the interplay of three hormones in the regulation of calcium and phosphate homeostasis.

The principal sources of energy in our diet are carbohydrates, protein, and fat. Nucleic acids have negligible nutritional value. Most food consists of polymers that must be hydrolyzed to monomers before metabolism can begin. This hydrolysis occurs during **digestion**.

8.1 DIGESTION

Preparation of food often involves processes that make food more digestible. For example, limited proteolysis occurs during the hanging (aging) of meat or by the action of meat tenderizer (protease), as well as during the manufacturing of cheese. The process of **cooking** denatures protein, thus making it a better substrate for hydrolyzing enzymes. Cooking also ruptures the starch granules, thus releasing their carbohydrate contents. Other processes include the emulsification of fat in milk and mayonnaise.

8.1.1 Digestive Processes in the Mouth

Digestion of starch (consisting of amylose, a linear α-1,4-glucan, and amylopectin, a branched glucan similar in structure to glycogen but containing fewer branch points) or glycogen begins in the mouth with the action of **salivary α-amylase**. This endoglucanase cleaves α-1,4-glucosidic bonds, producing small oligosaccharides from dietary starch. Hydrolysis usually stops at the disaccharide or trisaccharide level, and at branch points, since α-amylase is unable to hydrolyze 1,6-glucosidic bonds. The products of this digestion are maltose, maltotriose, somewhat larger branched oligosaccharides (average chain length of seven glucose residues), and a small amount of glucose. α-Amylase is most active at pH 7.0 and is quickly inactivated in the acidic environment of the stomach. It is stabilized by Ca^{2+} ions and requires Cl^- ions for activity.

In addition to α-amylase, saliva contains **mucin**, a proteoglycan with about 60% carbohydrate content, mostly *N*-acetyl-neuraminic acid **(sialic acid)** and *N*-acetyl-galactosamine. Mucin is produced throughout the alimentary canal and functions as a lubricating and protecting agent.

There is also a **lingual lipase** produced by the **Ebner's glands** of the tongue. Although it has a pH optimum of 4.0, it is active down to pH 1.8. This lipase becomes active only when the food passes into the acid environment of the stomach, where its action is responsible for the hydrolysis of about one third of all triacylglycerols.

Saliva contains many of the ions and small molecules present in blood and thus may be used for the noninvasive monitoring of certain blood components. One unusual ion found in saliva is thiocyanate (SCN^-), which arises from the detoxification of cyanide ions by thiosulfate (SSO_3^{2-}).

$$SSO_3^{2-} + CN^- \rightarrow SO_3^{2-} + SCN^-$$

The reaction is catalyzed by the iron-containing enzyme **thiosulfate sulfate transferase** (rhodanese), which is present in many types of cells.

8.1.2 Digestive Processes in the Stomach

The primary digestive process in the stomach is the hydrolysis of proteins by **pepsin.** Pepsin is produced by the **chief cells** of the stomach in the form of an inactive proenzyme. The conversion of pepsinogen to pepsin is initially catalyzed by pepsinogen itself, which is activated by the low pH in the stomach lumen. The resulting pepsin then catalyzes the activation of the remaining pepsinogen. In the process, a 42-amino-acid polypeptide is removed from the amino terminus of pepsinogen. This polypeptide is subsequently cleaved further into smaller fragments, some of which possess antitrypsin activity.

Pepsin is an endopeptidase with a relatively broad specificity. It cleaves peptide bonds on either side of aromatic (Phe, Tyr) and hydrophobic (Leu, Ile, Met) amino acid residues as well as a few other amino acids. For optimal activity, pepsin requires a pH of 2.0.

Gastric juice contains an additional endopeptidase, **gastrisin.** However, this enzyme has a pH optimum near 4.0, much higher than the prevailing stomach pH. It appears likely that it functions only in the stomach of the infant, where the pH is much higher than in the adult.

The pH of the adult stomach lies between 1.2 and 1.8. The hydrochloric acid responsible for this low pH is secreted by the **parietal cells** of the gastric mucosa. In this active transport process, hydrogen ions are pumped against a concentration gradient of $10^{-7}M$ inside the cell to about $10^{-0.4}M$ in the stomach lumen. The energy for this transport amounts to about 1.5 kcal/L of gastric secretion. (Parietal cells are especially rich in mitochondria.)

Despite extensive research, the mechanism of hydrogen secretion is still not completely understood. The process involves both an ATP-dependent H^+/K^+ exchange pump and an H^+ pump. Chloride ions probably leak out of the cell accompanied by K^+ ions, which are then exchanged for Na^+ ions (Figure 8.1).

Inhibitors of **carbonic anhydrase** reduce H^+ secretion by over 90%. This enzyme catalyzes the reaction

$$H_2O + CO_2 \rightleftharpoons H_2CO_3$$

The resulting carbonic acid dissociates immediately into bicarbonate and hydrogen ions:

$$H_2CO_3 \rightleftharpoons H^+ + HCO_3^-$$

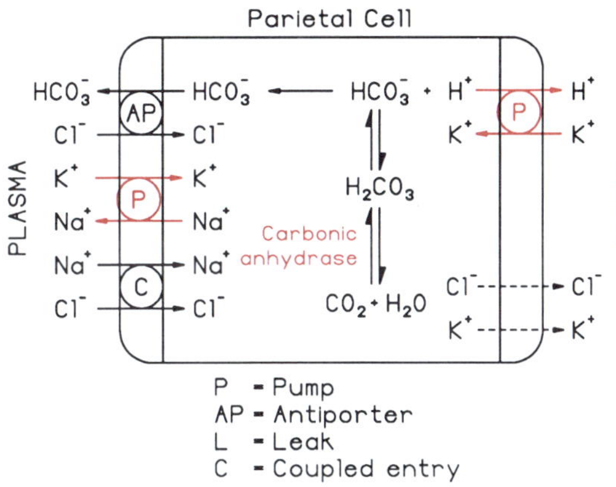

Figure 8.1.
Probable mechanism of HCl secretion in the stomach.

The reaction catalyzed by carbonic anhydrase thus provides an intracellular supply of hydrogen ions for subsequent secretion. The bicarbonate ion, created at the same time, is exchanged for a chloride ion by an antiporter system located in the cell membrane facing the plasma. The release of bicarbonate into the bloodstream after a large meal causes an **alkaline tide,** a brief rise in blood pH, which is followed by alkalinization of the urine (see Section 12.2).

HCl secretion is stimulated by **histamine** and by several oligopeptide hormones produced by the pyloric mucosa of the stomach, the **gastrins.** Histamine acts through the formation of cyclic adenosine monophosphate (cAMP), which, in turn, activates a protein kinase that phosphorylates carbonic anhydrase. The phosphorylated form of this enzyme is much more active than the unphosphorylated form. Inhibitors of phosphodiesterase (the enzyme that cleaves cAMP), such as caffeine or theobromine, cause intracellular cAMP levels to rise and thus stimulate HCl secretion. The action of gastrin may, in part, proceed through the activation of histidine decarboxylase, the enzyme that produces histamine.

The primary function of the hydrochloric acid in the stomach is to provide the proper pH for pepsin activity. In addition, the acid denatures proteins and also functions as a partial barrier against the passage of pathogenic microorganisms into the gastrointestinal tract.

Parietal cells also produce **intrinsic factor,** a glycoprotein required for the absorption of vitamin B_{12} from the intestine. Lack of intrinsic factor leads to **pernicious anemia.** Often this is accompanied by **achlorhydria** (lack of HCl production), resulting in low pepsin activity.

8.1.3 Digestive Processes in the Intestines

Completion of the hydrolytic processes occurs in the small intestine. Upon exit from the stomach, the well-mixed and partially digested bolus of food is rapidly neutralized by pancreatic, biliary, and intestinal secretions. These secretions have a high bicarbonate content and a pH value of about 8.0. Pancreatic and intestinal enzymes then complete the hydrolysis of carbohydrates, proteins, and nucleic acids to monosaccharides, amino acids, and nucleosides and free bases, respectively. Fats are only partially hydrolyzed to monoglycerides and free fatty acids.

Carbohydrate Digestion

Starch, predigested in the mouth, is further degraded by pancreatic α-amylase. The resulting mixture of α-1,4-disaccharides and α-1,4 trisaccharides, as well as oligosaccharides containing the α-1,6 branch point of glycogen or amylopectin, are then further hydrolyzed by a variety of **oligosaccharidases** and by **disaccharidases**, such as **maltase, isomaltase** (cleaves α-1,6 glucosidic bonds), **sucrase**, and **lactase**. Most of these enzymes are located on the brush border of the intestinal epithelium. Oligosaccharidases found free in the intestinal juice (succus entericus) are probably derived from desquamated epithelial cells. (The half-life of the intestinal epithelial cell is only about 3 days.)

Most adult humans have lost the ability to synthesize the enzyme lactase, which was produced in sufficient quantities during infancy. Thus, for them excessive ingestion of milk may cause digestive problems because unhydrolyzed lactose can pass into the lower intestine, where bacteria ferment it to organic acids causing water retention and diarrhea. Northern Europeans and their descendents retain the ability to synthesize lactase in adulthood.

Sucrase and isomaltase activities are expressed by a single polypeptide that also exhibits α-1,4-glucosidase activity. About 1% of the U.S. population lacks the **sucrase-isomaltase complex.** In people affected by this deficiency, ingestion of large amounts of sucrose causes diarrhea **(sucrose intolerance)** and may possibly lead to colitis and other diseases of the colon.

The monosaccharides glucose and galactose are absorbed from the intestine by cotransport with Na^+ ions. The carrier protein that transports the sugars into the intestinal cell functions only if both Na^+ and either glucose or galactose are bound to it. Apparently, the Na^+ concentration gradient across the cell membrane provides the driving force for the uptake of these sugars. Ouabain, a digitalis glycoside that inhibits the Na^+/K^+-exchange pump, also blocks the absorption of glucose and galactose. Fructose is taken up by a separate system (facilitated transport).

Protein Digestion

Protein digestion, initiated in the stomach, is completed in the small intestine by a variety of pancreatic and intestinal proteases. The **acinar cells** of the **pancreas** produce several proenzymes, **trypsinogen, chymotrypsinogen, procarboxypeptidases A** and **B**, and **proelastase.** The enzymes are released into the upper duodenum via the pancreatic duct. There, the intestinal enzyme **enteropeptidase** (old name: enterokinase) activates trypsinogen by cleaving the Lys_6-Ile_7 peptide bond. The resulting **trypsin** activates all other pancreatic pro-proteases. (Table 8.1).

The action of the pancreatic proteases produces a mixture of short peptides, mostly dipeptides and tripeptides, and a high proportion of free amino acids. Complete hydrolysis to amino acids is accomplished by several **oligopeptidases** and **aminopeptidases** with different specificities located on the brush border of the **intestinal epithelium**. Some of these enzymes will also cleave the prolyl bond, which is re-

Table 8.1. Specificities of Pancreatic Proteases

Enzyme	Activator	Bond Cleaved at Carboxyl Side of
Trypsin	Enteropeptidase	Lys, Arg
Chymotrypsin	Trypsin	Phe, Tyr, Leu, Met
Carboxypeptidase A	Trypsin	Phe, Tyr, Leu
Carboxypeptidase B	Trypsin	Arg, Lys
Elastase	Trypsin	Small hydrophobic amino acids

sistant to the action of pancreatic proteases. The released free amino acids are then absorbed by a variety of transport systems. As with the uptake of glucose, the transport of amino acids is coupled to Na^+ uptake.

Lipid Digestion

Lipids, consisting of triglycerides, phospholipids, and cholesterol esters, are first emulsified with the help of bile salts. Triacylglycerols are hydrolyzed to about 30% by lingual lipase in the acid environment of the stomach. Further hydrolysis is accomplished by a **pancreatic lipase.** Both processes yield monoglycerides and free fatty acids. The β-acyl bond is resistant to lipase, but the fatty acyl residue at the β position can migrate nonenzymatically to a free α position. Pancreatic lipase is activated in the duodenum by Ca^{2+} ions and by bile salts. Lipase itself is only partially active. For full activity, the enzyme combines with **colipase,** a 10 kd protein without enzymatic activity. Colipase enables lipase to bind to its lipid substrate.

Phospholipids are cleaved by **pancreatic phospholipase** (synthesized as a proenzyme and activated by trypsin). Lecithin, for example, is degraded to lysolecithin, a powerful detergent that, together with monoglycerides, free fatty acids, and bile salts, forms **micelles.** Although bile salts are not necessary for triglyceride hydrolysis, they are required for micelle formation. The products of lipid digestion, as well as undigested lipids, must be emulsified to micellar dimensions for absorption to occur. In the mucosal cells, triglycerides are resynthesized and assembled, together with other lipids and apolipoproteins, into **chylomicrons,** which then enter the circulation via the lymphatic system (see Section 3.3).

Digestion of Nucleic Acids

Nucleic acids are hydrolyzed by **pancreatic nucleases, nucleosidases,** and **phosphatases**. The nucleosides and free purine and pyrimidine bases can be taken up by the mucosal cells. However, the energy derived from the oxidation of purines to uric acid, and of pyrimidines to CO_2, is insignificant. Only the ribose and deoxyribose moieties of nucleic acids contribute to energy production. A diet that is rich in nucleic acids may elevate the uric acid pool of the system and thus contribute to the etiology of gout (see Section 5.5).

Table 8.2 summarizes the digestive enzymes of the gastrointestinal tract.

Table 8.2. Origin and Substrates of Digestive Enzymes

	Enzymes Digesting the Following Substrates			
Origin of Enzyme	Carbohydrates	Proteins	Lipids	Nucleic Acids
Mouth (saliva)	α-Amylase$_s$		Lingual lipase	
Pancreatic juice	α-Amylase$_p$	Trypsin Chymotrypsin Carboxypeptidases A and B Elastase	Lipase$_p$ Phospholipases Cholesteryl esterase	Nucleosidase Phosphatase
Intestinal mucosa	1,6-Glucosidase Maltase Saccharase Lactase	Aminopeptidases Dipeptidases	Phosphodiesterases	

Bulk and Fiber

In addition to the digestible components, our regular diet contains a variety of indigestible substances, generally called **bulk** or **fiber.** Bulk consists mainly of polysaccharides, including cellulose, hemicellulose, pentosans, lignins, and a large number of other indigestible polymers. Despite their indigestibility, these substances are beneficial because they are highly hydrophilic and thus aid in the retention of water in the intestinal contents, producing a softer stool. Some nutritionists hold that many of the diseases of the lower digestive tract, such as hemorrhoids, colon cancer, colon diverticula, and others, typically found in highly developed societies, stem from a low-bulk diet and hard stools.

The colon acts as a massive incubator for microorganisms. Generally, these symbionts are advantageous to the host. They represent an important source of certain vitamins. If supplied with excess substrate, however, such as lactose (in lactase deficiency), they produce organic acids that retain water in the feces (by osmotic forces), thus causing diarrhea. Certain types of food (e.g., beans) contain carbohydrates that resist digestion in the small intestine. In the large intestine, bacteria ferment these substances to methane and hydrogen.

8.1.4 Regulation of Synthesis of Digestive Enzymes

The production of digestive enzymes is controlled by both neural and hormonal stimuli. Filling of the stomach stimulates the cholinergic parasympathetic nervous system, which causes release of pepsin as well as of gastrins. Gastrins, in turn, stimulate HCl production in the stomach and enzyme secretion into the pancreatic juice.

The entering of the acidic stomach contents into the upper duodenum causes release of two duodenal hormones: **secretin** and **cholecystokinin**. Secretin, a 17-amino-acid oligopeptide, increases the flow of pancreatic juice without increasing enzyme activities. It acts via the second messenger cAMP. Cholecystokinin, on the other hand, stimulates the secretion of pancreatic proenzymes. This peptide hormone contains 33 amino acid residues and acts via cyclic guanosine monophosphate (cGMP). Its five amino-terminal amino acids are identical to those found in gastrin. The stimulatory effects of these two hormones can also be achieved directly through stimulation of the parasympathetic system (Figure 8.2).

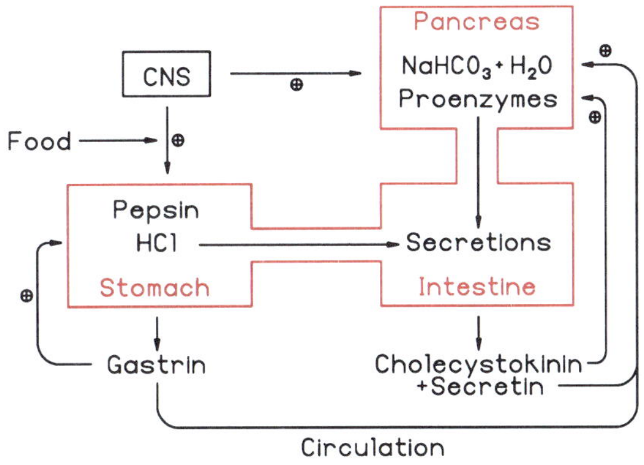

Figure 8.2.
Regulation of the synthesis of digestive enzymes.

8.2 NUTRITIONAL REQUIREMENTS OF HUMANS

The human body requires food for energy production as well as for the synthesis of its components, which are in constant turnover. Most of these components, although chemically rather different, can be synthesized in the body from the degradation products of the major food components: carbohydrates, protein, and fat. Only a few components cannot be synthesized and must be obtained from the diet; these include the essential amino acids and fatty acids as well as vitamins and minerals. The value of a diet, therefore, is determined not only by its caloric content but also by its content of essential components.

8.2.1 Caloric Value of Food

The energy content or caloric value of a particular food can be determined by **bomb calorimetry.** In this procedure, a known amount of food is completely combusted to CO_2 and water in a closed system, the calorimetric bomb, and the heat produced in the reaction is measured in kilocalories, often simply called "calories." The energy yield as measured by bomb calorimetry for carbohydrates is about 4 kcal/g and for fats is 9 kcal/g, values that agree well with their metabolic heat production. However, the value measured for proteins, 5.6 kcal/g, is too large. This is because the amino acid nitrogen is oxidized in the calorimetric bomb to nitrate (NO_3^-), whereas no such oxidation takes place during metabolism. The metabolic energy yield for proteins is only about 4.0 kcal/g, the same as that for carbohydrates.

8.2.2 Respiratory Quotient

Metabolic energy production can also be determined indirectly by measuring the amount of carbon dioxide produced and the oxygen consumed during a certain time period. The **respiratory quotient (RQ)**, defined as the ratio of CO_2 produced over O_2 consumed, varies with the kind of food a person metabolizes:

1. *Carbohydrates* (example: glucose):

$$C_6H_{12}O_6 + 6O_2 \rightarrow 6H_2O + 6CO_2$$

$$RQ = 6/6 = 1.0$$

2. *Fat* (example: trioleyl glycerol):

$$C_{57}H_{104}O_6 + 80O_2 \rightarrow 57CO_2 + 52H_2O$$

$$RQ = 57/80 = 0.71$$

3. *Protein*: The average RQ for protein metabolism is 0.81. The amount of protein that has been metabolized can be estimated from the urinary nitrogen excretion (N). On the average, the amount of protein metabolized is $6 \times N$.

In order to obtain the amount of fat and carbohydrate that has been metabolized, the CO_2 produced and O_2 consumed due to protein metabolism (obtained from urinary nitrogen) are subtracted from the total. This leaves a nonprotein RQ, which normally ranges from 0.71 to 1.0, allowing the calculation of the amount of energy that is produced by each major nutrient during its metabolism.

Table 8.3 illustrates how the human body, when subjected to prolonged physical exercise, switches from predominant carbohydrate metabolism to predominant fat metabolism. At the outset of running, most metabolic energy comes from the oxidation of glycogen (RQ = 0.97). As the exercise continues, however, the oxidation of fatty acids becomes the major energy source. Only during the spurts, when the speed is increased from 9.3 to 11.3 km/h, is glycogen metabolism temporarily increased. This is because glycogen degradation can be turned on much more quickly than fat mobilization.

Table 8.3. Energy Generation in a Marathon Runner

Speed (km/h)	Oxygen Used (L/min)	RQ	Blood Glucose (mg/100 ml)	Energy Produced (kcal)	Sources of Energy (kcal)		Percent Energy from Carbohydrate
					Carbohydrate	Fat	
0.0		100					
11.3	2.46	0.97	82	308	276	27	90
9.3	2.46	0.96	89	262	225	32	85
11.3	2.43	0.94	87	302	238	59	79
9.3	2.07	0.88	79	254	148	101	59
11.3	2.47	0.86	83	301	155	141	51
9.3	2.15	0.82	81	260	99	156	38
11.3	2.55	0.82	75	308	118	185	38
9.3	2.19	0.79	78	262	74	183	28
11.3	2.55	0.82	69	308	118	185	38
9.3	2.23	0.79	74	267	75	187	28
11.3	2.58	0.81	66	311	108	198	35
9.3	2.28	0.77	58	264	56	203	21

Note: The trained runner ran for 6 hours at speeds alternating between 9.3 and 11.3 km/h (25-minute intervals interrupted by 5-minute rest periods). Energy derived from protein metabolism amounted to 2% of the total energy need.
Source: After Edwards HT, et al. *Am J Physiol.* 1934; 108:203.

Respiratory quotients larger than 1.0 are observed when large amounts of glucose are converted into fat and deposited in adipose tissue.

8.2.3 Basal Metabolic Rate and Obesity

The basal metabolic rate (BMR) is the minimum energy requirement for the maintenance of body functions without any work by the major muscles. It consists of the basal metabolism of all cells, including the functioning of the central nervous system (20%), the cardiovascular system (10%), and the kidney (7%).

The BMR is measured after a night's sleep (12–14 hours after a meal). The person must be resting, lying down (but not sleeping), lightly clothed, at an ambient temperature of 28–30°C. Under these conditions, the RQ for most people is about 0.81, indicating that the major heat production comes from fatty acid oxidation. (Protein metabolism, probably negligible under these conditions, will not influence this RQ.) Hence, the only measurement necessary to determine the BMR under these conditions is oxygen consumption.

BMR is usually expressed in kilocalories per square meter of body surface per hour. Since surface area is difficult to determine, body weight to the ¾ power can be substituted as an approximation, or a nomogram can be employed.

BMR levels depend on age, sex and blood levels of catecholamines or thyroxine, two hormones that stimulate metabolism in general. (In **hyperthyroidism**, BMR is increased by over 15%; in **myxedema** [below normal levels of thyroid hormones], BMR is depressed.) An adult person typically has a BMR between 30 and 40 kcal • m^{-2} • h^{-1} (Figure 8.3).

The total caloric requirement of a person is determined by the BMR plus the level of activity. A sedentary type may require as little as 500 kcal over the BMR, whereas a decathlon athlete requires 10 times as much (5,000 kcal). Pregnancy and lactation raise the caloric requirement, partly related to growth. Low environmental temperatures also require more metabolism for heat production. Fever increases the BMR.

The variability of the caloric requirement demands a precise control of food intake. The appetite control center is located in the hypothalamus. It functions remarkably well if one considers that an extra pat of

Figure 8.3.
Dependence of basal metabolic rate on age and sex. Data are from the Mayo Foundation Normal Standards.

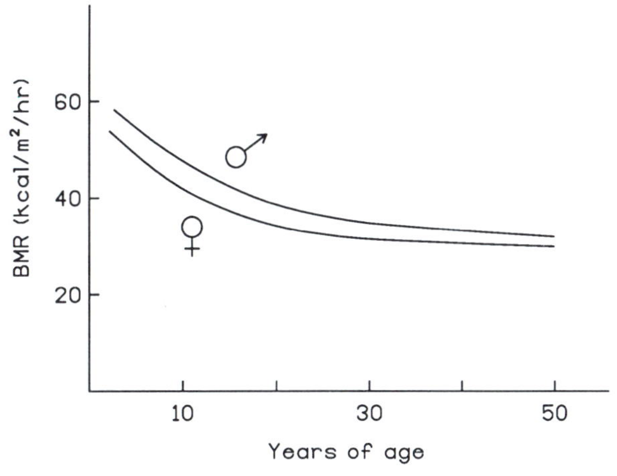

butter per day will lead to a weight gain of 6–10 pounds in a year's time.

Some people can eat as much as they want and never get fat; others gain weight on very little caloric intake. There is mounting evidence that there are mechanisms by which calories can be converted to heat rather than fat. In rats, **brown adipose tissue (BAT),** although comprising only 1% of the total body mass, contributes significantly to **nonshivering thermogenesis.** BAT is richly innervated by adrenergic fibers. In response to cold exposure or other factors, the generated norepinephrine activates adenylate cyclase in BAT that, in turn, initiates cAMP-mediated hydrolysis of cellular triglycerides by the protein kinase–dependent lipase. The thermogenic mechanism in BAT is localized in the inner mitochondrial membrane where an **uncoupling protein (UCP),** also called **thermogenin,** functions as a proton conductor that dissipates the electrochemical proton gradient generated by respiration. Long-chain fatty acids and their CoA esters have been shown to activate UCP. In addition, chronic administration of norepinephrine results in high levels of UCP mRNA.

In humans, the participation of BAT in diet-induced thermogenesis is questionable because of insufficient quantities of BAT and UCP. However, there is evidence of **non-BAT mitochondrial uncoupling**. The amino acid sequences of UCP and the adenosine diphosphate/adenosine triphosphate (ADP/ATP) carrier are similar, and both proteins are responsive to fatty acids and acyl CoA esters. It is speculated that under certain conditions, the ADP/ATP carrier might serve as a proton conductor similar to UCP.

It is also well established that the composition of the diet, especially the ratio of carbohydrate to fat, influences thermogenesis. It takes only 3% of the calories of ingested fat to convert it to adipose tissue, whereas 23% of the calories from carbohydrate and protein are expended during the conversion to triglyceride. The fraction of dietary calories that are used for thermogenesis is inversely proportional to the protein content of the diet. The thermogenesis that is induced by dietary carbohydrate is dependent on insulin function.

Individual variations in intake-induced thermogenesis can be part of the explanation why certain people remain thin whereas others gain weight on isocaloric diets.

Disorders of caloric intake include the following:

1. **Obesity**, a major disability in so-called advanced countries. The tendency for obesity may be related to infant feeding and to the number of fat cells in the body. According to one theory, the final number of fat cells is reached during early childhood; it does not increase later in life. Numerous other theories abound.

2. **Anorexia nervosa**, a serious psychogenic disorder characterized by inadequate food intake, often accompanied by voluntary induced vomiting. It is most frequently found in female teenagers.

3. **Marasmus**, a general deficiency in food intake (starvation) in small children. It merges indistinguishably with **kwashiorkor** (protein deficiency).

8.2.4 Specific Dynamic Action

Caloric determination generally overestimates the energy that can be extracted from the food during metabolism. This is because some of the metabolic steps require input of energy (in the form of ATP or reducing equivalents) before ATP-generating reactions can occur. This apparent loss of caloric content is called **specific dynamic action (SDA)**. Different types of food have different SDAs, and their SDAs are not additive.

1. **Fats** show a very low SDA of about 1% of their caloric content. Fatty acids are stored as triacylglycerides before they are mobilized for oxidation. The energy expense of triglyceride synthesis is about 1.6% of the energy available from fatty acid oxidation.
2. **Carbohydrates** exhibit intermediate SDA values. About 6% of the energy gained from the complete oxidation of glucose represents the energy expense for the synthesis of glycogen.
3. **Proteins** show relatively high SDA values of 20–30%, which stem from the extensive, energy-requiring processing many amino acids must undergo before entering energy-producing pathways.

For a mixed diet, the SDA is typically taken as 10% of the computed caloric content. Once the caloric requirements are determined, and a diet is constructed to meet them, an extra 10% is added to meet the SDA. If caloric contents of food were stated in ATP equivalents, there would be no SDA.

8.2.5 Nature and Function of Individual Diet Components

Carbohydrates, usually in the form of starch, are the major caloric source in most diets. (In so-called advanced countries, the average diet's content of sucrose can account for a significant part of caloric intake.) Digestible carbohydrates include the polysaccharides amylose and amylopectin (both in starch), glycogen, and the disaccharides sucrose, lactose, trehalose, maltose, and isomaltose. Monosaccharides that can be absorbed from the gut include glucose, fructose, mannose, and galactose.

No carbohydrate is essential, since all can be synthesized in the human body from other carbohydrates or amino acids. However, a diet that lacks carbohydrates completely will lead to the development of ketosis and acidosis because of the depletion of citric acid cycle intermediates. For this reason, at least 20% of the total energy need should be supplied in the form of carbohydrates. A typical satisfactory diet provides 40–60% of the energy need in the form of carbohydrates. Carbohydrates are also responsible for a **sparing action** on the amino acid requirement. (In the absence of carbohydrates, citric acid cycle intermediates must be synthesized from amino acids.)

Fats are the most efficient energy source in our diet. The presence of fat makes food more palatable. Linoleic acid and linolenic acid are essential fatty acids and must be provided in the diet. Linoleic acid serves as the precursor for the synthesis of arachidonic acid, from which prostaglandins are synthesized. Polyunsaturated fatty acids, found in many vegetable oils (exceptions are olive oil and coconut oil), have been reported to be responsible for lowering serum cholesterol levels and thus for reducing the chance of development of atherosclerosis.

Protein is required in the diet primarily as a source of essential amino acids. In addition, it provides amino groups for the synthesis of nonessential amino acids (by transamination).

Proteins from different sources often differ in their **biological values**, that is, their usefulness as sources of essential amino acids (Table 8.4). The biological value of a protein is determined by its ratio of essential amino acids. A protein with the correct proportion of all essential amino acids (reflecting the proportion found in human protein) is defined as having a biological value of 100. When such a protein is fed to a 70 kg man, nitrogen balance is observed on about 20 g of protein per day.

Nitrogen balance, however, may not be an accurate measure of protein requirement because a major portion of protein turnover, the desquamation of intestinal epithelium and the secretion of pancreatic enzymes, is drastically reduced under conditions of low protein intake. Higher amounts of dietary protein stimulate this protein turnover, causing increased nitrogen excretion. The optimal daily protein intake of a 70 kg man probably lies somewhere between 35 and 70 g.

The ratio of essential amino acids required to give a protein a biological value of 100 has been determined experimentally and is shown in Table 8.5. In the preparation of this table, the level of tryptophan was arbitrarily set at 1.0 because this amino acid is required in the lowest amount.

Chemical analysis of a protein usually defines its biological value. Some proteins, however, are difficult to digest; as a result, their biological values (as determined by chemical analysis) are overestimated. The same protein may also exhibit different biological values in different individuals because of variations in the efficiency of amino acid absorption.

The daily minimum requirement of 13.8 g of essential amino acids by a 70 kg man (see Table 8.5) is not sufficient for maintaining nitrogen balance. An additional 7 g of amino acids, essential or nonessential, are

Table 8.4. Biological Values of Proteins from Different Sources

Source	Biological Value	Protein Required per Day for N Balance (g)	Prepared Food (g)
Eggs	100	20	170
Beef (muscle)	100	20	120
Fish	87	23	75
Milk	85	24	700
Soybean meal (dry)	67	30	50
Potatoes	67	30	2000
Whole wheat bread	30	67	800

Table 8.5. Experimentally Determined Ratio of Essential Amino Acids in a Protein with a Biological Value of 100

Amino Acid	Proportion (molar ratio)	Amount Required per Day for 70 kg Man (g)
Trp	1.0 (7 $mg \cdot kg^{-1} \cdot day^{-1}$)	0.5 (7 mg × 70 = 490 mg)
Thr	5.5	1.6
His	2.0	0.8
Ile	5.5	1.8
Val	7.0	1.9
Lys	6.0	2.0
Leu	10.0	3.0
Met	2.0	0.7
Met + Cys	4.0	
Phe	3.5	1.5
Phe + Tyr	7.0	
Total		13.8

required to replenish the amino groups lost during irreducible amino acid catabolism.

The amino acid pairs Met + Cys and Phe + Tyr can be considered joint requirements. Although Cys is a nonessential amino acid, it derives its sulfur atom from Met. In other words, Cys exhibits a **sparing action** on the Met requirement. Similarly, Tyr (nonessential) is entirely derived from Phe. Consequently, Tyr in the diet has a sparing action on the Phe requirement.

If two proteins with low biological values are supplied at the same time, they supplement each other if the individual deficiencies differ. Such a supplementation has been utilized in the manufacturing of Incaparina, a mixture of corn, sorghum, and cottonseed with a high overall biological value. Incaparina was designed to combat protein deficiencies in underdeveloped countries but has met little acceptance because of its rather dull taste. (Incacola, a colalike soft drink, with a similar protein composition was received much more enthusiastically.) Breeding experiments with corn have produced genetic variants that contain higher proportions of lysine, an amino acid almost absent from regular corn.

Protein requirements during the period of growth are considerably higher than during maturity. The National Research Council suggests that a newborn (up to 2 months) should receive about 2.5 g of protein/kg of weight, and a 10 year old should receive 1.75 g/kg, compared with only 1 g/kg for the mature adult. During pregnancy and lactation, 1.3 g/kg are recommended.

8.2.6 Diet and Disease

Dieticians divide food into classes or so-called **exchange groups,** each containing food of similar character:

1. Vegetable group A (asparagus, cabbage, eggplant, mushrooms, radishes)
2. Vegetable group B (green peas, rutabaga, winter squash)

3. Fruits
4. Starch (bread, cereal, pasta, potatoes)
5. Protein (meat, fish, cheese, eggs)
6. Fat (butter, cream, salad dressing, nuts, avocado)
7. Milk

Equivalent amounts of individual foods can be exchanged within a group; they are measured in exchange units, or **exchanges.**

Once decisions about caloric intake, nutrient balance (e.g., high or low protein), and mineral contents (e.g., low sodium) have been made, tables are used to construct the basic diet. For example, a low-calorie, low-sodium diet for one day may consist of the following: unlimited amounts of vegetable group A exchange (low calories, low Na^+, high in vitamins), one vegetable group B exchange, three fruit exchanges, 10 starch exchanges, six protein exchanges, seven fat exchanges, and three milk exchanges, with total calories amounting to 2,200. Changes within each exchange group allow for reasonable variety in the diet.

The most striking example of a nutritional disturbance caused by inadequate protein intake is **kwashiorkor**, a disease occurring mainly in children in underdeveloped countries after weaning. The symptoms include patchy hair, gastrointestinal disturbances (which result in vitamin deficiencies), edema from low serum albumin levels, and muscle wasting. Treatment with a high-quality protein diet often comes too late to be of benefit, and even careful refeeding is dangerous.

Treatment of other diseases often can be supported by dietary restrictions. For example, maturity onset diabetes is frequently successfully treated by simple caloric restriction. On the other hand, purine restriction in gout patients has little effect on the course of the disease.

Debilitating diseases, fever, and surgery cause a negative nitrogen balance, which must be met by increased protein intake. Often this is difficult or impossible in a weak or apathetic patient. A solution to this problem is **total parenteral nutrition**, the infusion into a vein of a solution of glucose and amino acids, supplemented with fat emulsions.

High-protein diets designed for weight reduction must be supervised by a physician because of the ensuing ketosis and acidosis. The success of such a diet is mainly due to the fact that excess protein intake overloads the liver with the result that the patient feels sick and loses appetite.

8.3 VITAMINS

Vitamins are small organic molecules required by the human body. Most vitamins, or their derivatives, participate as coenzymes in enzymatic reactions. By definition, a vitamin cannot be synthesized in the human body but must be provided in the diet. Vitamins are present in small quantities in many types of food. An unbalanced diet may cause vitamin deficiencies. A number of vitamins, such as biotin, pyridoxine, vitamin K, and probably others are synthesized by intestinal bacteria and are easily absorbed from the gut so that deficiencies rarely develop.

Although highly diverse chemically, vitamins are usually divided into two groups, the fat-soluble and the water-soluble vitamins. Fat-soluble vitamins include vitamins A, D, E, and K; water-soluble vitamins include the B complex and vitamin C.

8.3.1 Fat-Soluble Vitamins

Vitamin A

The vitamin A_1 aldehyde, **retinal**, is derived from **β-carotene** (as well as from some other carotenes) present in most green and yellow vegetables (lettuce, spinach, carrots, sweet potatoes). In the intestinal mucosa, β-carotene is oxidatively cleaved by a dioxygenase into two molecules of retinal. The aldehyde is then reduced to vitamin A alcohol, **retinol** (Figure 8.4). This last reaction is inhibited by the end product retinol. Thus, in contrast to overdosage of vitamin A, excessive intake of carotene is not toxic.

Figure 8.4.
Oxidation of β-carotene to retinal.

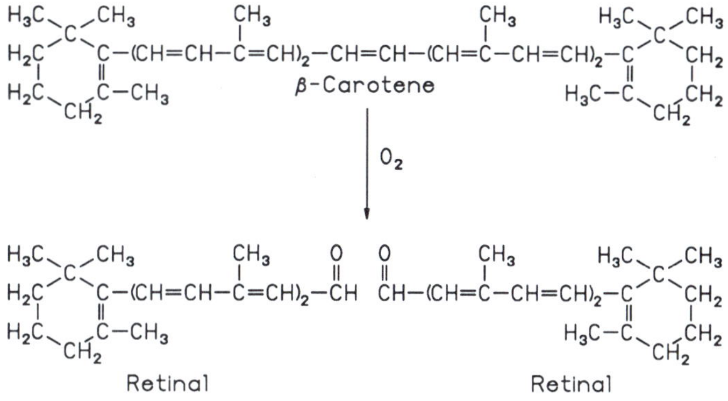

After uptake from the intestine, retinol is esterified to palmitic acid and transported via chylomicrons to the liver, where it is stored in the Kuppfer cells. Transport of free retinol in the blood occurs by the **retinol binding protein**, an α_1-globulin present in plasma (see Chapter 9).

Rhodopsin, the purple photoreceptor pigment of the rod cells in the retina, consists of the protein **opsin** linked via a Schiff base linkage to **11-*cis*-retinal**. A pulse of light isomerizes the 11-*cis*-retinal in rhodopsin to ***trans*-retinal**. This isomerization destabilizes rhodopsin (photoexcited rhodopsin), resulting in its cleavage into opsin and *trans*-retinal (through a series of intermediates). *Trans*-retinal is isomerized

Figure 8.5.
Interconversions of retinal and retinol in the visual cycle.

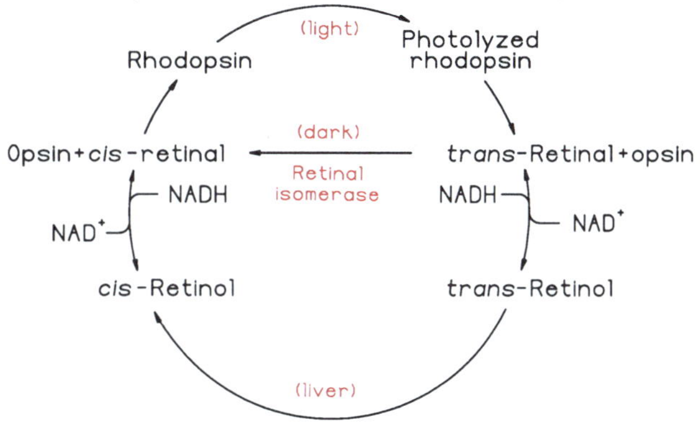

in the dark to 11-*cis*-retinal, which combines again with opsin to form rhodopsin. After longer exposure to light, *trans*-retinal is reduced to *trans*-retinol, which is reconverted to *cis*-retinol in the liver (Figure 8.5).

The mechanism by which the initial effect of a light pulse, namely the isomerization of 11-*cis*-retinal on rhodopsin, is converted into a neural signal is shown in Figure 8.6.

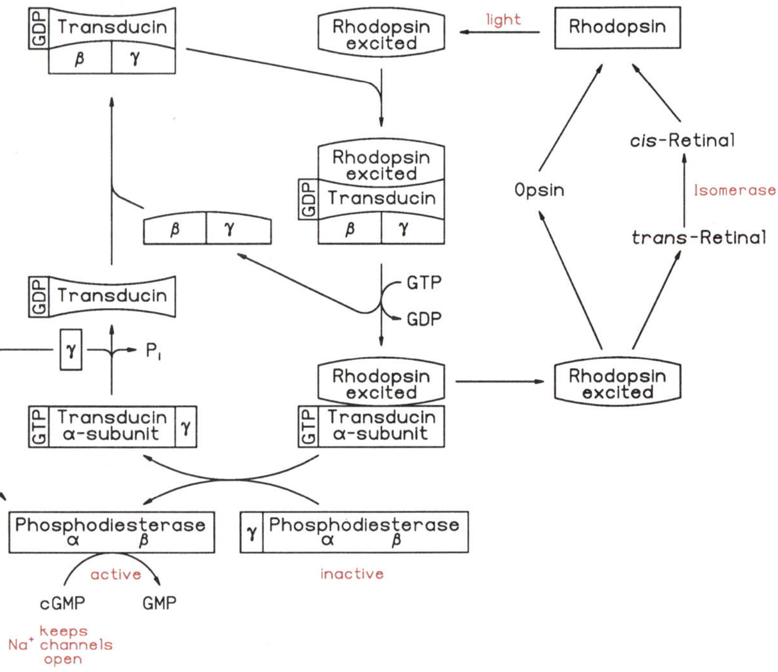

Figure 8.6.
Mechanism for the transmission of the light response in the retina.

Photoexcited rhodopsin binds to a regulatory protein, **transducin,** a G protein that has a guanosine diphosphate (GDP) molecule tightly bound to its α subunit. This binding results in the replacement of GDP by guanosine triphosphate (GTP) in the transducin complex, with the result that the complex becomes unstable and dissociates into free rhodopsin, α-transducin-GTP, and β,γ-transducin. α-Transducin-GTP then binds to **arrestin,** an inhibitory protein associated with phosphodiesterase. The removal of arrestin activates phosphodiesterase, which then cleaves cyclic GMP. Reduced levels of cGMP cause the sodium channels in the membrane of the rod cells of the retina to close, thus hyperpolarizing the cell.

The cell recovers from photoexcitation-induced hydrolysis of cGMP by resynthesizing cGMP, which reopens the sodium channels. Activation of **guanylyl cyclase** plays a key role in this recovery process. The activity of the enzyme is regulated by a calcium-binding protein, called **recoverin.** After the closure of the sodium channels, cytosolic Ca^{2+} concentration falls. The calcium-free recoverin then stimulates guanylyl cyclase. The rising level of cGMP opens the sodium channels, which in turn is followed by a rise in cytosolic Ca^{2+}. The calcium binds to recoverin and causes inhibition of guanylyl cyclase.

The cone cells of the retina, responsible for color vision, contain three different kinds of photoreceptor protein with absorption maxima in the blue, green, and red regions of the visible spectrum. Each of these receptors consists of a different opsin bound to retinal. **Color blindness** is the result of a defect in one of these opsins. Vitamin A deficiency causes **night blindness.**

In addition to its role in the visual process, vitamin A is also involved in the metabolism of mucopolysaccharides and glycoproteins. However, little is known about its mechanism of action there. (It has been speculated that vitamin A functions similarly to the dolichols in glycosylation.)

In the absence of vitamin A, the secretory epithelium of the eye becomes keratinized. The resulting **xerophthalmia** is a major cause of blindness in children in underdeveloped countries. Ingestion of **retinoic acid** can prevent this deficiency syndrome even though the human body is incapable of reducing retinoic acid to retinal. Retinoic acid has been demonstrated to function as a **morphogen** in development. Most of the details of this function are still obscure.

The recommended daily allowance for retinol is 600–750μg (2,000–2,500 IU) or twice this amount of β-carotene (less efficient absorption). Ingestion of large amounts of vitamin A (>500,000 IU/day) causes a variety of symptoms (headache, nausea, weakness, dermatitis, and others) because the human body cannot excrete excess quantities of this vitamin.

Vitamin D

There are several forms of vitamin D, or **calciferol.** The vitamin D_3 family can be synthesized in the human body. In a strict sense, therefore, vitamin D is not a vitamin but a steroid hormone.

Cholecalciferol, or **vitamin D_3,** is synthesized from 7-dehydrocholesterol, an intermediate in cholesterol biosynthesis. This synthesis occurs in the skin and is catalyzed by UV light. Vitamin deficiency occurs mainly in children who are not exposed to any sunlight.

Vitamin D_3 itself is not biologically active. It must be converted in the liver to **25-hydroxycholecalciferol** and further activated in the kidney to **1,25-dihydroxycholecalciferol** (Figure 8.7).

1,25-Dihydroxyvitamin D_3 is required for the absorption of Ca^{2+} and phosphate ions from the intestine. Specifically, 1,25-dihydroxyvitamin D_3 stimulates Ca^{2+} uptake by promoting the de novo synthesis of a calcium-binding protein in the brush border of the intestinal cells. Together with the hormones **parathyroid hormone** and **calcitonin**, vitamin D_3 regulates the overall homeostasis of calcium and phosphate (see Section 13.4).

Vitamin D is present in many foods of animal origin (eggs, fish liver oil, milk after UV irradiation). A deficiency during early childhood causes **rickets**; later in life, it leads to **osteomalacia** (demineralization of bones).

Minimum daily requirements for vitamin D are variable and depend on the exposure to UV irradiation. Excess vitamin D (10 times the daily allowance) may cause demineralization of bones, calcification of soft tissues, and formation of renal calculi (kidney stones). Higher doses are even more toxic.

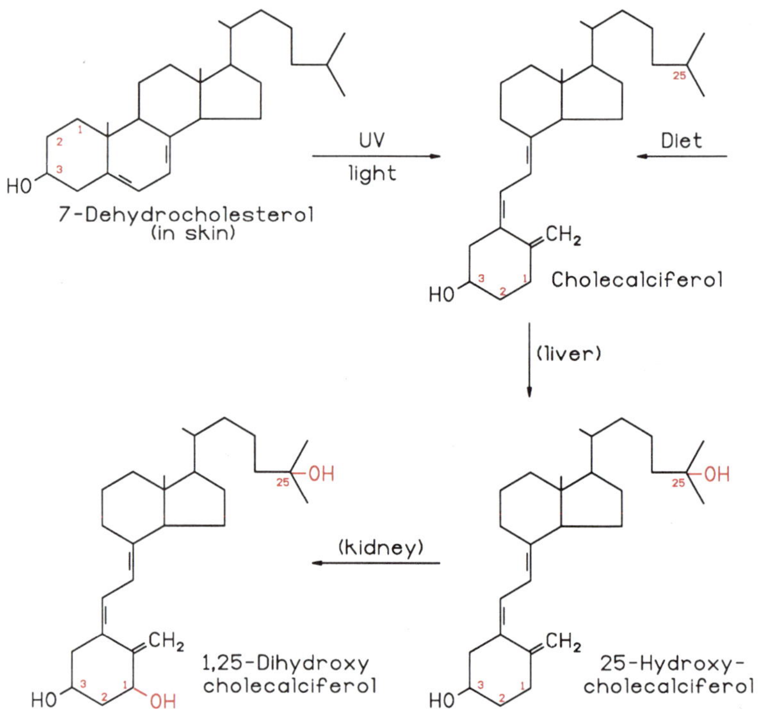

Figure 8.7.
Formation and activation of vitamin D3 (cholecalciferol).

Vitamin E

The vitamin E group consists of several forms of tocopherol, which are found in many foods of plant origin, especially in seed oils and wheat germ. **α-Tocopherol** is the most active component (Figure 8.8).

The molecular mechanism of vitamin E action in human metabolism is not known. In rats, vitamin E deficiency causes infertility. In humans, vitamin E deficiency has been recognized in newborns who were fed with a formula entirely lacking vitamin E. The deficiency resulted in skin lesions that responded specifically to vitamin E supplementation of the diet. Vitamin E also seems to facilitate the treatment of cardiac problems in kwashiorkor. Most likely, vitamin E functions as an antioxidant, protecting the double bonds in the unsaturated fatty acids of cellular membranes against oxidation. Vitamin E and selenium (see Section 8.4) act similarly in this respect, and their effects are synergistic. In adult humans, experimentally induced vitamin E deficiency resulted in increased excretion, via the lungs, of short hydrocarbons arising from the oxidative breakdown of polyunsaturated fatty acids.

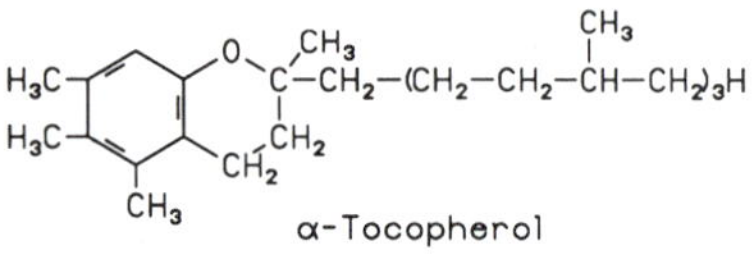

Figure 8.8.
Structure of α-tocopherol.

Vitamin K

All forms of vitamin K are derivatives of naphthoquinone. Plants contain **vitamin K_1** with a phytyl residue at position 3; bacteria contain **vitamin K_2** with a difarnesyl residue at the same position. The synthetic **menadione** also exhibits vitamin K activity (Figure 8.9).

Vitamin K functions as a coenzyme in the posttranslational carboxylation of glutamate residues in several factors of the blood-clotting system, of which **prothrombin** (factor II) is the most prominent. In prothrombin, the first 10 amino-terminal glutamate residues are carboxy-

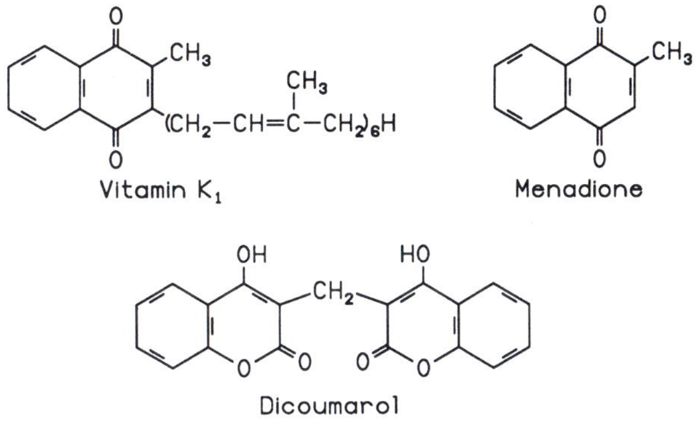

Figure 8.9.
Structures of vitamin K and some of its analogs.

lated to γ-carboxyglutamate residues, which appear to be required for the binding of Ca^{2+} ions (Figure 8.10). The exact mechanism of this reaction is still unclear.

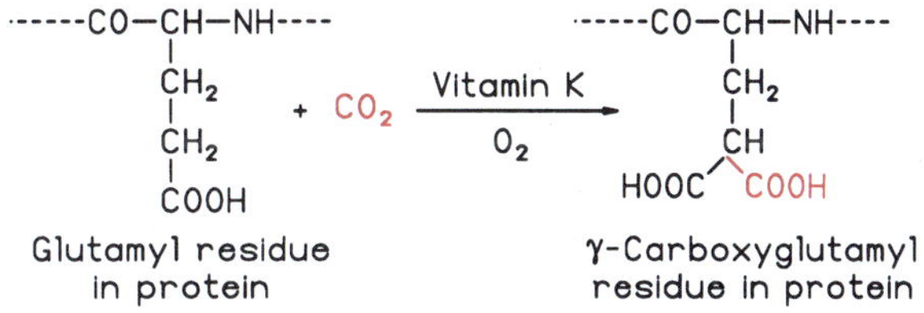

Figure 8.10.
Vitamin K–catalyzed carboxylation of glutamyl residues in proteins.

Vitamin K is synthesized by intestinal bacteria. Uptake occurs in the upper small intestine and requires the presence of bile salts. **Steatorrhea** (fatty stool), caused by defective lipid absorption, can lead to vitamin K deficiency, which is characterized by blood-clotting problems (hemorrhages).

Dicoumarol is a vitamin K antagonist that is used clinically as an anticoagulant. A related vitamin K antagonist, **warfarin,** is widely used as rat poison.

8.3.2 Water-Soluble Vitamins

Water-soluble vitamins consist of vitamin C and the vitamins of the B complex, which includes thiamine (B_1), riboflavin (B_2), niacin, pyridoxine (B_6), pantothenic acid, biotin (vitamin H), folic acid, inositol, and cyanocobalamin (B_{12}). The functions of many of these vitamins as coenzymes are discussed in the chapters on metabolism. The discussion here will be concerned mainly with their nutritional importance.

Vitamin C

Vitamin C, or **ascorbic acid**, is found abundantly in fresh fruits, green leafy vegetables, tomatoes, etc. Cooking destroys much of this vitamin. Most animals can synthesize ascorbic acid from D-glucuronic acid via L-gulonic acid. Primates have lost this capability.

Ascorbic acid participates in several biological oxidation reactions, such as the hydroxylation of proline and lysine residues in collagen and the hydroxylation of γ-butyrobetaine to carnitine. The claim that ingestion of large quantities of vitamin C aids the body in fighting viral infections such as the common cold has not yet been substantiated. Excess vitamin C (>50 mg/day) is excreted mostly unchanged in the urine. Oxalic acid, a degradation product of ascorbic acid, may lead to the formation of kidney stones.

Scurvy, the clinical manifestation of ascorbic acid deficiency, is characterized by a weakening of connective tissue (loosening of teeth, receding gums, joint pain, subcutaneous hemorrhages). The disease is rare in developed countries.

Thiamine

Thiamine, or **vitamin B_1**, is the antiberiberi substance found in the outer layers of rice and other seeds. **Beriberi**, a severe thiamine deficiency syndrome with neurologic and cardiovascular manifestations, is encountered predominantly in populations that depend on polished rice for their main diet. The function of thiamine as a coenzyme in pyruvate dehydrogenase and transketolase is discussed in Sections 1.22 and 2.8.

Riboflavin

Riboflavin, or **vitamin B_2**, was originally demonstrated as a coenzyme of Warburg's yellow enzyme (reduced nicotinamide-adenine dinucleotide [NADH] dehydrogenase). It is found in the form of flavin mononucleotide (**FMN**) or flavin adenine dinucleotide (**FAD**) in many dehydrogenases and oxidases.

Animals cannot synthesize riboflavin and must obtain it from plants or bacterial sources. Deficiencies are often found in patients with beriberi, pellagra (see below), and kwashiorkor.

Niacin

Niacin, or **nicotinic acid**, forms the basis of the important nicotinamide adenine nucleotides. It can be synthesized from tryptophan in the human body and, therefore, is not a true vitamin, although tryptophan is an essential amino acid (Figure 8.11). (Note that the first steps of niacin biosynthesis are identical to those of tryptophan catabolism.)

Pellagra, a dietary deficiency of tryptophan and niacin as well as of other members of the vitamin B complex, is characterized by a variety of symptoms (dermatitis, stomatitis, atrophy of the tongue, diarrhea, neural disorders).

Pyridoxine

Vitamin B_6 includes the alcohol **pyridoxine,** the aldehyde **pyridoxal,** and the amine **pyridoxamine.** This vitamin functions as a coenzyme in several reactions in amino acid metabolism (transamination, decarboxylation, deamination). A deficiency of vitamin B_6 in humans usually does not develop because intestinal bacteria provide it abundantly.

Pantothenic Acid

Pantothenic acid, a constituent of both coenzyme A and the acyl carrier protein, is essential to human nutrition. The vitamin is abundantly found in many types of food of both animal and plant origin.

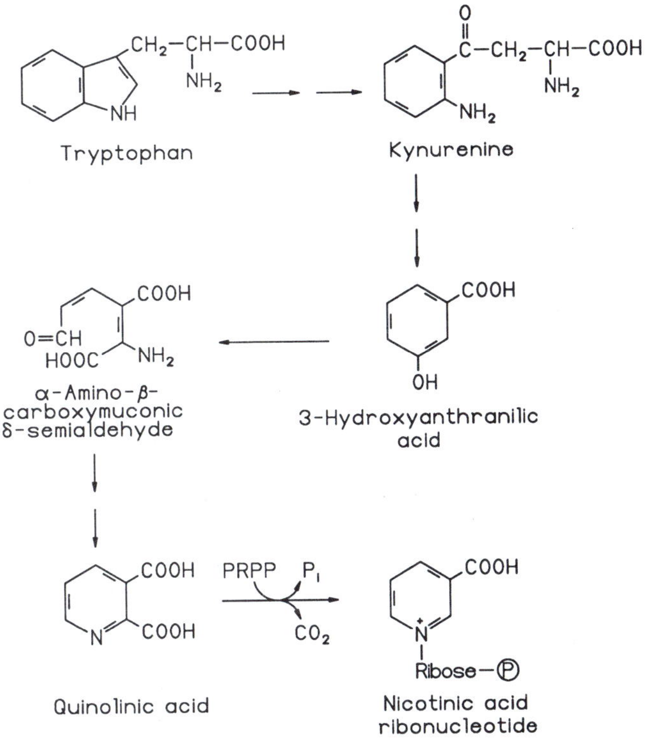

Figure 8.11.
Biosynthesis of nicotinic acid from L-tryptophan.

Biotin

Biotin functions as a coenzyme in most carboxylation reactions. It is normally provided in sufficient quantities by intestinal bacteria. A deficiency can be induced, however, by ingestion of large quantities of raw egg white. Egg white contains **avidin**, a glycoprotein that binds to biotin, and thus prevents its absorption from the gut.

Folic Acid

The role of folic acid derivatives in the metabolism of one-carbon fragments is discussed in Chapter 4. Folic acid is present in many types of food of both animal and plant origin. The richest sources of this vitamin are green lettuce and vegetables, kidney, and liver. Deficiencies of folic acid are known in humans. **Megaloblastic anemia** may develop in pregnant women because of their increased need for this vitamin and in alcoholics because of their poor dietary habits.

Sprue is a chronic intestinal disorder caused by an intolerance to **gluten**, the protein present in wheat. The accompanying changes in the intestinal mucosa lead to impaired absorption of many vitamins and nutrients. Insufficient uptake of folic acid, as well as of vitamin B_{12}, leads to megaloblastic anemia.

Vitamin B_{12}

Vitamin B_{12}, or **hydroxycobalamin** (cyanocobalamin contains a cyanide group that is introduced during the isolation of the vitamin), is the precursor of the cobalamin coenzymes that function in transmethylation and molecular rearrangement reactions discussed in Chapter 4. The vitamin is absent from higher plants but found in small quantities in food of animal origin. (Animals derive their vitamin B_{12} from bacterial sources.) Liver and kidney are the richest animal sources. All vitamin B_{12} is ultimately derived from bacteria that can synthesize this vitamin.

Pernicious anemia, a vitamin B_{12} deficiency syndrome, is usually caused by the absence of **intrinsic factor**, a stomach protein required for the intestinal absorption of vitamin B_{12} (see Section 8.1). This absorption occurs in the upper intestine. Thus the vitamin B_{12} that is produced by the bacteria in the lower intestines cannot be absorbed.

8.4 MINERAL METABOLISM

In addition to organic molecules, the human body contains a large number of inorganic ions whose concentrations are subject to precise homeostatic regulation. Because of their involvement in the maintenance of water and acid-base balance, the major ions of the body fluids, Na^+, K^+, Cl^-, and bicarbonate, are discussed in Chapter 12. The discussion here will be limited to the metabolism of calcium, phosphate, magnesium, iron, and copper as well as the trace elements, which are found in very small quantities but serve vital functions in the human body.

8.4.1 Calcium and Phosphate Metabolism

Calcium represents the largest mineral component of the human body. Almost all of it exists in bones in the form of **hydroxyapatite** [$Ca_{10}(PO_4)_6(OH)_2$]. The plasma concentration of Ca^{2+} is about 2.5 m*M* (10 mg/100 ml), half of it in free ionized form, the other half bound to serum proteins, especially albumin. Intracellular concentrations of free Ca^{2+} are in the micromolar range; most intracellular calcium is tightly bound to Ca-binding proteins, such as **calsequestrin,** a 46 kd protein found in muscle tissue. Another Ca-binding protein, **calmodulin** (16 kd), is found in every cell of the body. In the presence of Ca^{2+} ions, it undergoes a conformational change, enabling it to bind to and activate a variety of enzymes such as phosphodiesterase, the enzyme that cleaves cyclic nucleotides, or protein kinase (see Section 11.3).

Ca^{2+} ions are required in many enzymatic reactions, in muscle contraction, in the blood-clotting mechanism, and in the release of neural transmitters at nerve endings. Because of its effect on the transmembrane potential, the plasma Ca^{2+} concentration must be maintained within rather narrow limits. A lowering of Ca^{2+} causes depolarization of the cell and leads to tetany; an increase in Ca^{2+} produces hyperpolarization that may cause muscle weakness and cardiac failure.

Calcium is absorbed from the small intestine by an active transport system and carried via the bloodstream to the bones, where it is deposited. Part of the large calcium pool in bone is easily exchangeable and turns over continuously. Excretion of Ca^{2+} ions is accomplished in part through the bile and pancreatic juice released into the intestine. The remainder is excreted in the urine and sweat. The kidney normally reabsorbs 99% of the Ca^{2+} that is filtered through the glomeruli (Figure 8.12).

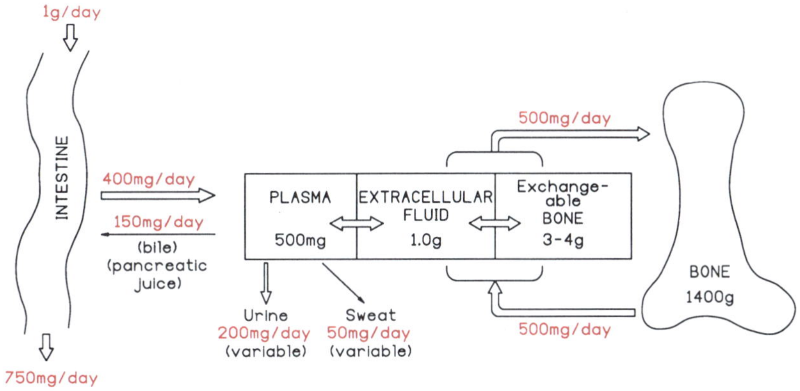

Figure 8.12.
Calcium homeostasis.

Calcium homeostasis is accomplished through the action of three hormones: **parathyroid hormone (PTH), calcitonin,** and **1,25-dihydroxyvitamin D_3.** If the concentration of plasma Ca^{2+} falls, PTH is released from the parathyroid gland. This hormone stimulates the activity of osteoclasts, which catalyze the solubilization (resorption) of hydroxyapatite from bone. It also increases the reabsorption of Ca^{2+} in the kidney and stimulates the hydroxylation of vitamin D_3. 1,25-Dihydroxyvitamin D_3 triggers the synthesis of Ca-binding proteins that are required, in addition to the ATP-driven Ca^{2+} pump, for intestinal absorption of calcium, for the reabsorption of filtered Ca^{2+} in the kidney, and for bone resorption. The combined actions of PTH and 1,25-dihydroxyvitamin D_3 thus raise the level of Ca^{2+} ions in plasma. Calcitonin, on the other hand, lowers the plasma Ca^{2+} by increasing the activity of osteoblasts, which deposit calcium in the bone. As with PTH, calcitonin stimulates the hydroxylation of vitamin D_3 in the kidney.

Calcium deficiency is rare but occurs occasionally in childhood, in pregnancy, and during lactation. Milk products are an excellent source of calcium; they contain **casein,** a phosphoprotein complexed with calcium. Grain products are also rich in calcium and phosphate, but the presence of phytic acid in some grains prevents the absorption of Ca because it forms insoluble salts.

The **homeostasis of phosphate** is closely linked to that of calcium. Most of the phosphate is deposited in bone. Absorption occurs in the intestine by active transport. In the plasma, phosphate exists as HPO_4^{2-} and $H_2PO_4^-$ ions in the ratio 4:1 (total concentration: 4mg/100 ml). Excretion of phosphate occurs mainly in the kidney. After filtration through the glomeruli, phosphate is largely reabsorbed. The rate of reabsorption is lowered by PTH, thus causing a reduction in the plasma concentration of phosphate. In contrast, calcitonin increases the rate of reabsorption of phosphate in the kidney and promotes the deposition of phosphate in bone (together with calcium).

Phosphate deficiency is extremely rare. However, it may occur in ulcer patients undergoing treatment with the antacid aluminum hydroxide because the antacid converts phosphate into soluble $AlPO_4$, which cannot be absorbed.

8.4.2 Magnesium Metabolism

The human body contains much less Mg than Ca, most of it concentrated in bone (total Mg content = 20 g). The Mg concentration in plasma is between 0.6 and 1.7 m*M*. Absorption and excretion of Mg resembles that of calcium. However, renal excretion of Mg^{2+} is controlled together with that of K^+ rather than Ca^{2+} (aldosterone increases both K^+ and Mg^{2+} excretion). The reason for this may be the fact that both Mg^{2+} and K^+ are predominantly intracellular, whereas Ca^{2+} and Na^+ are extracellular.

Magnesium ions are essential for the functioning of many enzymes and participate in the formation of structural components of the body. Magnesium deficiency is fairly common but often not recognized. Usually, it is associated with other diseases such as sprue, renal disease, and alcoholism, representing disorders in absorption, excretion, and nutrition, respectively.

8.4.3 Iron Metabolism

Iron functions as an oxygen carrier in hemoglobin and myoglobin and as an electron carrier in many oxidases. The homeostasis of iron in the body is unique in that the iron level is maintained solely by control of iron uptake from the intestine. There is no mechanism for the removal of excess iron. The total iron content of the human body amounts to about 4 g. Only 1–2 mg of iron is lost every day by uncontrolled processes such as sluffage of epidermal cells, menstruation, and minor but continual hemorrhage into the lower gut where reabsorption does not occur. These small irretrievable losses are balanced by iron uptake.

Iron uptake from the intestine is precisely controlled. Under normal conditions (i.e., no iron deficiency), there exists a **mucosal block** against the uptake of iron. In response to a large iron loss, such as after a severe hemorrhage, iron uptake is increased gradually until, after 3 days, maximal absorption is reached. This time interval coincides with the average lifetime of an intestinal mucosal cell. It appears, therefore, that a mucosal cell is set for a specific rate of iron uptake at cell division and maintains that rate until its death.

Iron is absorbed from the intestine as Fe^{2+} ion. Upon crossing the brush border of the mucosal cell, Fe^{2+} binds to a carrier protein that delivers Fe^{2+} either to **apoferritin**, a 450 kd protein with 24 identical subunits, for storage as **ferritin** or to the plasma side of the cell, where the plasma protein **transferrin** takes up the iron and transports it to other tissues.

Before combining with apoferritin, Fe^{2+} is oxidized to Fe^{3+}. The resulting ferritin is a large spherical particle made from many apoferritin molecules. It contains a mixture of ferric hydroxide and ferric phosphate in its center in the form of mixed micelles. Each ferritin particle contains several thousand iron atoms (iron content: 35%). Release of iron from ferritin is accomplished by **ferritin reductase**, which regenerates Fe^{2+} to be transported across the cell membrane (Figure 8.13).

Transferrin, a β_1-globulin, is a large single-peptide glycoprotein (molecular weight = 76,000) that contains two Fe^{3+} binding sites. Before binding to transferrin, Fe^{2+} must be oxidized to Fe^{3+} by the action of either one of two serum **ferroxidases**. One of these enzymes is **ceruloplasmin**, the carrier protein for Cu^{2+} ions. The other ferroxidase is identical to **xanthine oxidase**. In the presence of CO_2, transferrin has a very high affinity for Fe^{3+}, with a bicarbonate ion

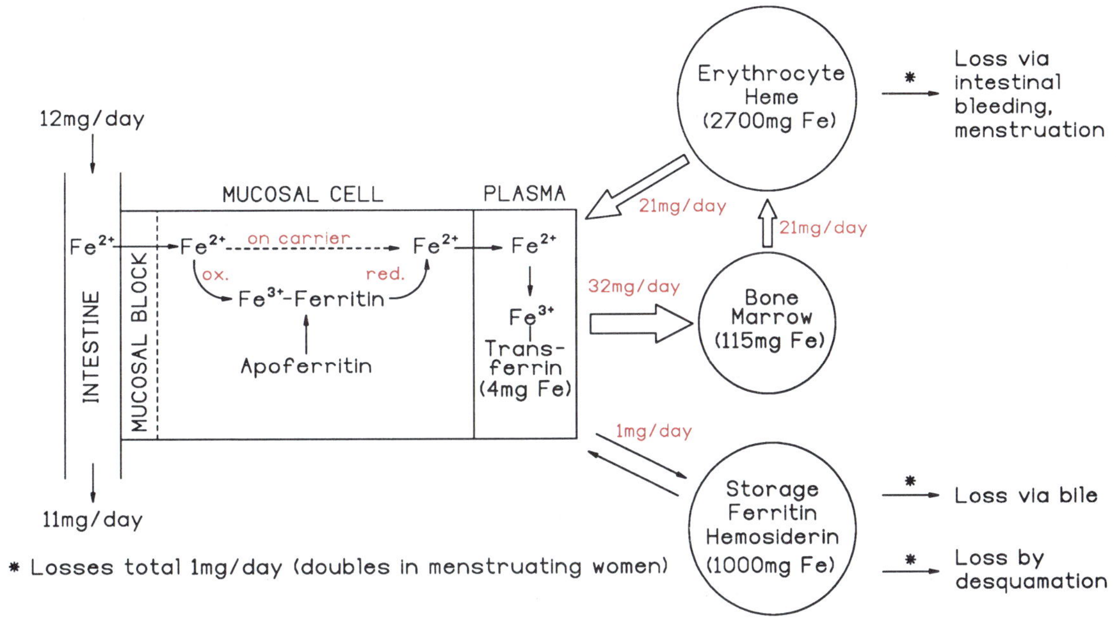

Figure 8.13.
Iron homeostasis.

functioning as a coligand to Fe^{3+}. Under normal conditions, plasma transferrin is only 30% saturated with iron. The total binding capacity of plasma transferrin is about 4 mg of iron. Yet during one day about 40 mg of iron are transported by transferrin, suggesting rapid exchange. Transferrin also binds Zn^{2+} and Mn^{3+} ions and may function in the transport of these ions.

After being transported to the tissue where the iron is needed (reticulocytes, liver, spleen, bone marrow), an Fe^{2+} ion is again transferred across the cell membrane to an intracellular carrier protein. Because of the tight binding of Fe^{3+} to transferrin, this transfer may involve reduction of Fe^{3+} to Fe^{2+}, removal of bicarbonate from transferrin by carbonic anhydrase, or allosteric modification of transferrin by an unknown effector in order to reduce transferrin's affinity for Fe^{3+}. The exact mechanism is still not known.

For storage of larger quantities of iron in the liver or spleen, many ferritin particles combine and form even larger granules, called **hemosiderin,** which contain up to 40% iron. In **primary hemochromatosis,** excess iron uptake from the gut leads to large deposits of hemosiderin, which distort the architecture and function of the liver and pancreas, causing cirrhosis and diabetes mellitus. Symptoms usually appear at 40 years of age. **Secondary hemochromatosis** may result from excessive transfusions. Repeated phlebotomy (500 ml/week) may slowly (over several years) reduce the excessive levels of stored iron to normal. The use of chelators, such as **deferoxamine,** a small molecule that can be excreted in the urine, has proven less successful in reducing iron stores; daily injections of large quantities of the chelator are required, which may produce side effects.

The regulation of cellular iron metabolism is mostly controlled post-transcriptionally. The mRNAs of the two major proteins involved in iron homeostasis contain regulatory sequences, called **iron-responsive elements (IREs).** Ferritin mRNA has one IRE in its 5′-untranslated segment, whereas the mRNA of the transferrin receptor (Tfr) contains multiple IREs within its 3′-untranslated region. The cell also contains an **IRE-binding protein (IRE-BP)** that binds to IRE. This binding represses the translation of ferritin mRNA and also inhibits the otherwise rapid degradation of Tfr mRNA. The expression of IRE-BP is regulated by the availability of iron, increasing under conditions of iron deprivation and decreasing when iron is available.

Iron deficiency anemia may result from dietary restrictions, especially in women after reaching puberty and during pregnancy. The presence of large quantities of inorganic phosphate in the diet interferes with the uptake of iron.

8.4.4 Copper Metabolism

Copper functions in several enzyme systems, notably in cytochrome oxidase of the electron-transport chain. Together with iron, Cu is required for hemoglobin synthesis. Uptake from the gut requires a specific transport system. A copper deficiency that seems to result from the congenital absence of this transport system is known as **Menkes' kinky hair syndrome**. Parenteral treatment with Cu^{2+} salts is possible.

Most of the copper in plasma is tightly bound to the α-globulin **ceruloplasmin**. A small amount is also carried on albumin, a carrier that may function in the transfer of Cu^{2+} to and from ceruloplasmin. In **Wilson's disease**, free or loosely bound Cu^{2+} ions are present in the blood and are excreted in the urine. The free Cu^{2+} ions react with many tissue proteins and cause excessive deposits of copper in the eye, brain, liver, and other tissues. This results in psychosis, cirrhosis, and diabetes. The exact defect in Wilson's disease in not known. In most, but not all, patients, plasma levels of ceruloplasmin are lower than normal. Treatment with copper-binding chelators, such as **penicillamine**, is possible.

8.4.5 Trace Elements

The list of trace elements in human nutrition has grown rapidly in recent years, mainly due to advances in analytical techniques. These techniques made it possible to conduct feeding studies with diets from which even traces of a particular element were absent with certainty. The functions of most trace elements in human metabolism are known; only a few remain obscure. Some have only been shown to be essential in rats.

Iodine has long been recognized as essential in human nutrition. Iodine deficiency causes **goiter**, a hyperplasia of the thyroid gland, where the iodine-containing hormone thyroxine is synthesized. Natural sources of iodide (I^-) are water and seafood. So-called goiter belts are geographical areas, distant from the ocean, with low soil and water iodine content. In the United States, goiter is very rare because table salt is supplemented with NaI. Secondary iodide deficiency may develop after frequently eating large amounts of cabbage, which contains goitrogenic compounds that interfere with tyrosine iodination (see Section 13.3).

Fluorine functions in the development of bone and teeth. It specifically serves to strengthen bone and the enamel of teeth. Supplementation of drinking water in areas with low fluoride content markedly reduced the incidence of caries. Lack of fluoride causes **osteomalacia** and poorly developed teeth. Excessive intake of fluoride produces mottled teeth as well as symptoms of toxicity.

Silicon appears to function in bone formation by participating in the cross-linking of the matrix before calcification. It is plentiful in all diets, and no human deficiencies are known.

Selenium has long been suspected to be a required element. It is now known to function in **glutathione peroxidase**, the enzyme that reduces peroxide and thus prevents the oxidation of the polyunsaturated fatty acids in the cell membranes. In the process, glutathione (GSH) is oxidized to the disulfide form

$$2GSH + H_2O_2 \rightarrow GS—SG + 2H_2O$$

Selenoproteins contain **selenocysteine,** which is incorporated into the polypeptide chain by a specific transfer RNA (tRNA). This tRNA binds to the codon UGA, normally used as a stop codon. The presence of an in-frame UGA codon has been demonstrated by sequence analysis of the complementary DNA (cDNA) of a selenoprotein.

The reduced form of NADPH serves to regenerate glutathione. The source of NADPH is the pentose phosphate pathway. This explains why red blood cell membranes are very fragile when the enzyme glucose 6-phosphate dehydrogenase is defective (see Chapter 9).

Zinc is found as a prosthetic group in many enzymes: for example, DNA polymerase, carboxypeptidase, and carbonic anhydrase. Zinc deficiencies are probably more common than is generally recognized. A diet based mainly on grain that is harvested in a region where the zinc content of the soil is low may produce a deficiency because Zn-phytate is insoluble and cannot be absorbed. Symptoms of zinc deficiency include loss of appetite, decreased sense of taste, and impaired wound healing.

Zinc binds tightly to **metallothionein,** a small protein 6.5 kd that also binds to other metal ions such as Cu or Cd. The protein is found in most tissues and can be induced by the presence of metal ions or stress. Its main functions include the detoxification of Cd ions and the control of Zn and Cu absorption.

Chromium's exact role in human nutrition is still not known. In rats, chromium deficiency leads to decreased glucose tolerance. It has been suggested that chromium participates in the binding of insulin to cell membranes. Chromium may also function in lipid metabolism. Deficiency (in rats) causes high serum cholesterol levels and increased formation of atherosclerotic plaques.

Manganese is a component of the enzymes arginase (urea cycle), α-keto acid oxidases, phosphoglucomutase, and others. No human deficiency is known.

Molybdenum functions in xanthine oxidase. No human deficiency is known.

Tin participates in the enzymatic breakdown of heme (heme oxidase).

Other candidates for required trace elements include **nickel, vanadium, arsenic, antimony,** and **lead.** Their functions are unknown. **Cobalt** is found only as part of vitamin B_{12}, which is taken up in the diet as a complete molecule.

Further References

Brenner I, Beattie JH. Metallothionein and the trace minerals. *Annu Rev Nutr.* 1990; 10:63–83.

Dizhoor AM, Ray S, Kumar S, Niemi G, Spencer M, Brolley D, et al. Recoverin: A calcium sensitive activator of retinal rod guanylate cyclase. *Science.* 1991; 251:915–918.

Lardy H, Shrago E. Biochemical aspects of obesity. *Annu Rev Biochem.* 1990; 59:689–710.

Rouault TA, et al. Cloning of the cDNA encoding an RNA regulatory protein—the human iron-responsive element-binding protein. *Proc Natl Acad Sci USA.* 1990; 87:7958–7962.

Stadtman TC. Selenium biochemistry. *Annu Rev Biochem.* 1990; 59:111–128.

Suda T, Shinki T, Takahashi N. The role of vitamin D in bone and intestinal differentiation. *Annu Rev Nutr.* 1990; 10:195–212.

Ziboh VA, Miller CC. Essential fatty acids and polyunsaturated fatty acids. *Annu Rev Nutr.* 1990; 10:433–450.

CHAPTER

9 The Constituents of Blood

OBJECTIVES

After studying this chapter, the student should be able to

1. Describe the life cycle of an erythrocyte.
2. Explain why ATP concentrations are much lower in erythrocytes than in other cells.
3. Explain the significance of the pentose phosphate pathway in preventing oxidative processes in the erythrocyte.
4. Name the two enzymes involved in the reduction of methemoglobin and describe their relative significance.
5. Understand the molecular basis of drug-induced hemolytic anemia.
6. Name three major proteins of the erythrocyte membrane and describe their functions.
7. Name the subunit composition of the normal human hemoglobins and state their physiologic significance.
8. Draw the chromosomal arrangement of α-like and β-like globin genes.
9. Describe some of the deficiencies that cause thalassemia.
10. Explain the significance of the Hill coefficient of hemoglobin binding to oxygen.
11. Give a molecular explanation for the sigmoidal shape of the hemoglobin-oxygen saturation curve.
12. List the conditions and effectors that influence the Hb-O_2 saturation curve.
13. Recognize the major plasma proteins and describe their functions, if known.
14. Explain the contribution of the Gibbs-Donnan equilibrium to the total osmotic pressure of plasma.
15. Understand the term "acute phase reactants."
16. Explain why α_1-antitrypsin deficiency causes emphysema.
17. Explain the function of the haptoglobins.
18. Draw the subunit structure of an immunoglobulin molecule and specify the functions of its parts.
19. Describe the clonal selection theory of antibody production.
20. Outline the arrangement of immunoglobulin genes and explain how a limited number of structural genes can give rise to an almost infinite number of immunoglobulins of different specificity.
21. Describe the genomic events that occur during the differentiation of a precursor cell into a lymphocyte.
22. Explain the function of the major histocompatibility complex proteins in the activation of T lymphocytes.
23. List the sequence of events that lead to primary hemostasis and explain why aspirin interferes with the formation of a platelet plug.
24. Describe the events that activate the intrinsic coagulation pathway and those that activate the extrinsic pathway.
25. Name the coagulation factor that is activated by both the intrinsic and the extrinsic pathway and describe the function of this activated factor.

A major function of blood is to carry oxygen from the lungs to the tissues and to return to the lungs with the carbon dioxide derived from oxidative processes. In addition, blood transports ions and metabolites and generally serves as the communication link between different parts of the body. Blood also carries many of the cells that participate in the immunosurveyance of the human body, thereby facilitating the transport of lymphocytes and macrophages to sites of infection.

When blood is subjected to centrifugation (in the presence of a coagulation inhibitor, e.g., heparin or citrate), the blood cells sediment into a pellet. Most of this pellet consists of red cells (1 mm^3 of blood contains about 5×10^6 erythrocytes and $5–10 \times 10^3$ white blood cells). The percentage of the original blood volume taken up by the sedimented erythrocytes is the **hematocrit** (normal values are 46% for males and 44% for females). The supernatant is called **plasma** and contains all extracellular blood proteins, including the clotting factors and fibrinogen. If clotting is not prevented, the erythrocytes precipitate together with fibrin. The supernatant of clotted blood is called **serum.** Serum contains most plasma proteins except fibrinogen, the soluble precursor of fibrin.

9.1 THE ERYTHROCYTE

Erythrocytes are small disc-shaped cells that have lost their nucleus and are filled with a 34% solution of hemoglobin, the protein molecule that carries oxygen and also functions as a buffer for carbonic acid (see Section 12.1).

9.1.1 The Life Cycle of the Erythrocyte

As with all blood cells, erythrocytes are derived from a **pluripotential stem cell** in the bone marrow (Figure 9.1). This process of **hematopoiesis** is regulated by several hematopoietic growth factors. Under the influence of different growth factors, stem cells can differentiate into four different types of committed progenitor cells with restricted differentiation potentials. These progenitor cells then proliferate and differentiate into the mature blood cell types. The growth factors are named **colony stimulating factors (CSF)** because of their ability to make a progenitor cell grow into a colony of cells. Their precise mech-

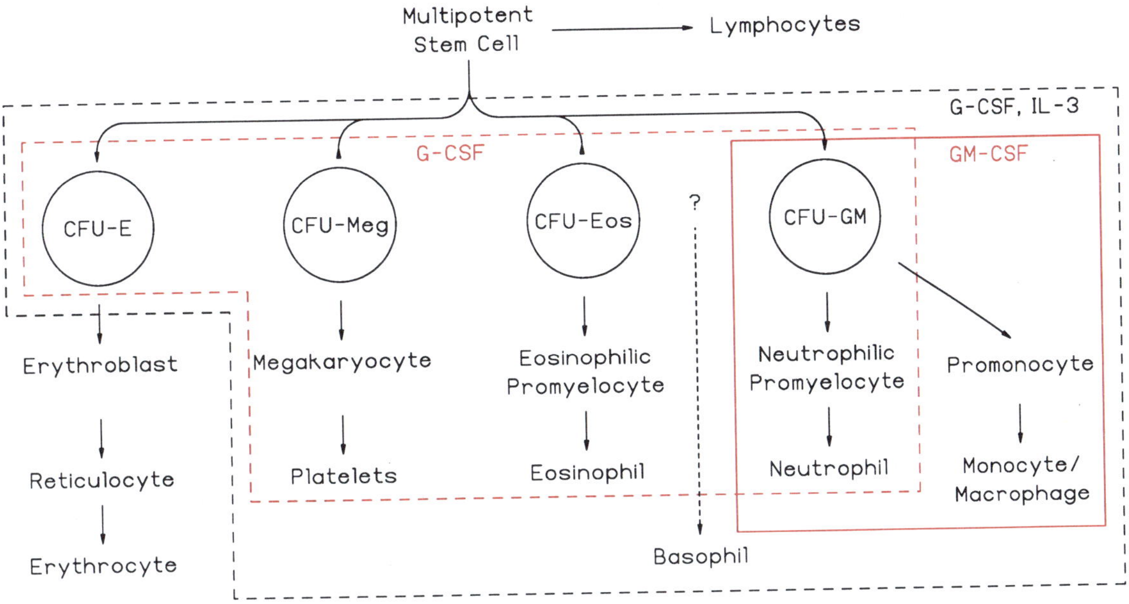

Figure 9.1.
Differentiation of a pluripotential stem cell under the influence of different growth factors. (CFU = colony forming unit.) Enclosed areas shown the range of effect of different colony-stimulating factors (CSFs).

anism of action is still little understood. **Erythropoietin** is a peptide hormone produced by the kidney and acts directly upon the erythroid cell lineage. **Reticulocytes,** the immediate precursors of erythrocytes, lose their nuclei before being released into the bloodstream where they mature into red blood cells.

The average life span of a normal erythrocyte is about 120 days. During this time period, the rate of glycolysis, the only ATP-generating pathway of the erythrocyte, slowly decreases, and the cell becomes mechanically more and more fragile. It is not known which single factor is responsible for the final sequestration of old erythrocytes in the reticuloendothelial systems (RES) of spleen, liver, and bone marrow where the old cells are destroyed and their contents degraded. (The degradation of heme is discussed in Section 10.3.)

9.1.2 Erythrocyte Metabolism

In the erythrocyte, as in other cells, the major intracellular cation is K^+. Its level is maintained by an ATP-dependent Na^+/K^+ exchange pump that removes Na^+ ions from the cell in exchange for K^+ ions against a steep concentration gradient. If the influx of Na^+ ions, due to leakage, is increased more than twofold, the cell begins to swell and may burst. This is because glycolysis cannot accelerate by more than a factor of two. Intracellular concentrations of ATP are low (1–1.5 mM) when compared with those observed in cells that possess mitochondria. A fraction of the 1,3-diphosphoglycerate generated by glycolysis is shunted via 2,3-diphosphoglycerate to 3-phosphoglycerate. This **Rapaport-Luebering shunt** does not produce ATP (Figure 9.2); it serves to generate **2,3-diphosphoglycerate,** an important effector in the control of the affinity of hemoglobin for oxygen (see below).

In addition to the glycolytic pathway, erythrocytes possess a functional **pentose phosphate pathway.** The NADPH generated in this pathway is required for the reduction of oxidized glutathione. Glutathione participates in the detoxification of hydrogen peroxide and, in

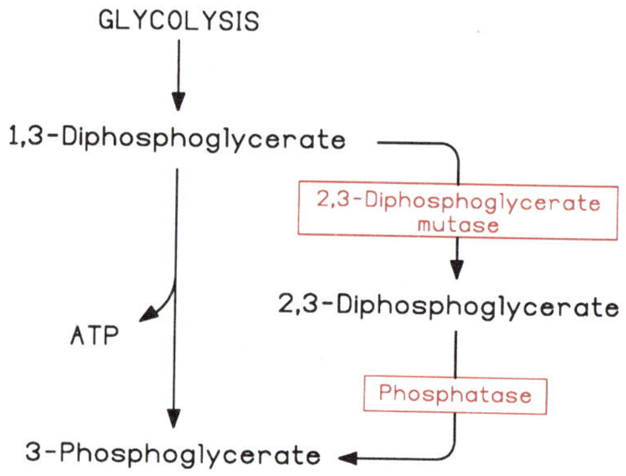

Figure 9.2.
The Rapaport-Luebering shunt for the generation of 2,3-diphosphoglycerate.

general, prevents the oxidation of SH groups in hemoglobin and membrane proteins. In order to function as oxygen carrier, hemoglobin must maintain its heme iron in the divalent state (Fe^{2+}). Oxidation to **methemoglobin,** containing Fe^{3+}, occurs continuously even under normal conditions. (Normal blood contains about 2% methemoglobin.) There are two enzymatic mechanisms for the reduction of methemoglobin. The first, and physiologically the only significant one, is catalyzed by the enzyme **NADH diaphorase** (methemoglobin reductase).

$$Fe^{3+}\text{ heme} + \text{NADH} \rightarrow Fe^{2+}\text{ heme} + \text{NAD}^+$$

The second mechanism has no physiologic function; it uses NADPH as a reducing agent but requires, in addition, an external electron donor such as **methylene blue.** This second **methemoglobin reductase** is much more active than the first one and can be utilized for the rapid treatment of drug-induced methemoglobinemia (see below) by injecting methylene blue, a relatively innocuous dye. Oxidation of hemoglobin to methemoglobin produces the superoxide ion O_2^-.

$$\text{Hb} + O_2 \rightleftharpoons \text{metHb} + O_2^-$$

The erythrocyte enzyme **superoxide dismutase** converts two superoxide ions into molecular oxygen and hydrogen peroxide.

$$O_2^- + O_2^- + 2H^+ \rightarrow H_2O_2 + O_2$$

The resulting peroxide is then detoxified either by the heme-containing enzyme **catalase,**

$$2H_2O_2 \rightarrow 2H_2O + O_2$$

or by the selenium-containing enzyme **glutathione peroxidase** (see Section 8.4).

9.1.3 Hemolytic Anemias Caused by Metabolic Defects

Premature lysis of erythrocytes is called **hemolytic anemia.** The most common cause for this anemia is a deficiency in the enzyme **glucose**

6-phosphate dehydrogenase (G6PD). In normal situations, the deficiency is asymptomatic. Certain drugs, however, can induce hemolytic episodes. The antimalarial drug **primaquine,** for example, and a long list of other drugs, including sulfonamides, nitrofurans, antipyretics, and analgesics, all have the property of oxidizing hemoglobin (by formation of disulfide bonds), causing it to precipitate inside the erythrocyte. The resulting clumps are known as **Heinz bodies** and can be seen in a light microscope. Their presence indicates an erythrocyte at risk. In the presence of a functioning pentose phosphate pathway, the initial oxidative damage is reversible because of the cell's capacity to synthesize enough NADPH (the pentose phosphate pathway can accelerate up to 30-fold). The NADPH then acts via glutathione to reduce the disulfide bonds. In the absence of G6PD, however, the pentose phosphate pathway is inoperative. This causes oxidative destruction of hemoglobin, leading to hemolysis.

There are many different forms (variants) of G6PD deficiency characterized by enzymes of different degrees of instability. G6PD deficiency is X-linked and recessive. Most common is the African variant; it is found in 10% of American black males. A Mediterranean variant, found mostly in Sardinians, Sephardic Jews, and Arabs, causes especially severe hemolysis. In Caucasians, the disease is rare. It may be precipitated by the ingestion of Fava beans, which produce an allergic reaction causing extreme hemolysis **(Favism)** if G6PD is genetically deficient.

Glutathione reductase deficiency is another rare genetic disorder that also sensitizes the patient to drug-induced hemolysis.

Genetic deficiencies in most of the enzymes of the glycolytic pathway have been reported. **Pyruvate kinase deficiency** is the second most common erythrocyte enzyme deficiency. Homozygotes exhibit moderate to severe hemolysis (even in the absence of drugs), which can be alleviated, in some cases, by splenectomy. The rationale for this operation is that splenectomy is usually followed by reticulocytosis (overproduction of reticulocytes); since reticulocytes still contain mitochondria, they do not depend on glycolysis for ATP generation, as do mature red cells.

9.1.4 The Erythrocyte Membrane

Study of the erythrocyte membrane has contributed significantly to our knowledge of cell membrane structure. When erythrocytes are subjected to osmotic shock, their membranes rupture and can be separated as **ghosts** from the cytoplasmic proteins by centrifugation. SDS gel electrophoresis of the membrane proteins reveals more than eight prominent bands when stained with Coomassie Blue (a protein-specific stain) and four additional bands when stained with periodic acid Schiff's base (PAS) reagent, a carbohydrate stain (Figure 9.3).

The protein present in the largest quantity in the erythrocyte membrane is **spectrin** (molecular weight = 220,000–250,000). This protein (bands 1 and 2) is located on the inner surface of the erythrocyte membrane and appears to support the structure of the membrane by binding to actin (band 5), the main component of microfilaments (see Section 11.1). Another major protein in the erythrocyte membrane, commonly called the **band-3 protein,** has a molecular weight of about 90 kd and represents the **anion channel** that allows chloride and bi-

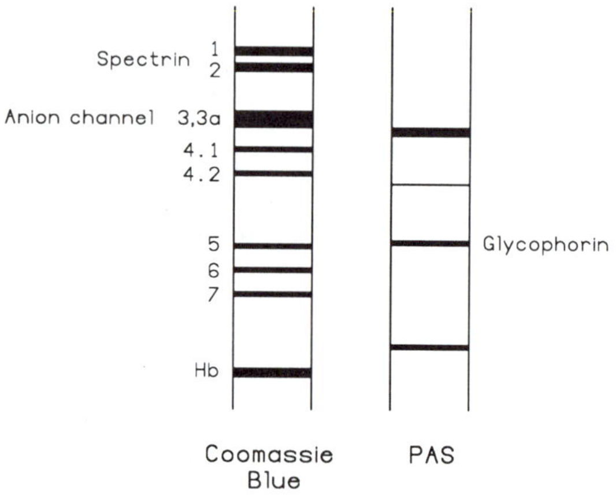

Figure 9.3.
SDS gel of red cell membrane proteins.

carbonate ions to pass through the membrane at very high rates. The third major protein, **glycophorin,** also crosses the cell membrane (as does band-3 protein). Its N-terminus is very rich in threonine and serine residues, which are thought to protrude from the external membrane surface. A total of 16 oligosaccharide chains are attached to these amino acid residues, making glycophorin a **sialo-glycoprotein** with 60% carbohydrate content. The surface carbohydrate residues of glycophorin carry the MN blood group antigens. (Blood group antigens of the ABO type are part of another membrane glycoprotein.)

Hereditary spherocytosis (hereditary hemolytic jaundice) is an autosomal recessive disorder in which the erythrocyte membrane is defective. As a result, the red cells are spherical, osmotically fragile, and are prematurely removed from the circulation in the spleen. The metabolism of the red cell appears to be normal. The disorder can be cured by splenectomy.

In **hereditary elliptocytosis,** an autosomal dominant disorder, between 40% and 60% of the red cells are ellipsoidal in shape and osmotically fragile, causing hemolysis.

An acquired membrane defect, possibly arising from a mutated clone of abnormal cells, is **paroxysmal nocturnal hemoglobinuria** (PNH). Serum complement factors, which normally absorb to the erythrocyte surface without consequences, cause lysis of the cells in these patients whenever the blood pH drops (as it does during sleep). It should be noted that mature erythrocytes have lost the ability to synthesize lipids. Therefore, repair of their phospholipid membranes relies on uptake of lipids from plasma lipoproteins, with lecithin-cholesterol acyltransferase (LCAT) playing a major role.

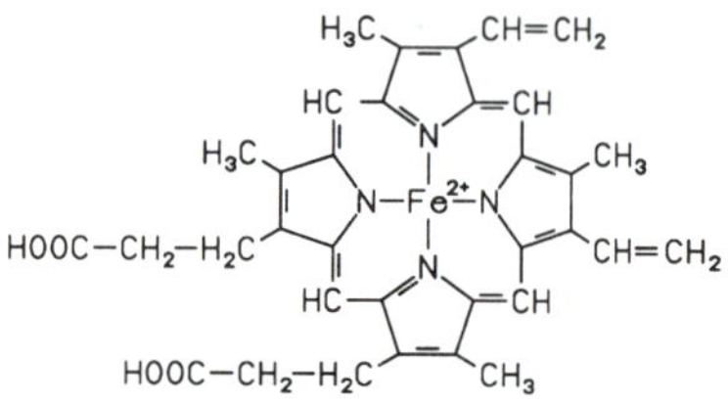

Figure 9.4.
The structure of heme (ferro-protoporphyrin IX).

9.2 THE HEMOGLOBINS

9.2.1 Hemoglobin Biosynthesis

Normal adult human hemoglobin, **hemoglobin A (HbA),** is a tetramer consisting of two α-globin and two β-globin polypeptide chains. Each hemoglobin chain is coordinated to a ferroprotoporphyrin IX complex, called heme (Figure 9.4).

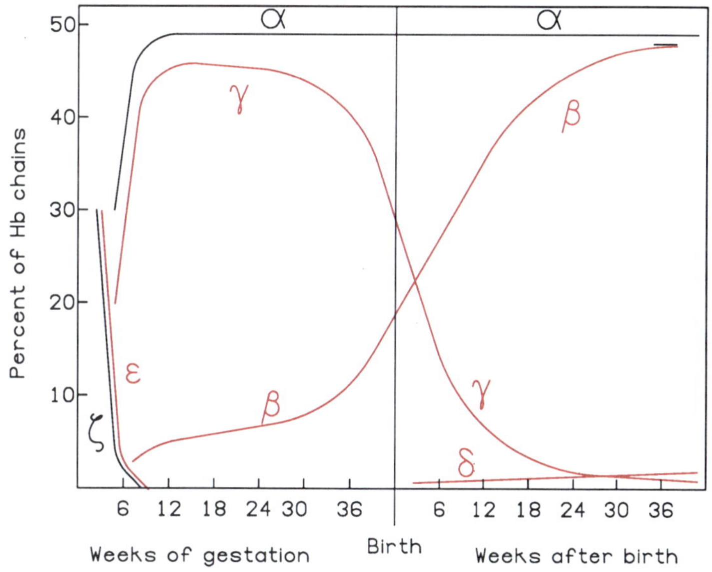

Figure 9.5.
Distribution of Hb chains during human development. β-Like globins are drawn in red. After Huehns ER, et al. *Cold Spring Harbor Symp Quant Biol.* 1964; 29:327.

During development, several different forms of hemoglobin are made (Figure 9.5). All hemoglobins consist of two α-like and two β-like globin chains. Table 9.1 lists the different hemoglobins and their time of expression.

During the first 3 months of life, Hb Gower 1, Hb Gower 2, and Hb Portland are made. The ϵ and ζ chains are probably synthesized in the yolk sack. At birth, the major hemoglobin species is HbF. It is replaced during the first year of life by HbA.

9.2.2 Gene Structure and Expression

The haploid human genome contains five α-like Hb genes: two identical α, and ζ, one pseudo-ζ, and one pseudo-α gene. (Pseudogenes are not expressed; they often contain mutations in their promoters.) The α-like

Table 9.1. Human Hemoglobins

Hemoglobin	Structure	Percent of Normal Adult Hemolysate	Increased
A	$\alpha_2\beta_2$	92	
A_2	$\alpha_2\delta_2$	2.5	β-Thalessemia
A_{1a}	Not known	<1	Diabetes mellitus
A_{1b}	Not known	2	Diabetes mellitus
A_{1c}	$\alpha_2(\beta$—*N*—Glucose$)_2$	5	Diabetes mellitus
F	$\alpha_2\gamma_2$*	<1	Fetal erythrocytes, β-thalassemia, sickle cell anemia
F_1	$\alpha_2(\gamma$—*N*—Acetyl$)_2$	<1	
Gower 1	$\zeta_2\epsilon_2$	0	Early embryo
Gower 2	$\alpha_2\epsilon_2$	0	Early embryo
Portland	$\zeta_2\gamma_2$	0	Early embryo
H	β_4	0	α-Thalassemia
Bart's	γ_4	0	α-Thalassemia

*In humans, two genetic forms of HbF, $\alpha_2\gamma(A)_2$ and $\alpha_2\gamma(G)_2$ are expressed simultaneously.

hemoglobin genes are clustered on chromosome 16. The β-like genes are all closely linked on the short arm of chromosome 11. There are a total of six β-like hemoglobin genes: one ε, two γ, one δ, one β, and one pseudo-β gene (Figure 9.6).

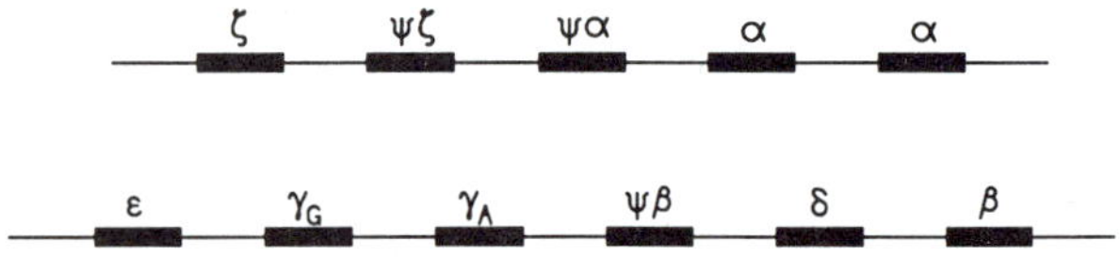

Figure 9.6.
Arrangement of α-like and β-like hemoglobin genes on chromosomes 16 and 11, respectively.

As with the α-like genes, the β-like genes appear to have evolved from a single ancestral gene. The two γ chains, for example, differ by only one amino acid at position 136 (Gly in γ^G and Ala in γ^A). The δ and β chains show more differences but still share large homologous regions. All hemoglobin genes consist of three exons that are interrupted by two introns. The length of the introns is variable, but their positions are identical in all α-like and β-like globin genes. Although α- and β-globin genes are expressed at high levels and are regulated coordinately in adult bone marrow cells, their promoter structures are surprisingly different. When globin genes are introduced by transfection into other types of mammalian cells, the α-globin gene is expressed without difficulty because it carries an enhancer as part of its coding sequence. In contrast, the β-globin gene requires a strong viral enhancer, such as the viral SV40 enhancer, for expression.

Several different transcription factors are required for the expression of globin genes in erythroblasts where hemoglobin transcription begins. In humans, the switch from embryonic hemoglobin to fetal hemoglobin occurs after about 5 weeks of gestation, and the site of hemoglobin synthesis changes from the yolk sac to the fetal liver. Embryonic erythrocytes are nucleated and very large, whereas fetal liver erythrocytes are already anucleate and small, similar to adult erythrocytes. At about 12 weeks of gestation, the primary site of hematopoiesis switches from the liver to the spleen and the bone marrow, which at the time of birth becomes the main site of hematopoiesis. The mechanism of hemoglobin switching is little understood. It probably involves the sequential expression of different sets of transcription activation factors. It has been proposed that a developmental clock determines the onset of expression of the different hemoglobin genes. The model has found some support from transplantation experiments. When adult bone marrow cells are injected into fetal sheep in utero, the erythroid cells were found to synthesize adult-type hemoglobin. Similarly, when fetal liver cells are injected into lethally irradiated adult sheep, the hemoglobin synthesized initially was HbF and the time of switch to adult-type hemoglobin was dependent on the gestational age of the donor fetus. On the other hand, when fetuses were hypophysectomized, the switch from HbF to HbA was delayed, indicating the participation of an endocrine factor in hemoglobin switching. In addition, DNA methylation seems to be involved in the control of hemoglobin gene expression. Genes that are being expressed are undermethylated, whereas genes that are inactive are fully methylated.

Reticulocytes (immature red blood cells) contain about 40% more α-globin messenger RNA (mRNA) than β-globin mRNA. However, synthesis of the α- and β-globin chains occurs at equal rates. **Hemin** is a potent stimulator of globin synthesis. In the absence of hemin (oxidized form of heme), a protein kinase phosphorylates, and thereby inactivates, the initiation factor eIF-2 (the letter "e" refers to eukaryotic). Hemin inactivates the protein kinase, allowing a protein phosphatase to restore activity to eIF-2. When hemin levels are depressed, as found in iron deficiency, α chain synthesis is more depressed than β chain synthesis, presumably because of a less efficient ribosome binding site on α-globin mRNA. This leads to unbalanced synthesis of globin chains as seen in forms of α-thalassemia (see below).

9.2.3 Genetic Defects in Hemoglobin Synthesis

Thalassemia syndromes are genetic defects in the globin genes causing the absence, or reduced expression, of either α-globin (α-thalassemia) or β-globin chains (β-thalassemia). Since the normal diploid human genome has four copies of α-globin genes, one distinguishes four different degrees of severity of α-thalassemia depending on how many of these genes are nonfunctional.

A person who has only one α-globin gene deleted is a silent, that is, phenotypically normal, carrier of the **α-thalassemia** trait. Deletions in two α-globin genes cause the α-thalassemia syndrome, deletions in three genes cause the severe **hemoglobin H (β_4) disease,** and deletion of all four α-globin genes is never found, presumably because the defect is lethal.

In **β-thalassemia** the β-globin genes are deleted or otherwise defective. β-Thalassemias are usually less severe than α-thalassemias because the fetal γ-hemoglobin continues to be produced into adult life, replacing the missing β-hemoglobin. For example, in homozygous δβ-thalassemia, neither δ- nor β-globin is produced, but fetal hemoglobin F is present. (The disorder is only moderately severe.) The Lepore and anti-Lepore hemoglobins contain altered β chains that are in fact hybrids between the β and γ chains (Figure 9.7). The hybrid chains arise through unequal crossing-over during meiosis. (Similar crossing-over events between β and γ genes lead to the **Kenya** and **anti-Kenya** genotypes.) **Lepore** globin is synthesized at very low rates, causing a β-thalassemia phenotype. **Anti-Lepore** hemoglobin is also synthesized very inefficiently. However, intact β- and δ-globin genes are being made as well, so that no thalassemia results.

In **β^+-thalassemia,** only a small amount of β-globin mRNA is made, whereas excessive amounts of pre-mRNA are found in the nucleus. The defect has been shown to involve a mutation that maps in one of the splice sites of the pre-mRNA, thus preventing the proper maturation and transport of the mRNA. In one case of β^+-thalassemia, a mutation was found in the 3′-polyadenylation site. Apparently the poly A-less β-globin mRNA was too unstable to be used for efficient translation.

β^0-Thalassemia is another β chain defect in which the β-globin genes are present but no β-globin chains are synthesized. One type of

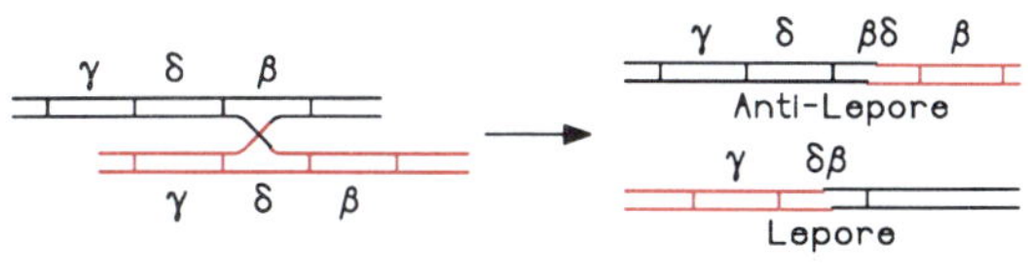

Figure 9.7.
Unequal crossing-over between globin genes leading to hybrid globin genes.

β^0-thalassemia has also been shown to be caused by a nonsense mutation in the β-globin gene. β^0-Thalassemia have been shown to be caused by promoter mutations and splice site mutations.

β-Thalassemia patients have been treated experimentally with the cytidine analog 5-azacytidine. 5-Azacytidine interferes with the methylation of DNA and causes an increase in γ-globin synthesis. DNA isolated from bone marrow cells of patients treated with 5-azacytidine was found to be undermethylated in the γ-globin genes. This observation suggests that methylation is one of the mechanisms responsible for the inactivation of genes during development. The activation of fetal hemoglobin genes in thalassemia patients can also be accomplished by injection of genetically engineered erythropoietin. This therapy appears considerably safer than using 5-azacytidine.

Many **α-globin variants** (mutants) are known. Some of these mutants produce longer than normal α-globin chains because of a mutation in the chain termination codon or a frameshift mutation. Most of these mutant globins are made inefficiently, causing an α-thalassemia phenotype. The most common hemoglobin mutant is the hemoglobin variant HbS, which is due to a mutation in the β-globin gene. This mutation causes **sickle cell anemia,** a condition in which the hemoglobin molecules aggregate to form intracellular filaments that distort the shape of the erythrocyte (sickle shape). This makes the red cell fragile and eventually leads to cell destruction, causing severe anemia. The mutation involves the change of a glutamate, the sixth amino acid from the amino terminus of the molecule, into valine. This mutation makes the deoxygenated hemoglobin molecule much less soluble. As long as HbS is fully oxygenated, the hemoglobin functions normally. The heterozygous carrier of the HbS allele makes about 65% HbA ($\alpha_2\beta_2^A$) and 35% HbS ($\alpha_2\beta_2^S$). This condition is relatively benign and is called sickle cell trait. Sickling occurs only under conditions of stress, such as low oxygenation at high altitudes. Our understanding of how the molecular defect affects Hb solubility still does not explain all observed phenomena such as the stickiness of sickling erythrocytes and the variability of painful crises in homozygotes.

There are three other hemoglobin variants (all affecting the β chain) that also have altered solubility. **HbC** (6Glu→Lys) is found, like HbS, predominantly in African populations; **HbE** (26Glu→Lys) is found in Southeast Asia; and **HbD** (121Glu→Gln) is found in India and Pakistan. Heterozygous carriers of these variants, including the sickle cell trait, appear to be less susceptible to malaria and thus have a selective advantage over the normal HbA carriers. Homozygotes, however, are subject to periods of severe anemia. About 9% of the American black population carries the HbS allele. The treatment of sickle cell anemia and related anemias is similar to that of thalassemias involving the activation of fetal hemoglobin genes by erythropoietin.

9.2.4 Structure and Function of Oxygen-Binding Proteins

The oxygen-binding protein in erythrocytes is hemoglobin. It is a tetramer consisting of two identical dimeric αβ subunits that are connected by salt bridges. Each of the subunits carries one heme group with an Fe^{2+} ion at its center. Thus, a hemoglobin molecule carries four heme groups and can bind four molecules of oxygen. Dissolved oxygen diffuses from the plasma into the erythrocyte, where it is bound to hemoglobin (Hb). This process is called **oxygenation,** the reverse process is called deoxygenation. Oxygenated Hb is called **oxyhemoglobin,** deoxygenated Hb is called **deoxyhemoglobin.** The average amount of Hb in erythrocytes in 100 ml of blood is 15 g. This amount of Hb can carry up to 20 ml of O_2, considerably more than the 0.31 ml of O_2 that can be dissolved the same volume of blood plasma.

In contrast to hemoglobin, **myoglobin,** the oxygen-binding protein in muscle, is a single polypeptide. Its primary structure is closely related to that of hemoglobin. Each myoglobin molecule binds one molecule of oxygen. It binds oxygen much tighter than hemoglobin. The partial oxygen pressure, pO_2, required for 50% saturation with O_2 for myoglobin is only 1 mmHg. The corresponding value for Hb is 26 mmHg (see Figure 9.8). Cytochrome oxidase exhibits an even higher affinity for oxygen. In humans, muscle myoglobin acts as a carrier during the diffusion of O_2 from the capillary membrane into the muscle. In diving mammals, myoglobin also functions in O_2 storage. (In humans, myoglobin is only abundant in heart muscle.) The kinetics of oxygen binding of hemoglobin and myoglobin are quite different. Myoglobin exhibits a hyperbolic oxygen-binding curve as would be expected from a direct one-to-one reaction between myoglobin and oxygen.

$$Mb + O_2 \rightleftharpoons MbO_2 \tag{9.1}$$

The fractional saturation of myoglobin, Y, is defined as

$$Y = \frac{[MbO_2]}{[MbO_2] + [Mb]} \tag{9.2}$$

and K_d, the dissociation constant of the myoglobin-oxygen complex, is defined as

$$K_d = \frac{[Mb][O_2]}{[MbO_2]} \tag{9.3}$$

Combining the two equations and substituting pO_2, the partial pres-

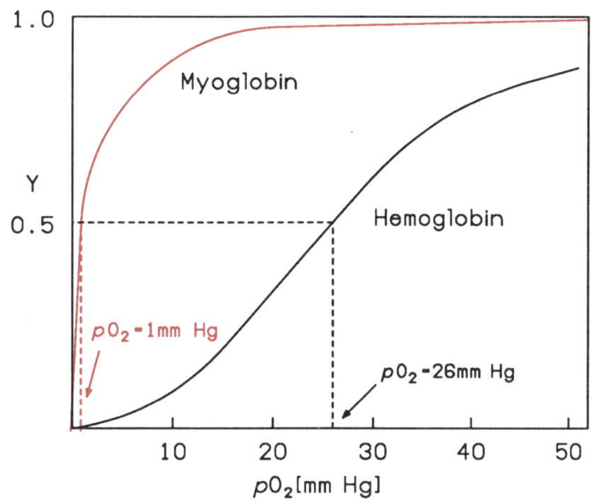

Figure 9.8.
Oxygen saturation curves for hemoglobin and myoglobin.

sure of oxygen which is proportional to the concentration of O_2, for $[O_2]$ we obtain:

$$Y = \frac{pO_2}{pO_2 + K_d} \tag{9.4}$$

Plotting Y versus pO_2 gives the oxygen saturation curve (Figure 9.8) for myoglobin, a hyperbolic curve similar to the Michaelis-Menten curve (see Chapter 1).

In further analogy with the Michaelis-Menten constant, the dissociation constant K_d equals the pO_2 value at which half the myoglobin is oxygenated. Thus, the ratio of oxygenated to free myoglobin, $Y/(1 - Y)$, is

$$\frac{Y}{1 - Y} = \frac{pO_2}{P_{50}} \tag{9.5}$$

or, in logarithmic form,

$$\log \frac{Y}{1 - Y} = \log pO_2 - \log P_{50} \tag{9.6}$$

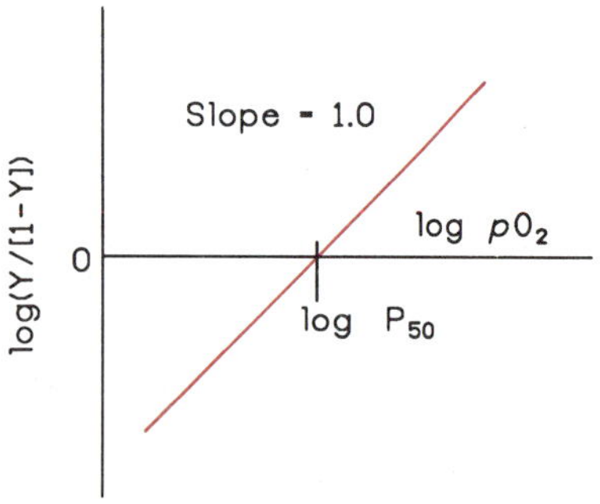

Figure 9.9.
Hill plot for the binding of oxygen to myoglobin.

This is the **Hill equation.** When $\log [Y/(1 - Y)]$ is plotted against $\log pO_2$, myoglobin gives a straight line with a slope of $n = 1.0$, the Hill coefficient (Figure 9.9).

In contrast, the oxygen saturation curve for hemoglobin is sigmoidal, suggesting that the binding of oxygen occurs in a cooperative manner. The initial binding of an oxygen molecule to a hemoglobin subunit induces a conformational change in the other subunits with the result that additional oxygen molecules are bound more easily than the first one. This **allosteric behavior of hemoglobin** is apparent in a sigmoidal Hill plot (Figure 9.10). The Hill coefficient of Hb is $n = 2.9$.

Figure 9.10.
Hill plot for the binding of oxygen to hemoglobin. The Hill coefficient is the slope of the curve at P_{50}.

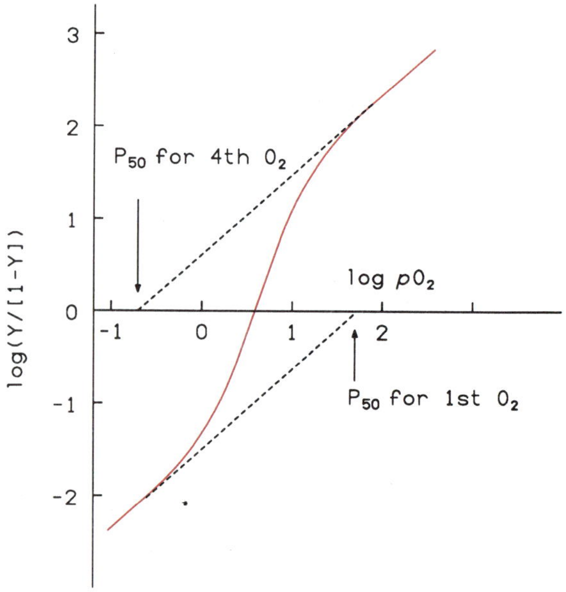

The more general form of the Hill equation is

$$\frac{Y}{1 - Y} = \left(\frac{pO_2}{P_{50}}\right)^n \qquad (9.7)$$

In **deoxyhemoglobin,** the four subunits of hemoglobin are tightly bound together by **salt bridges,** electrostatic interactions between the carboxyl termini and positively charged groups on the other chains. The terminal amino groups of the α chains also participate in these interactions. Altogether, there are eight such salt bridges in deoxyhemoglobin. Oxygenation causes the disruption of these salt bridges. The Fe^{2+} ion in the center of the heme ring has four bonds to the nitrogens of the porphyrin ring system, a fifth bond to a histidine side chain in the F helix of globin, and an unpaired sixth coordination site. Because of its four unpaired electrons, the Fe^{2+} in deoxyhemoglobin is paramagnetic and too large to fit into the plane of the porphyrin ring. When oxygen binds to this sixth site, the Fe^{2+} ion becomes diamagnetic, decreases in size, and moves into the porphyrin plane. This movement pulls the F8 histidine toward the heme plane and causes a conformational change of the entire Hb molecule, with the result of loosening up its structure (Figure 9.11). (Recent magnetic measurements, however, throw some doubt on the actual movement of the Fe^{2+} ion.)

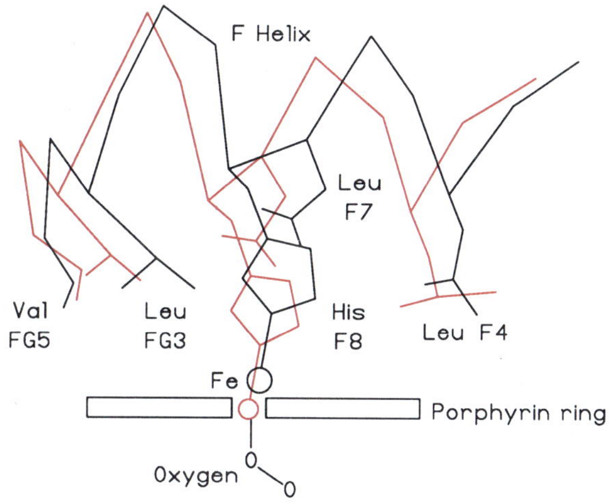

Figure 9.11.
The movement of the Fe^{2+} ion into the plane of the porphyrin ring, pulling with it the F8 histidine residue in hemoglobin. Adapted from Baldwin J, et al., *J Mol Biol.* 1979; 129:175.

The cooperative binding of oxygen to Hb is most easily visualized by postulating two conformational states for the Hb subunits: the T (for tight) and the R (for relaxed) state. As oxygen binds to one subunit, it converts it from the T state to the R state, breaking salt bridges to its two neighbor subunits. The conversion of the next subunit into its R state will be easier because salt bridges to only one other subunit must be broken (Figure 9.12).

The **Hb-O_2 saturation curve** is shifted downward and to the right by an increase in pCO_2, $[H^+]$, temperature, or the concentration of

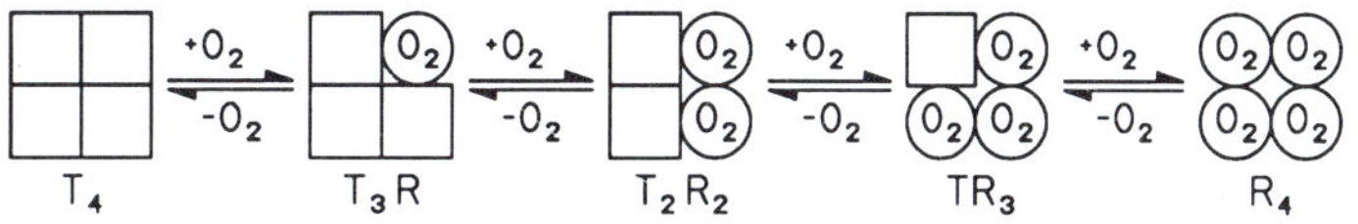

Figure 9.12.
Sequential model of allosteric heme-heme interaction without postulating a direct communication between the heme groups.

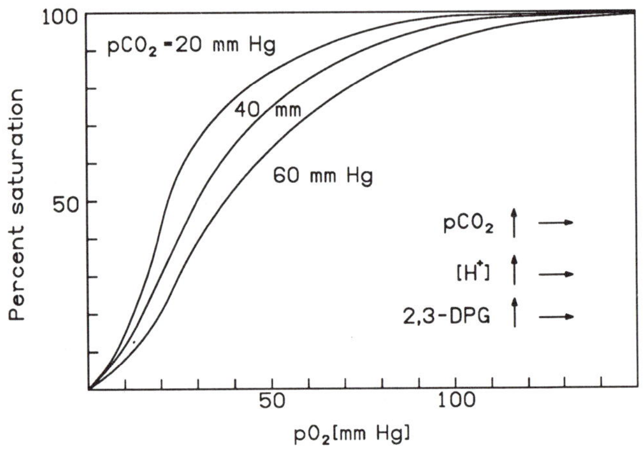

Figure 9.13.
Hemoglobin-oxygen saturation curves at different values of pCO_2.

2,3-diphosphoglycerate (2,3-DPG). The result is a reduced affinity of hemoglobin for oxygen but increased affinity for H^+ ions (Figure 9.13).

The shift in the Hb-O_2 saturation curve caused by the presence of CO_2 or protons is called the **Bohr effect.** It facilitates O_2 release in the tissues, where levels of CO_2 and H^+ are higher because of metabolic activity. The reverse effect occurs in the lungs, where the increased pO_2 drives off H^+ and CO_2 **(Haldane effect).** The Bohr effect involves three pairs of proton-binding groups, the terminal α-amino group of the α chains, 2 Arg residues and two His residues, all of which, upon deoxygenation, increase their pK values because of the increased proximity of negatively charged groups.

2,3-Diphosphoglycerate binds to Hb and reduces its affinity for oxygen. It stabilizes deoxyhemoglobin by cross-linking the β chains. Carrying four negative charges, it is bound within a central cavity to Lys and His residues on the β chains. Upon oxygenation, 2,3-DPG is extruded because the central cavity becomes too small. The 2,3-DPG level in the erythrocyte (4.5 m*M*) is apparently regulated by the atmospheric pO_2. At high altitude, the 2,3-DPG level increases, allowing the Hb to release O_2 more easily. 2,3-DPG is rapidly lost from erythrocytes when blood is stored in citrate. This may cause complications during transfusions. Addition of inosine prevents this loss of 2,3-DPG. Hemoglobin F (fetal Hb) binds 2,3-DPG less strongly than does HbA, the normal adult Hb. Consequently, fetal Hb has a higher affinity for O_2, facilitating the transfer of O_2 from maternal to fetal blood in the placenta.

9.3 PLASMA PROTEINS

Normal plasma contains between 6 and 8 g of protein per 100 ml. These proteins are usually classified according to their electrophoretic mobilities. There are (in order of decreasing mobility) five major classes of plasma proteins: albumin (59%), α_1-globulins (5%), α_2-globulins (8%), β-globulins (12%), and γ-globulins (16%) (Figure 9.14).

The majority of the plasma proteins are synthesized in the liver. Major exceptions are the immunoglobulins (γ-globulin fraction), which are synthesized by the lymphoid tissues. Apolipoproteins A and B are also

Figure 9.14.
Electrophoretogram of plasma proteins.

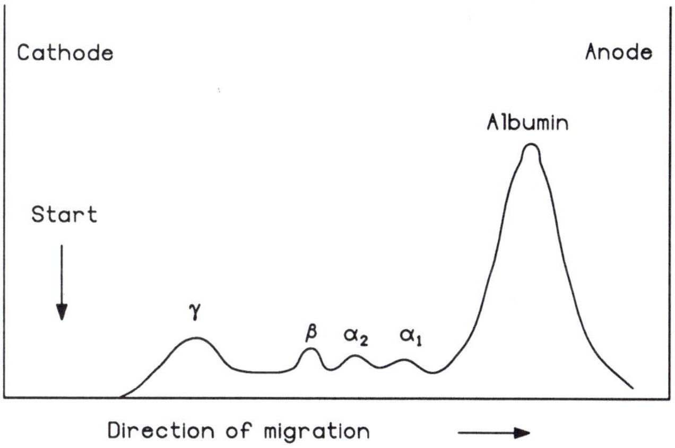

Table 9.2. Major Human Plasma Proteins

Protein	Molecular Weight	Serum Concentration (mg/100 ml)	Function	Disorders
Prealbumin	55,000	10–40	Binding and transport of thyroxin, RBP	
Albumin	66,000	3500–4500	Oncotic pressure, transport	Decreased in liver disease or protein deficiency
α_1-Globulins		300–600		
α_1-Acid glycoprotein	40,000	55–140	Unknown	Acute phase reactant
α_1-Antitrypsin	54,000	200–400	Protease inhibitor	Acute phase reactant, reduced in pulmonary emphysema or hepatic cirrhosis
α_1-Fetoglobulin	64,000	0.001	Unknown	Present in the fetus, and in liver carcinoma
Transcortin	52,000	3–3.5	Binding and transport of corticosteroids	
Thyroxine-binding globulin	58,000	1–2	Binding and transport of thyroxine	
Transcobalamin I	—	—	Minor carrier for vitamin B_{12}	
α_2-Globulins		400–900		
α-Macroglobulin	725,000	150–420	Protease inhibitor	
Inter-α-trypsin inhibitor	160,000	20–70	Protease inhibitor	
Cls inhibitor	—	20–30	Inhibits first component of complement system	Genetic deficiency (only heterozygotes)
Haptoglobins	100,000	380–780	Binding of Hb	Acute phase reactant
Ceruloplasmin	151,000	15–60	Cu transport	Decreased in Wilson's disease
β-Globulins		600–1100		
Transferrin	76,500	200–300	Fe transport	Increased in pregnancy and iron deficiency
Hemopexin	57,000	50–100	Binds heme	
Transcobalamin II	—	—	Major vitamin B_{12} carrier	
β_2-Microglobulin	12,000	0.2	HLA antigen	
C-reactive protein	118,000	1	Unknown	
Lipoproteins			Discussed in Chapter 3	
γ-Globulins	150,000	700–1500	Antibodies	
Cryoglobulins	220,000		Unknown	

RBP = retinol-binding proteins, HLA = histocompatibility locus antigen.

synthesized in the intestinal mucosa. (Apolipoproteins are discussed in Chapter 3.) The major human plasma proteins are listed in Table 9.2.

Most plasma proteins turn over rapidly. They are continually being removed from the circulation in the liver and kidney. Plasma glycoproteins from which the carbohydrate moiety (usually sialic acid) has been removed have a greatly reduced half-life.

9.3.1 Serum Albumin

Serum albumin is a single-polypeptide-chain protein (molecular weight [MW] = 69,000) containing 17 intrachain disulfide bonds and one free SH group. About 50% of the molecule is α-helical in nature. There are no carbohydrates linked to albumin. The molecule can be divided into three clearly recognizable domains of similar structure. It has been suggested that albumin evolved from the 80 carboxyl-terminal amino acids of hemoglobin through a series of gene duplications (Figure 9.15).

One of the main functions of albumin is to act as a nonspecific carrier for a variety of different molecules and ions. For example, Ca^{2+} ions are bound to albumin, and fatty acids are transported almost exclusively by albumin. (The solubility of free fatty acids in water is only 10^{-8} *M*, whereas in serum it is increased to 5×10^{-4} *M*.) Albumin also binds to steroid hormones, bilirubin, Cu^{2+} and Zn^{2+} ions, as well as to many drugs such as sulfonamides, penicillin, aspirin, and others.

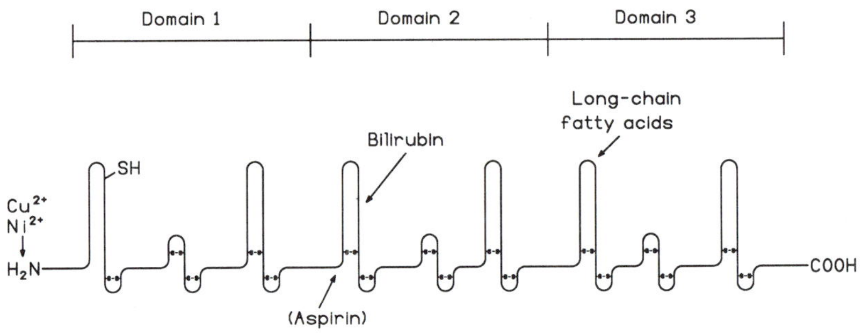

Figure 9.15. Model of the primary structure of human serum albumin showing its disulfide-bonded loops and three similar domains. The binding sites for some ligands are indicated.

In addition to its carrier function, albumin is responsible for maintaining the **oncotic pressure** of plasma. (Oncotic pressure is the osmotic pressure due to nondiffusible ions.) Because of the high concentration of albumin in plasma and its relatively low molecular weight compared with that of the other plasma proteins, over 75% of the oncotic pressure of blood is due to albumin. The total osmotic pressure of plasma is even higher because of the contribution of the **Gibbs-Donnan equilibrium.** At pH 7.5 (normal pH of blood), albumin carries 18 negative charges that must be balanced by positively charged counter ions, mainly Na^+. These "trapped" Na^+ ions, as well as chloride ions that enter the blood vessel, contribute to the osmotic pressure of plasma. The extent of this contribution can be calculated as follows. Suppose two fluid compartments are separated by a semipermeable membrane that allows passage of small molecules and ions but not macromolecules such as proteins. Initially, the inside compartment contains only a negatively charged protein (P^-) and Na^+ and ions, whereas the outside compartment contains Na^+ and Cl^- ions. Because of the concentration gradient of chloride ions across the membrane, Cl^- ions

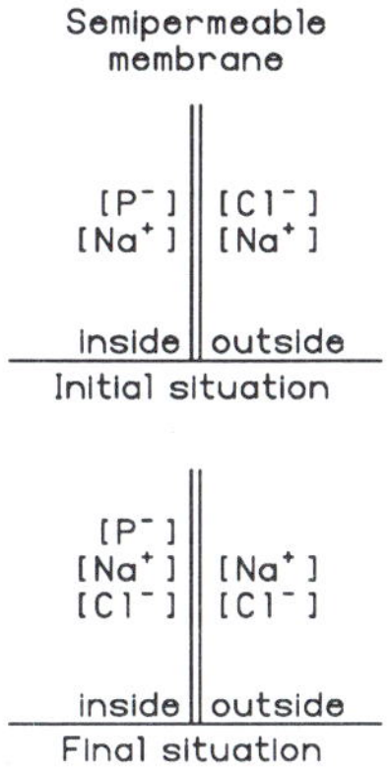

Initial Concentrations	Final Concentrations qt Equilibrium
	$(P^-)_i = z$
$(P^-)_i = z$	$(Na^+)_i = z + x$
$(Na^+)_i = z$	$(Cl^-)_i = x$
$(Na^+)_O = y$	$(Cl^-)_O = y - x$
$(Cl^-)_O = y$	$(Na^+)_O = y - x$

Figure 9.16.
Gibbs-Donnan equilibrium.

will soon pass into the inner compartment accompanied by additional Na^+ ions to preserve electrical neutrality. (P^- is restricted to the inner compartment.) This net movement of NaCl into the inside compartment is opposed by the increasing osmotic pressure in the inside compartment and comes to a halt when the two forces equal each other. This state is called the Gibbs-Donnan equilibrium (Figure 9.16). The subscripts "i" and "o" refer to the inside or outside compartment, respectively.

Gibbs-Donnan equilibrium is established when the following relationship exists for the diffusible ions:

$$(Na^+)_i\,(Cl^-)_i = (Na^+)_o\,(Cl^-)_o \tag{9.8}$$

By substituting x, y, and z, we obtain

$$x\,(z + x) = (y - x)^2 \quad \text{or} \quad x = \frac{y^2}{z + 2y} \tag{9.9}$$

From equation (9.9) it can be easily seen that the number of diffusible ions present in the inside compartment (e.g., a blood vessel) must exceed the number of diffusible ions in the outside compartment (e.g., interstitial fluid). A mathematical theorem states that if the product of two unequal numbers equals the square of a third number, the sum of the two unequal numbers is always greater than two times the third number. Thus, the sum of $(z + x) + x$, representing the inside ions, must be larger than $2(y - x)$, the number of outside ions.

The excess of diffusible ions in the vascular compartment over the diffusible ions in the outside compartment adds to the osmotic effects of the nondiffusible proteins (P^-). Serum albumin levels are decreased in kwashiorkor (severe dietary protein deficiency), nephrotic syndrome, glomerulonephritis, hepatic cirrhosis, and protein-losing enteropathy, all of which may cause edema. **Analbuminemia,** a rare genetic deficiency of albumin, is accompanied by impaired lipid transport and elevated levels of cholesterol, phospholipids, and lipoproteins. Surprisingly, there is only minimal edema. An interesting feature of this disorder is the presence of proteins with lower molecular weights than albumin, which seem to compensate for the absence of the osmotic effects normally exerted by albumin.

Albumin is synthesized in the liver. Its half-life is about 19 days. It continually leaks out of the vascular compartment into the interstitial space. Most of it is returned to the bloodstream by the lymphatic system. Some of it, however, is taken up by cells and degraded.

9.3.2 The Prealbumin Fraction

Some plasma proteins migrate faster than albumin. Because of their low concentration, they are usually not discernible by routine electrophoresis. **Thyroxine-binding prealbumin (TBPA)** is one of two transport proteins for thyroid hormone. (The other is thyroxine-binding globulin.) TBPA, sometimes called tryptophan-rich prealbumin, is a tetrameric protein (MW = 54,000) with one binding site (per tetramer) for either thyroxine (T4) or triiodothyronine (T3). TBPA also has binding sites for retinol-binding protein. This association does not affect the binding of thyroid hormones.

Retinol-binding protein (RBP), the specific plasma carrier of retinol, is normally 75% saturated with retinol. It has a rather low molecular weight (21 kd), which would allow it to pass through the glomeruli of the kidney and be excreted in the urine were it not bound to the larger TBPA. (In free form, RBP is found in the α_1-globulin fraction.)

9.3.3 α_1-Globulins

α_1-Acid glycoprotein, also known as orosomucoid, is a glycoprotein with 42% carbohydrate content. Although much is known about its structure and genetic variations, its major biological role remains unclear. (One known function of this protein is its capacity to inactivate circulating progesterone.) It belongs to a group of plasma proteins called **acute phase reactants** because their plasma concentrations increase dramatically in response to injury or surgical trauma, rheumatoid arthritis, and some malignancies. (Other members of this group of proteins are found in the α_1- and β-globulin fractions.)

α_1-Antitrypsin (α_1-AT) is the major protease inhibitor in plasma. It is a glycoprotein with considerable medical significance. As an acute phase reactant, its normal plasma concentration can quadruple under stress conditions. α_1-AT's principal function appears to be the inhibition of elastase released by leukocytes when combating an infection. The inhibition is irreversible, and the adduct elastase–α_1-AT is degraded. α_1-AT contains a methionine residue at its active center. During the formation of an abscess, leukocytes release oxidizing compounds, such as the superoxide ion O^{2-}, which kill cells and at the same time inactivate α_1-AT by oxidizing its $-S-CH_3$ group to $-SO-CH_3$, thus allowing local tissue destruction. However, this process of tissue destruction by leukocyte elastase is limited to the center of the abscess because of the presence of active α_1-AT in the surrounding tissue.

α_1-Antitrypsin deficiency has been shown to be associated with **emphysema,** although not in all cases. Homozygotes are more likely to develop the disease than heterozygotes, who have about 60% of the normal α_1-AT levels. The mutant allele occurs with a frequency of about one in 20. This incidence is comparable to that of cystic fibrosis, one of the most frequent inborn metabolic disorders in Caucasians. Children with hepatic cirrhosis have α_1-AT deficiency. In these children, an α_1-antitrypsin–like material, possibly deficient in sialic acid residues, is deposited in the liver.

Transcortin is a cortisol-binding protein with a single site for cortisol or corticosterone.

Thyroxine-binding globulin (TBG) is present at much lower concentrations than thyroxine-binding prealbumin. However, it exhibits a much higher affinity for the hormone.

Transcobalamin I is a minor carrier for vitamin B_{12}.

α_1-Fetoglobulin (α_1-F) is found in the serum of the fetus. Its structure closely resembles that of albumin. It is normally absent from adult serum. It is found at low concentrations in the serum of pregnant women reaching much higher levels if the fetus is abnormal (e.g., Down's syndrome, spina bifida). The protein is also found in the plasma of a large percentage of patients with primary liver carcinoma, and in patients with germ cell tumors of the ovary or testis.

9.3.4 α_2-Globulins

α_2-Macroglobulin (α_2-M) is a very large glycoprotein (MW = 820,000), comprising about 33% of the α_2-globulin fraction. Little is known about its function. It may also be involved in the transport of some hormones and metal ions. (One quarter of the Zn^{2+} in plasma is found associated with α_2-M.) In addition, α_2-M appears to act as a "trap" for various proteases such as trypsin, plasmin, and thrombin. It may, therefore, also have a role in blood coagulation. The binding of α_2-M to proteases is irreversible and affects only the protease activity against large protcins, whereas lower molecular weight substrates are still degraded. This residual activity cannot be inhibited by macromolecular inhibitors (e.g., soybean trypsin inhibitor). The probable explanation for this mode of action is that the protease to which α_2-M binds cleaves a peptide bond within α_2-M, causing a conformational change of the molecule that leads to the irreversible trapping of the protease within α_2-M. Small substrate molecules can still reach the active protease inside the α_2-M cage. All four classes of endopeptidases bind to α_2-M and can be inactivated: serine proteases, thiol proteases, metal proteases, and carboxy proteases. The spent α_2-M protein is taken up and destroyed by the reticuloendothelial system.

C1s inhibitor specifically inhibits C1s, the activated form of the first component of the complement system. Genetic deficiency of this inhibitor is found only in heterozygous form, suggesting that the complete absence of C1s inhibitor is incompatible with life. Heterozygotes suffer from angioedema, a disorder characterized by transitory localized painless swelling of the face, hands, feet, and genitals or by severe visceral edema. Laryngeal edema is a frequent cause of death.

Haptoglobins (Hp) are a family of glycoproteins that specifically form complexes with hemoglobin. Upon hemolysis of old erythrocytes in the spleen and other organs of the reticuloendothelial system, the oxygenated tetrameric hemoglobin dissociates into two dimers: $\alpha_2\beta_2 \rightarrow 2\alpha\beta$. Because of their small size, these Hb dimers would be lost through the kidney, which would result in a loss of iron. To prevent this loss, as might occur under conditions of excessive hemolysis, the Hb is complexed to the large-molecular-weight haptoglobin. The hemoglobin-haptoglobin complex is then metabolized in the reticuloendothelial system. The haptoglobin molecule consists of two pairs of nonidentical chains connected by disulfide bonds. The larger β chains (MW = 42,600) are identical in all haptoglobins, whereas there are two different α chains. The larger α_2 chain appears to have evolved from a duplication of the gene for the α_1 chain. Combinations of different α chains with β chains produce three genetic variants of haptoglobin: Hp1-1, Hp2-1, and Hp2-2 (Figure 9.17). (The numbers indicate the types of α chains found in the tetramer.) Since the α_2 chain carries one more SH group than the α_1 chain, and thus can engage in additional intermolecular disulfide bond formation, Hp2-1 and Hp2-2 form a series of oligomers that produce multiple bands on electropherograms.

Each Hp tetramer binds (via its β chains) two halves ($\alpha\beta$) of a hemoglobin molecule. Upon binding to Hp, the normally rather weak peroxidase activity of hemoglobin is greatly enhanced. This phenomenon is being exploited in the analytical determination of haptoglobins.

Figure 9.17.
Structures of haptoglobins.

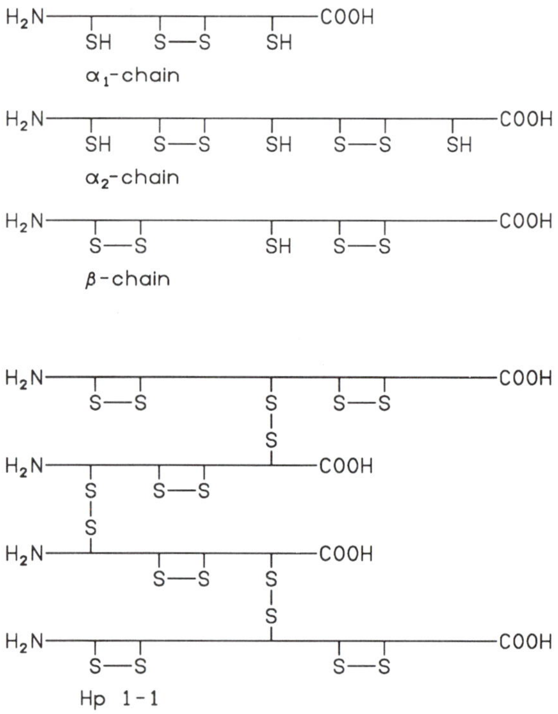

Levels of circulating haptoglobin are elevated in inflammation and reduced in hemolytic anemia. Very low, or absent, levels of haptoglobin are diagnostic of accelerated red cell destruction.

Ceruloplasmin, a blue glycoprotein, is the major plasma carrier of Cu^{2+} ions. Each molecule can carry eight Cu ions (four Cu^{2+} and four Cu^{+}). Ceruloplasmin's activity as a ferroxidase in the oxidation of Fe^{2+} to Fe^{3+} (see Section 8.4) appears to be only incidental. In Wilson's disease or hepatolenticular degeneration, genetic disturbances of Cu metabolism, the plasma levels of ceruloplasmin are greatly reduced or absent. This is associated with abnormal copper deposits in liver, brain, and cornea.

9.3.5 β-Globulins

Transferrin, the specific iron-binding protein, is described in Section 8.4. Its level in plasma decreases in many acute and chronic diseases but rises in pregnancy and chronic iron deficiency.

Hemopexin binds free heme. In this function, it competes with albumin but has a higher affinity for heme. Levels of hemopexin are reduced in hemolytic diseases. Hemopexin delivers the heme to the liver, where specific receptors recognize the hemopexin-heme complex.

Transcobalamin II is the major carrier of vitamin B_{12}. It binds to the vitamin in the ileum and transports it to the tissues. It is required for cellular uptake of the vitamin.

β_2-Microglobulin is a very small protein (MW = 12,000) that is excreted in the urine. It represents the light chain of the class I major histocompatibility complex (see Figure 9.27) and is homologous to the immunoglobulins.

C-reactive protein (CRP), so named because it precipitates pneumococcal C polysaccharide, is normally present at very low concentrations. It increases significantly after acute infections, binding to the membranes of microorganisms that contain phosphoryl choline and in the form of such a complex activating the complement system.

9.3.6 γ-Globulins

Immunoglobulins (antibodies) make up the large majority of γ-globulins. Their structure and biosynthesis are discussed in the next section. **Cryoglobulins** are serum proteins that become insoluble and form gels or crystals when the temperature is lowered. One of the cryoglobulins is **fibronectin,** a protein secreted by fibroblasts. Fibronectin may serve as a scaffold for the formation of connective tissue.

9.4 THE IMMUNE RESPONSE

When a foreign body or substance invades the human body, several defense mechanisms are initiated. One of these defense mechanisms is the immune response. There are two basic types of immune response, the **humoral** and the **cell-mediated immune response.** Both responses are brought about by circulating lymphocytes, the **B lymphocytes,** which develop in bone marrow and fetal liver, and the **T lymphocytes,** which originate in the thymus.

In the humoral response, differentiated B lymphocytes, called **plasma cells,** synthesize and release large quantities of **immunoglobulins (Igs)** or **humoral antibodies** that specifically recognize and bind to the invading **antigen.** Antibodies form complexes with the antigen and thereby inactivate it. By binding to the surface of bacterial cells, they mark these for subsequent destruction by phagocytosis.

Cytotoxic T lymphocytes are responsible for the cell-mediated immune response. They recognize foreign antigens on the surface of the body's own cells and then kill such cells. This mechanism is mainly responsible for stopping viral infections.

Only macromolecules (proteins, polysaccharides, nucleic acids) are recognized as **immunogens** and as such are capable of triggering an immune response. Small molecules, called **haptens** (half-antigens), are not immunogenic by themselves but become so when linked to a macromolecule. There they form the antigenic determinant against which the immune response is directed.

The human body contains about 2×10^{12} lymphocytes, each carrying on its surface identical copies of a single type of immunoglobulin molecule capable of binding to one **epitope** (antigenic determinant of an antigen). There are literally millions of different clones of lymphocytes with antibodies on their surfaces that can bind to any conceivable antigen. When an antigen combines with the surface immunoglobulin of a **B lymphocyte**, it causes this cell to divide and multiply into a large clone of lymphocytes, all of which are capable of synthesizing the same type of antibody directed against the antigen. After a certain number of divisions, the stimulated B cells differentiate terminally into **plasma cells,** which then begin to secrete large quantities of the immunoglobulin that was originally carried on the surface of their B lymphocyte precursors. This mechanism of **clonal selection** of antibody-producing cells by the antigen was originally proposed by Sir MacFarlane Burnet.

Actually, the process is a little more complicated, since T lymphocytes also participate in this clonal stimulation of B lymphocytes. There are three types of T lymphocytes (Figure 9.18). The **cytotoxic T lymphocytes** are responsible for the cell-mediated immune response and will be discussed later. The large majority of the T lymphocytes, on the other hand, are regulatory in nature. These are the **helper T lymphocytes** and **suppressor T lymphocytes.** Their participation in the humoral immune response will be discussed in the section on cell-mediated immunity.

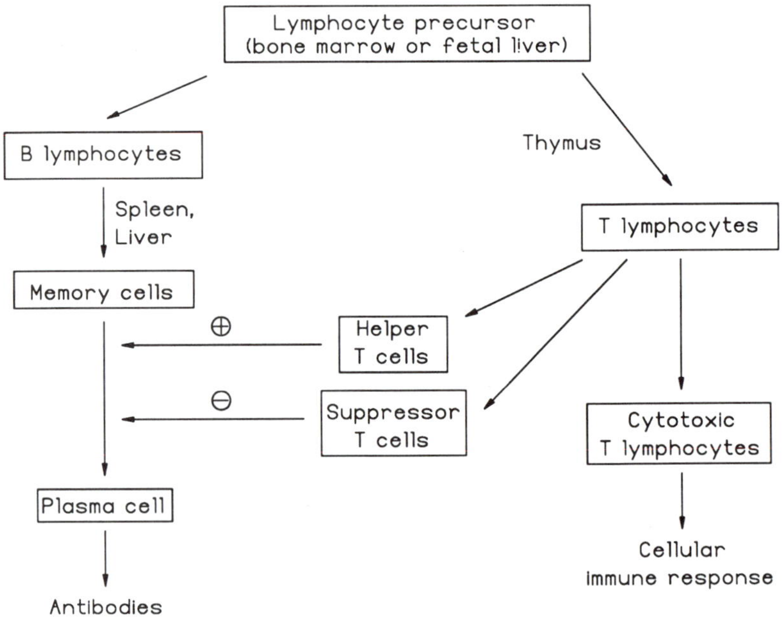

Figure 9.18.
Origin and differentiation of lymphocytes.

9.4.1 B Lymphocytes and Antibody Synthesis

When a mature virgin B cell (that which has IgM and IgD on its surface but which has not yet seen antigen) is exposed to an antigen, it divides and differentiates into both active B cells (plasma cells), which secrete antibody, and **memory cells.** These memory cells respond much more quickly to a challenge with the same antigen that does a virgin cell. Therefore, the secondary antibody response is much quicker than the first one. It is also stronger, because there are more memory cells of a specific clone than virgin cells and the antibodies carried on their surface exhibit a higher affinity to the antigen than those found on virgin cells (see class switching below).

Structure of Immunoglobulins

Isolation of large enough quantities of identical immunoglobulin molecules for structural analysis and amino acid sequence determination first became possible when it was realized that **multiple myeloma** is a neoplasm of an antibody-producing plasma cell. Cells isolated from such tumors represent a single clone of plasma cells, all producing the same immunoglobulin. Some patients with this disease excrete in their urine large quantities of immunoglobulin light chains **(Bence Jones proteins).**

Another source for pure antibodies in quantities large enough for structural analysis are the **hybridoma cells.** Hybridoma cells are generated when antibody-producing B lymphocytes are fused with cells derived from a B cell tumor. Since the tumor cells are immortal, the hybrid cells continue to divide but also continue to secrete antibody. Thus, large quantities of **monoclonal antibodies** can be obtained from hybridoma cultures.

An immunoglobulin molecule consists of four polypeptide chains, two light chains (L) and two heavy chains (H), linked together by disulfide bonds (Figure 9.19). Each chain can be divided into a **variable region (V),** in which the amino acid sequence varies considerably from one Ig molecule to another, and a **constant region (C),** which shows the same amino acid sequence in different Ig molecules. It is the variable regions of both heavy and light chains that form the antigen-binding site of the Ig molecule. The constant region interacts with other molecules, such as the complement system, or with mast cells or macrophages, thereby exerting the effector function of the antibody. Amino acid analysis indicates that the constant region of the heavy chain consists of three domains (C_H1, C_H2, C_H3) of similar sequence that appear to have evolved from a single ancestral gene. (The μ and ϵ chains are larger and contain four similar domains in their constant regions.)

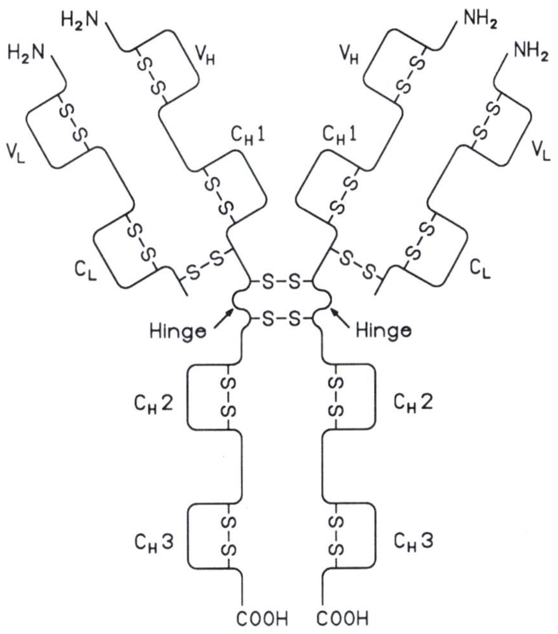

Figure 9.19. Immunoglobulin structure.

The protease **papain** cleaves the Ig molecule into an **Fc fragment** and two **Fab fragments.** Only the Fab fragments bind antigen (Figure 9.20).

There are five major classes of immunoglobulins: **IgM, IgD, IgG, IgA,** and **IgE,** containing μ, δ, γ, α, and ϵ heavy chains, respectively. Each class may contain either two κ light chains or two λ light chains (Table 9.3).

The first antibody that appears in the serum in response to antigen stimulation is IgM. IgMs are pentamers of four-chain subunits linked together via their Fc segments to form rosettelike structures. Such structures are closed by the covalent insertion of an additional protein called **J chain** between two adjacent Fc segments. Each IgM thus contains 10 antigen-binding sites that allow it to form large insoluble complexes with antigen.

IgM is soon replaced by IgG, the most prevalent type of antibody in serum. Interestingly, this switch from IgM to IgG affects only the constant segments of the heavy immunoglobulin chains. The variable segments remain unchanged, thus allowing IgG to bind to the same antigen as the original IgM. IgA is the major antibody present in body secretions, such as saliva, tears, bronchial mucus, and intestinal secretions. Reactions between antigens and IgE, bound to the surface of mast cells, are responsible for allergic reactions. IgD is found almost exclusively bound to the surface of B lymphocytes.

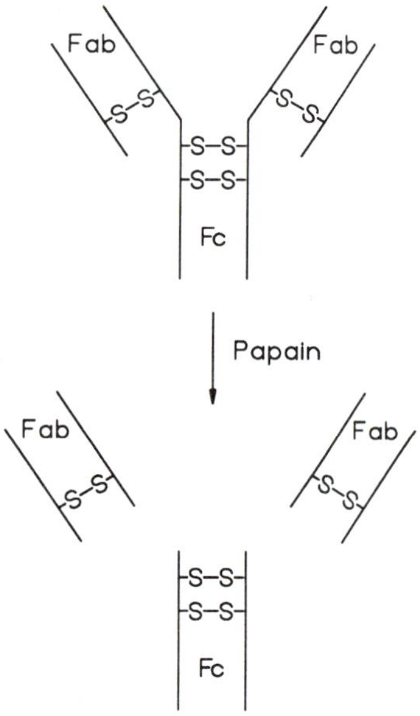

Figure 9.20.
Proteolytic cleavage of immunoglobulin by papain.

Biosynthesis of Immunoglobulins

The number of different antigenic determinants against which specific antibodies can be synthesized is almost limitless. Since the human genome is limited in size, the question arises how a limited number of structural genes can code for an almost infinite number of variable region sequences. The answer came with the discovery that variable and constant regions of the same Ig molecule are encoded by different structural gene segments, the V and the C gene segments. Furthermore, there are multiple copies of the V gene segment, and the joining of the V and C segments is subject to somatic mutational and recombinational alterations, thus generating a very large number of different somatic genes, all derived from a limited number of germ-line genes. The sequence of the C segment of an immunoglobulin is not subject to such alterations. The next question arising from these observations was how a variable region is connected to a constant region, and how the switch from one class of Ig to another class of Ig is accomplished. The answer came with the advent of recombinant gene technology.

In embryonic cells, the V_L **gene segment,** coding for amino acids

Table 9.3. Classes of Human Immunoglobulins

Class	Chain Composition	Molecular Weight	Serum Concentration (mg/100 ml)	Remarks
IgG	$\lambda_2\gamma_2$ or $\kappa_2\gamma_2$	[illegible]50,000–300,000	600–1700	Crosses placenta
IgA	$(\lambda_2\alpha_2)_n$ or $(\kappa_2\alpha_2)_n$		140–420	n = 1, 2, or occasionally 4
IgM	$(\lambda_2\mu_2)_5$ or $(\kappa_2\mu_2)_5$	950,000	50–190	Strongest activator of complement
IgD	$\lambda_2\delta_2$ or $\kappa_2\delta_2$	160,000	3–40	Present on cell surface, involved in B cell activation
IgE	$\lambda_2\epsilon_2$ or $\kappa_2\epsilon_2$	200,000	0.01–0.14	Binds to mast cells

1–98 of the light chain, is located on the chromosome at a site distant from the C_L gene segment. During differentiation into an antibody-producing cell, a recombinational event brings the V_L gene segment into direct contact with one of several **J segments** (J for joining). The J segment codes for amino acids 99–112 of the variable region and is located about 1–2 kb upstream (5′) of the **C_L gene.** This chromosomal rearrangement activates the immunoglobulin light chain promoter that is located just upstream of the V_L segment because of the influence of a powerful enhancer element, located downstream of the J segments.

Transcription of the rearrangement immunoglobulin gene begins with a short leader sequence, followed by a short intron, then the V_L segment, the J segment, another intron (which may include additional J segments that did not recombine with the V_L segment), and finally the C_L gene segment. After processing, which includes the removal of the two introns by RNA splicing, the mRNA is exported into the cytoplasm and translated into a complete immunoglobulin light chain. The leader peptide is removed in the endoplasmic reticulum before export from the cell (Figure 9.21).

The genes involved in the synthesis of mouse κ light chains have been analyzed. They include approximately 300 different V_L gene segments, four J gene segments, and one C_κ gene segment. This allows for 1200 different combinations to form a functional light chain. However, the actual number is probably about 15 times larger, since the recombinational events that occur during this immunoglobulin gene assembly are not precise. Often nucleotides are lost or gained, causing **junctional diversification** at one of the **hypervariable regions** of the immunoglobulin chain.

The organization of the gene segments for the heavy chains is similar although slightly more complex. In addition to V_H, J_H, and C_H segments, there are D_H segments (D for diversity) located between V_H and J_H (Figure 9.22). In the mouse, a pool of about 1000 V segments can be joined to 12 D segments and four J segments, allowing for 48,000 dif-

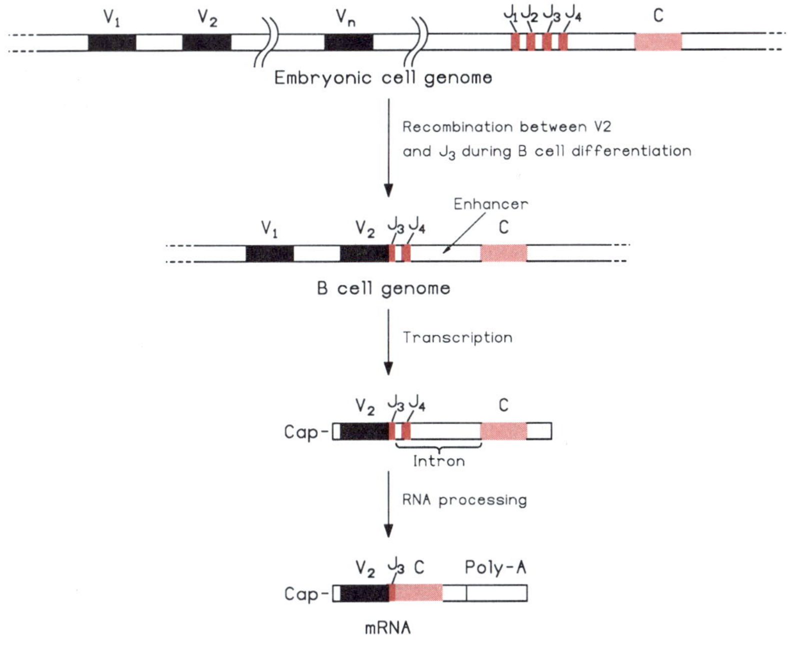

Figure 9.21.
Differentiation and expression of immunoglobulin genes (example: mouse κ light chain).

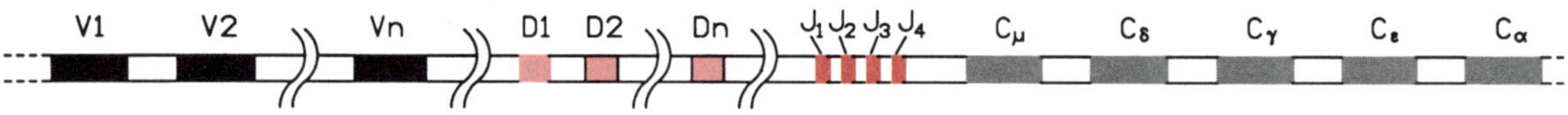

Figure 9.22.
Gene arrangement for the immunoglobulin heavy chain.

ferent heavy chain combinations. Again, junctional diversification brings this figure to about 10^6. Thus, the total number of possible mouse Ig combinations may be $10^4 \times 10^6 = 10^{10}$, considerably larger than the total number of lymphocytes in the mouse.

Since antigen provides a selective pressure for the replication of B cells with better binding antibodies, the lymphocytes that produce antibodies in response to secondary immunizations often acquire somatic mutations in the variable regions of their Ig genes, allowing for the synthesis of even better fitting antibodies. **Somatic hypermutation,** selected for by antigen, leads to the accumulation of immunoglobulins with mutations in the three hypervariable regions (the regions that bind antigen). This somatic hypermutation is therefore the chief contributor to **affinity maturation.**

As mentioned before, the first antibody synthesized by a B cell is IgM. It is anchored in the lymphocyte cell membrane by its hydrophobic carboxy-terminal segment. It is this membrane-bound antibody that is first recognized by the antigen. The switch from making membrane-bound IgM to secreting IgM is accomplished by differential splicing of the pre-mRNA (Figure 9.23). Depending upon the site of polyadenylation of the precursor RNA, either the membrane-bound or secreted form of the antibody will be produced. If the RNA is polyadenylated at the first site, then no acceptor splice site (further downstream) will be available, and as a consequence secreted IgM will be formed. The first stop codon comes at the end of a hydrophilic amino acid sequence. As the B cell matures into a plasma cell, a second form of secreted/membrane-bound regulation takes place. Now transcription is terminated immediately 3′ to the first stop codon. In both cases, translation of the hydrophobic carboxy terminus, which normally anchors the protein in the cell membrane, is prevented.

At later stages, as the B cell matures into a memory cell or a plasma

Figure 9.23.
Differential splicing of IgM pre-mRNA to produce either membrane-bound or secreted protein.

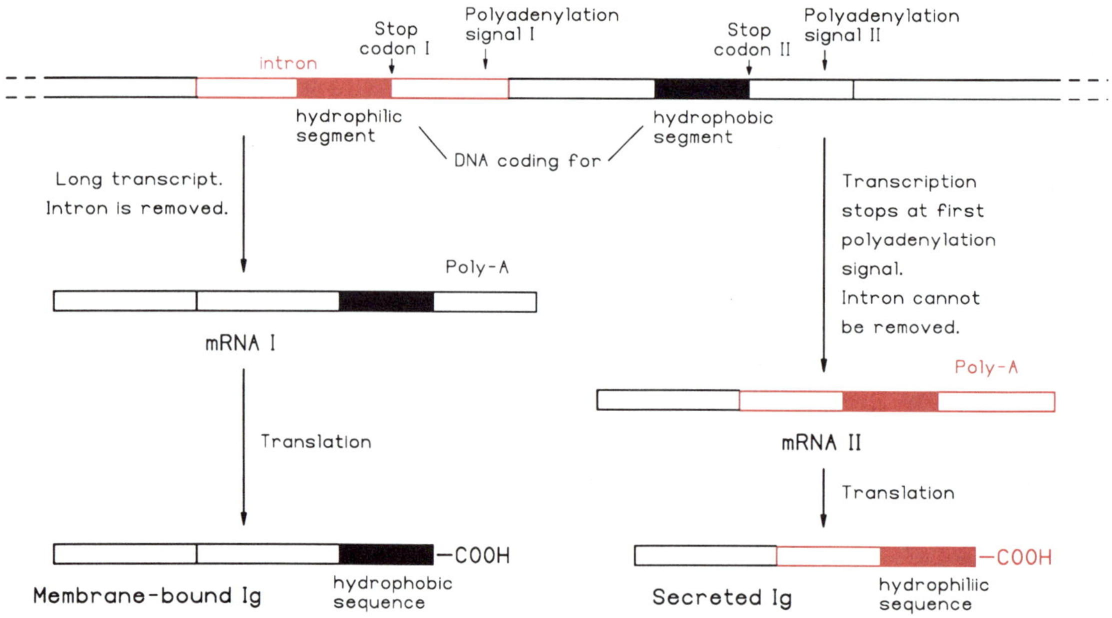

cell, it switches to the synthesis of other immunoglobulins carrying different constant regions in their heavy chains. (The variable regions remain unchanged.) This **class switching** also involves DNA recombination and, in the mature plasma cell, is accomplished by the deletion of a DNA segment, bringing the J_H region near a C_γ (there are actually four C_γ regions), C_ϵ, or C_α segment, all of which are located downstream from C_μ. The recombinational event occurs within a special **switching region** that precedes every constant region with the exception of the C_δ gene. (The switch from IgM to IgD is accomplished by alternative RNA splicing.) As an example, the switch from IgM synthesis to IgG synthesis is illustrated in Figure 9.24.

Figure 9.24.
Example of Ig class switching.

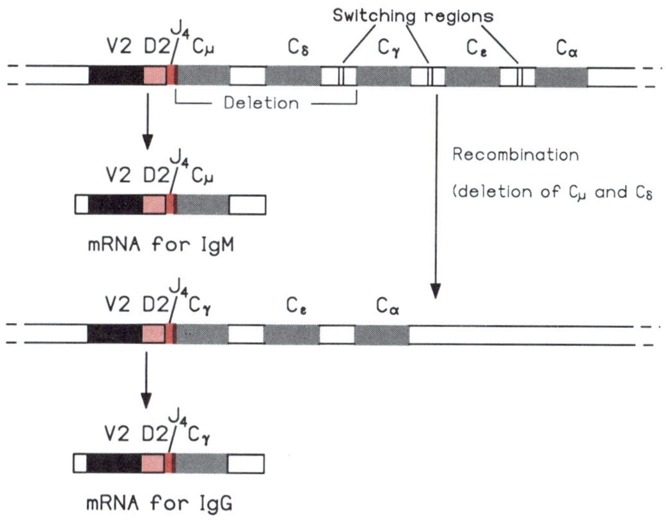

As mentioned before, each plasma cell produces only a single type of antibody. Since the cell is diploid, only one of the two parental alleles can be expressed. This **allelic exclusion** is most likely due to the fact that the presence of a completed immunoglobulin on the cell surface feedback inhibits Ig recombination.

The antibodies that are generated in response to an invading antigen are themselves antigenic. Their antigenic determinants lie in the variable regions of their H and L chains. The antibodies produced against these epitopes **(idiotypes)** are called **anti-idiotype antibodies.** In turn, the anti-idiotypic antibodies induce the generation of anti-anti-idiotype antibodies, and so on. It has been proposed that the immune response to a single antigen is not limited to the production of specific antibodies but represents a perturbation of a complex **network** of interreacting lymphocytes and antibodies.

9.4.2 T Lymphocytes and Cell-Mediated Immunity

There are three major classes of T cells: the regulatory **suppressor T cells** and **helper T cells** and the **cytotoxic T cells,** which can kill virus-infected cells directly. T cells are activated by direct contact with the surface of another cell, where they bind to a complex of foreign

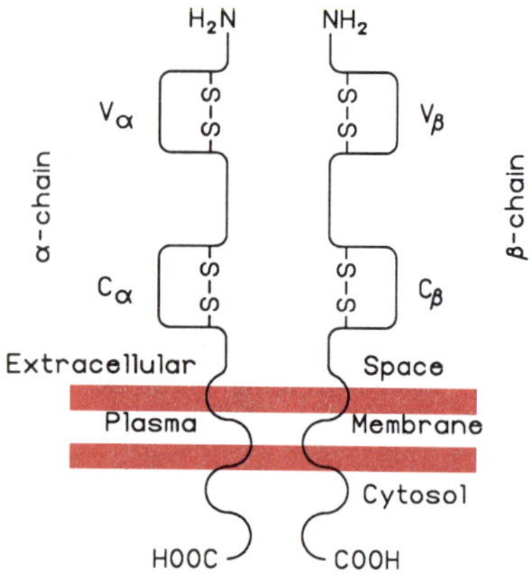

Figure 9.25.
The structure of the T cell receptor.

antigen and cellular glycoprotein called the **major histocompatibility complex (MHC) protein.** The **T cell receptor** that recognizes these antigen-MHC complexes is an immunoglobulinlike molecule anchored in the T cell membrane. It is a heterodimer consisting of an α chain and a β chain. The genes for the T cell receptor chains are located on different chromosomes and consist of separate V, D, J, and C segments that must be brought together by a recombinational event during the maturation of the T cell. Thus, a very large pool of different T cell receptors can be formed. In contrast to immunoglobulins, T cell receptors are **monovalent** (Figure 9.25).

In T cells of the gut, epidermis, and thymus, another type of receptor, containing γ and δ chains, is found. Its significance is not known. All T cell receptors are found associated with the same group of membrane proteins, called the **T3** or **CD3 complex,** which mediate signal transduction of the antigen to the cell.

The mechanism by which cytotoxic T cells kill their target cells is not known. They are activated by binding to viral proteins associated with MHC proteins on the virus-infected cell. In tissue culture, some cytotoxic T cell lines have been shown to release a protein called **perforin** whose structure is related to the complement component C9 (see below). Perforin molecules assemble in the membrane of the target cell, forming a transmembrane channel through which cytoplasmic components of the cell leak out, thereby killing the cell.

The Structure and Function of MHC Proteins

The mechanism of T cell activation involves interaction with the proteins of the histocompatibility complex. The presence of these proteins in the cell membranes of all cells is responsible for the **graft rejection** that often occurs when tissue is transplanted from one individual to another. In humans, the MHC proteins are called **HLA-antigens** (human leukocyte–associated antigens; Figure 9.26).

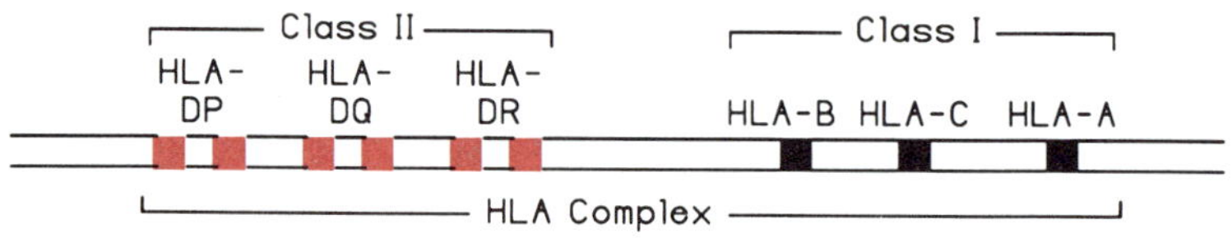

Figure 9.26.
The human HLA-A complex on chromosome 6.

One distinguishes two classes of MHC molecules: the **class I** and the **class II MHC molecules.** The genes for the MHC proteins in humans are located in a cluster on chromosome 6. They are highly **polymorphic,** which means that there exist many (>100) different alleles in the population. The class I MHC glycoprotein consists of a transmembrane α chain with three extracellular domains and an extracellular β_2-microglobulin chain. The class II MHC glycoprotein also contains an α and a β chain, both of which span the cell membrane (Figure 9.27).

Class I MHC molecules are found on the membranes of essentially all cells. In combination with a foreign antigen, they function in the activation of cytotoxic T cells. On the other hand, class II MHC molecules are limited to the membranes of cells of the immune system; they are recognized by helper T cells and play a role in their activation.

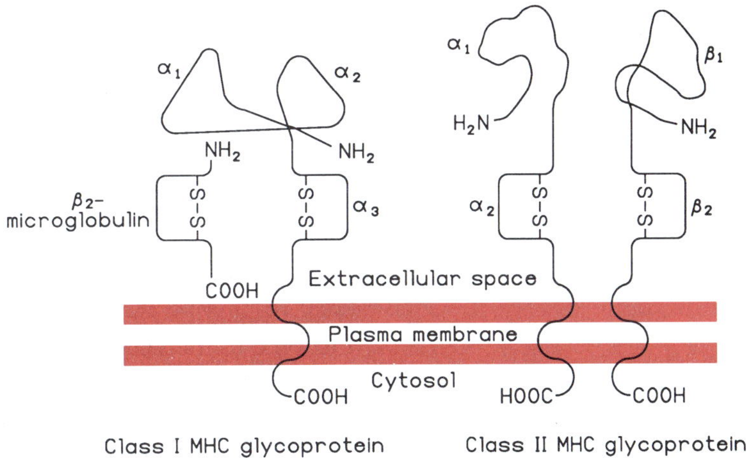

Figure 9.27.
The structures of class I and class II MHC proteins.

The Mechanism of T Cell Activation

In contrast to B cells, which bind complete antigenic molecules via their membrane-bound immunoglobulin molecules, **cytotoxic T cells** recognize only short fragments of foreign antigens (10–20 amino acids long), for example, a peptide from a virus protein, on the surface of a cell. The original antigenic molecule appears to be degraded inside the cell, and the degradation products are then somehow transported to the cell surface. However, the antigenic peptide is only recognized if the cell expresses the same **class I MHC glycoproteins** as the T cells. The peptide appears to be complexed in a groove between the α_1 and α_2 domains of the MHC molecule.

Helper T cells recognize degradation products of foreign antigens that are complexed to **class II MHC glycoproteins** on the surface of specialized **antigen-presenting cells.** Such cells include B cells, thymus epithelial cells, dendritic cells of lymphoid tissues, Langerhans cells in the skin, and macrophages. Only cells expressing class II MHC glycoproteins on their surface can function as antigen-presenting cells. The process of helper T cell activation is not completely understood. It involves the release of several kinds of **interleukins,** locally acting growth factors. In response to the interaction with a helper T cell, the antigen-presenting cell releases **interleukin-1 (IL-1)** which, in turn, stimulates the T cell to secrete **interleukin-2 (IL-2)** and, at the same time, express on its surface IL-2 receptors. In response to this autocrine stimulation by interleukin-2, the T cell proliferates.

The mechanism by which activated helper T cells cause the proliferation and maturation of antibody-secreting B cells is similar. After the antigen is bound to the membrane-bound antibody on the surface of the B cell, the antibody-antigen complex is ingested by endocytosis. Inside the cell, the antigen is degraded into shorter peptides that are recycled to the cell surface bound to class II MHC molecules. (These are the same complexes that the T cell originally recognized on the surface of antigen-presenting cells.) The interaction between helper T cell and B cell causes the T cell to release several other interleukins. IL-4 initiates B cell activation, IL-5 stimulates B cell proliferation, and IL-6 causes maturation of a B cell into a plasma cell.

Interleukins, also called **lymphokines,** belong to the class of **cytokines,** proteins and glycoproteins produced by activated T cells and macrophages that mediate immune and inflammatory responses. The list of cytokines also includes the interferons (see Section 7.8) and the **tumor necrosis factors TNF-α** and **TNF-β.** TNF-α is produced by activated mononuclear phagocytes and causes necrosis of certain tumors when injected into tumor-bearing animals. It has general antiproliferative effects and is responsible for weight loss, fever, and acute phase reaction that is associated with neoplasia. For this reason TNF-α is also called **cachectin.** TNF-α and TNF-β are homologous proteins that bind to the same cellular receptor. Table 9.4 lists the functions of the known cytokines.

The interaction between antigen-MHC complexes and T cell receptors is strengthened by the participation of **adhesion proteins.** These are glycoproteins whose extracellular domains show homology to similar domains in immunoglobulins. The best known T cell adhesion proteins are **CD4,** found on regulatory T cells binding to class II MHC proteins, and **CD8,** present on cytotoxic T cells which must bind to cells expressing class I MHC proteins.

Little is known about the mechanism of action of suppressor T cells, and in fact there is some question as to whether they are a separate T cell lineage or not. It has been proposed that they act indirectly by suppressing helper T cells. It has also been suggested that they participate in the phenomenon of **immunological tolerance,** which is the body's ability to "tolerate" antigenic determinants on its own proteins without mounting an immune response. The mechanism of self-tolerance is not completely understood. However, it appears that most tolerance is based on the elimination of those lymphocyte clones early during development that express antibodies against the body's own proteins. Perhaps suppressor T cells play a secondary role in this self-tolerance.

9.4.3 The Complement System

When an antibody binds to the antigenic determinants on the surface of a cell, such as a bacterium, the cell is eventually lysed. The sequence of events that leads to this lysis is catalyzed by a group of plasma proteins called the **complement system.** It is composed of serum proteins that are sequentially activated by a cascade of proteolytic reactions. The central component of the complement system is the **C3 protein.** When this protein is cleaved by **C3 convertase,** one of the proteolytic products (C3b) catalyzes the assembly of proteins C5, C6, C7, C8, and C9 into a **membrane attack complex** that creates a hole in the target cell membrane.

The activation of C3 can occur by two different pathways: the **classical pathway** and the **alternative pathway.** The components of the classical pathway, C1, C2 and C4, are activated by antibodies bound to the surface of the invading cell. The alternative pathway, containing factors B and D, on the other hand, can be activated directly by polysaccharides located in the bacterial cell wall. The story is slightly more complex because the two pathways are interconnected.

Table 9.4. Functions of Some Cytokines

Interleukin	Other Name	Source	Target	Action
IL-1	—	Antigen-presenting cell	Helper T cells	Activates T cell
IL-2	T cell growth factor	T_H 1 helper cells	Activated T cells and B cells	Stimulates B and T cell proliferation
IL-3	Colony-stimulating factor	T_H 1 cells	Several hematopoietic cells	Stimulates proliferation
IL-4	B cell stimulating factor (BSF-1)	T_H 2 cells	B cells, T cells, mast cells	Helps activation of B cells, increases class II MHC molecules on B cells, induces IgE synthesis by B cells
IL-5	B cell growth factor 2 (BCGF-2)	T_H 2 cells that also make IL-4	B cells, eosinophils	Promotes proliferation and maturation and IgA secretion (in mouse)
IL-6	B cell stimulating factor 2 (BSF-2)	Helper T cells and macrophages	Activated B cells, T cells	Promotes B cell maturation into plasma cell
γ-Interferon	—	T cells that make IL-2	B cells, macrophages, endothelial cells	Induces class II MHC molecules and activates macrophages
IL-7	—	Bone marrow, stroma	Pre-B cells, thymocytes, immature T cells	Stimulates proliferation of pre-B cells, thymocytes, and T cells
IL-8	—	T cells, monocytes, endothelial cell	Neutrophils	Neutrophil degranulation and chemotaxis (a proinflammatory mediator)
IL-9	P40	T cells	T helper cells, mast cells	T-helper cell growth (antigen independent)
IL-10	Cytokine synthesis inhibitory factor (CSIF), B cell–derived T cell growth factor (BTCGF)	B cells	T_H 1 cells, mature and immature thymocytes	Inhibits release of γ-interferon from T_H 1, does not inhibit proliferation, increases proliferation of thymocytes in presence of IL-2 and IL-4
TGF-β	Transforming growth factor β	Endothelial cells	B and T cells, endothelial cells	Inhibits proliferation of epithelial, endothelial, T and B cells; stimulates IgA production, chemotactic for fibroblasts, growth-inhibiting for most other cells
TNF-α	Tumor necrosis factor α	Macrophage	Thymocytes, B cells, T cells	Increases proliferation of thymocytes (in presence of IL-2); inhibits cell-mediated responses

9.5 HEMOSTASIS, BLOOD COAGULATION, AND FIBRINOLYSIS

Upon injury to a blood vessel, the initial loss of blood is quickly stemmed by the formation of a mechanical plug consisting of **platelets.** This primary hemostasis is followed by coagulation, that is, the formation of a **fibrin clot,** which reinforces and stabilizes the original plug (Figure 9.28). The reactions that trigger the clotting mechanism must be limited to the site of injury. This is accomplished by the presence in plasma of inhibitors of the clotting factors as well as by feedback inhibition of the sequence of clotting reactions. Finally, after repair of the injured vessel, the fibrin clot is dissolved by the proteolytic enzyme **plasmin.**

9.5.1 Hemostasis

Primary hemostasis is achieved by the formation of a **platelet** plug. Platelets are very small cells without nuclei. They are derived from **megakaryocytes,** large polyploid cells that form long filopodia (2.5×120 μm) that invade the capillaries in the marrow where they are fragmented into mature platelets. Normally, blood contains about $2–3 \times 10^5$ platelets/mm^3 (compared with 5×10^6 red cells/mm^3).

Platelets contain many canaliculi that are continuous with their plasma membrane. This **open canalicular system** gives the platelet cell a huge membrane surface. Platelets contain mitochondria and three types of **secretory granules:** the α and the dense granules and the lysosomes. α **granules** contain the coagulation factor V, von Willebrand factor, and fibrinogen as well as platelet-derived growth factor (PDGF) and platelet factor 4. The **dense granules** contain Ca^{2+} and Mg^{2+} ions, ATP, adenosine diphosphate (ADP), and serotonin.

When the vascular endothelium is damaged and the underlying **collagen** fibers are exposed, platelets adhere to these fibers. This adherence is dependent on the presence of **von Willebrand factor (vWF).** It is thought that vWF first binds to collagen and thereby changes its configuration so that it can now bind to the vWF receptors on the platelet membranes. Congenital lack of vWF **(von Willebrand disease)**

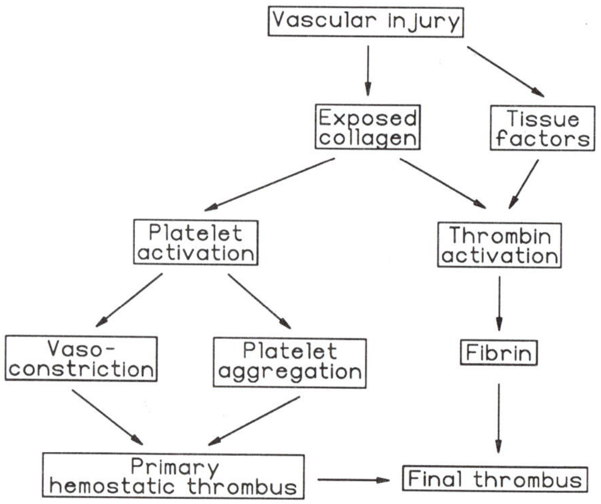

Figure 9.28.
Overview of blood clot formation.

or platelet receptor **(Bernard-Soulier syndrome)** causes spontaneous mucocutaneous bleeding.

Von Willebrand factor is a 250 kd single-chain glycoprotein with a tendency to form multimers (via disulfide bonds). It is synthesized by endothelial cells and megakaryocytes. About 15% of the circulating vWF is bound to the surface of platelets. Levels of vWF are increased under conditions of physiologic stress. vWF has two main functions: it is a carrier of factor VIII and it promotes platelet adhesion to collagen.

After adhering to collagen, the platelets release **ADP** and **thromboxane A_2,** which leads to the production of **thrombin** (see later). Both ADP and thrombin activate other platelets causing them to adhere to each other **(platelet aggregation),** thereby forming a plug at the site of injury. The release of Ca^{2+} ions during platelet activation causes phosphorylation of myosin light chains by protein kinase C. This causes contraction of the platelet. Furthermore, a membrane phospholipase is activated that cleaves arachidonic acid from platelet membrane phospholipids. Another platelet enzyme, **cyclooxygenase,** converts arachidonic acid to a cyclic peroxide, which is further transformed into thromboxane A_2. Thromboxane A_2 is a powerful vasoconstrictor (see Chapter 13) that causes constriction of the vessel at the site of injury. Aspirin, an inhibitor of cyclooxygenase, interferes with the formation of the platelet plug.

Endothelial cells produce **prostacyclin (PGI_2),** which raises intracellular cAMP levels and thereby tends to inhibit platelet activation. Intact endothelial cells also inhibit thrombosis via their cell surface protein **thrombomodulin.** This protein binds thrombin, and the complex activates **protein C** to become a serine protease that cleaves coagulation factor Va, which is required for the activation of prothrombin to thrombin (see later).

Primary hemostasis is independent of the coagulation mechanism. Its efficiency is measured by the **bleeding time.** Normally, bleeding from a 1 mm deep, 3 mm long cut in the forearm stops within 6 minutes. Bleeding time is increased in **thrombocytopenia** (decreased number of platelets), **von Willebrand disease,** and some disorders of platelet function.

9.5.2 Blood Coagulation

Blood coagulation is a cascade mechanism in which a small stimulus is amplified, through a series of reactions of increasing size and intensity, into a very large response. There are two separate pathways that lead to the formation of **thrombin,** the enzyme that converts **fibrinogen** into **fibrin** and thus causes blood coagulation:

1. The **intrinsic pathway,** which is activated by the exposure of blood to the subendothelial tissue at the site of vascular injury
2. The **extrinsic pathway,** which depends on a tissue factor for activation.

Figure 9.29 illustrates the series of reactions that lead to the formation of a fibrin network at the site of vascular injury.

The **intrinsic pathway** consists of four reaction sequences that lead

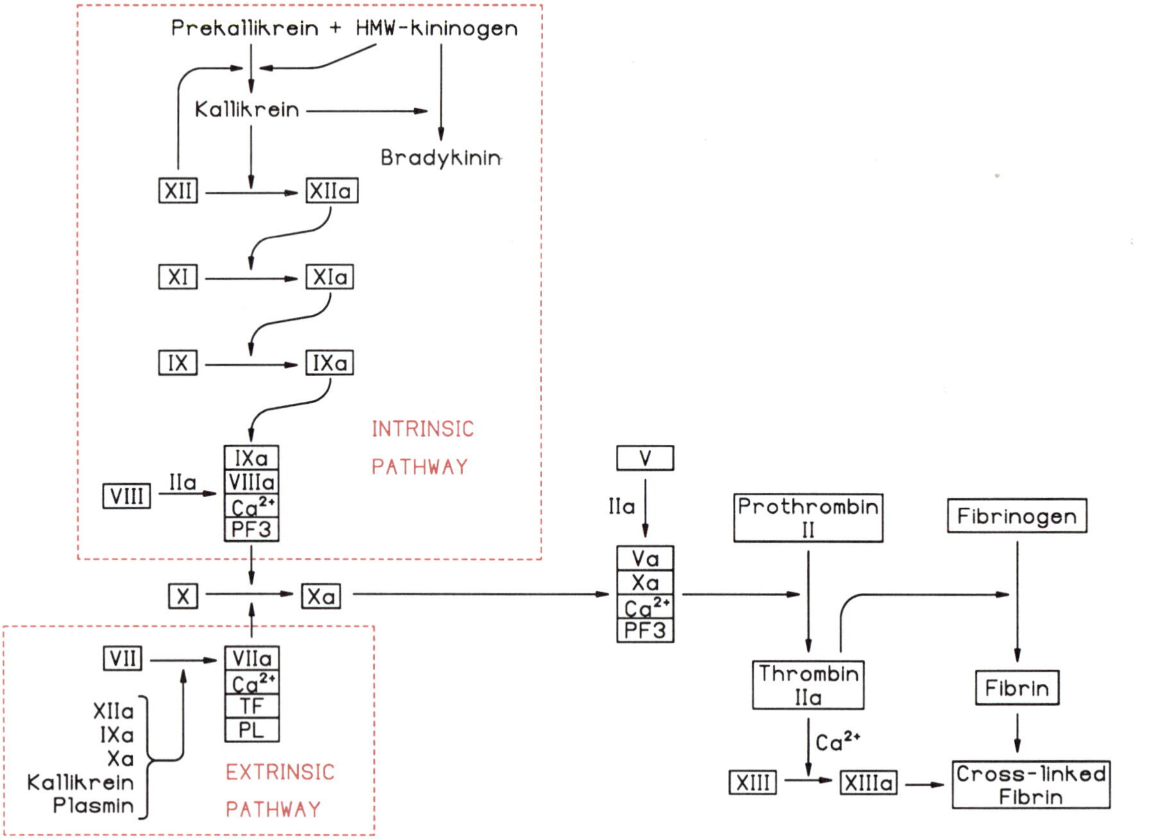

Figure 9.29.
Overview of the intrinsic and extrinsic coagulation pathways.

to the activation of factor X, the enzyme required for the formation of thrombin:

1. The plasma **factor XII,** also called the **Hageman factor,** binds to exposed subendothelial structures. This interaction triggers a conformational change that allows the surface-bound factor XII to convert **prekallikrein** to the active proteolytic enzyme **kallikrein.** Prekallikrein circulates in the plasma as a complex with **high molecular weight kininogen (HMW kininogen),** a cofactor required in the activation reaction. In a feedback reaction, kallikrein then cleaves factor XII into an active 28 kd serine protease, called **factor XIIa.** (This reaction also requires the participation of HMW kininogen.) There are two side reactions that occur after the activation of factor XII: first, factor XIIa participates in the activation of plasminogen to plasmin (see Figure 9.34), the enzyme that degrades fibrin; second, kallikrein cleaves HMW kininogen to yield **bradykinin,** a powerful vasodilator.
2. Factor XIIa then activates **factor XI** to **XIa.** Factor XI also circulates in a complex with HMW kininogen, and its activation is stimulated by HMW kininogen.
3. Factor XIa, still surface-bound, converts **factor IX (Christmas factor)** to **IXa.** This two-step reaction requires Ca^{2+} ions.
4. Factor IXa, together with Ca^{2+}, phospholipid (PL), and another cofactor, **factor VIIIa,** cleaves **factor X (Stuart factor)** to **Xa.**

Factor VIII (antihemophilic factor) is a very large protein ($MW = 1.5\text{–}2 \times 10^6$) composed of several identical subunits. Thrombin (factor IIa) initially potentiates the activity of factor VIII but, upon longer exposure, inactivates it. The phospholipid required for the activation of factor X is also called **platelet factor 3 (PF-3).** It is provided by the surface membrane of platelets that have contracted and are aggregated. Circulating platelets do not possess the appropriate membrane configuration.

The **extrinsic pathway** consists of **factor VII (proconvertin),** a circulating glycoprotein, which is activated after interaction with **tissue factor (TF, thromboplastin),** a glycoprotein present on the membranes of endothelial and other cells. The tissue factor must be complexed with a phospholipid to be able to function as cofactor for factor VII. Possibly, factor VII forms a complex with the tissue factor and phospholipid via its γ-carboxyglutamate residues and Ca^{2+} ions. The single-polypeptide factor VII protein is only partially active when bound to tissue factor. It can be activated further by a variety of clotting factors that convert it to a two-polypeptide chain protein, VIIa. These factors include factor XIIa, kallikrein, plasmin, factor IXa, and factor Xa. It is not known which of these is the primary activator. Prolonged action of factor Xa inactivates factor VIIa (autoregulation).

Although factors of the intrinsic pathway can activate factor VII, the key factor of the extrinsic pathway, the two pathways function physiologically independently. Patients who are deficient in components of the intrinsic pathway, such as factor VIII, the classical hemophiliac protein, bleed despite the fact that they have normal levels of factor VII. **Factor VII deficiency** also causes bleeding, although the intrinsic pathway is intact. It appears that both pathways are required for optimal activation of factor X.

In the **final common pathway,** factor Xa first cleaves **prothrombin** into two fragments, an amino terminal fragment (domains F_1 and F_2), containing 10 γ-carboxyglutamyl residues and a single β-hydroxy aspartate residue, and the carboxy terminal **prethrombin 2** fragment. (In a side reaction, the removal of F_1 and F_2 occurs in two steps with the formation of prethrombin 1 as an intermediate.) An additional cut within prethrombin 2 then produces the active serine protease **thrombin,** in which the two polypeptide chains are linked by a disulfide bond (Figure 9.30).

The structures of prothrombin and factors IX and X are rather similar. However, in contrast to thrombin, factor Xa retains the aminoterminal domain of its zymogen, linked to the carboxyterminal domain via disulfide bonds, and thus remains bound via its γ-carboxyglutamyl residues to the platelet surface, the site of factor X activation. Thrombin, on the other hand, has lost the γ-carboxyglutamyl residues and thus can diffuse and act on fibrinogen at some distance from its site of activation.

The activation of prothrombin is greatly accelerated by **factor Va,** a large protein that binds to the phospholipid membrane of platelets and there provides part of the receptor site for factor Xa. Factor V itself is activated by thrombin. However, as the concentration of thrombin increases, thrombin degrades factor V, thus regulating its own production.

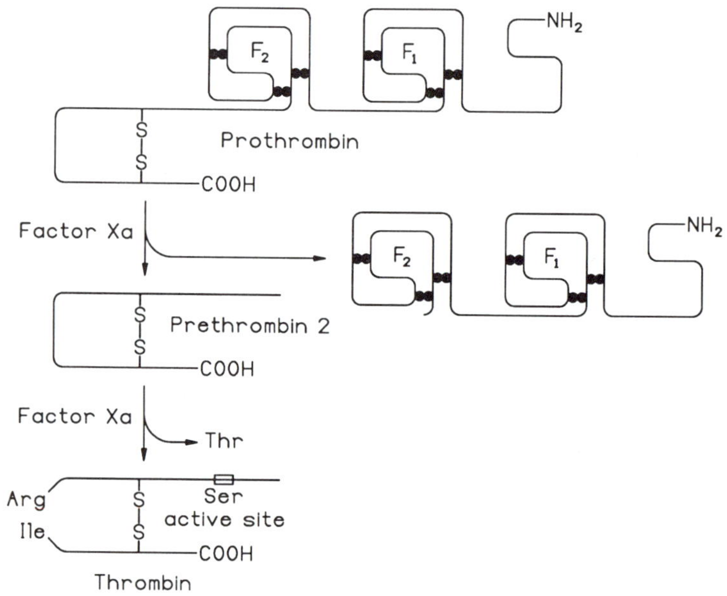

Figure 9.30.
The activation of prothrombin to thrombin. The "Kringle" structures F_1 and F_2 contain three disulfide bonds each (indicated by filled circles).

Furthermore, thrombin can cleave prethrombin, yielding more thrombin, but at higher concentrations it may also cleave the F_1 and F_2 domains from free prothrombin, thus preventing prothrombin from binding to the platelet surface. Platelet activation appears to be required for the activation of prothrombin. In addition to platelets, the membranes of peripheral white cells and endothelial cells can also serve as sites for the activation of prothrombin.

Fibrinogen is a large glycoprotein (MW = 330,000) consisting of dimers with three polypeptide chains each $(\alpha A, \beta B, \gamma)_2$, which are linked together by disulfide bonds (Figure 9.31). Thrombin cleaves peptides A and B from the αA and βB chains of fibrinogen. (The cleavage occurs at Arg-Gly bonds.) The removal of these peptides appears to allow fibrin molecules to aggregate into soft clots, which can still be solubilized by 3*M* urea.

The last step in blood coagulation is the cross-linking of fibrin (Figure 9.32). It is catalyzed by **factor XIIIa,** a transglutaminase that forms ϵ-(γ-glutamyl)-lysine cross-links between γ and α chains. (Activation of factor XIII to factor XIIIa is catalyzed by thrombin.)

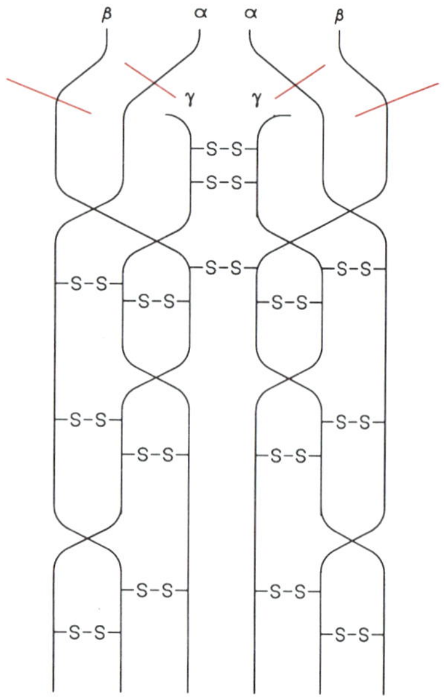

Figure 9.31.
Schematic structure of fibrinogen. All three chains are wound around each other in a coiled coil. The thrombin cutting sites are indicated by red dashed lines.

Regulation of Blood Coagulation

The coagulation cascade must be precisely regulated so that it is restricted to sites where vascular damage has occurred. There are two anticoagulation mechanisms that accomplish this, the **heparin-antithrombin mechanism** and the **protein C-thrombomodulin mechanism.**

The heparin-antithrombin mechanism. A circulating α_2-globulin in plasma, **antithrombin III,** is the principal inhibitor of circulating thrombin. Antithrombin III forms a 1 : 1 complex with thrombin. In this complex, an arginine residue interacts, and thereby interferes, with the active site serine of thrombin. In the absence of heparin, this interaction occurs only slowly. Heparin binds to lysyl residues on antithrombin III and thereby accelerates its rate of complex formation with thrombin.

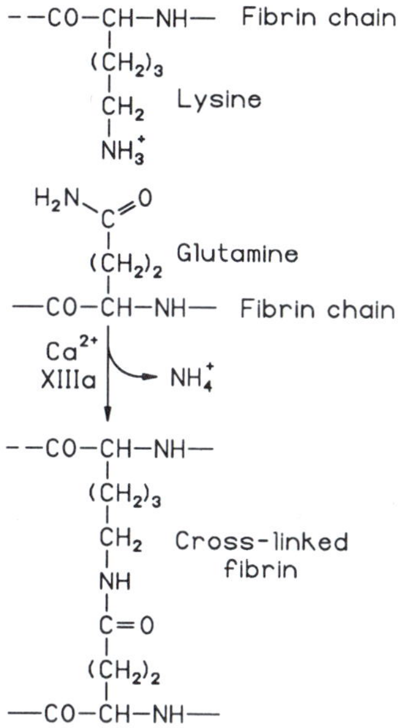

Figure 9.32.
Cross-linking of fibrin chains.

The other activated coagulation factors of the intrinsic pathway, factors IXa, Xa, XIa, and XIIa, are inactivated by the antithrombin III–heparin complex in a similar way.

Heparin is a sulfated polysaccharide synthesized by mast cells (Figure 9.33). It is synthesized in the form of a proteoglycan with side chains containing alternating residues of *N*-acetylglucosamine and glucuronic acid. During maturation, the glucuronic acid is epimerized to iduronic acid, and all sugar residues are sulfated to various degrees at the C3 and C6 positions. A small fraction of plasma antithrombin III is found associated with heparinlike molecules located on the membranes of endothelial cells. This appears to protect the endothelial surface from thrombus formation.

The protein C–thrombomodulin mechanism. Protein C is a vitamin K–dependent plasma glycoprotein (MW = 56,000; concentration 4 μg/ml) containing two polypeptide chains that are linked by a disulfide bond. In order to function as an anticoagulant, protein C must be activated by thrombin. However, only thrombin that is bound to thrombomodulin on the surface of platelets is capable of activating protein C to Ca at a reasonable rate. Activated protein Ca is a serine protease that inactivates factor Va. Another plasma protein, **protein S,** is required as a cofactor for this degradation of factor Va.

In summary, the small amounts of factor Xa and factor Va that are

Figure 9.33.
The biosynthesis of heparin begins with the attachment of mucopolysaccharide side chains to a proteoglycan. The carbohydrate side chains are then modified to heparin.

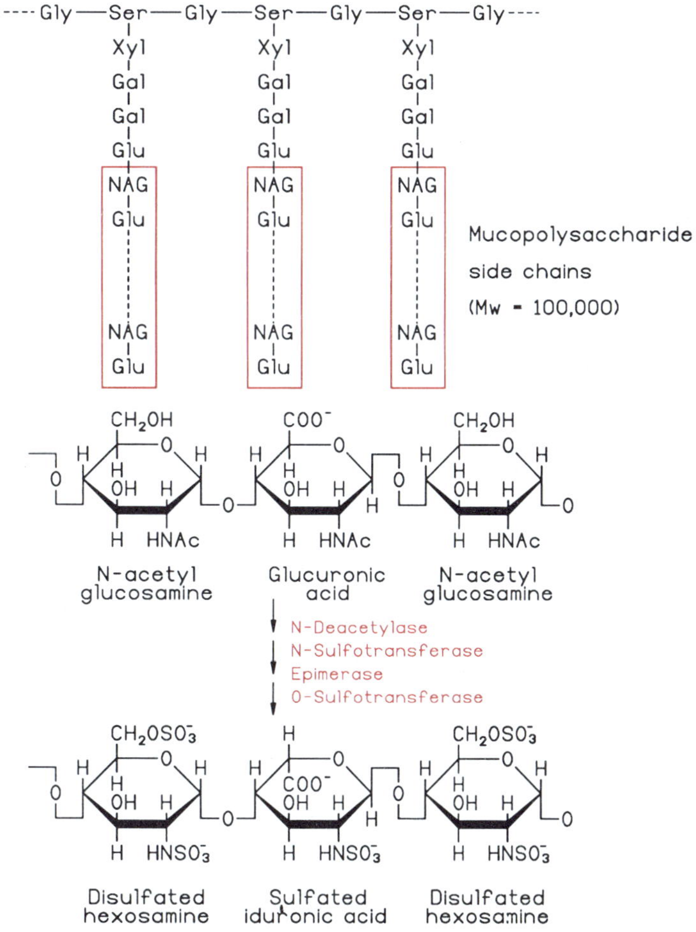

continuously formed within the vascular system are capable of binding to their receptors on the platelet surface where they form small amounts of thrombin. The two anticoagulation mechanisms discussed above prevent the formation of the **prothrombinase complex** (Va, Xa, PF-3, Ca^{2+}) and inactivate free thrombin. Furthermore, the production of prostacyclin (PGI_2) by endothelial cells tends to suppress platelet activation. This natural suppression of coagulation is interrupted when large amounts of platelets are activated via collagen (damaged endothelium); large amounts of platelet factor 4 are released that inhibit the heparin-antithrombin mechanism.

Disorders in Blood Coagulation

The most common forms of **hemophilia** are caused by genetic deficiencies in factor VIII **(classical hemophilia)** or factor IX **(Christmas disease).** Both hemophilias are inherited as X-linked recessive traits. The genes for all other coagulation factors are located on autosomes.

Factor XII deficiency **(Hageman trait)** does not cause excessive bleeding. (As a matter of fact, the patient Mr. Hageman, after whom the disease is named, died at old age of thrombosis.) On the other hand, **von Willebrand's disease,** one of the most common hereditary bleeding disorders, is characterized by both an abnormally long bleeding time and a partial deficiency of factor VIII. This is expected from its known functions as carrier protein for factor VIII and participant in platelet activation.

Chelation of Ca^{2+} ions by citrate prevents clotting of blood very effectively. This fact is used when blood is collected for analysis or transfusion. Clotting factors VII, IX, and X as well as thrombin contain γ-carboxyglutamyl residues that are required for the binding of Ca^{2+} ions. The posttranslational carboxylation reaction requires vitamin K as a cofactor (see Section 8.3). Vitamin K deficiency, therefore, leads to bleeding problems. The vitamin K antagonist **dicoumarol** is used as an oral anticoagulant drug.

Several tests of coagulation function are being used.

1. **Prothrombin time (PT)** measures the extrinsic pathway. Blood is collected in the presence of citrate and centrifuged, tissue factor is added to the plasma, and the time to form a fibrin clot is measured after readdition of Ca^{2+}. PT measures the activities of factors VII, X, V, and prothrombin.
2. **Partial thromboplastin time (PTT)** measures the intrinsic pathway (factors XII, XI, IX, VIII, X, V, and prothrombin). Plasma is first incubated with diatomaceous earth (surface activation of factor XII) and phospholipids (platelet membrane substitute), and then Ca^{2+} ions are added. The time required for clot formation is measured.
3. **Thrombin time** tests the activity of fibrinogen. A solution of thrombin is added to plasma, and the time required for clot formation is measured.

Fibrinolysis

Degradation of fibrin is catalyzed by the serine protease **plasmin** (fibrinolysin), which circulates in the blood as **plasminogen.** Conversion

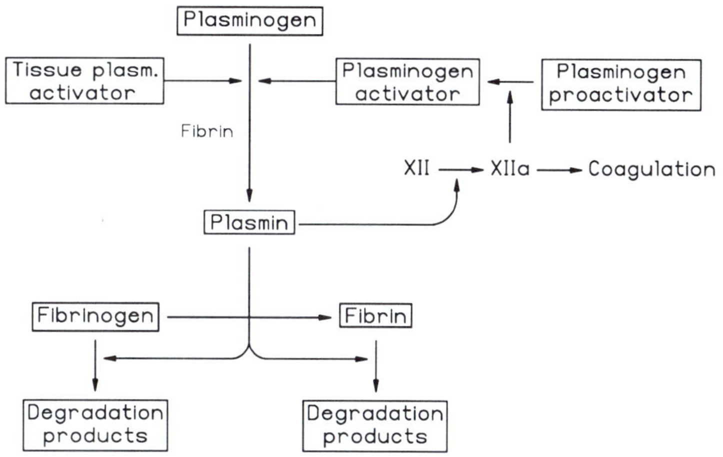

Figure 9.34.
The activation of plasminogen.

of plasminogen to plasmin is accomplished by two mechanisms (Figure 9.34):

1. **Intrinsic activation** involves components of the intrinsic pathway (factor XII, prekallikrein, and HMW kininogen). The significance of this activation mechanism is not known, since patients lacking components of this pathway show no symptoms.
2. **Tissue-type plasminogen activator (t-PA)** is a serine protease released into the bloodstream by cells. The mechanism of this secretion is not clear. In plasma, t-PA occurs in free form or complexed with α_2-antiplasmin and α_1-antitrypsin. The activation of plasminogen by t-PA requires the presence of fibrin. Apparently, by binding to fibrin, plasmin is protected from inactivation by α_2-antiplasmin. Free plasmin is immediately inactivated.

There are two other known plasminogen activators: **urokinase,** found in normal urine, and **streptokinase,** a nonenzymatic protein produced by streptococcus bacteria that directly activates plasminogen by binding to it in a 1:1 complex. Neither one of these plasminogen activators binds to fibrin and thus will activate free circulating plasminogen. The resulting plasmin is then immediately neutralized by α_2-antiplasmin. Thus t-PA, which has recently been cloned and expressed in bacteria, appears to be the preferred activator for use in the thrombolytic therapy of heart infarction. However, this issue is still controversial.

For a more extensive discussion of constituents of blood, consult the following textbooks:

Mathews et al. (1990): Chapters 7, 10, 35

Stryer (1988): Chapters 7, 11

Further References

Arai K, Lee F, Miyajima A, Miyatake S, Arai N, Yokota T. Cytokines: Coordinators of immune and inflammatory responses. *Annu Rev Biochem.* 1990; 59:783–836.

Bjorkman PJ, Parham P. Structure, function, and diversity of class I major histocompatibility complex molecules. *Annu Rev Biochem.* 1991; 59:253–288.

Brodsky FM, Guagliardi L. The cell biology of antigen processing and presentation. *Ann Rev Immunol.* 1991; 9:707–744.

Davis MM. T cell receptor gene diversity and selection. *Annu Rev Biochem.* 1990; 59:475–496.

Evans T, Felsenfeld G, Reitman M. Control of globin gene transcription. *Ann Rev Cell Biol.* 1990; 6:95–124.

Golde DW, Gasson JC. Hormones that stimulate the growth of blood cells. *Sci Am.* 1988; 259 (1):62–70.

Grey HM, Sette A, Buus S. How T cells see antigen. *Sci Am.* 1989; 261:56–64.

Massagúe, J. The transforming growth factor-β family. *Ann Rev Cell Biol.* 1990; 6:597–642.

Nepom GT, Erlich H. MHC class-II molecules and autoimmunity. *Ann Rev Immunol.* 1991; 9:493–526.

Rothbard JB, Gefter ML. Interaction between immunogenic peptides and MHC proteins. *Ann Rev Immunol.* 1991; 9:527–566.

Smith KA. Interleukin-2. *Sci Am.* 1990; 262:50–57.

Stamatoyannopoulos G, Nienhuis AW, Leder P, Majerus PW. The molecular basis of blood diseases. Philadelphia PA: WB Saunders; 1987. (An excellent monograph on the molecular biology of all blood components.)

Staudt LM, Lenardo MJ. Immunoglobulin gene transcription. *Ann Rev Immunol.* 1991; 9:373–398.

Wagner DD. Cell biology of the von Willebrand factor. *Ann Rev Cell Biol.* 1990; 6:217–246.

Vitetta ES, Berton MT, Burger C, Krepton M, Lee WT, Yin X. Memory B and T cells. *Ann Rev Immunol.* 1991; 9:193–218.

CHAPTER

10 Liver Functions

OBJECTIVES

After studying this chapter, the student should be able to

1. Describe the fate of glucose absorbed after an average meal and explain how the liver maintains an adequate blood glucose level during periods of fasting.
2. Write a balanced equation for the oxidation of ethanol to acetyl CoA.
3. Name the metabolic disorders that can cause excessive deposition of fat in the liver and explain their probable mechanisms of action.
4. Name several liver function tests that evaluate the capacity of the liver to metabolize glucose and store it in the form of glycogen, to synthesize plasma proteins, or to conjugate and excrete foreign substances.
5. Give at least two examples of hepatic biotransformation reactions and include details of the two steps involved.
6. Describe, in general terms, the biochemical steps involved in the degradation of heme and name their cellular locations.
7. State the significance of early labeled bilirubin.
8. Explain the significance of the direct and indirect van den Bergh reaction.
9. Name conditions that would lead to unconjugated or conjugated hyperbilirubinemia, respectively.

The liver is the central organ that coordinates the synthesis and metabolism of carbohydrates, fat, and protein throughout the body. It also synthesizes a variety of important compounds. Specific liver functions include

1. **Metabolic reactions** concerned with the maintenance of glucose homeostasis, gluconeogenesis, amino acid catabolism and the synthesis of urea, triacylglycerides, phospholipids, cholesterol, and most plasma proteins
2. **Biotransformation reactions** that chemically modify, and thereby facilitate subsequent excretion of, a large variety of foreign chemicals that enter the body as, for example, drugs and toxins
3. The **processing of bilirubin**
4. The **secretion of bile,** which contains conjugated bilirubin, cholesterol, and bile salts

10.1 METABOLIC FUNCTIONS

The liver regulates the supply of energy in the body by maintaining a constant level of glucose in the bloodstream. This is accomplished primarily by storing excess glucose in the form of glycogen and then releasing it during periods of fasting.

For example, after an average meal about 100 g of glucose are absorbed, over a 3 hour period, from the gut. Only 25 g of this glucose is used directly for energy production, mainly by the cells of the central nervous system (CNS) and by erythrocytes. (The uptake of glucose by these cells is not dependent on insulin.) Another 20 g of glucose are stored as glycogen and triacylglycerides in muscle cells and adipocytes, respectively. (These latter reactions are dependent on insulin.) The majority of the absorbed glucose, 55 g or 75% of the excess, is taken up by the liver and either stored as glycogen or converted to triacylglycerides and exported to adipose tissue in the form of very low density lipoprotein (VLDL; see Section 3.4).

Dietary fat is transported first to adipose tissue via chylomicrons. Only the triacylglycerides contained in the chylomicron remnants reach the liver, where they are hydrolyzed and, after modification of the fatty acids, resynthesized and exported to adipose tissue via VLDL.

In contrast to glycogen and triacylglycerides, there is no specific protein that serves to store excess amino acids. To a limited extent, essentially all proteins participate in this function. The degradation of amino acids and their conversion to glucose (gluconeogenesis) with the concomitant production of urea occurs predominantly in the liver.

During periods of **fasting,** the liver maintains blood glucose levels by first degrading its glycogen. If that is depleted, glucose is synthesized from amino acids, glycerol (from triacylglycerides), or lactic acid (generated by erythrocytes). Under normal conditions (no starvation), the CNS requires about 145 g of glucose per day, whereas erythrocytes require another 35 g. The glycogen content of the liver (about 70 g) can satisfy only part of this demand if no additional dietary glucose is provided. The remaining glucose is derived from gluconeogenesis.

When fasting persists over a period of 10 days **(starvation)**, the CNS partially adapts to using ketone bodies for energy (see Section 3.9), thus reducing the overall requirement for glucose. Erythrocytes continue to depend on glucose as their sole source of energy. Their glycolytic metabolism, however, stops at lactate, which is reconverted to glucose in the liver and recycled to the erythrocytes.

All other cells in the body are capable of metabolizing fatty acids and ketone bodies. During prolonged periods of fasting, the depletion of body proteins for gluconeogenesis is reduced to a minimum of about 20 g/day.

Gluconeogenesis initially takes place predominantly (over 90%) in the liver. In starvation, however, up to 40% of the glucose may be manufactured in the kidney. This is because the kidney degrades amino acids in order to produce ammonia, which it excretes in the urine. Ammonia neutralizes urinary hydrogen ions and thus facilitates further excretion of H^+ ions, which is necessitated by the ketoacidosis that develops during starvation (see Section 12.2).

10.1.1 Disturbances in Liver Metabolism

Alcoholism

Alcohol is rapidly absorbed through the gastric mucosa and upper small intestine and enters the liver via the portal vein. There it is oxidized to **acetaldehyde** by two different enzymes: **alcohol dehydrogenase** and a **microsomal mixed-function oxidase.**

1. **Alcohol dehydrogenase (ADH)** uses NAD^+ to oxidize alcohol.

$$CH_3{-}CH_2{-}OH + NAD^+ \xrightarrow{ADH} CH_3{-}CHO + NADH + H^+$$

   The human genome has five genes for ADH with a multiplicity of alleles. Orientals often express an isozyme that is very active and thus produces larger amounts of acetaldehyde very quickly. This makes the individual more sensitive to drinking alcohol (flushing reaction). Alcohol dehydrogenase is a constitutive enzyme.

2. **The microsomal ethanol oxidizing system** (P_{450} enzyme

system) also produces acetaldehyde but uses NADPH and O_2 as cosubstrates.

$$CH_3\text{—}CH_2\text{—}OH + NADPH + H^+ + O_2 \rightarrow CH_3\text{—}CHO + NADP^+ \; 2H_2O$$

In contrast to ADH, the microsomal system is inducible. The induction is rather long-lived. After induction, it accounts for 30–50% of the total alcohol oxidation. It has a higher K_m for ethanol than ADH.

The induction of the P_{450} system by alcohol is important clinically because it affects the metabolism of many drugs. Heavy drinkers, even after they have stopped drinking, show an increased metabolism of drugs. This must be considered when treating chronic alcoholics with certain medications.

The acetaldehyde is further oxidized to acetyl-CoA by **aldehyde dehydrogenase.**

$$CH_3\text{—}CHO + NAD^+ + CoA \rightarrow \text{acetyl CoA} - NADH$$

Thus, the metabolism of alcohol produces large amounts of acetyl CoA and reducing equivalents. This leads to fatty acid synthesis and the formation of triacylglycerides (see the discussion of fatty liver below). The glycerol moiety is derived from glucose metabolism if drinking is accompanied by eating. However, if the drinker does not eat, as is frequently the case in chronic alcoholics, the glycerol is derived from gluconeogenesis. This is why chronic alcoholics, who derive 70–80% of their total caloric intake from alcohol, often exhibit severe muscle atrophy that can include the heart muscle.

The increased requirement for pyridine nucleotides in the oxidation of ethanol stimulates purine biosynthesis and turnover (adenine is part of NAD), which may increase levels of uric acid. Hyperuricemia is frequently associated with overindulgence in alcohol (see gout, Chapter 5).

Acetaldehyde is toxic and can form adducts with a variety of different molecules. For example, its interaction with tetrahydrofolic acid traps folic acid and thereby leads to folic acid deficiency. This deficiency is exacerbated by the poor nutritional habits of alcoholics whose food intake does not include many sources of folic acid.

Interaction of acetaldehyde with thiamine pyrophosphate (TPP) may lead to the **Wernicke-Korsakoff syndrome,** a neurological disorder associated with functional thiamine deficiency. The function of TPP in the transketolase reaction is apparently important in certain functions of the central nervous system.

Fatty Liver

The fat content of the normal liver amounts to only about 5% of its weight. This can be increased to 50% in certain pathological conditions, as for example excessive consumption of alcohol, diabetes mellitus, overnutrition, and undernutrition (kwashiorkor).

A single heavy consumption of **alcohol** causes lipolysis in adipose tissue (via sympathetic stimulation) and deposition of the released fatty acids as triacylglycerides in the liver. This occurs in addition to the de novo fatty acid synthesis that results from alcohol oxidation. Normally, this fat deposit in the liver is redistributed into adipose tissue within 24–48 hours. In **chronic alcoholism,** however, the alcohol may interfere with the hepatic synthesis of lipoproteins and thus the transport of triacylglycerides into adipose tissue.

In **diabetes mellitus** the absence of insulin causes excessive lipolysis in adipose tissue. Since β-oxidation of the released fatty acids in the liver cannot accelerate beyond a certain level, the excess fatty acids are deposited as triacylglycerides.

Overnutrition causes excessive deposition of triglycerides not only in adipose tissue but also in the liver. **Undernutrition,** especially protein deficiency (kwashiorkor), interferes with the synthesis of lipoproteins that are required for the export of triacylglycerides from the liver as VLDL. Since adipose lipolysis is increased in starvation, more fatty acids are available for deposition in the liver.

Several drugs (e.g., tetracyclines) and toxins (e.g., carbon tetrachloride) also interfere with lipoprotein biosynthesis and may cause fatty liver.

Fatty liver may lead to **cirrhosis,** the proliferation of fibrous tissue that replaces the normal liver cells.

10.1.2 Tests for Liver Metabolism and Disease

The ability of the liver to metabolize carbohydrates and to synthesize proteins can be tested in several ways.

In the **glucose tolerance test,** a patient is given a dose of glucose (about 1 g/kg of body weight) after an overnight fast of 12 hours. Blood glucose levels are determined at hourly intervals for 3–4 hours after the ingestion of glucose. Normally, the level increases from about 80 mg/100 ml of blood to about 130 mg/100 ml and returns to normal levels within 2–2.5 hours. In the absence of diabetes mellitus, this test evaluates hepatic function in glucose metabolism.

The **galactose tolerance test** (oral intake of 0.5 g of galactose/kg of weight after 12 hours of fasting) gives information on the perfusion and metabolic capacity of the liver. The test can be used to distinguish intrahepatic jaundice (liver cell destruction) from obstructive jaundice (blockage of hepatic or biliary bile canals). Normally, blood galactose levels are zero 75 minutes after ingestion. In intrahepatic jaundice, blood galactose values are greater than 20 mg/100 ml at this time, whereas in obstructive jaundice, galactose values are less than 20 mg/100 ml.

The **epinephrine** (or glucagon) **tolerance test** measures the capacity of the liver to store glycogen. A patient is given a high-carbohydrate diet for 3 days and is then injected with epinephrine after an overnight fast. Normally, blood glucose levels rise within 1 hour to 40 mg/100 ml *above* the fasting level (80 mg/100 ml) observed before the injection of epinephrine. The rise in glucose, reflecting the presence of stored liver glycogen, is followed by a rise in lactate (glucagon gives no rise in lactate).

Low serum albumin levels **(hypoalbuminemia)** are usually observed in liver disease. An imbalance of synthesis of serum proteins in patients with liver cell damage can be recognized by several flocculation tests. The **cephalin-cholesterol flocculation test,** the **thymol turbidity test,** and the **zinc turbidity test** are used most frequently. Flocculation or turbidity is observed when serum protein factors are imbalanced.

Prothrombin time (see Section 9.5), the time required for clotting of plasma, is increased up to 10-fold in liver disease owing to the reduced levels of prothrombin produced by the liver. This test is usually employed when the progress of a patient with liver disease is monitored. Prothrombin time is also increased in vitamin K deficiency that may result from biliary obstruction.

The serum levels of several enzymes have been used in the diagnosis of liver disease. The most important enzymes in this respect are **serum glutamate oxaloacetate transaminase (SGOT)** and **serum glutamate pyruvate transaminase (SGPT).** Both enzymes are released into the bloodstream from damaged liver cells. Such damage occurs in hepatitis, cirrhosis, or intoxication with carbon tetrachloride. Elevated transaminase levels are an early indication of developing icterus. (Myocardial infarction also raises transaminase levels, predominantly SGOT.)

Serum alkaline phosphatase levels usually increase in obstructive jaundice. Osteoblasts (in children) and placental tissue (in pregnant women) release different isozymes of alkaline phosphatase that can be distinguished from the liver enzyme by electrophoresis or by their different heat sensitivities.

The sulfobromophthalein excretion test is used to evaluate the capacity of the liver to conjugate and excrete foreign substances. Sulfobromophthalein is conjugated to cysteine and excreted into the bile. In normal conditions, over 90% of an injection of 5 mg of the dye/kg of body weight is cleared from the blood within 30 minutes.

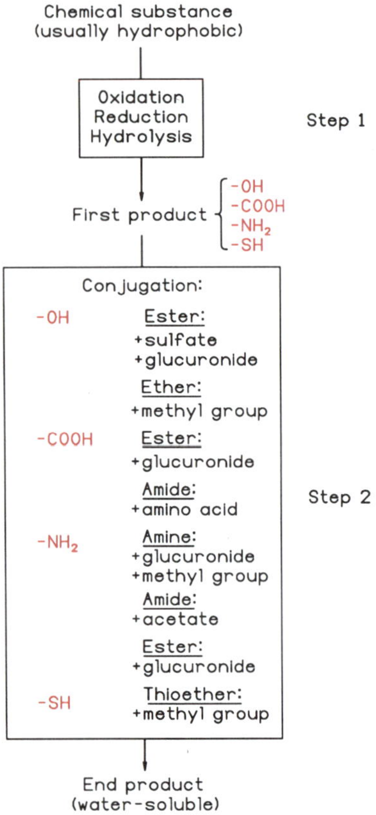

Figure 10.1.
Overview of biotransformation reactions.

10.2 BIOTRANSFORMATION AND CONJUGATION

Many organic chemicals that enter the human body, either as components of the diet or in the form of drugs, are chemically modified in the liver. This **biotransformation** occurs on the endoplasmic reticulum and involves two steps:

1. At first, the compound is transformed by **oxidation, reduction,** or **hydrolysis.**
2. In a second step, called **conjugation,** the substance is made more water-soluble by attachment of a variety of hydrophilic groups (Figure 10.1).

Biotransformation may alter the biological activity of a substance in several ways. Many drugs are inactivated by biotransformation; other drugs become more water-soluble and, as a consequence, are excreted more readily (e.g., pentobarbital). However, biotransformation does not always make a substance less active. The insecticide parathion, for example, is itself inactive. Biotransformation in the liver converts parathion to paroxon, a very toxic acetylcholinesterase inhibitor blocking

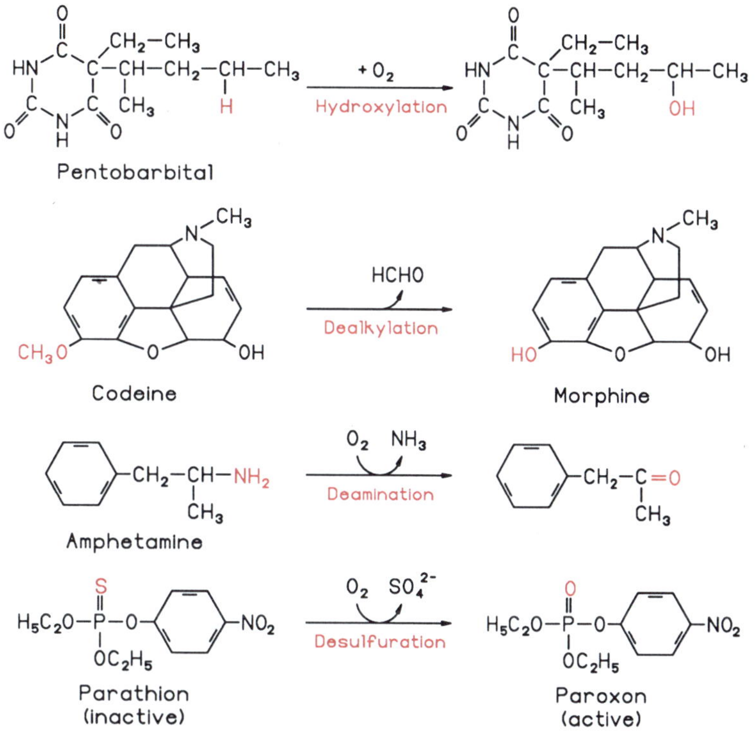

Figure 10.2.
Examples of biotransformations.

synaptic transmission. Other compounds are made mutagenic or carcinogenic. Several examples of biotransformation are illustrated in Figure 10.2.

10.2.1 Transformation

The most prevalent type of biotransformation is **hydroxylation.** It is catalyzed by membrane-bound (endoplasmic reticulum) **cytochrome P_{450}**–containing monooxygenases. (Cytochrome P_{450} derives it name from the absorption maximum at 450 nm of its carbon monoxide adduct.) The mechanism of the hydroxylation reaction is as follows. After cytochrome P_{450} binds to the substrate, its heme-Fe^{3+} is reduced to Fe^{2+}. The electron is provided by the flavoprotein NADPH–cytochrome P_{450} reductase. (This enzyme contains one molecule of flavin adenine dinucleotide [FAD] and one molecule of flavin mononucleotide [FMN].) The cytochrome P_{450} (Fe^{2+})-substrate complex then binds

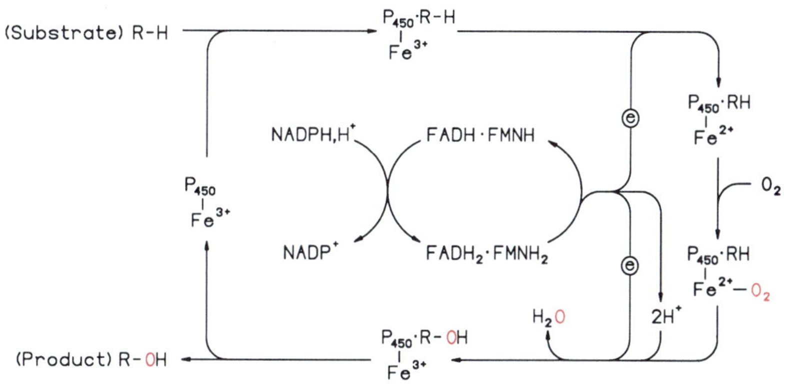

Figure 10.3.
Mechanism of hydroxylation by cytochrome P_{450}.

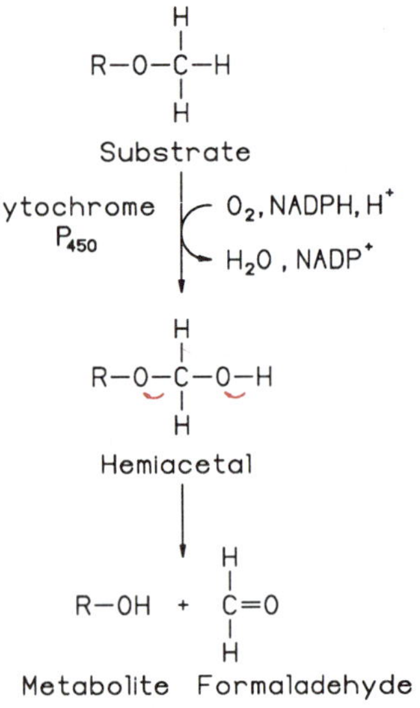

Figure 10.4.
Mechanism of dealkylation by cytochrome P_{450}.

molecular oxygen (see the analogy to hemoglobin, Section 9.2) and accepts the second electron generated by the cytochrome P_{450} reductase. One oxygen atom subsequently accepts two hydrogen ions to form water while the other oxygen forms the hydroxyl group on the substrate (Figure 10.3).

Cytochrome P_{450} is an unusual *b*-type cytochrome that has as the sixth ligand of the Fe atom a protein (bound via a histidinyl residue) that is easily displaced by CN^-, CO, O_2, or a variety of other substrate molecules.

There are several different forms of cytochrome P_{450}, each with a rather wide substrate specificity. This explains why so many different chemical compounds can be hydroxylated by this system. Different forms of the enzyme are induced by different groups of chemical agents.

Although the cytochrome P_{450}–catalyzed hydroxylation is a relatively slow reaction (0.1–50 molecules/minute), the hydroxylation capacity of the liver is large, since cytochrome P_{450} constitutes up to 10% of the membrane-bound proteins in the liver. The liver is specially designed for metabolizing foreign substances, since it receives 25% of the total cardiac output and directly obtains, via the portal vein, all substances absorbed from the gut.

Dealkylation, another form of biotransformation, is also initiated by

Figure 10.5.
Examples of conjugation reactions.

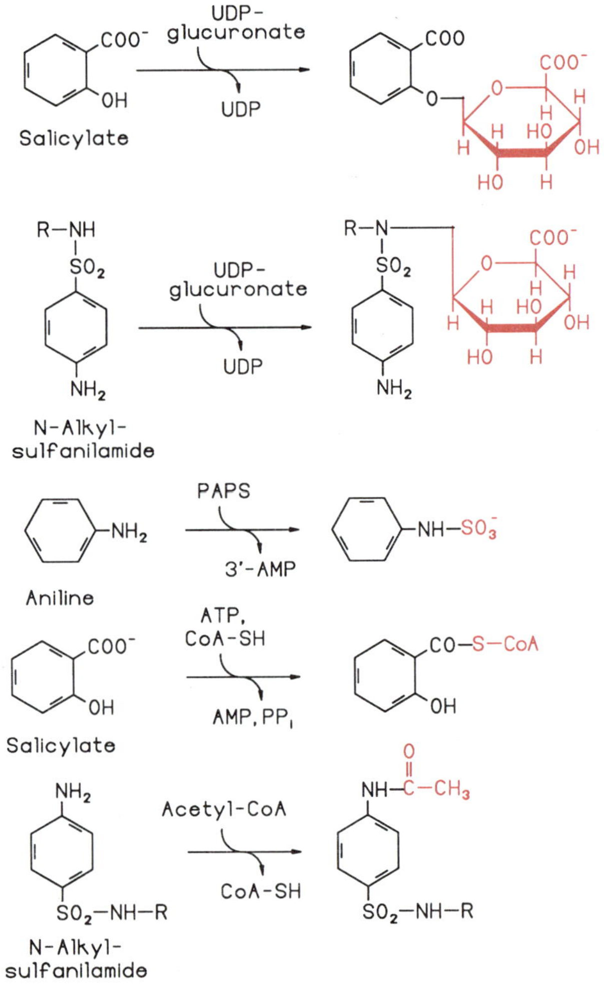

hydroxylation. Hydroxylation of an *O*-, *N*-, or *S*-linked carbon atom produces an unstable hemiacetal, which dissociates spontaneously into an aldehyde and the corresponding OH, SH, or NH_2 group (Figure 10.4).

***N*-oxidation, *S*-oxidation, deamination,** and **desulfuration** are probably also catalyzed by cytochrome P_{450} monooxygenases. **Reductions** are catalyzed by membrane-bound NADPH- and NADH-dependent reductases. The nitro group (—NO_2), for example, is reduced to the carcinogenic nitroso group (—NO). **Hydrolysis** of esters and amides occurs in the blood and is catalyzed by nonspecific hydrolases.

The main purpose of all these reactions is to create free OH, NH_2, and SH groups, which can be conjugated to a hydrophilic compound in the second step of biotransformation.

10.2.2 Conjugation

The second step in the biotransformation of organic chemicals is their conjugation to form water-soluble, and thus rapidly excretable, derivatives. Several endogenous substrates function as conjugation partners for activated (first step of biotransformation) molecules: glucuronate, sulfate, amino acids (Gly, Gln, Cys), and acetate. Examples of some conjugation reactions are shown in Figure 10.5.

Conjugation enzymes are also inducible and are located on the membranes of the endoplasmic reticulum.

10.3 DEGRADATION OF HEME

Old and defective red blood cells are removed from the circulation by the phagocytic cells of the reticuloendothelial system in the spleen, liver, and bone marrow. After lysis of the erythrocytes in these cells, the hemoglobin is oxidized to methemoglobin, the globin chains are degraded to amino acids, and the hemin (Fe^{3+}) is oxidized to **biliverdin.** This oxidation occurs in the endoplasmic reticulum and is catalyzed by **heme oxygenase,** a cytochrome P_{450}–containing mixed-function oxidase. The heme ring is cleaved at the α-methene group, which is oxidized to carbon monoxide (Figure 10.6). This is the only enzymatic reaction in the human body that generates CO. Iron-free porphyrins are not attacked by the enzyme.

The green-colored tetrapyrrol biliverdin is reduced to the orange-red **bilirubin.** This reaction takes place in the cytoplasm of the cell and requires NADPH.

About 85% of the bilirubin (300 mg/day) arises from the degradation of erythrocytes in the spleen. The remainder is produced in the liver and the bone marrow. It is derived from both heme and other sources such as cytochromes, catalase, and myoglobin.

Two different types of cells are involved in heme degradation in the liver. A subpopulation of **Kupffer cells** (reticuloendothelial system) produces bilirubin from old erythrocytes. In addition, **parenchymal cells,** also called **hepatocytes,** take up from the bloodstream either hemoglobin bound to haptoglobin or heme bound to hemopexin. Following a splenectomy or in hemolytic anemia, when substantial quantities of hemoglobin are released into the bloodstream, the heme oxy-

Figure 10.6.
Formation of bilirubin.

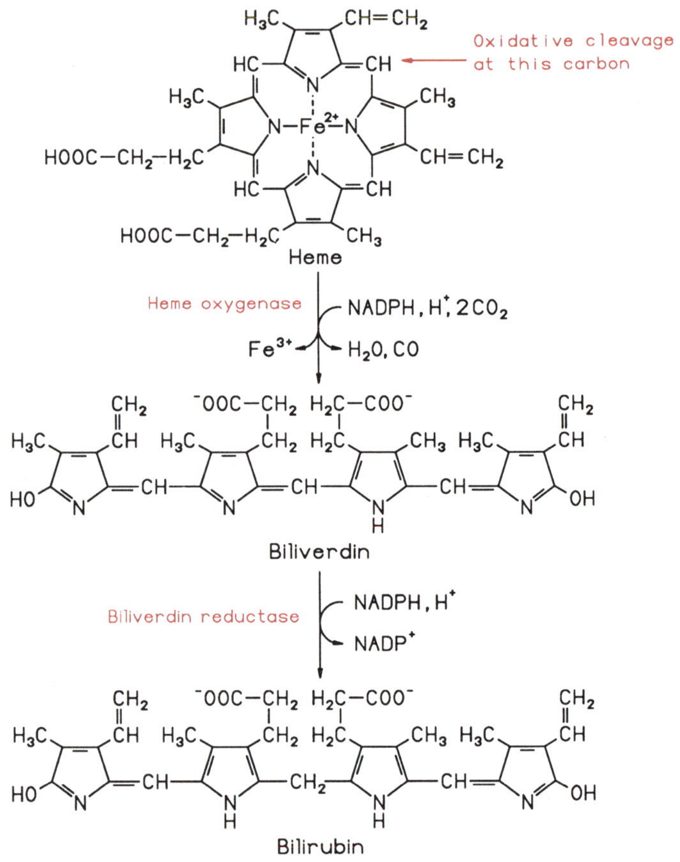

genase activity in the liver parenchymal cells is significantly elevated. Bilirubin produced in bone marrow macrophages is derived from abortive hemoglobin synthesis in immature red blood cells and from the turnover of heme-containing enzymes. It is called **early labeled bilirubin** because a short pulse of radioactive label is found in bilirubin within only a few days. This early labeled bilirubin fraction normally amounts to about 15% of the total incorporated label. The majority of the label is released from hemoglobin after 120 days, the normal lifetime of the erythrocyte. In certain disease processes, however, the early labeled bilirubin fraction can range from 30% to 80% of the total incorporated label.

All bilirubin, formed at different locations in the body by the reticuloendothelial cells, is carried to the liver bound to serum albumin. Each albumin molecule can carry up to two molecules of bilirubin. The binding to albumin is necessary because unconjugated bilirubin is almost insoluble at physiologic pH. The large plasma concentration of albumin ensures that this transport system is never overtaxed even in patients with jaundice (yellow discoloration of tissues and body fluids by bile pigments).

At the hepatocyte cell membrane, bilirubin dissociates from albumin, enters the cell, and is bound intracellularly by two proteins, called **Y protein** (ligandin) and **Z protein.** (Z protein is the intracellular sterol carrier protein, a general carrier for hydrophobic molecules.) The significance of these proteins in the uptake and storage of bilirubin is still unclear.

Inside the hepatocyte, bilirubin is conjugated at its two propionic acid side chains to glucuronic acid. The enzyme **bilirubin–UDP glucuronyl transferase** is an inducible enzyme located on the membrane of the smooth endoplasmic reticulum. After conjugation, bilirubin is actively secreted into the bile, where it forms complexes with cholesterol, phospholipids, and bile salts. Once released into the intestine, bilirubin diglucuronide is hydrolyzed, and the free bilirubin is reduced by the bacterial flora through a series of colorless **urobilinogens** to **stercobilinogen.** Urobilinogens are partly reabsorbed into the enterohepatic circulation, and a very small amount (1% of the total bile pigment formed) is excreted in the urine. There it is oxidized by air to yellow **urobilin.** Stercobilinogen is also oxidized by air to form **stercobilin** (Figure 10.7) as well as a series of dipyrrol compounds.

10.3.1 Hyperbilirubinemias

Normal serum bilirubin levels range from 0.1 to 1.0 mg/100 ml. Most of this bilirubin is unconjugated; only a small fraction (less than 0.25 mg/100 ml) is conjugated. It arises by reabsorption from the bile canaliculi, which occurs even under normal conditions. Both forms of bilirubin are bound to albumin.

Clinical determination of free and conjugated bilirubin in serum is based on the **van den Bergh reaction.** In this reaction, diazotized sulfanilic acid (Ehrlich's reagent) is coupled to bilirubin, forming a

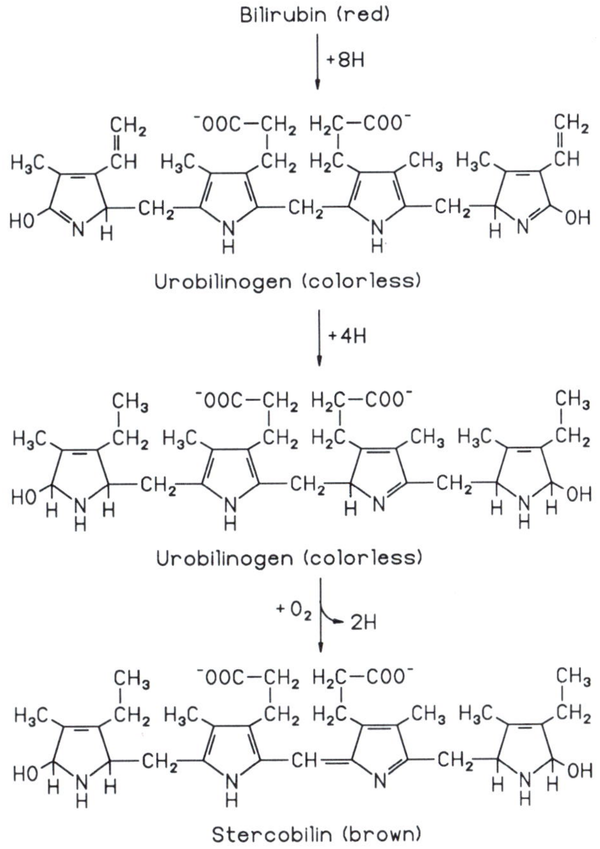

Figure 10.7.
Intestinal metabolism of bilirubin.

purple azocompound that can be quantitated photometrically. At first, bilirubin is cleaved at the central methylene group to form two dipyrrols, which are then coupled to diazosulfanilic acid. If the reaction is performed in an aqueous milieu, only the soluble conjugated bilirubin **(direct-reacting bilirubin)** reacts. If alcohol is added to the reaction mixture, both types of bilirubin contribute to the color formation. The difference between the two reactions gives the serum concentration of unconjugated **(indirect-reacting)** bilirubin.

Hyperbilirubinemia exists when serum levels exceed 1 mg/100 ml. Levels above 2 mg/100 ml cause a yellow discoloration of skin and sclera, called **icterus** or **jaundice.** Depending on the site and nature of the defect in bilirubin metabolism, the hyperbilirubinemia may be due to unconjugated or conjugated bilirubin. The van den Bergh test allows distinction between the two.

Unconjugated hyperbilirubinemia may be due to an overproduction of bilirubin, occurring, for instance, in hemolytic anemias. The liver normally has a limited reserve capacity to conjugate and excrete bilirubin; if the red cell lifetime is reduced by more than 50%, unconjugated hyperbilirubinemia (hemolytic jaundice) develops. Abnormal erythropoiesis may also cause bilirubin levels to rise.

Gilbert's disease is a relatively benign defect in the uptake of bilirubin by hepatocytes. It produces intermittent jaundice.

In newborn babies, the hepatic glucuronyl transferase is not yet fully induced. This enzyme is not needed before birth because unconjugated bilirubin can pass through the placenta to the mother's liver. After birth, when excretion through the mother's liver is no longer possible, plasma levels of unconjugated bilirubin temporarily rise until the baby's hepatic glucuronyl transferase is fully induced. This relatively harmless, physiologic **neonatal jaundice** usually disappears by itself after the first 2 weeks of life. It can be treated by daily irradiation with blue light, which causes photooxidation of the bilirubin to water-soluble products that can be excreted in the urine. However, if bilirubin levels exceed 20 mg/100 ml, the jaundice becomes life-threatening because at these levels bilirubin affects basal ganglia in the brain. This situation is known as **kernicterus** and arises primarily from hemolysis caused by Rh incompatibility. The condition can be treated by exchange transfusion.

The rare **Crigler-Najjar syndrome** (congenital nonhemolytic jaundice) is characterized by a defective glucuronyl transferase. There is a severe form (type I) that leads to kernicterus and death and a milder form (type II) that responds to treatment with phenobarbital (induction of transferase).

Conjugated hyperbilirubinemia arises from a defect in the excretion of conjugated bilirubin into the bile. A genetically defective mechanism of bilirubin secretion is present in both the **Dubin-Johnson syndrome** and the **Rotor's syndrome.** As a consequence, conjugated bilirubin enters the bloodstream and is excreted in the urine. Both syndromes are rather harmless.

Certain steroid drugs (e.g., methyltestosterone) cause the bile canaliculi and hepatocytes to become excessively permeable so that conjugated bilirubin escapes into the blood and urine. **Extrahepatic obstruction,** caused by stones or neoplasms of the bile duct, pancreas, or gall bladder, can be recognized by the presence of bilirubin **diglucuronide** in blood and urine.

10.4 METABOLISM OF BILE ACIDS

The biosynthesis from cholesterol of the primary bile acids, cholic acid, and chenodesoxycholic acid as well as their conversion to secondary bile acids by intestinal bacteria are discussed in Section 3.12. Essentially all bile acids are conjugated to either glycine or taurine. The rate-limiting step in bile acid synthesis is the hydroxylation at the 7α position. In liver disease, the biosynthesis of bile acids is decreased. As a consequence, the ratio of cholesterol to bile acids increases, causing cholesterol to precipitate and form biliary stones. Ingestion of chenodesoxycholic acid sometimes facilitates resolubilization of the cholesterol stones.

For a more extensive discussion of liver functions, consult the following textbooks:

Mathews et al. (1990): Chapter 18

Stryer (1988): Chapter 24

Further References

Arias IM, Frenkel M, Wilson JHP. *The Liver.* Annual/6. Elsevier Science Publishers, New York, 1987. (A series of critical surveys of the international literature. An excellent review on metabolism and hepatic injury by drugs and toxic chemicals, as well as other liver-related topics.)

Rubin, E. ed. Alcohol and the cell. *Ann New York Acad Sci.* 1987; 492. (A volume containing many articles on alcohol metabolism and the effects of alcohol on liver and other cells.)

Stolz A, Takikawa H, Ookhtens M, Kaplowitz N. The role of cytoplasmic proteins in hepatic bile acid transport. *Ann Rev Physiol.* 1989; 51:161–176.

CHAPTER

11 Special Structures and Functions

OBJECTIVES

After studying this chapter, the student should be able to

1. List the three major types of intracellular fiber structures that make up the cytoskeleton.
2. Give an example for each of the four membrane transport mechanisms: simple diffusion, carrier-mediated diffusion, active transport, and coupled transport.
3. Describe how macromolecules are transported through the endothelial cells of capillaries.
4. Explain the functions of tight junctions and of gap junctions.
5. Describe the nature and function of adhesion molecules.
6. Explain how tropocollagen molecules are arranged in the collagen fiber to produce the characteristic 670 Å cross-striation pattern observed in the electron microscope.
7. Outline the sequence of posttranslational events that are required for the formation of a stable collagen fiber.
8. Describe some of the biochemical reactions involved in the cross-linking of α chains in collagen.
9. Name two genetic disorders of collagen metabolism and describe the defects in biochemical terms.
10. Describe the biosynthesis of an elastin fiber.
11. Describe the basic structure of proteoglycans, including their glycosaminoglycan side chains.
12. Draw the basic structure of the proteoglycan complex found in cartilage.
13. Name the bands, lines, and zones of sarcomeres observed in the electron microscope and explain their relation to the molecular substructure of the sarcomere.
14. Describe the molecular structures of the thick and the thin filaments of the sarcomere.
15. Explain the sequence of events that leads to muscle contraction.
16. Explain why colchicin inhibits cell division.

11.1 THE PLASMA MEMBRANE

The plasma membrane separates intracellular from extracellular space. Its importance as an organizing unit becomes apparent when one considers the fact that life is impossible without cell membranes. In addition, the plasma membrane functions in

1. the interaction with fibers of the cytoskeleton to give the cell mechanical stability,
2. the transport of small molecules and ions into and out of the cell, and in
3. cell recognition and interaction with other cells

11.1.1 Interaction with the Cytoskeleton

The basic structure of a biological membrane has been discussed in Section 3.2. In order to give the plasma membrane mechanical support and also to allow the cell to change its shape, certain integral membrane proteins are connected on the inner membrane surface to three major types of intracellular fiber structure: microfilaments, microtubules, and intermediate filaments. Together these fibers make up the cytoskeleton.

Microfilaments (diameter of 8 nm) consist of two strands of polymerized actin subunits wound around each other. These fibers are noncontractile but can change their length through polymerization and depolymerization of **actin subunits**. During their polymerization into microfilaments, actin subunits are tightly associated with adenosine triphosphate (ATP). The conformational change that occurs during assembly causes the hydrolysis of ATP to adenosine diphosphate (ADP). ATP hydrolysis lowers the affinity of one actin subunit to its neighbors in the microfilament, thus increasing its chance for depolymerization. Microfilaments are **polar;** that is, they possess a **plus end** and a **minus end.** Monomers are added or removed more rapidly at the plus end because the monomer has to undergo a conformational change before being capable of binding to the minus end (Figure 11.1).

The rates of polymerization and depolymerization are determined by the concentrations of free monomers at the two ends of the microfilament. (About 50% of the actin in a cell is in the monomeric state.) Because of the different rates of assembly at their two ends, microfilaments often reach a steady state at which polymerization at the plus end equals depolymerization at the minus end. In this condition, the

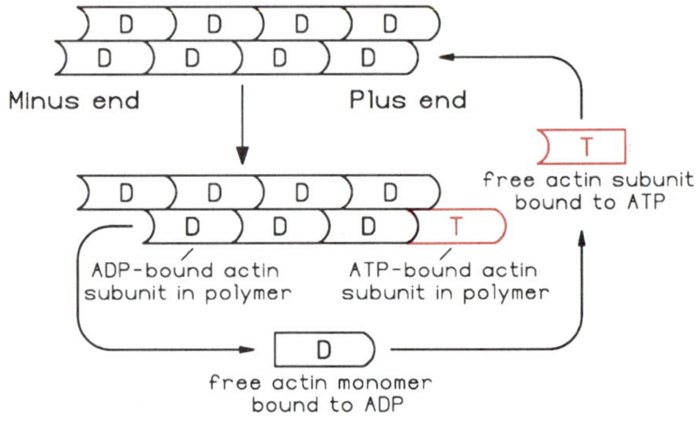

Figure 11.1.
The process of polymerization of actin subunits into microfilaments.

microfilament maintains its length but moves through the cytoplasm. This situation is known as **treadmilling.**

Cytochalasin B, a molecule excreted by certain molds, inhibits polymerization of actin into microfilaments. This inhibition also paralyzes the cell; it is no longer capable of ameboid motion, phagocytosis, or cytokinesis, processes that are based on microfilament polymerization and depolymerization. **Phalloidin,** a mushroom poison, has the same effect: it inhibits depolymerization of actin subunits and therefore stabilizes microfilaments.

Microfilaments often cross the entire cytoplasm and may thus be responsible for the three-dimensional shape of the cell. There are several proteins that interact with microfilaments and anchor them to other cellular structures. Table 11.1 lists some of these proteins.

Microtubules (diameter of 25 nm) are assembled from two types of globular tubulin molecules: **α-tubulin** and **β-tubulin** (MW = 55,000 each). Microtubules consist of hollow tubes with 13 tubulin dimers arranged around the circumference. The polymerization of the tubulin dimers (αβ) into microtubules is similar to the polymerization of actin monomers into microfilaments (Figure 11.1). The dimers are tightly bound to guanosine triphosphate (GTP), which is hydrolyzed to GDP after polymerization. Like microfilaments, microtubules have plus and minus ends. However, minus ends of microtubules are tightly bound to cellular structures called **microtubule-organizing centers** and are thus inactive. Polymerization and depolymerization usually occur only at the plus end. This end usually carries a GTP-associated tubulin dimer that favors further growth rather than depolymerization. Guanosine diphosphate–bound (GDP-bound) dimers dissociate much

Table 11.1. Actin-Binding Proteins and Their Functions

Protein	Function
Tropomyosin	Strengthens filament by binding to it along its entire length
Spectrin	Attaches the sides of a filament to the plasma membrane
Filamin	Cross-links microfilaments
α-Actinin	Associates with microfilaments to form bundles
Fimbrin	Bundles microfilaments similar to α-actinin
Vinculin	Binds to α-actinin and thereby attaches filaments to plasma membrane
Myosin	Causes sliding of microfilaments and is responsible for cytoplasmic streaming

more readily than GTP-bound dimers. Thus, microtubules depolymerize about 100 times faster if they are capped at their plus ends by a GDP-bound tubulin dimer. This leads to a situation called **dynamic instability,** in which most microtubules grow and only a few are getting smaller at any moment in time.

Microtubules grow from microtubule-organizing centers. One example of such a center is the **centrosome,** located near the cell nucleus; another is the **basal body** of a ciliary axoneme. Both these structures contain two **centrioles** from which microtubules grow outward. However, not all microtubule-organizing centers contain centrioles. The plus ends of microtubules can be capped, which stabilizes them. As with microfilaments, microtubules participate in cell movement by polymerization and depolymerization. In addition, there are two microtubule-binding proteins that move vesicles and organelles along microtubules. **Kinesin** catalyzes movement outward and **cytoplasmic dynein** inward along microtubules. Both proteins hydrolyze ATP in the process. **Colchicin** inhibits tubulin polymerization and depolymerization and thus interferes with cell division (movement of chromosomes along spindle fibers) and cell motility in general. It is used in the treatment of **gout** because it interferes with the endocytosis of urate crystals by macrophages and thus eliminates the painful inflammatory process.

Intermediate filaments (diameter of 8–10 nm) form meshlike structures in the cytoplasm. They are especially frequent in regions of the cell that are subjected to mechanical stress, such as the desmosomal junctions (see later) between epithelial cells. There are four types of intermediate filaments, each consisting of different protein subunits (Table 11.2).

The polypeptides that make up the different types of intermediate filaments belong to the same gene family. They all contain a homologous central α-helical domain that is interrupted at three locations by short non–α-helical segments. The α-helical segments are capable of forming so-called **coiled coil** structures in which two α helices are wound around each other. These dimeric polypeptides then assemble into much larger filaments in which the non–α-helical domains within the central region and at the termini interact.

Some intermediate filaments are anchored in the cell membrane to **desmosomes.** Desmosomes are points of intercellular contact (see Figure 11.6) and are responsible for the mechanical connection of neighboring cells, increasing the resistance of tissue to tear. They are found predominantly in skin and other epithelia.

Table 11.2. Types of Intermediate Filaments

Type	Protein Components	Cell Types
I	Acidic keratins	Epithelial cells
	Neutral and basic keratins	
II	Vimentin	Mesenchymal cells
	Desmin	Muscle cells
	Glial fibrillary acidic protein	Glial cells
III	Neurofilament proteins	Neurons
IV	Nuclear lamins (3 types: A, B, and C)	All cells

11.1.2 Membrane Permeability and Transport

The lipid bilayer of the cell membrane provides an impermeable barrier for most water-soluble and polar molecules. However, the cell must be able to take up nutrients and ions from the surrounding medium and to discard waste products. This is accomplished by specific transport mechanisms.

Basically there are four mechanisms for the passage of small molecules and ions across a biological membrane:

1. Simple diffusion
2. Facilitated diffusion (e.g., carrier-mediated)
3. Active transport
4. Coupled transport

Only nonpolar substances follow the laws of **simple diffusion** and can cross biological membranes without the aid of catalysts. A few polar molecules, such as CO_2, water, or urea, can cross the membrane by diffusion. All these molecules are very small and uncharged.

Facilitated diffusion involves specific membrane transport proteins that act like enzymes. As with simple diffusion, it is a passive transport process that depends on a concentration gradient of the transported substance across the membrane. The process exhibits Michaelis-Menten kinetics; that is, at high substrate concentrations (on one side of the membrane), the transport system becomes saturated, and the rate of transport reaches a maximum (V_{max}). As observed in competitive enzyme inhibition, structurally similar molecules compete with the substrate. The molecular mechanism may involve shielding of the polar groups of the substrate followed by a conformational change of the carrier on the other side of the membrane (Figure 11.2). Some carrier proteins may also function by forming channels through the plasma membrane similar to the ionophore gramicidin A (see Section 2.6).

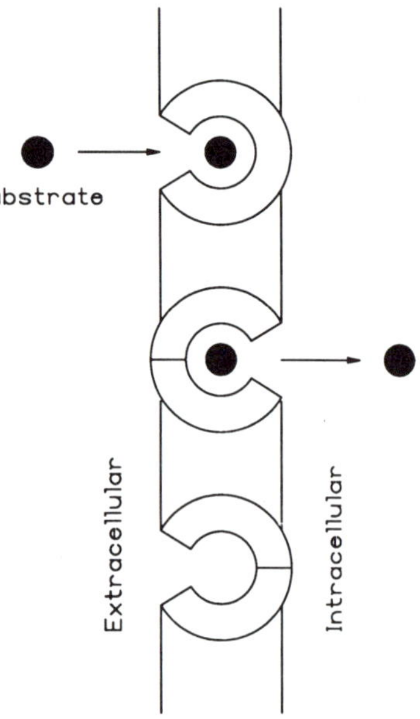

Figure 11.2. Hypothetical mechanism of facilitated diffusion by a carrier protein.

Glucose uptake into erythrocytes and many other cells, with the exception of epithelial cells, occurs by facilitated diffusion. It is called a **uniport system,** since it involves the transport of only one substance at a time. Once inside the cell, glucose is immediately phosphorylated to glucose 1-phosphate. Thus, the glucose concentration gradient across the cell membrane is maintained.

Active transport requires the expenditure of energy in the form of ATP. In contrast to simple diffusion and carrier-mediated diffusion, it usually functions "uphill" or against a concentration gradient. In mammalian cells, there are several **primary active transport systems,** or "pumps," whose function is directly coupled to the hydrolysis of ATP. Several members of the **P class of ion pumps,** so called because they form phosphorylated intermediates during the transport process, have been studied extensively:

1. The first system, also called **Na^+/K^+ exchange pump** or Na^+, K^+-dependent ATPase, has been studied most extensively. It is found in all vertebrate cells and is responsible for maintaining the high intracellular K^+ concentration and, as a consequence, the cell's negative membrane potential. The overall stoichio-

metry of the reaction is

$$2K_o^+ + 3Na_i^+ + MgATP_i \rightarrow MgADP_i + P_i + 3Na_o^+ + 2K_i^+$$

The subscripts i and o refer to intracellular and extracellular locations, respectively.

The molecular mechanism is illustrated in Figure 11.3. After binding three Na^+ ions on the intracellular surface, the membrane-bound transport protein is phosphorylated by ATP, which causes a conformational change of the protein, resulting in the release of the Na^+ ions on the outside of the cell membrane. Following the binding of two extracellular K^+ ions, the protein is dephosphorylated and returns to its original conformation, releasing the K^+ ions into the cytoplasm. The cycle then starts all over again.

The **Na^+/K^+ exchange pump** is an **antiport system** that only works when Na^+ export is coupled to K^+ entry. Vanadium ions (V^{5+}) are specific inhibitors of the system when applied intracellularly; **ouabain,** a **digitalis glycoside,** inhibits when applied extracellularly.

2. The **Ca^{2+} pump,** a **uniport system,** is found in the plasma membranes of most cells and in the membranes of the sarcoplasmic reticulum of muscle cells. The Ca^{2+} that is pumped into the sarcoplasmic reticulum is tightly bound to the protein **calsequestrin.** This protein (MW = 43,000) has many aspartate and glutamate residues that can bind up to 43 Ca^{2+} ions.

$$2Ca_i^{2+} + MgATP_i \rightarrow MgADP_i + P_i + 2Ca_o^{2+}$$

3. The K^+/H^+ exchange pump is found in the parietal cells of the stomach where it functions in the secretion of HCl into the lumen of the stomach.

$$K_o^+ + H_i^+ + MgATP_i \rightarrow MgADP_i + P_i + H_o^+ + K_i^+$$

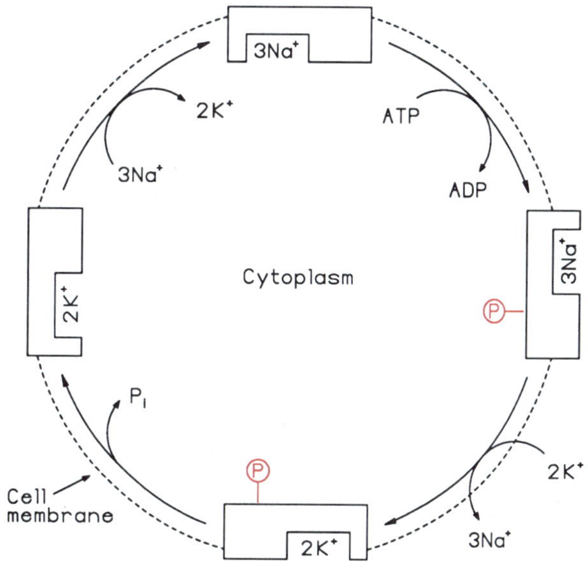

Figure 11.3.
Schematic model of the Na^+/K^+ exchange pump.

The members of the P class of ion pumps are related proteins. They contain nearly identical phosphorylation sites in which an aspartate residue becomes phosphorylated. Other classes of ion pumps also hydrolyze ATP, but the membrane protein is not phosphorylated in the process.

Coupled transport may be called a **secondary active transport** system because it depends on the chemical concentration gradient across the membrane that is generated by a primary pump. For example, the uptake of **glucose** or **amino acids** from the intestinal lumen into epithelial cells, against a concentration gradient, is coupled to the entry of Na^+ ions along a concentration gradient. The Na^+ ion concentration gradient is, of course, created by the Na^+/K^+ exchange pump. It has been postulated that this **Na-coupled glucose transport** system employs a carrier protein with two binding sites, one for glucose and another for a monovalent cation. The occupation of both of these sites by glucose and Na^+ on the extracellular membrane surface triggers a conformational change of the protein resulting in the transport of glucose and Na^+ to the cytoplasmic side of the membrane. Because of the intracellular prevalence of K^+ ions, the Na^+ ion is then replaced by K^+, which results in the dissociation of glucose from the carrier protein and the reformation of the original protein conformation.

Since glucose and Na^+ are transported in the same direction, the transport is called a **symport** system. It is spatially separated from the Na^+/K^+ pump on the membrane. The transport of many other molecules involves symport with Na^+ ions. However, coupled transport does not imply the existence of a symport system. The transport of many substances across the inner mitochondrial membrane (see Figure 2.28) involves **antiport** systems that appear to be driven by concentration gradients. Table 11.3 lists some of the known transport processes.

Table 11.3. List of Cellular Membrane Transport Processes

Type of Diffusion	Examples	Driving Force	Comments	Classification
Simple diffusion	H_2O, O_2, CO_2, NH_3, urea, hydrophobic compounds	Osmotic gradient, concentration gradient	No saturation	Passive
Facilitated diffusion	Glucose uptake in erythrocytes, muscle, adipocytes, liver cells	Concentration gradient	Michaelis-Menten kinetics, saturation	Passive uniport
	Transport of ATP/ADP, malate/α-ketoglutarate, citrate/malate, pyruvate/OH^-, and others		Mitochondrial, saturation kinetics	Passive antiport
Active transport	Na^+/K^+ exchange pump	ATP hydrolysis	Occurs against concentration gradient	Primary active transport
	H^+/K^+ exchange pump	ATP hydrolysis	Occurs against concentration gradient	Primary active antiport
	Ca^{2+} pump	ATP hydrolysis	Occurs against concentration gradient	Primary active uniport
	Cl^- pump	ATP hydrolysis	Ascending loop of Henle	?
Coupled transport	Uptake of glucose or amino acids into intestinal epithelial cells and renal tubular cells	Na^+-gradient dependent	Occurs against concentration gradient	Secondary active symport

Larger molecules (macromolecules) and small particles are transported across the plasma membrane by **endocytosis** (called **pinocytosis** if soluble molecules are involved) and **exocytosis.** For example, proteins (liver), enzymes (pancreas), neurotransmitters (nerve endings), or hormones are packaged into small intracellular vesicles that fuse with the cell membrane upon demand (exocytosis), releasing their contents on the outside. Thereafter, the vesicular part of the plasma membrane becomes coated on its cytoplasmic side by proteins, called **clathrins.** This clustering of clathrins in the plasma membrane **(coated pits)** at the site of vesicle-membrane fusion seems to facilitate the reformation of the vesicle. It appears that clathrin is required for the formation of vesicles that function in the endocytosis of proteins. Clathrin-coated endocytotic vesicles have a latticelike appearance under the electron microscope.

Many soluble macromolecules pass through the endothelial cells of blood capillaries via vesicles (Figure 11.4).

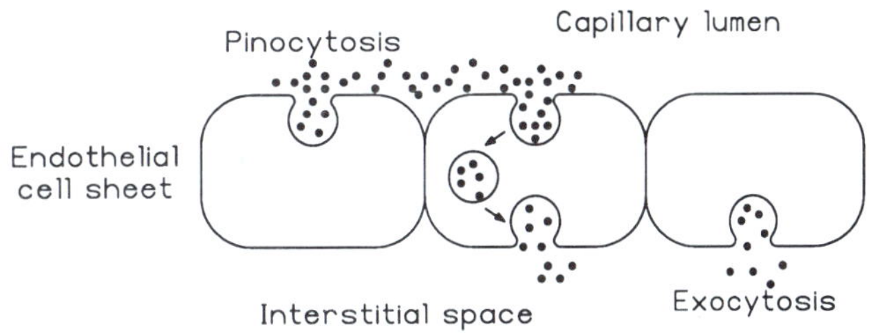

Figure 11.4. Passage of macromolecules through an endothelial cell.

11.1.3 Intercellular Communication

Different cells in a multicellular organism must be able to communicate with one another. This is accomplished by three basic mechanisms:

1. Chemical signaling. One cell releases a chemical compound that binds to a receptor on the target cell where it initiates a specific action. Examples of this type of signaling are hormonal regulation and the release of transmitters at nerve terminals (synapses).
2. Direct interaction between membrane proteins of different cells. Examples are the adhesion molecules found in the membranes of most cells.
3. Exchange of small molecules and ions between cells via gap junctions, which are assemblies of cell-cell channels that directly connect the cytoplasms of two cells.

In this chapter, we will discuss only the last two types of communication. Communication by hormones is discussed in Chapter 13.

Cells of a specific type recognize each other because they carry on their membrane surface-specific markers, usually represented by oligosaccharide side chains of glycoproteins or glycolipids embedded in the membrane. For example, when liver cells and kidney cells are mixed, the liver cells, recognizing only those surface markers present on other liver cells, aggregate only with other liver cells, whereas the kidney cells associate with other kidney cells. The membrane proteins responsible for this cell-cell recognition are the **adhesion molecules.** Each cell type appears to express a different type of adhesion molecule.

For example, neurons express in their membranes **neural cell adhesion molecules (N-CAM).** N-CAMs have been isolated from retinal cell membranes. They consist of several types of membrane proteins with identical extracellular domains. They are produced by differential splicing of a single pre-mRNA that generates different carboxy terminal domains. Their extracellular segment consists of five immunoglobulin-like domains (Figure 11.5). These domains bind by homophilic interaction to identical domains on the surface of another cell thereby causing cell-cell adhesion.

A different class of adhesion molecules are the **cadherins.** It now appears that cadherins are involved in the general phenomenon of cell adhesion between cells of the same type. Cadherins require calcium ions for activity. There are several subclasses of cadherins, all of which share the same basic structure. So far, **E cadherin** (epithelial), **P cadherin** (placental), **N cadherin** (neural), and **L cadherin** (liver) have been characterized biochemically. Cadherins are integral membrane proteins with a conserved cytoplasmic carboxy terminal domain that interacts with the cytoskeleton and an extracellular amino terminal domain with three repeats that contain several amino acid clusters of the type DXNDN, DXD and DRE (X = any amino acid). It is thought that the aspartate residues are involved in Ca^{2+} binding that allows the protein to assume the configuration required for interaction with identical molecules on the neighboring cell. The extracellular domains differ in different cadherins; they provide the specificity for cell-cell adhesion.

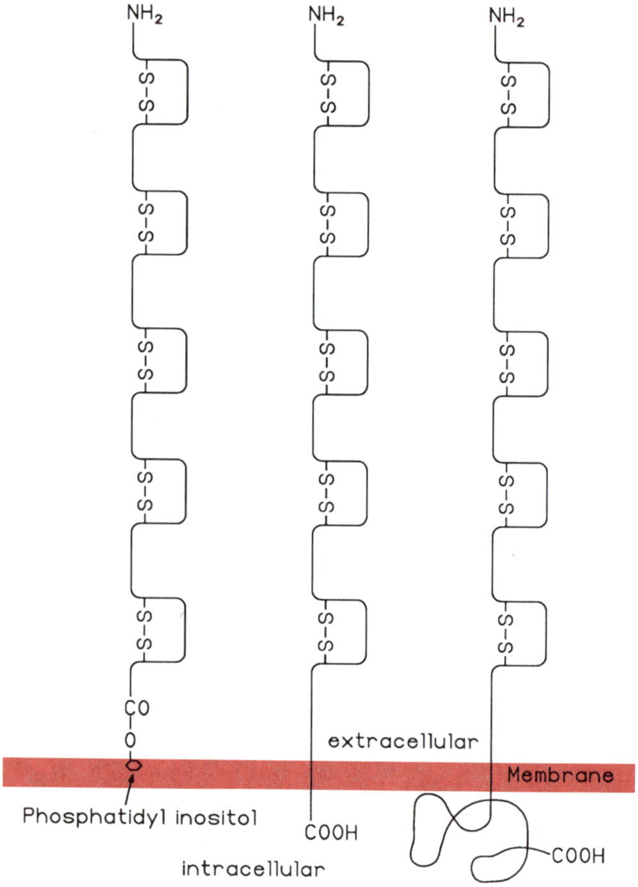

Figure 11.5.
Structures of three types of N-CAMs. The left-most molecule has no cytoplasmic domain. Its carboxyl terminus is linked to phosphatidyl inositol in the membrane.

There is some evidence that metastatic tumor cells, that is, tumor cells that invade other tissues, have lost the expression of their cadherins.

Many types of cells form small cytoplasmic channels with their neighbor cells through which small molecules and ions can pass. A **cell-cell channel** consists of two sets of six transmembrane proteins, called **connexins,** circularly arranged in the membranes of two neighboring cells. The complete channel thus contains 12 connexin subunits. Usually cell-cell channels occur in the form of large clusters. They are called **gap junctions** because of their typical appearance in the electron microscope. Different cells often express different types of cell-cell channels. Because of the highly homologous extracellular domains of all connexin proteins, hybrid channels consisting of two types of hemichannel sometimes can form between two different types of cells.

Cell-cell channels are **gated,** i.e., their **open-closed state** can be controlled by effectors such as ligands or transjunctional voltage. For example, cell-cell channels in heart muscle are sensitive to Ca^{2+} ions. Upon a rise in the intracellular Ca^{2+} concentration, as might occur after cell damage, the channels close. This mechanism seems to protect undamaged cells from the influx of Ca^{2+} ions that might occur in a neighboring cell. Embryonic cell-cell channels are very sensitive to protons; upon a slight decrease in the cytoplasmic pH the channels close. Embryonic channels are also sensitive to transmembrane voltage (voltage-gated).

Gap junctions are found in the membranes of most cells. They are especially abundant in heart muscle where they couple neighboring cells electrically, thus allowing an action potential to spread through the muscle mass. Smooth muscle cells of the uterus in both virgin and pregnant females have few gap junctions in their membranes. However, just before parturition, gap junctions appear throughout the muscle, allowing action potentials to pass through the uterus muscle, thus causing synchronous contractions.

Some channels in the nervous system are asymmetrically sensitive to transjunctional voltage; that is, they close when cell 1 is positive in

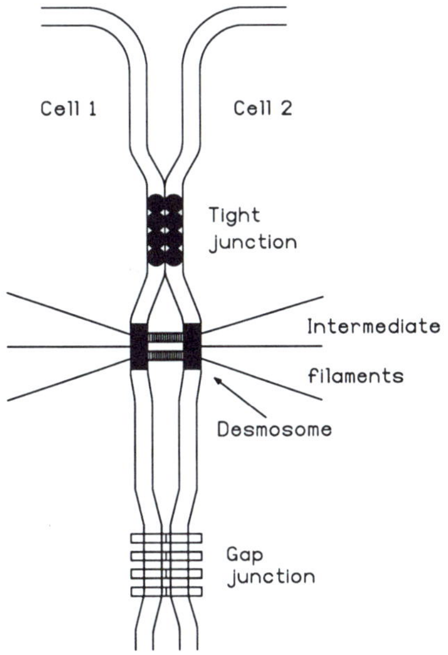

Figure 11.6.
Schematic drawing of the three different types of cell-cell junctions.

relation to cell 2 but remain open when the polarity of the transjunctional voltage is reversed. Such channels function as rectifiers, passing current only in one direction. It is likely that such **rectifying electrical synapses** are hybrid channels consisting of two different types of hemichannel: one being voltage-gated, the other not.

Another type of cell-cell junction is the **tight junction.** Tight junctions bring the membranes of adjacent cells into very close contact, thereby providing a seal through which extracellular fluid cannot pass. They also restrict the lateral diffusion of membrane proteins and thus confine certain proteins to one or the other side of the cell (e.g., mucosal and serosal side). They are found most abundantly in renal tubular epithelium and in the capillary endothelium of the central nervous system (CNS; blood-brain barrier). Figure 11.6 illustrates the different types of cell-cell junctions.

11.2 CONNECTIVE TISSUE

Connective tissue consists of four major components: collagen, elastin, proteoglycans, and glycoproteins. Tissues requiring high tensile strength, for example, skin, bone, and tendons, contain **collagen** as the major structural protein. Elastic tissues that are subject to rapidly changing tension, for example, the walls of large arteries, contain **elastin,** a highly cross-linked protein with rubberlike properties. **Proteoglycans** represent the ground substance that fills the extracellular space not occupied by fibrous or elastic structures. Because of their hydrophilic nature, proteoglycans bind large amounts of water and thus facilitate the diffusion of water-soluble molecules through the tissue. Glycoproteins are involved in interconnections among cells and the other matrix components.

11.2.1 Collagen

Collagen Structure

Collagen is by far the most abundant protein in the human body, accounting for roughly 25% of all proteins. Collagen fibers are constructed from many individual α **chains,** forming structures of increasing complexity: α chains → tropocollagen → collagen fiber.

Collagen α chains (do not confuse with α helix) contain about 1,050 amino acids. Every third amino acid is a glycine residue; proline and hydroxyproline account for another 30% of the amino acids. Because of its high proline content (no rotation possible between C_α and N atoms), each α chain assumes the form of a left-handed helix. Three such α chains assemble spontaneously into a **superhelix** in which individual α chains are wound around one another. This triple-stranded **tropocollagen** molecule is 3050 Å in length. It is assembled into much longer and thicker **collagen fibers** in such a way that the amino terminals of individual tropocollagen molecules are staggered against each other by roughly one-fifth of their length. Furthermore, there is a 400 Å gap between the carboxyl terminus of one tropocollagen molecule and the amino terminus of the next. This precise alignment of tropocollagen molecules is responsible for the cross-striations of collagen fibers at 670 Å intervals that are observed under the electron microscope (Figure 11.7).

Biosynthesis of Collagen

The synthesis of α chains occurs on the rough endoplasmic reticulum (ER) of fibroblasts. The native **pre-pro–α chain** contains a 70-amino-acid signal sequence at its amino terminus that is responsible for the transport of the molecule into the cisternae of the ER. There, the signal peptide is cleaved off and the resulting **pro–α chain** is further modified by three posttranslational processes:

1. Proline is hydroxylated to hydroxyproline.
2. Lysine is hydroxylated to hydroxylysine.
3. Specific hydroxylysine residues are glycosylated to glucosyl-galactosyl-hydroxylysine.

After these modifications, the C-terminal propeptides of three α chains come together, forming disulfide bonds between them. This aligns the α chains for the formation of the triple helix. The resulting dumbbell-shaped **procollagen** molecule is transported to the Golgi apparatus, where oligosaccharides are added to the C-terminal peptides, and is subsequently exported from the cell via exocytotic vesicles (Figure 11.8).

In the extracellular space, two **procollagen peptidases** (metalloproteases) cleave off the C- and N-terminal peptides, producing the final **tropocollagen** molecule. (These peptide domains prevent the procollagen molecule from assembling into tropocollagen while still inside the cell.) Tropocollagen molecules assemble spontaneously into collagen fibers. Tropocollagen still carries small **telopeptide** regions

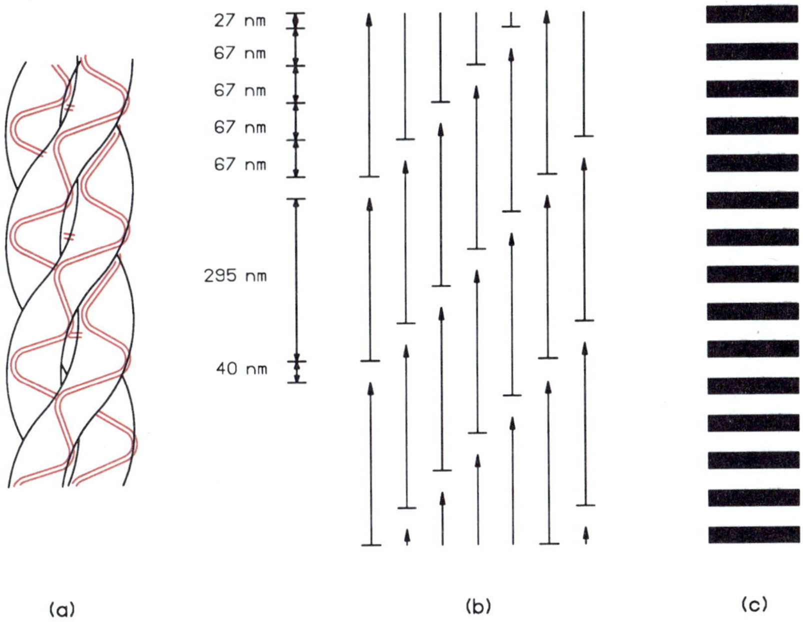

Figure 11.7.
Structure of collagen: (a) tropocollagen superhelix, (b) staggered alignment of tropocollagen in the collage fiber, (c) repeat pattern observed under the electron microscope.

Figure 11.8.
Biosynthesis of collagen.

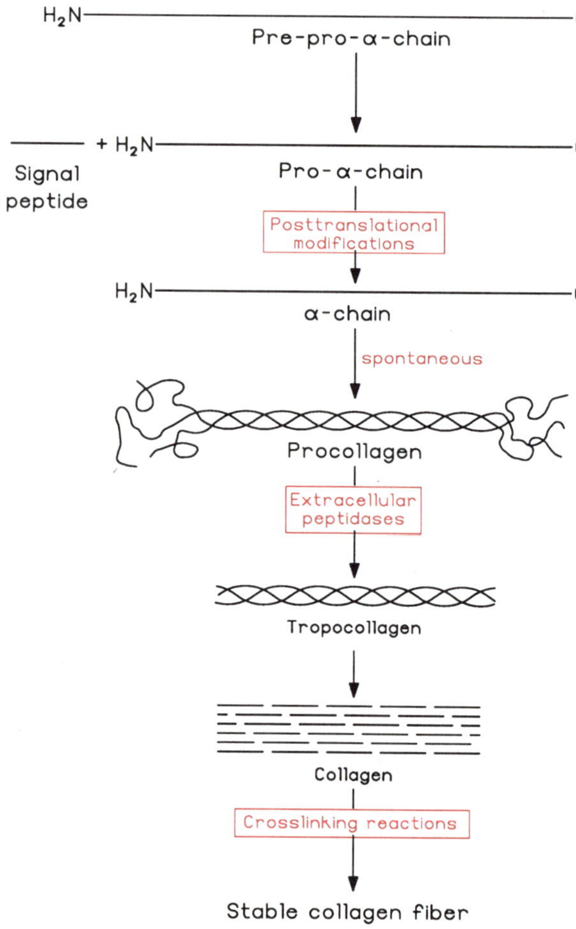

at each end that are not part of the triple helix. Presumably, they direct the alignment of the tropocollagen molecules within the collagen fiber structure and participate in the cross-linking between molecules. The final size of the collagen fibril appears to be determined by **dermatan sulfate proteoglycans** (see next section) that coat the surface of the fibril.

In order to stabilize the still rather weak collagen fiber, several **cross-linking** reactions occur that covalently interconnect α chains within individual tropocollagen molecules as well as in different tropocollagen molecules. Cross-linking is initiated by **lysyl oxidase,** a Cu-containing enzyme, which converts the ϵ amino group of a lysyl residue into an aldehyde **(allysine)** by oxidative deamination. Two basic types of cross-linking reactions can now take place. Either two allysines undergo an aldol condensation reaction or an allysine reacts with the ϵ amino group of an unmodified lysine to form a Schiff base. More complex cross-links are derived from these two basic structures (Figure 11.9). Most cross-links occur between the telopeptide of one molecule and the helical region of a neighboring molecule. However, helical-helical cross-links are also formed. Without these cross-links the collagen fiber would possess little or no tensile strength.

So far, more than 22 different genes for collagen have been identified. There are four major types of collagen in the human body; they are listed in Table 11.4. Another eight minor types have been identified to

Figure 11.9.
Types of cross-links between lysine (or hydroxylysine) residues of different α chains.

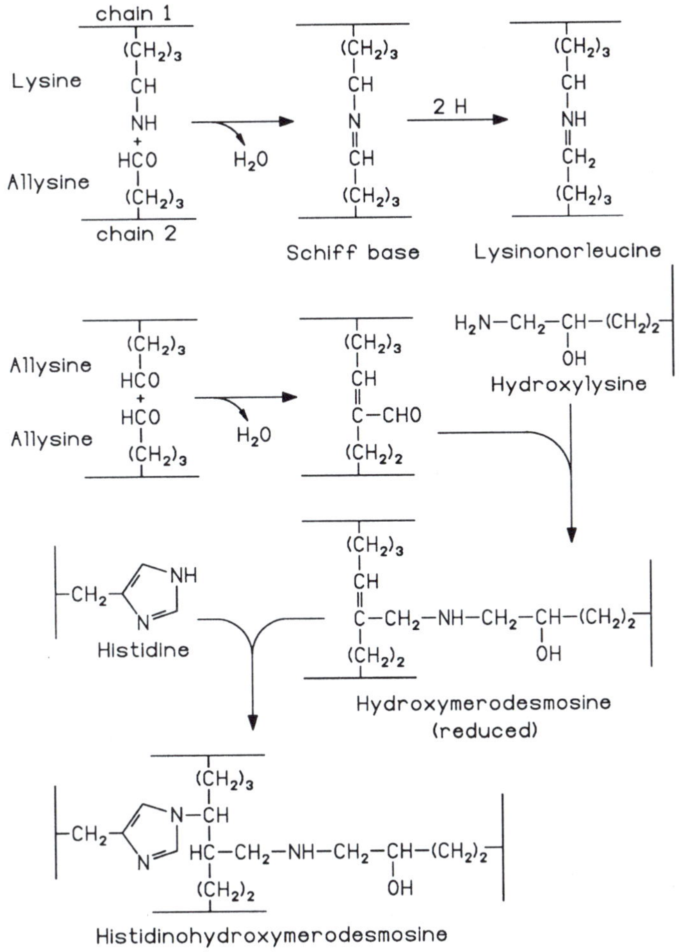

date. Depending on the type of tissue, collagen may exist in the form of fiber bundles or as a fiber network.

Collagen Degradation

Collagen is an extremely stable protein with a very long half-life. There are some instances, however, where collagen fibers are degraded rather

Table 11.4. Major Types of Human Collagen and Their Properties

Type	Composition	Form	Tissue Source
I	$\alpha1(I)_2$, $\alpha2(I)$, $\alpha1(I)_3$	Fibril	Skin, bone, tendon, ligaments, cornea (accounts for 90% of all collagen in human body)
II	$\alpha1(II)_3$	Small fibril	Cartilage, vitreous body of the eye, intervertebral disk
III	$\alpha1(III)_3$	Small fibril	Skin, blood vessels, uterus
IV	$\alpha1(IV)_2$, $\alpha2(IV)$	Network	Basal lamina

rapidly. Examples are the postpartum involution of the uterus, bone remodeling, or the development of decubitus ulcers. A defect of collagen degradation is responsible for the development of keloids or in scleroderma. There are two pathways of collagen degradation:

1. The triple helix can be degraded extracellularly by **collagenases,** which are produced by many cell types, into smaller peptide fragments that are subsequently degraded by unspecific proteases.
2. Fragments of collagen fibrils may be phagocytosed by macrophages. Inside the cell, the fragments enter the lysosomes where they are degraded by **cathepsin B.**

Both pathways probably are required and function in sequence.

The synthesis of collagenase, a Zn^{2+}-containing metalloprotease, is tightly controlled. It only occurs in response to specific stimulation by factors such as interleukin 1 or retinoic acid. The enzyme is secreted as an inactive proenzyme that must be activated by the action of another protease. Furthermore, tissues contain a specific collagenase inhibitor. In the bloodstream, collagenase is inactivated by the general protease inhibitor α_2-macroglobulin.

11.2.2 Elastin

Elastin fibers are found in structures such as the walls of large arteries, for example, the aortic arch, which are constantly changing their shape. Despite their fibrous appearance, the inner structure of elastin fibers is rubberlike and amorphous. Individual **tropoelastin** molecules crisscross the fiber in all orientations and are connected by many cross-links to form a three-dimensional network. When this structure is stretched, it snaps back to its original shape as soon as the stretching force disappears.

Elastin is synthesized by smooth muscle cells. The process begins with the synthesis of a **glycoprotein** rich in cysteinyl residues. This protein aggregates outside the cell into **bundles of microfibrils** by forming intermolecular disulfide bridges. The same cells also synthesize and excrete **tropoelastin,** a 74 kd protein rich in glycine, alanine, and valine. Tropoelastin contains a small amount of hydroxyproline and 33 residues of lysine, of which about 24 may be involved in cross-linking.

The tropoelastin molecules form disulfide bonds to the microfibrils and then undergo an extensive cross-linking process. The process of cross-linking is similar to that occurring in collagen. Lysine side chains are oxidized to allysine, which then reacts with other allysine and lysine residues. α-Helical regions of one tropoelastin molecule, with two allysine residues on adjacent turns, interact with similar regions on another tropoelastin molecule to form four-membered ring structures called **desmosines** (Figure 11.10). A desmosine is derived from three allysine side chains and one lysine. It joins two tropoelastin molecules together. Multiple desmosines, as well as lysinonorleucine bridges, in one tropoelastin molecule join it to several other molecules, thus generating a three-dimensional network. The overall biosynthetic pathway for elastin fibers is shown in Figure 11.11.

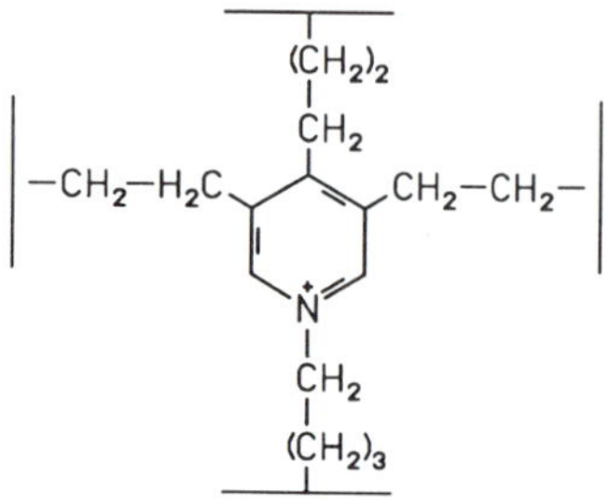

Figure 11.10.
Structure of a desmosine.

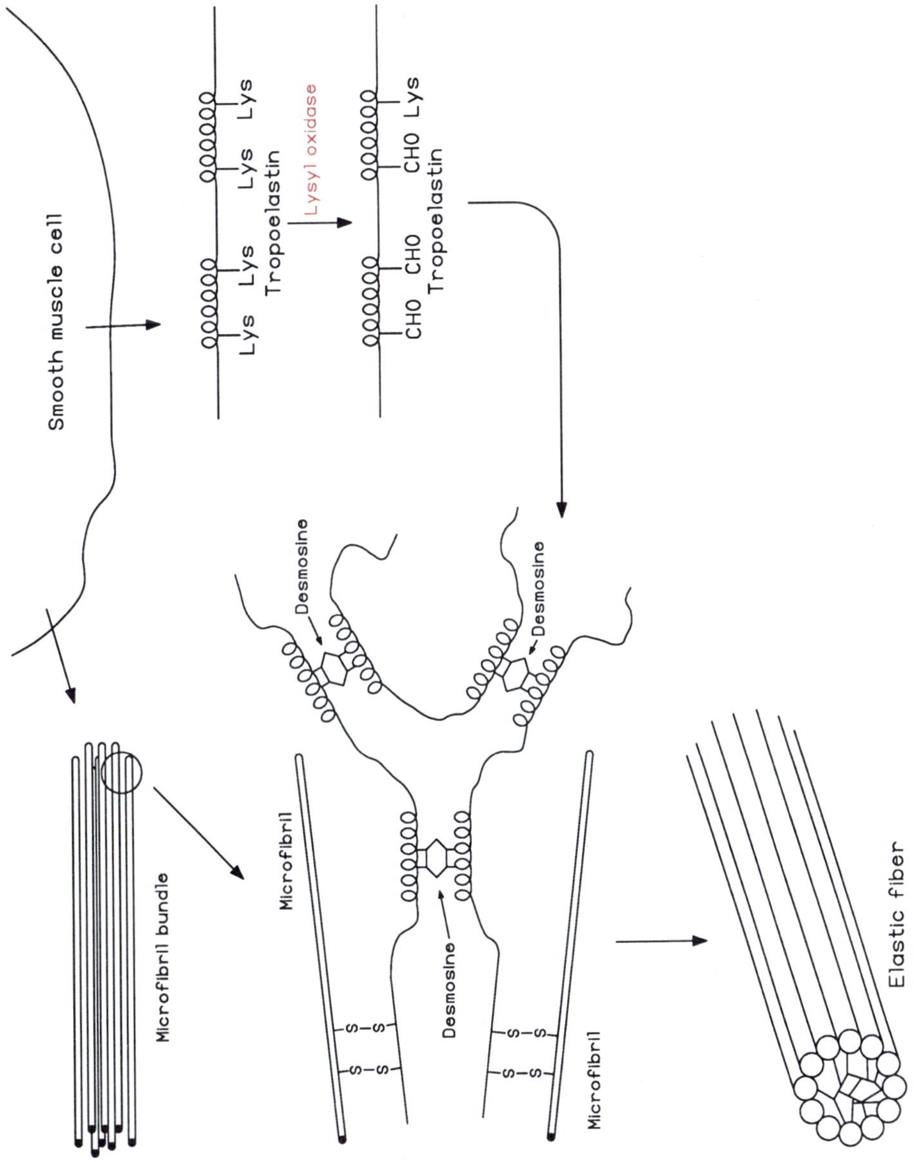

Figure 11.11.
Biosynthesis of elastin fibers.

11.2.3 Proteoglycans

The space between collagen and elastin fibers is occupied with different types of proteoglycans, depending on the type of connective tissue. A proteoglycan molecule consists of a **core protein** to which up to 100 polyanionic glycosaminoglycan side chains are covalently linked via glycosidic bonds to serine and threonine residues. The linkage usually occurs through a specific trisaccharide with the sequence

Core protein-Ser-xylose-galactose-galactose-[disaccharide repeat]$_n$

Glycosaminoglycans (GAG), earlier called **mucopolysaccharides,** are linear polysaccharides with repeating disaccharide units (20–50 units per chain). One of the sugars always is a glucosamine or galactosamine with a sulfate or acetyl substituent. The other sugar may be a uronic acid or a simple hexose. **Hyaluronic acid** is unique among the GAGs in that it contains no sulfate groups. The structures of the other GAGs, **keratan sulfate, chondroitin 4-sulfate, chondroitin 6-sulfate,** and **dermatan sulfate** are shown in Figure 11.12. **Heparan sulfate** is similar to heparin, except that it contains more *N*-acetyl groups and fewer *O*-sulfate groups (see Figure 9.33).

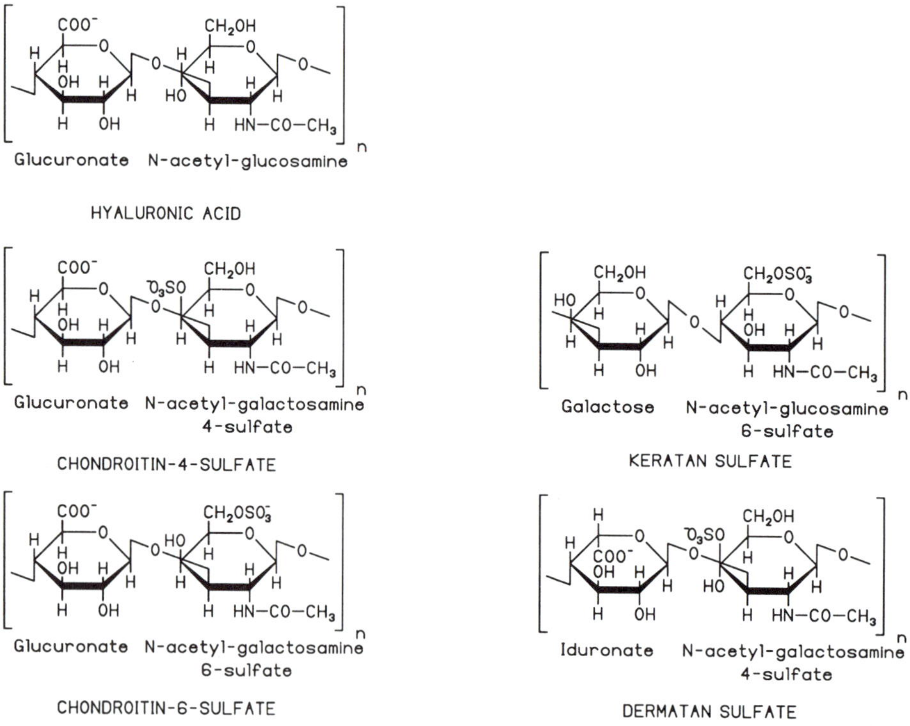

Figure 11.12.
Disaccharide units found in glycosaminoglycans.

The **mucus** produced by epithelial cells also contains glycoproteins but includes mostly neutral sugars without sulfate and carboxyl groups.

In **cartilage**, proteoglycan molecules are further organized into a giant **proteoglycan complex.** In this complex, several core proteins (20–30), together with their GAG side chains, are connected noncovalently with the aid of a **link protein** to a linear chain of hyaluronic acid (Figure 11.13). Table 11.5 shows the composition of different types of connective tissue.

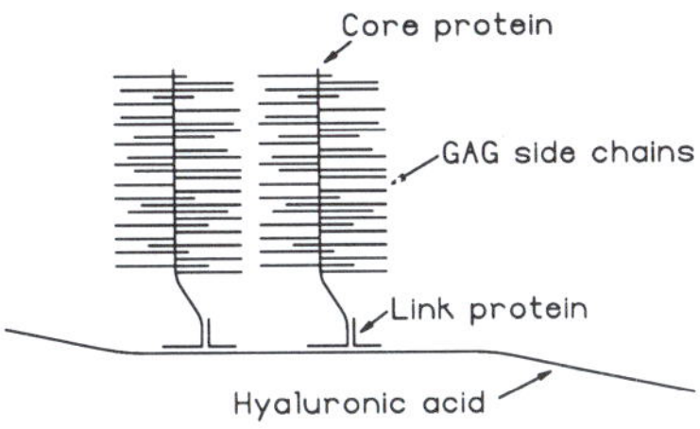

Figure 11.13.
Structure of the proteoglycan complex found in cartilage.

11.2.4 Architecture of the Extracellular Matrix

In addition to its major components, collagen, elastin, and proteoglycans, the extracellular matrix contains more than 30 other glycoproteins that participate in the interconnection of cells and extracellular structures. Only a few of these components can be mentioned here.

Fibronectin functions in the adhesion between cells and extracellular matrix. It consists of two similar subunits that are connected at their C-termini by disulfide bonds. The polypeptide chains contain several domains specific for binding to collagen, fibrin, and heparin. In addition, it has a cell-binding domain that contains the critical sequence **Arg-Gly-Asp (RGD)** also found in other glycoproteins that bind to a cell surface receptor called **integrin.** The fibronectin gene contains 47 exons; its pre-mRNA is subject to alternate splicing that produces a variety of different forms of fibronectin.

Laminin is a large glycoprotein (Mw ~ 850,000) consisting of three polypeptide chains: A, B_1, and B_2. It functions in the attachment of epithelial cells to the **basal lamina.** In electron micrographs, laminin appears as a crosslike structure (Figure 11.14). The molecule has a triple helix domain (coiled coil α-helical domain) in which three polypeptide chains are wound around each other. In the center of the molecule, the two B chains separate from the A chain, forming the two side arms of the cross. At the center of this cross is the cell-binding domain, the base binds to heparan sulfate, and the arms bind to type IV collagen, the major collagen in the basal lamina.

Basal laminae are sheetlike meshworks, consisting of a collagen type IV network filled with glycoproteins and proteoglycans. Basal laminae separate cells from the surrounding connective tissue. They function as a scaffolding device for the organization of cells. For example, muscle cells are individually wrapped in sheets of basal lamina, and epithelial

able 11.5. Composition of Different Types of Connective Tissue

		Types of Glycosaminoglycans					
issue	Type of Fibers	Hyaluronic Acid	Keratan Sulfate	Chondroitin 4-Sulfate	Chondroitin 6-Sulfate	Dermatan Sulfate	Heparan Sulfate
Jmbilical cord		+ + +					
itreous humor (eye)	Collagen (N)	+ + +					
ornea	Collagen (N)		+ + +	+ +		+	
kin	Collagen (F) Elastin	+ +				+ + +	+ +
essels	Collagen (N,F) Elastin	+ +	+		+ + +	+	
ung	Collagen (N)		+ +			+	+
iver	Collagen (N)					+	+
endons	Collagen (F)	+			+ +	+ +	
ntervertebral disc	Collagen (F)		+ +	+ +			
leart valves	Collagen (N)	+ + +			+ +	+ +	
artilage	Collagen (F) Elastin	+	+	+ + +	+		
one	Collagen (F)		+	+	+		

I = network, F = fiber.

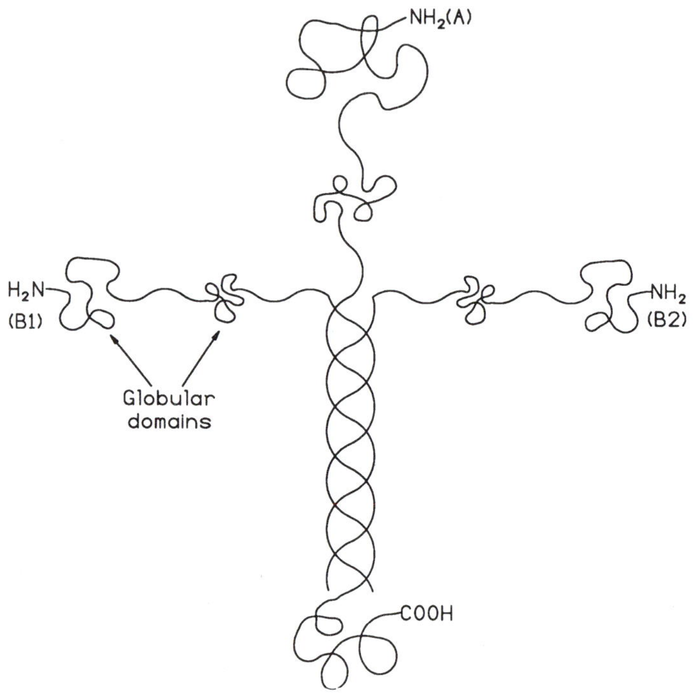

Figure 11.14.
Schematic representation of the structure of laminin.

cells grow on the surface of basal laminae. In the glomerulus of the kidney, basal laminae separate the endothelial cells of capillaries from the epithelial cells of the tubular cells. Here the basal lamina functions as a filter for the filtration of plasma.

11.2.5 Disorders of Connective Tissue Metabolism

Defects in several steps of collagen biosynthesis cause the **Ehlers-Danlos syndrome.** Several clinically distinct types of the syndrome can be distinguished, although they all produce similar symptoms characterized by fragile, elastic skin, hyperextensible joints causing frequent dislocations, and weakness in vessel walls. Enzyme defects that have been identified in Ehlers-Danlos syndrome include lysyl oxidase, lysyl hydroxylase, and procollagen peptidase. It is clear that in all these cases the stability of the collagen fiber structure must be severely affected.

Marfan syndrome is a disorder whose cause is still debated. It has been shown that the number of desmosomes in elastin is reduced, causing a general weakness of connective tissue. Recent reports point to a defect in **fibrillin,** an elastin-associated glycoprotein. Patients with Marfan syndrome frequently die of aortic aneurisms.

Osteogenesis imperfecta is a group of inherited disorders characterized by brittle bones that break frequently. It involves mutations in the $\alpha 1(I)$ gene. Heterozygotes with one normal and one mutant $\alpha 1(I)$ gene produce reduced amounts of type I collagen that results in the formation of bones that are structurally weak.

There are several diseases, the **mucopolysaccharidoses,** in which the metabolic turnover of GAGs is impaired because of defective hydrolytic (lysosomal) enzymes. Such a defect usually leads to the accumulation of the undegraded GAGs within the lysosomes of various

Table 11.6. Some Mucopolysaccharidoses

Syndrome	Type	Clinical Characteristics	Biochemical Findings	Enzyme Defect	Genetics
Hurler	I	Mental retardation, skeletal deformations, corneal opacity, somatic changes	Dermatan sulfate and heparan sulfate in urine and tissues, dermatan sulfate in fibroblasts		Autosomal recessive
Hunter's	II	Mental retardation, skeletal deformations, deafness, somatic changes	Dermatan sulfate and heparan sulfate in urine and tissues, dermatan sulfate in fibroblasts	α-L-iduronidase, iduronate sulfatase	X-linked recessive
Sanfilippo's (two types)	III	Mental retardation, some skeletal changes	Heparan sulfate in urine and tissues, dermatan sulfate in fibroblasts	Heparan *N*-sulfatase, *N*-acetyl-α-D-glucose aminidase	Autosomal recessive
Morquio's	IV	Skeletal deformations, spondylepiphyseal dysplasia	Keratan sulfate and chondroitin sulfate in urine	*N*-acetyl-D-galactosamine 6-sulfatase	Autosomal recessive
Scheie's	V	Severe corneal opacity	Dermatan sulfate in urine	α-L-iduronidase	Autosomal recessive
Maroteaux-Lamy	VI	Skeletal deformations, corneal opacity	Dermatan sulfate in urine	*N*-acetyl-D-galactosamine 4-sulfatase	Autosomal recessive

cells. It may cause brain damage, retardation, and skeletal deformations (see Table 11.6).

11.3 MUSCLE CONTRACTION

The human body contains three types of muscle: **striated** or **skeletal muscle, smooth muscle,** and **cardiac muscle.** All three contain the same basic **actomyosin** contractile system.

The functional subunit of a muscle fiber (muscle cell) is the **myofibril,** a 1 μm-thick bundle of contractile proteins surrounded by an excitable cell membrane, the **sarcolemma.** Each myofibril extends over the entire length of the muscle fiber and is composed of 2.3 μm-long repeating subunits, the **sarcomeres.** Each muscle cell contains many myofibrils. The characteristic pattern of cross-striations of skeletal muscle, as seen under the electron microscope, is due to the arrangement of the sarcomeres (Figure 11.15).

The myofibril contains two types of interdigitating protein filaments: the **thick filaments** (150 Å) composed of **myosin,** and the **thin filaments** containing mostly **actin** and smaller amounts of **tropomyosin** and **troponin.** During muscle contraction, the thin filaments slide into the space between the thick filaments, thus reducing the length of the sarcomere by 25–50% (Figure 11.15b).

Thick filaments consist of several hundred myosin molecules arranged in a bundle (see Figure 11.16b). Each myosin molecule (Mw = 460,000) is constructed from two identical heavy chains (Mw = 200,000) and four light chains of two types, A-1 (Mw = 21,000) and A-2 (Mw = 17,000). Trypsin cleaves myosin in the middle (at the hinge region) to yield two fragments, **light meromyosin (LMM)** and **heavy meromyosin (HMM).** HMM can be further cleaved by papain into an **S2 fragment** and two **S1 fragments.** Both LMM and S2 are largely α-helical in nature, whereas the S1 heads are globular (Figure 11.16a).

The two myosin heads (S1) have actin-binding properties and also exhibit ATPase activity. They reach out from the thick filament to interact with the thin filaments. In the center of the thick filament there

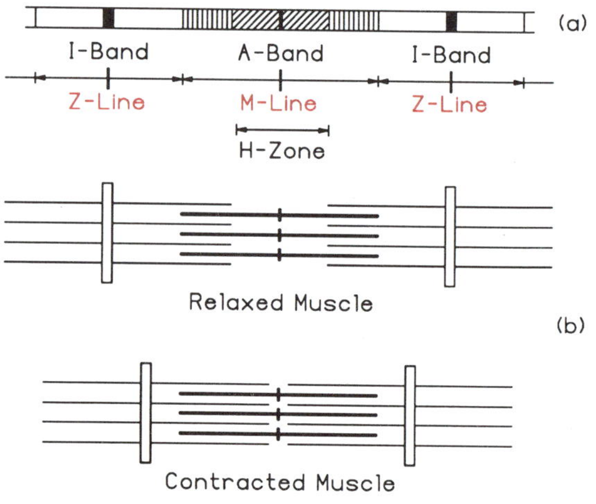

Figure 11.15.
The sarcomere: (a) EM pattern, (b) arrangement of thin and thick filaments.

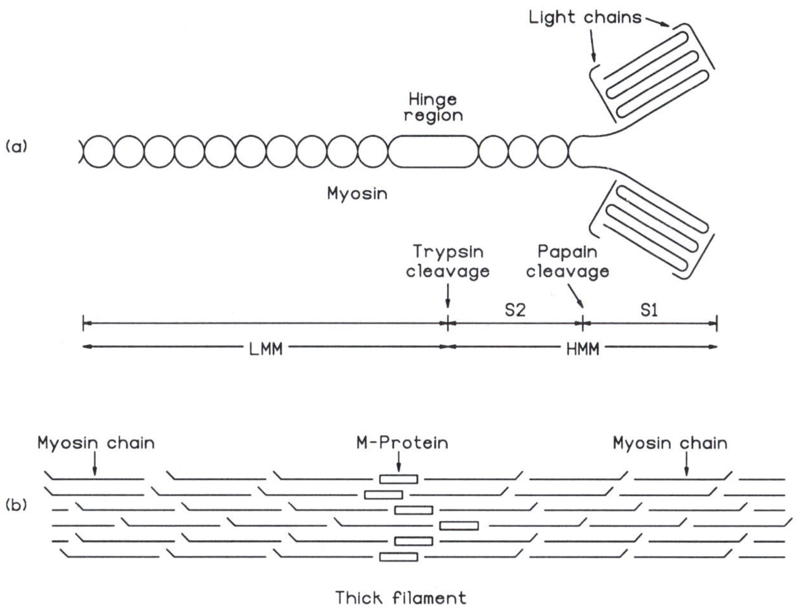

Figure 11.16.
Structure of the thick filament: (a) structure of myosin, (b) arrangement of myosin molecules in the thick filament.

is a bare zone that consists only of myosin tails that are linked to a central **M protein**. The arrangement of the myosin molecules within the thick filament is such that their heads point away from the center and outward from the surface of the filament.

Thin filaments consist primarily of actin. The globular **G actin** subunits (MW = 42,000) form a polymer, **F actin,** in which two strings of G actin molecules are wound around each other to produce a double helix (Figure 11.17).

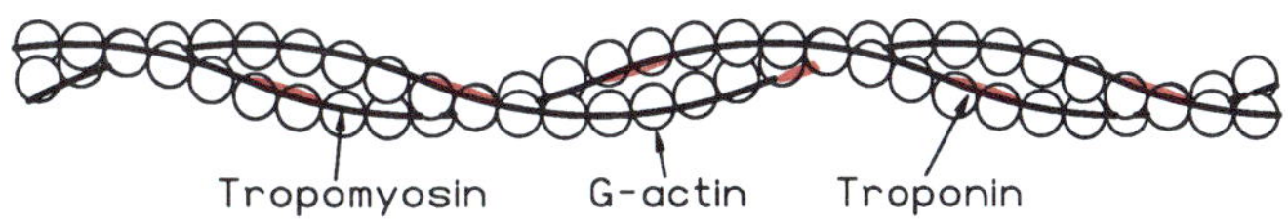

Figure 11.17.
Structure of the thin filament.

Analogous to the bipolar arrangement of myosin molecules in the thick filaments, the two halves of the thin filaments, extending from the Z line (see Figure 11.15), also exhibit opposite polarity. This can be visualized under the electron microscope when separated myosin heads (S1) are complexed to thin filaments. The interaction of F actin filaments at the Z disc involves the protein **α-actinin.**

In addition to G actin, thin filaments contain two regulatory proteins, **tropomyosin** and **troponin.** Tropomyosin is a long α-helical molecule consisting of two nonidentical polypeptide chains that form a continuous helix that is wrapped around the F actin helix. Troponin is a globular protein of three subunits, **troponin C** (C for Ca^{2+} binding), **troponin I** (I for inhibitory), and **troponin T** (T for tropomyosin binding).

In the absence of Ca^{2+} ions, the troponin-tropomyosin complex inhibits the interaction of actin subunits with the myosin heads sticking out from the surface of the thick filaments. This interaction between actin and myosin is required for the release of ADP and inorganic phosphate from the ATPase site of the myosin head. The release of ADP and

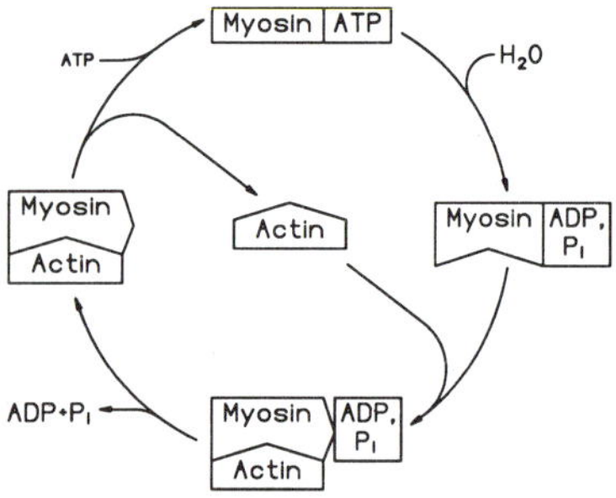

Figure 11.18.
ATP-catalyzed interaction of actin and myosin.

P_i causes a conformational change of the myosin molecule, resulting in a tilting of the myosin head.

Myosin has a very high affinity for ATP. As soon as ATP binds to myosin, the S1 head dissociates again from the actin molecule that was bound to it. ATP is then hydrolyzed by the ATPase but the reaction products, ADP and P_i, remain bound to the site, thus preventing further ATP binding and hydrolysis until the S1 head again interacts with another G actin molecule (Figure 11.18).

The sequence of events that causes the sliding of thick filaments onto the thin filaments, thus producing a contraction of the myofibril, is illustrated in Figure 11.19 and can be summarized as five distinct steps:

1. Muscle contraction is initiated by an action potential that depolarizes the muscle cell membrane and causes the release of Ca^{2+} ions from the sarcoplasmic reticulum into the myofibrils.
2. The Ca^{2+} ions bind to troponin C, causing a conformational change of the troponin-tropomyosin complex on the thin filament. This activates actin for interaction with the S1 heads of myosin.
3. Upon combining with actin, the ADP-P_i-myosin complex releases ADP and P_i, and the S1 head, still bound to the G actin subunit of the thin filament, tilts backward, causing a movement of the thick filament in relation to the thin filament. ADP is rephosphorylated to ATP by creatine kinase, which transfers phosphate from creatine phosphate, the storage form of high-energy phosphate in muscle (see Section 4.11).
4. ATP binds to the S1 head of myosin, causing it to dissociate from the actin and assume its original configuration. ATP is then hydrolyzed, and the resulting ADP-P_i-myosin complex can again interact with another G actin subunit as long as Ca^{2+} ions are still present, as is the case during a prolonged muscle contraction.
5. Relaxation of the muscle begins when no new action potentials occur, so that the intracellular levels of Ca^{2+} fall due to the action of the Ca^{2+} pumps. This restores the inhibition of actin-myosin interaction by the troponin-tropomyosin complex.

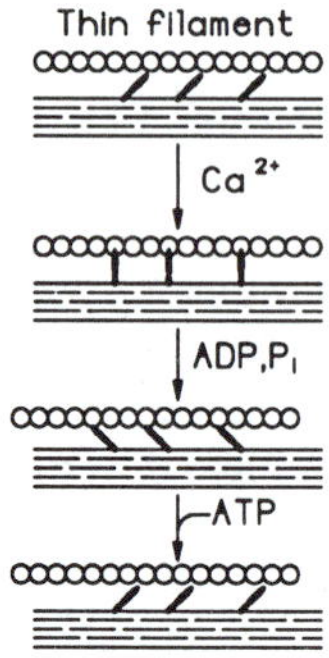

Figure 11.19.
The mechanism of muscle contraction.

It is important to note that the hydrolysis of ATP by itself does not produce muscle contraction. It is only the release of the hydrolysis products, ADP and inorganic phosphate, from the myosin head that triggers the conformational change necessary for the sliding of thick filaments in relation to the thin filaments.

In **tetany** (rapid succession of action potentials), the high intracellular concentration of Ca^{2+} ions prevents muscle relaxation. **Rigor mortis** (muscle contraction after death) is caused by the depletion of intracellular stores of ATP. Since ATP is required for the dissociation of actin from the myosin heads, the muscle remains in a contracted state.

In smooth muscle, the myofibrils are not as tightly packed and organized as in skeletal muscle. There are also differences in the structures of myosin and tropomyosin. These filaments do not contain any troponin. Smooth muscle contraction is regulated by phosphorylation and dephosphorylation of the light chains located in the myosin heads. In unphosphorylated form, these chains prevent the interaction of the myosin head with actin. As in skeletal muscle, smooth muscle contraction is initiated by the release of Ca^{2+} ions from intracellular stores. In a complex with **calmodulin,** Ca^{2+} activates a **myosin kinase,** which phosphorylates the myosin light chains, thus initiating the interaction between myosin and actin. Myosin is inactivated again by a phosphatase that removes the phosphate from the light chains. Myosin kinase itself can be inactivated by cAMP-dependent phosphorylation, which prevents binding of Ca^{2+}-calmodulin (Figure 11.20). This is how epinephrine relaxes intestinal and vascular smooth muscle by raising intracellular cAMP levels through interaction with β receptors (see Section 13.6). In general, smooth muscle contracts more slowly than skeletal muscle. Other variations of the actomyosin system are found in **cardiac muscle.**

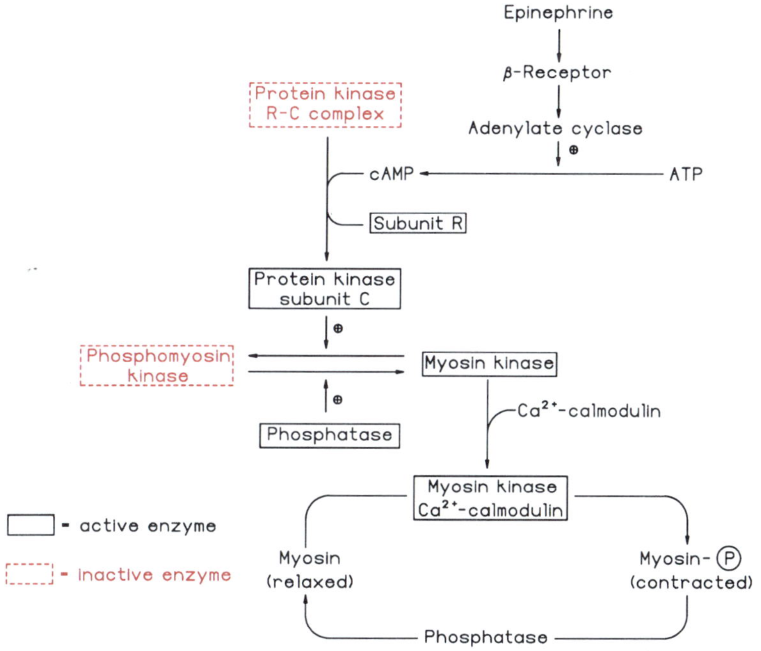

Figure 11.20. Regulation of smooth muscle contraction by epinephrine.

11.3.1 Actomyosin in Other Cell Types

Essentially all cells of the human body contain, in addition to actin (microfilaments), small amounts of myosin. The resulting actomyosin system, possibly interacting with the ends of microfilaments and other cytoskeletal structures, is involved in a variety of cellular processes such as the **secretion** of enzymes, hormones, or neurotransmitters by exocytosis, endocytosis, **ameboid motion,** cytoplasmic streaming, and mitosis. The fact that both **cytochalasin B,** which blocks polymerization of actin, and **phalloidin,** which inhibits depolymerization of microfilaments, inhibit cell motility suggests that cell motility depends to a large part on the polymerization and depolymerization of microfilaments.

11.3.2 Cilia and Flagella

Certain cells in the mucosal epithelium of various organs carry cilia on their surface that beat rhythmically and thus move liquid (mucus) unidirectionally over the surface of the epithelium. Nonsessile cells, such as spermatozoa, possess flagella that by their beating action propel the cell. Both cilia and flagella contain a complex motile system that also includes microtubules. Nine doublets of microtubules are arranged in a ring with a pair of singlet tubules in the center. The structure is called the **9 + 2 axoneme** (Figure 11.21). Each outer doublet contains a regular tubulin cylinder with 13 tubulin molecules per ring and is attached to a larger diameter partial cylinder containing 10 tubulin monomers. Attached to the regular outer tubule cylinder are two arms containing the protein **dynein.** Dynein is an ATPase, and it appears that the arms interact with the subunits of the neighboring tubule doublet in a manner analogous to the interaction of the myosin head with actin. This would cause a sliding of the outer tubule doublets against each other, thus producing the bending movement of the cilium or flagellum.

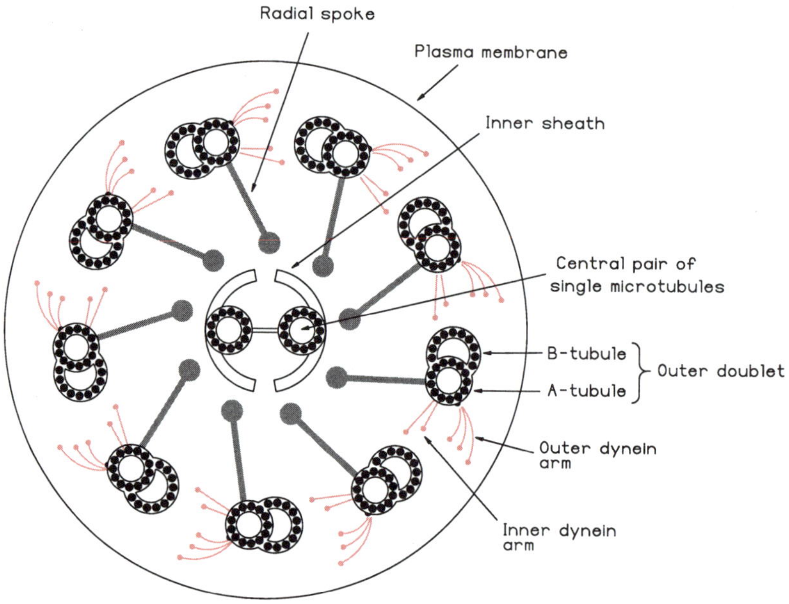

Figure 11.21.
Cross section of an axoneme, found in cilia and flagella.

For a more extensive discussion of special structures and functions, consult the following textbooks:

Mathews et al. (1990): Chapters 11, 36, 37

Stryer (1988): Chapters 6, 9, 29

Further References

Burgeson RE. New collagens, new concepts. *Annu Rev Cell Biol.* 1988; 4:551–577.

Carafoli E. The calcium pumping ATPase of the plasma membrane. *Ann Rev Physiol.* 1991; 53:531–547.

Cooke R. The mechanism of muscle contraction. *CRC Crit Rev Biochem.* 1986; 21:53–118.

Cooper JA. The role of actin polymerization in cell motility. *Ann Rev Physiol.* 1991; 53:585–605.

Edelman GM. Cell-adhesion molecules: a molecular basis for animal form. *Sci Am.* 1984; 250:118–129.

Gerisch G, Noegel AA, Schleicher M. Genetic alteration of proteins in actin-based motility systems. *Ann Rev Physiol.* 1991; 53:607–628.

Hassel JR, Kimura JH, Kascall VC. Proteinglycan core protein families. *Annu Rev Biochem.* 1986; 55:539–567.

Horisberger JD, Lemas V, Kraehenbühl JP, Rossier BC. Structure–function relationship of Na,K-ATPase. *Ann Rev Physiol.* 1991; 53:565–584.

Lazarides E, Woods C. Biogenesis of the red cell membrane skeleton and the control of erythroid morphogenesis. *Ann Rev Cell Biol.* 1989; 5:427–452.

Martin GR, Timpl R, Muller PK, Kühn K. The genetically distinct collagens. *Trends Biochem Sci.* 1985; 10:285–287.

Pollard TD, Cooper JA. Actin and actin-binding proteins. A critical evaluation of mechanisms and functions. *Annu Rev Biochem.* 1986; 55:987–1035.

Pollard TD, Doberstein SK, Zot HG. Myosin–I. *Ann Rev Physiol.* 1991; 53:653–681.

Porter ME, Johnson KA. Dynein structure and function. *Ann Rev Cell Biol.* 1989; 5:119–151.

Prockop DJ, Kivirillo KI. *N Engl J Med* 1984; 311:376–386.

Ruoslahti E. Fibronectin and its receptors. *Annu Rev Biochem.* 1988; 57:375–414.

Rutishauser U, Goridis C. N-CAM: the molecule and its genetics. *Trends Genet.* 1986; 2:72–76.

Schroer TA, Sheetz MP. Functions of microtubule-based motors. *Ann Rev Physiol.* 1991; 53:629–652.

Smith KA. The interleukin 2 receptor. *Ann Rev Cell Biol.* 1989; 5:397–426.

Takeichi M. Cadherins: a molecular family important in selective cell-cell adhesion. *Annu Rev Biochem.* 1990; 59:237–252.

Takeichi M. Cadherin cell adhesion receptors as a morphogenetic regulator. *Science.* 1991; 251:1451–1455.

Warrick HM, Spudich JA. Myosin structure and function in cell motility. *Annu Rev Cell Biol.* 1987; 3:379–421.

CHAPTER

12 Respiration, Water Balance, and Acid-Base Control

OBJECTIVES

After studying this chapter, the student should be able to

1. Explain Dalton's law of partial pressures in gas mixtures.
2. State the partial pressures of CO_2 and H_2O in alveolar air.
3. Describe the molecular events responsible for the Bohr effect.
4. Describe the mechanism of CO_2 transport by blood and explain the terms isohydric shift and chloride shift.
5. Explain why venous blood hematocrit is slightly higher than arterial hematocrit.
6. Write the Henderson-Hasselbalch equation for the dissociation of carbonic acid.
7. State the normal values of human plasma bicarbonate and α P_{CO_2} in millimolar.
8. Draw the pH-bicarbonate diagram with buffer lines for whole blood and separated plasma and P_{CO_2} isobars, and indicate on it the nine possible acid-base states.
9. Describe two methods for the determination of the acid-base status of a patient.
10. Name the main fluid compartments of the human body and give a method for determining their relative sizes.
11. Give a brief description of the parts of a nephron and describe their functions.
12. Describe the molecular mechanism of tubular sodium reabsorption.
13. State the fraction of the filtered sodium whose reabsorption is under the control of aldosterone.
14. Explain the interplay of aldosterone and vasopressin in the control of volume and osmolality of extracellular fluid.
15. Calculate the minimum amount of H^+ ions that have to be secreted daily to reabsorb the filtered bicarbonate.
16. Explain why carbonic acid anhydrase inhibitors cause diuresis (increased urine flow).
17. Explain how arterial P_{CO_2}, plasma potassium concentration, plasma chloride concentration, and aldosterone affect the renal bicarbonate threshold.
18. Calculate the ratio of HPO_4^{2-} to $H_2PO_4^-$ in plasma at pH 7.4 and in urine at pH 6.2 (pK of $H_2PO_4^-$ dissociation is 6.8).
19. Outline, in principle, the mechanism of renal production of ammonia.
20. Describe the mechanism of potassium homeostasis.
21. Explain why most acid-base disturbances lead to hypokalemia.
22. Describe the mechanism of renal compensation of respiratory and metabolic acid-base disturbances.

12.1 RESPIRATION

Respiration is defined as the exchange of oxygen and carbon dioxide between body fluids and the surrounding atmosphere. It occurs in two stages. First, in the lungs, carbon dioxide dissolved in venous blood is released into the alveolar air, and oxygen is taken up by the blood. Both gases diffuse along their concentration gradients. Because of the extreme thinness of the walls separating the blood in the pulmonary capillaries from the alveolar air, this diffusion is extremely fast and reaches equilibrium within a fraction of a second.

The second stage of respiration occurs in the tissues. Oxygen is delivered to the cells for oxidative processes, and carbon dioxide is taken up. Here the diffusion of gases occurs only between liquid phases but is also driven by their concentration gradients. There is no active transport involved.

12.1.1 Composition of Alveolar Air

The pressure of alveolar air is identical to the barometric pressure (the standard barometric pressure is 1 atmosphere or 760 mmHg).

According to **Dalton's law,** the total pressure of a gas mixture equals the sum of the partial pressures of the individual components. Taking into account that alveolar air is saturated with water vapor, which at 37°C is 47 mmHg, the partial pressures of the atmospheric gases in aveolar air are:

O_2	101.2	(159.0)	mmHg
CO_2	40.0	(0.3)	mmHg
N_2	571.8	(597.0)	mmHg
H_2O	47.0	(760.0)	mmHg

Note: The values in parentheses are those of atmospheric air.

Note that because of the 47 mmHg of water vapor pressure (value is independent of the barometric pressure) and because of the 40 mmHg of CO_2 pressure (value of arterial blood), the partial pressures of O_2 and N_2 in alveolar air are lower than those found in the atmosphere.

Since equal volumes of gases contain the same number of molecules (1 mol = 22.4 L), the **mole fraction** of a gas is directly **proportional** to its partial volume (v) or its **partial pressure** (P).

$$\text{Mole fraction} = \frac{P_{gas}}{P_{total}} = \frac{V_{gas}}{V_{total}}$$

12.1.2 Solution of Oxygen in Plasma

Henry's law states that the concentration of a gas in physical solution is directly proportional to its partial pressure:

$$\text{Concentration of dissolved gas} = \alpha P_{gas}$$

The solubility coefficient α is also called the **Bunsen coefficient.** For O_2 dissolved in plasma at 37°C (reduced to standard conditions):

$$\alpha_{oxygen} = 0.023\ [\text{atm}^{-1}]$$

For example, the amount of oxygen (in milliliters) dissolved in 1 ml of plasma at normal alveolar pressure is

$$O_{2diss} = 0.023 \times P_{O_2} = 0.023\ \frac{101.2}{760} = 0.0031\ \text{ml}$$

In other words, 100 ml of plasma can carry only 0.31 ml of oxygen in dissolved form. This is far too little to satisfy the O_2 requirement of the body's cells. In contrast, the hemoglobin (Hb) contained in 100 ml of blood (about 15 g) carries 20 ml of oxygen. (The transport of oxygen by hemoglobin has been discussed in section 9.2.4.)

12.1.3 Transport of CO_2 and Buffering of H^+

Carbon dioxide is about 20 times more soluble in plasma than oxygen. It rapidly diffuses from the tissues into the capillaries, where it is hydrated to carbonic acid. This hydration of CO_2 to H_2CO_3 takes place inside the erythrocyte and is catalyzed by the enzyme **carbonic anhydrase.** Carbonic acid spontaneously dissociates into a bicarbonate ion and a proton.

$$CO_2 + H_2O \rightleftharpoons H_2CO_3 \rightleftharpoons HCO_3^- + H^+$$

Due to the increased P_{CO_2} in tissue, Hb is deoxygenated (Bohr effect). As discussed in Chapter 9, deoxygenation increases the pK of Hb. Some of the hydrogen ions produced by the dissociation of carbonic acid can thus be accommodated without a decrease in the pH. This effect is known as the **isohydric shift** and can best be illustrated graphically.

Figure 12.1 shows the titration curves of both reduced (Hb) and oxygenated Hb (HbO_2). The addition of 0.5 mEq of H^+ to 1 mEq of HbO_2 (point A) lowers the pH by about 0.06 unit (point B). Simultaneous deoxygenation, however, reduces the pH by only 0.02 unit (point C) because of the increased pK value of Hb. In other words, about 0.3 mEq of H^+ ions can be accommodated by 1 mEq of Hb without a pH change. (In vivo, the isohydric shift can accommodate less H^+ because Hb is never completely oxygenated.) The remaining

Figure 12.1.
Demonstration of the isohydric shift.

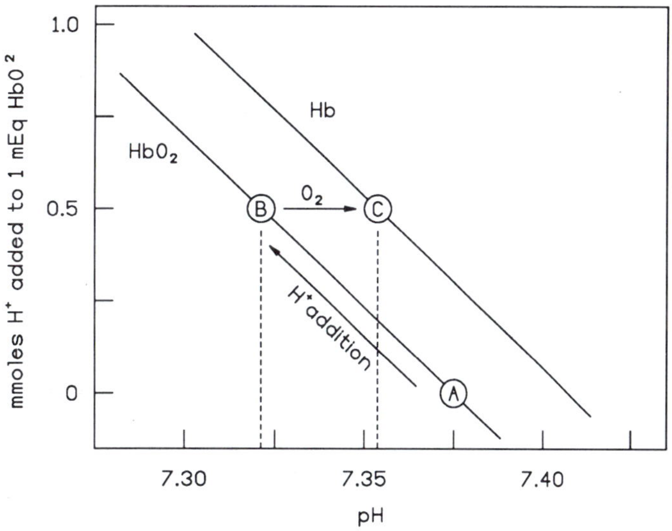

H^+ ions are mostly bound to His residues in which Hb is relatively rich.

Aside from its buffering action, Hb binds CO_2 chemically. The CO_2 reacts with the terminal α amino groups of all four chains to form **carbamino Hb.**

$$R{-}NH_2 + CO_2 \rightleftharpoons RNH{-}COOH \rightleftharpoons R{-}HN{-}COO^- + H^+$$

The formation of carbamino groups reduces the theoretically maximal isohydric shift because it removes potential proton acceptors. Carbamino acid formation also occurs on plasma proteins, but it contributes less to the CO_2 transport, partly because the protein concentration in plasma is lower.

The bicarbonate ions produced in the erythrocyte diffuse into the plasma. To conserve electroneutrality, chloride ions move from the plasma into the erythrocyte **(chloride shift).** As a result, most of the CO_2 is actually carried in the plasma in the form of bicarbonate (27 m*M*), although bicarbonate formation and buffering of H^+ ions occur inside the erythrocyte. The movement of bicarbonate against an apparent concentration gradient can be explained by the fact that the intracellular pH is about 0.5 pH unit lower than the extracellular pH. Since the pH determines the ratio of bicarbonate to dissolved CO_2 (see next section) and since the P_{CO_2} is the same in both compartments, the intracellular equilibrium concentration of bicarbonate (~12.5 m*M*) is only about one-half its extracellular concentration.

The extent of the chloride shift is determined by the Gibbs-Donnan equilibrium, which requires that the ratio of bicarbonate to chloride ions inside the cell is identical to that outside the cell.

As negative charges on Hb are neutralized through the binding of H^+ ions, the main intracellular cation, K^+, has to be neutralized by HCO_3^- or Cl^-. These additional ions raise the intracellular osmolarity, which, in turn, causes an influx of water. Venous hematocrit, therefore, is slightly higher than arterial hematocrit.

12.2 ACID-BASE BALANCE

12.2.1 The CO_2-Bicarbonate Buffer System

In contrast to carbonic acid, which is buffered mainly by Hb, other acids, called **fixed acids,** or bases are buffered by the bicarbonate-CO_2 system. It is an open system because one of its components, CO_2, is in equilibrium with alveolar air and is being adjusted continuously.

The pK value for the dissociation of carbonic acid into bicarbonate and H^+ is 6.1. According to the **Henderson-Hasselbalch equation** (see Section 1.1), the pH of plasma is determined by the ratio of the concentration of bicarbonate ions to that of undissociated carbonic acid:

$$pH = 6.1 + \log \frac{[HCO_3^-]}{[H_2CO_3]} \qquad (12.1)$$

As carbonic acid is in equilibrium with CO_2, the term $[H_2CO_3]$ includes both carbonic acid and dissolved CO_2. According to Henry's law, the amount of dissolved CO_2 (including the 0.2% that is hydrated to H_2CO_3) is determined by the partial pressure of CO_2, P_{CO_2}:

$$[\text{Dissolved } CO_2 + H_2CO_3] = \alpha P_{CO_2} \qquad (12.2)$$

where the solubility coefficient of CO_2 in plasma, α, is 0.03, when CO_2 is expressed in millimolar and P_{CO_2} in millimeters of mercury. Combining equations (12.1) and (12.2), we obtain

$$pH = 6.1 + \log \frac{[HCO_3^-]}{0.03\ P_{CO_2}} \qquad (12.3)$$

Usually the concentration in plasma of **total CO_2** rather than bicarbonate is determined:

$$[\text{Total } CO_2] = [CO_2] + [HCO_3^-] + [H_2CO_3]$$

Equation (12.3) may be rewritten:

$$pH = 6.1 + \log \frac{[\text{total } CO_2] - 0.03\ P_{CO_2}}{0.03\ P_{CO_2}} \qquad (12.4)$$

Equation (12.4) contains three unknown variables: pH, total CO_2, and P_{CO_2}. If any two are measured, the third can be calculated. At pH 7.4, the normal pH value of blood, the ratio of $[HCO_3^-]/0.03P_{CO_2}$ equals 20:1. Blood pH values that are compatible with life range from 7.0 to 7.8

On addition of H^+ ions, bicarbonate is converted to CO_2, which is eliminated during respiration.

$$HCO_3^- + H^+ \rightarrow CO_2 \uparrow + H_2O$$

The buffering of fixed acids, therefore, consumes plasma bicarbonate. To reestablish the 20:1 ratio of $[HCO_3^-]/0.03P_{CO_2}$, the rate of respiration is normally increased by a reflex mechanism. This lowers alveolar P_{CO_2} and is known as **compensation.** Usually, respiratory compensation is not sufficient to restore the original pH completely. Restoration of the original pH requires restoration of the original bicarbonate concentration. This can only be accomplished by renal mechanisms, which take considerably longer (see Section 12.4).

Addition of alkali increases plasma bicarbonate:

$$OH^- + CO_2 \rightarrow HCO_3^-$$

12.2.2 The pH-Bicarbonate Diagram

The interrelationship between pH, $[HCO_3^-]$, and P_{CO_2} is usually illustrated in the **pH-bicarbonate diagram** (Figure 12.2).

The $[HCO_3^-]$ and pH values corresponding to a particular P_{CO_2} are calculated by the Henderson-Hasselbalch equation and are represented by the P_{CO_2} isobars. The normal P_{CO_2} of alveolar air, and therefore also of arterial plasma, is 40 mmHg. Venous P_{CO_2} is about 46 mmHg. The normal bicarbonate concentration of arterial plasma is 24 m*M*. Thus, with αP_{CO_2} = 1.2 m*M*, the ratio $[HCO_3^-]/\alpha P_{CO_2}$ is 24/1.2 or 20.

When blood is titrated with carbonic acid (by increasing the P_{CO_2}), the **buffer line** is obtained, as indicated in Figure 12.3. The slope of the buffer line represents the **buffer value** of the system for carbonic acid and is defined as $\Delta[HCO_3^-]/\Delta pH$. Plasma that is separated from the erythrocytes (separated plasma) is a poor buffer when compared with whole blood because most of the buffering in whole blood takes place within the erythrocytes (via Hb).

In vivo, the buffer line is usually different from the one observed in vitro. This is because all cells in the body participate in the buffering

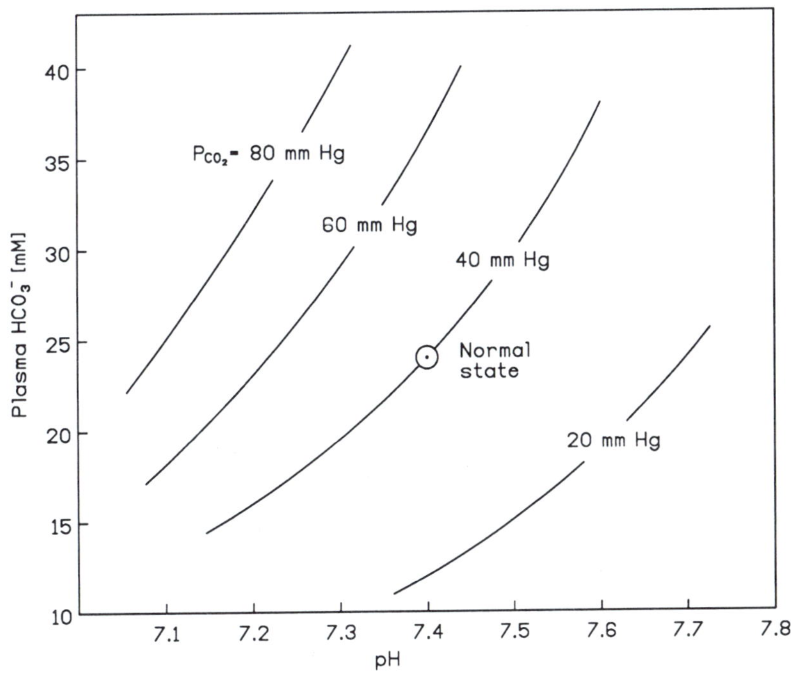

Figure 12.2.
The pH-bicarbonate diagram.

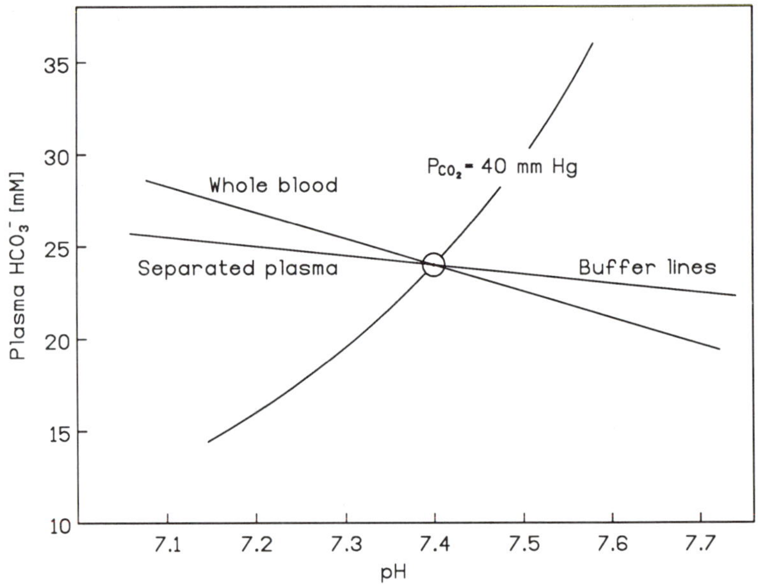

Figure 12.3.
Titration of blood and separated plasma with CO_2.

of hydrogen ions. The clinician must be careful when trying to correct a base excess or deficit, since the whole body buffering capacity varies and makes it difficult to estimate the extent of the correction necessary.

The buffering of fixed acids (such as HCl) by whole blood is best illustrated graphically (Figure 12.4). Let us assume that the addition of a certain amount of HCl to whole blood lowers the pH to 7.25 (point A) if the P_{CO_2} is kept constant at 40 mmHg. Part of the added acid is buffered by bicarbonate (by simply eliminating an equivalent amount of bicarbonate from the plasma), another part is buffered by proteins. The amount of acid added to the blood can be determined by back-titrating along a line parallel to the buffer line to the original pH by lowering P_{CO_2} (point B). The distance O → B should represent the amount of added HCl (base deficit). (Actually, the base deficit thus

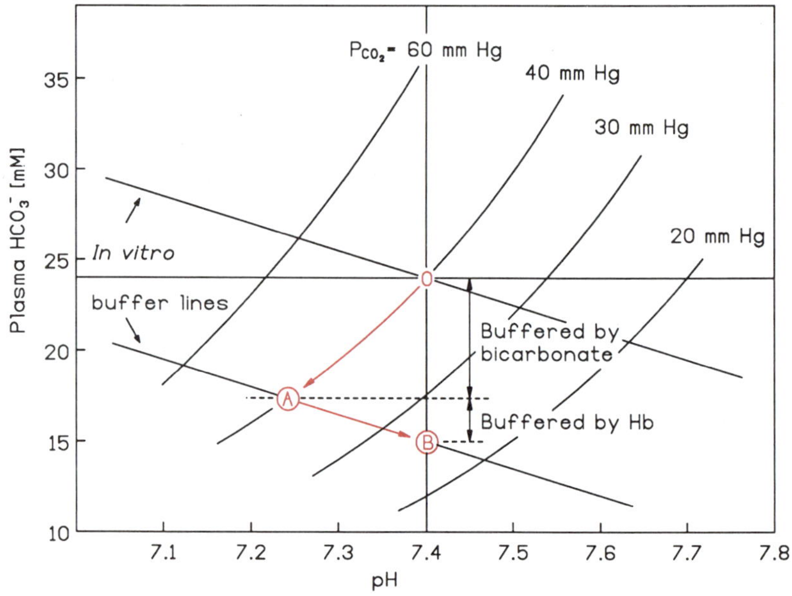

Figure 12.4.
Buffering of "fixed acids" by blood.

Table 12.1. Acid-Base Disturbances and Their Compensation

Disturbance	HCO_3^-	P_{CO_2}	Total CO_2	Compensation
Metabolic acidosis	Large decrease	Small decrease	Large decrease	Alveolar hyperventilation
Metabolic alkalosis	Large increase	Small increase	Large increase	Alveolar hypoventilation
Respiratory acidosis	Small increase	Large increase	Small increase	Renal HCO_3^- retention and H^+ elimination
Respiratory alkalosis	Small decrease	Large decrease	Small decrease	Renal H^+ retention and HCO_3^- elimination

determined must be multiplied by about 0.8 to correct for the lower concentration of HCO_3^- in erythrocytes as compared with plasma.)

12.2.3 Acid-Base Disturbances

Plasma pH may be affected by a change in either the bicarbonate concentration or the P_{CO_2} [see equation (12.3)]. When the primary change is in the P_{CO_2}, the disturbance is called **respiratory;** when it is in the bicarbonate concentration, it is called **metabolic.** Most primary disturbances elicit a secondary response (compensation), which partially normalizes the pH.

Compensation involves the system *not* involved in the primary response and is seldom complete (Table 12.1). The renal mechanisms for raising or lowering bicarbonate concentrations will be discussed later.

Abnormal acid-base states may be due to the simultaneous presence of two disturbances: one respiratory and one metabolic in nature. The nine possible acid-base states (including the normal state) are best illustrated in a pH-bicarbonate diagram (Figure 12.5). As a rule, a metabolic disturbance moves the acid-base status along the P_{CO_2} isobar; a respiratory disturbance moves it along the buffer line.

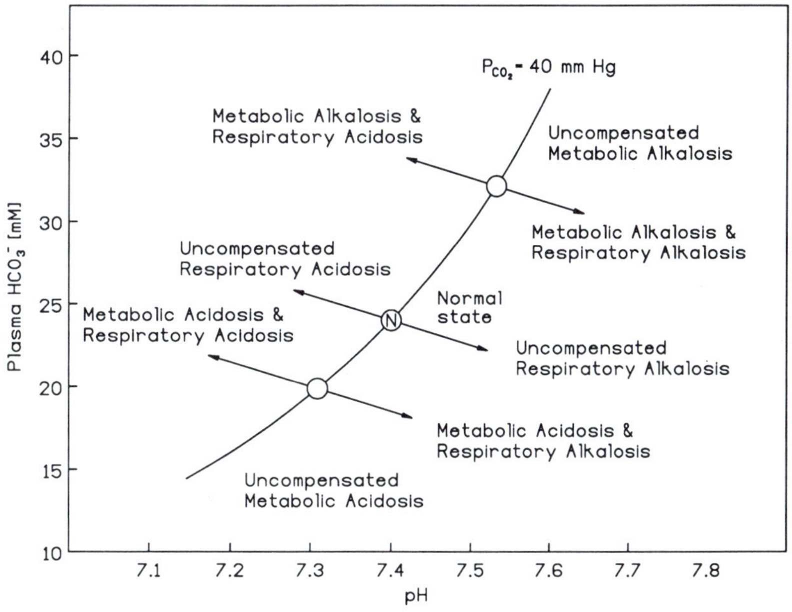

Figure 12.5.
Acid-base disturbances. Redrawn with permission from Davenport HW. *The ABC of Acid-Base Chemistry*, 6th ed. Chicago, Ill: University of Chicago Press; 1974, p. 69.

12.2.4 Determination of Acid-Base Status

At least two independent measurements are required for determining a patient's acid-base status. The **pH** is easily measured on arterial (or arterialized) blood using a glass electrode. This measurement has to be done as quickly as possible in order to avoid changes in pH caused by a loss of CO_2 or the production of lactic acid by the erythrocytes.

The bicarbonate concentration is not measured directly. Instead, the **total CO_2 content** of plasma is determined on plasma that has been separated anaerobically. Total CO_2 measures the sum of HCO_3^-, dissolved CO_2, and H_2CO_3. The bicarbonate concentration can then be calculated or determined from a nomogram (using the measured pH value). The **P_{CO_2}** can either be calculated from the total CO_2 content or it can be measured directly by a P_{CO_2} electrode.

Another method (developed by Astrup and Siggaard-Andersen), less subject to errors in measurement, equilibrates an oxygenated blood sample with CO_2 at a specified pressure, for instance 75 mmHg, and another sample at a different pressure, for instance 20 mmHg. After equilibrium, the pH values of the samples are measured and plotted in a log P_{CO_2} versus pH diagram (Figure 12.6, points A and B). This produces a line equivalent to the buffer line in the pH-bicarbonate diagram. The original pH of the blood sample, measured before CO_2 equilibration, is then marked on the log P_{CO_2}-pH line (point C). This point gives the P_{CO_2} of the original blood sample and allows the calculation of the original bicarbonate concentration. The slope of the line depends on the hemoglobin content of the blood and can be checked by measuring the hemoglobin concentration of the blood independently.

Various nomograms have been designated in which the log P_{CO_2}-pH line of Figure 12.6 intersects several scales that yield values for buffer base, actual bicarbonate, standard bicarbonate, and base excess concentrations. These commonly used terms are defined as follows:

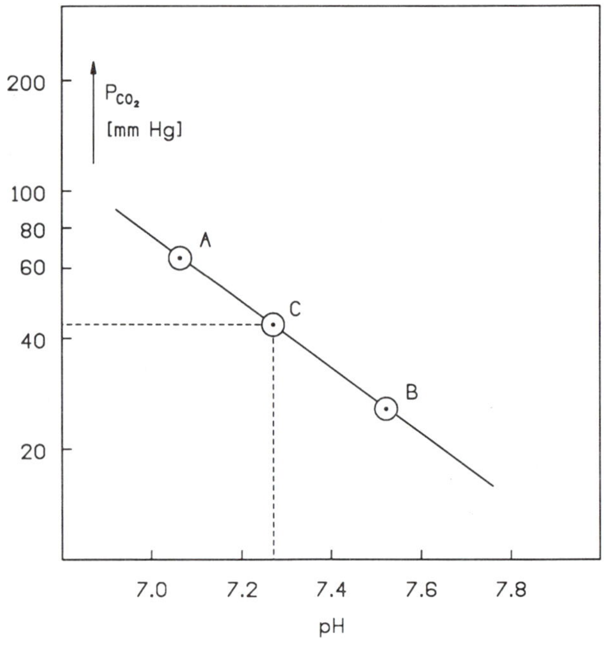

Figure 12.6.
P_{CO_2}-pH diagram. After Astrup P, et al. *Lancet.* 1960; ■:1035.

Buffer base: The sum of the concentrations of the buffer anions (in mEq/l) of whole blood (HCO_3^-, inorganic phosphate, and proteins in plasma; HCO_3^- and Hb in red cells); normal range: 45–50 mEq/L.

Standard bicarbonate: The plasma bicarbonate concentration of whole blood that has been equilibrated to a P_{CO_2} of 40 mmHg at 37°C; Units: mEq/l.

Base excess: The base concentration (in mEq/l) of whole blood as measured by titration with strong acid or base to pH 7.40 at a P_{CO_2} of 40 mmHg at 37°C. Negative values indicate "base deficit." The normal value of base excess (i.e., for normal blood with a pH of 7.40, P_{CO_2} of 40 mmHg) is 0 mEq/L (range: −2.5 to +2.5).

12.3 RENAL CONTROL OF WATER BALANCE

12.3.1 Body Fluid Compartments

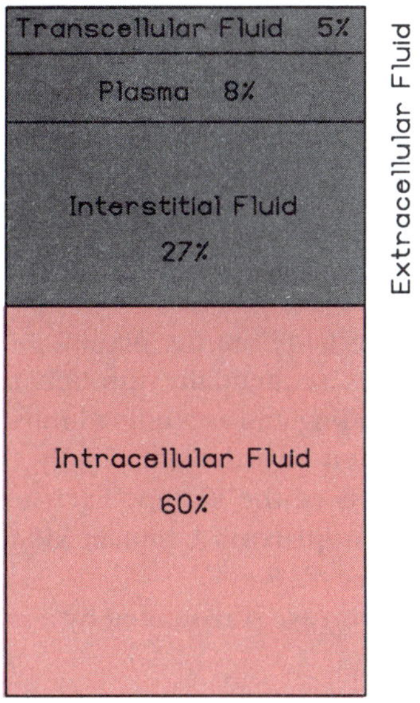

Figure 12.7. Composition of total body water.

The water content of the human body amounts to 50–70% of its total weight (the lower figure pertains to obese persons). **Total body water (TBW)** is distributed between two compartments: the **intracellular fluid (ICF)** and the **extracellular fluid (ECF).** ICF comprises about 60% of TBW. ECF has three compartments: **plasma** (8% of TBW), **interstitial fluid** (27% of TBW), and **transcellular fluid** (cerebrospinal, peritoneal; 5% of TBW) (Figure 12.7).

The size of a fluid compartment can be measured by injecting a known amount of a substance whose distribution in the body is limited to this compartment and determining its concentration after equilibration. Table 12.2 lists the most frequently employed methods.

Table 12.3 compares the electrolyte compositions of plasma and intracellular fluid. Note that the principal extracellular cation is Na^+. The electrolyte concentration (in mEq/l) of interstitial fluid is slightly lower than that of plasma. Interstitial and intracellular fluids are iso-osmolar, i.e., they contain the same number of osmotically active ions (mOsm/l). This is because intracellular fluid contains proteins that act as single osmotically active particles with multiple negative charges on them. The osmolarity of plasma is about 5% higher than that of interstitial

Table 12.2. Methods for Determining Body Fluid Compartments

Fluid Compartment	Method of Determination
TBW	Antipyrine, D_2O, HTO (tritiated water)
ECF	Inulin, sucrose, SCN^-, ^{35}S-sulfate
Plasma volume	^{131}I-albumin, Evans blue (T-1824), ^{51}Cr-labeled erythrocytes
Interstitial fluid	ECF minus plasma
ICF	TBW minus ECF

Table 12.3. Composition of Extracellular and Intracellular Fluid

	Ion	ECF (plasma; mEq/L)	ICF (muscle cell; mEq/L)
Cations	Na^+	145	12
	K^+	5	155
	Mg^{2+}	2	15
	Ca^{2+}	2 (free)	2 (sequestered)
Anions	Cl^-	110	8
	HCO_3^-	27	8
	$Protein^-$	15	60
	$Phosphate^-$	2	90
	Other ions		18
Total ions		308	368

fluid mainly due to the Gibbs-Donnan effect. The resulting osmotic pressure is balanced by the intravascular blood pressure.

12.3.2 Functional Anatomy of the Kidney

The kidneys enable the body to eliminate waste products (e.g., urea, uric acid), adjust the concentrations of important plasma components, and maintain water balance. Under normal conditions, about 20–25% of the cardiac output flows through the kidney. This amounts to 1200 ml/min of blood or 660 ml/min of plasma. About 20% of this plasma flow is filtered in the **glomeruli.** The **glomerular filtration rate** is 125 ml/min. Since the glomerular membrane is permeable to all small ions and molecules, the glomerular filtrate has the same electrolyte composition as plasma. Only proteins are retained in the plasma.

From the glomerulus the filtrate passes sequentially through the **proximal convoluted tubule,** the descending and ascending limbs of the **loop of Henle,** the **distal convoluted tubule,** and finally the **collecting duct,** which leads to the pelvis of the kidney. Each unit originating from a glomerulus is called a **nephron.** A human kidney contains about 1.3 million nephrons (Figure 12.8).

The tubules and ducts within each nephron are surrounded by a rich

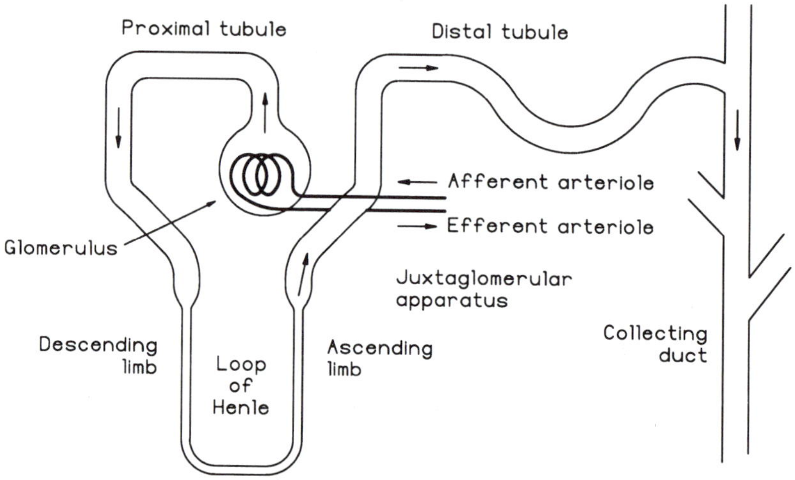

Figure 12.8. Schematic drawing of a nephron.

network of capillaries that allow the secretion and reabsorption of molecules and ions into and out of the tubular filtrate. For example, glucose and amino acids are quantitatively reabsorbed from the glomerular filtrate by the cells that line the proximal tubules. Na^+, K^+, and Cl^- ions are reabsorbed throughout the nephron.

Sodium chloride is the principal osmotically active solute in ECF. Since the volume of ECF is regulated primarily through the maintenance of its osmotic pressure, the extent of reabsorption of Na^+ ions from the glomerular filtrate becomes the most important factor that directly determines ECF volume. (Chloride ions mostly follow Na^+ ions passively.)

12.3.3 Mechanism of Reabsorption of Sodium and Chloride Ions

Reabsorption of Na^+ ions is an active, energy-consuming process that takes place along the entire nephron with the exception of the loop of Henle (see below). The mechanism of Na^+ reabsorption is illustrated in Figure 12.9. Na^+ ions diffuse passively through sodium channels located in the brush border of the tubular cell and are then actively transported out of the cell into the intercellular space by Na^+/K^+ exchange pumps located in the tubular cell membrane facing the intercellular space. Cl^- ions, to which the tubular cell membrane is permeable, follow Na^+ passively along the electrochemical gradient created by the Na^+ transport. The same applies to water, which follows the reabsorbed NaCl through osmotic forces.

The active (energy-requiring) nature of the Na^+ transport can be demonstrated during **osmotic diuresis.** Normally, there exists no transtubular (lumen-peritubular space) concentration gradient for Na^+, and the transtubular electric gradient is small (0 to -4 mV, tubular

Figure 12.9.
Mechanism of tubular reabsorption of Na^+ and water.

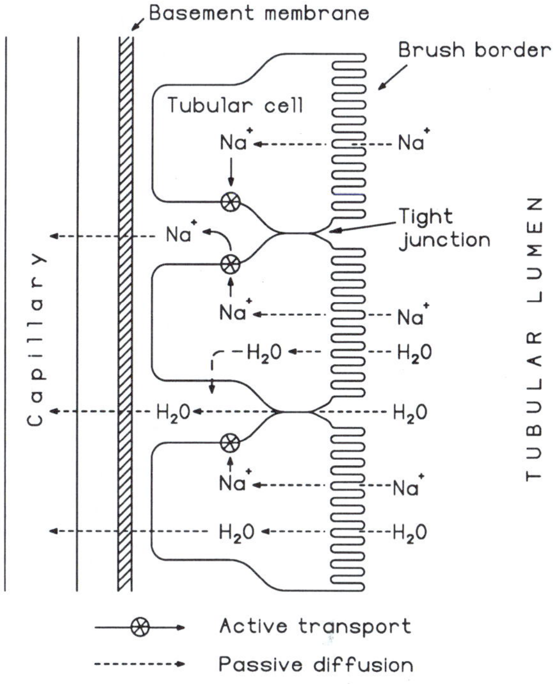

side negative). (By contrast, the electric gradient across distal tubular cells is −40 to −60 mV.) To produce osmotic diuresis, a substance like mannitol is administered, which cannot be reabsorbed and, therefore, tends to retain water in the tubules (through osmotic forces). As a result, the Na^+ concentration in the tubular lumen falls below its plasma concentration while Na^+ continues.

Osmotic diuresis occurs in disease processes, such as diabetes mellitus, when filtered glucose exceeds the capacity of its reabsorption mechanism. This will cause not only the excretion of larger volumes of urine but also the excretion of Na^+ and Cl^- ions, because the developing concentration gradient described above can only approach a certain limit at which active transport and passive back-diffusion are in equilibrium.

12.3.4 The Concentration of Urine

About 65% of the total filtered Na^+, Cl^-, and water are reabsorbed within the proximal tubules. As the tubular fluid enters the loop of Henle, it has the same osmolarity as plasma (about 300 mOsm/L). Concentration of urine is accomplished by a **countercurrent system** consisting of the loop of Henle, the collecting ducts, and a salt gradient that develops in the renal interstitial fluid surrounding these structures.

The **descending limb of the loop of Henle** is relatively impermeable to ions but freely permeable to water. In contrast, the **ascending limb of the loop of Henle** is impermeable to water but actively transports chloride ions into the interstitium. Na^+ follows passively along the developing electrochemical gradient. As a result of these diffusion and transport processes, a concentration gradient of NaCl develops in the interstitium with the highest concentration at the bottom of the loop (Figure 12.10). This gradient is not disturbed by the renal circulation because long capillaries, called **vasae rectae,** run parallel to the limbs of the loop of Henle deep down into the renal medulla and back up again. Plasma osmolarity in the vasa recta follows that of the surrounding interstitial fluid.

It should be mentioned that urea, which easily diffuses through the

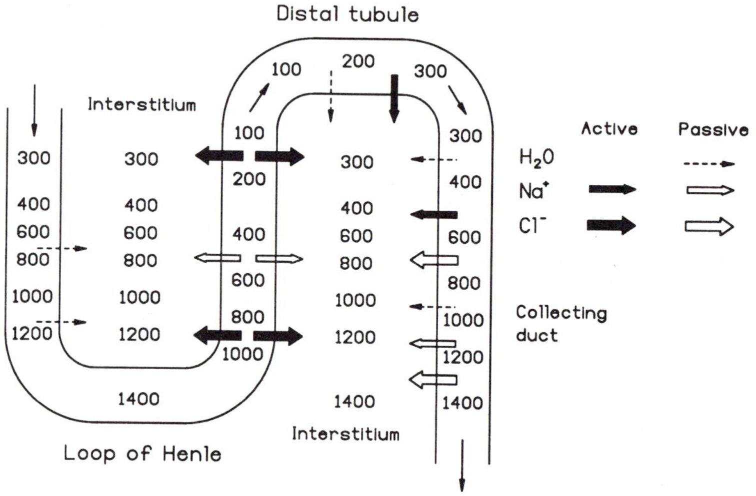

Figure 12.10.
Transport of Na^+, Cl^-, and water in the loop of Henle and the collecting duct generating the interstitial salt concentration gradient. The numbers indicate osmolalities.

tubular cell membranes, is an important participant in the generation of the interstitial concentration gradient.

The NaCl that is reabsorbed in the ascending limb of the loop of Henle amounts to another 25% of the total that is filtered. The fluid entering the distal convoluted tubules is hypoosmolar compared with plasma. Salt and water reabsorption are completed in the distal tubules and the collecting ducts. This part of the nephron is under the control of **vasopressin,** also called **antidiuretic hormone (ADH)**. Vasopressin regulates the water permeability of these structures. There is some evidence that vasopressin promotes the insertion of new water channels in the cell membranes of the distal tubules and collecting ducts. The water channels allow water to passively follow the reabsorption of Na^+ ions. This results in the formation of small volumes of highly concentrated urine. In the absence of the hormone, the water channels are internalized with the result that the urine becomes very dilute since the reabsorption of Na^+ ions continues throughout these structures (Figure 12.11).

The maximal urinary concentration achieved by the human kidney is 1400 mosm/L. Since daily waste products (urea, sulfate, phosphate, and others) excreted in the urine amount to about 600 mosm, the minimum daily urine volume is 600/1400 = 0.44 L. In addition to this urinary loss of about 0.5 L of water, the human body loses under normal conditions approximately 1 L of water/day through the skin and lungs (insensible loss) and in sweat and feces. Since the human body produces about 350 ml of metabolic water, the minimum daily intake of water for maintaining a positive water balance amounts to about 1.2 L.

12.3.5 Control of Sodium and Water Reabsorption

In the absence of the steroid hormone **aldosterone,** 98% of the filtered Na^+ ions are reabsorbed in the kidney. Changes in the glomerular filtration rate are accompanied by corresponding changes in the rate of

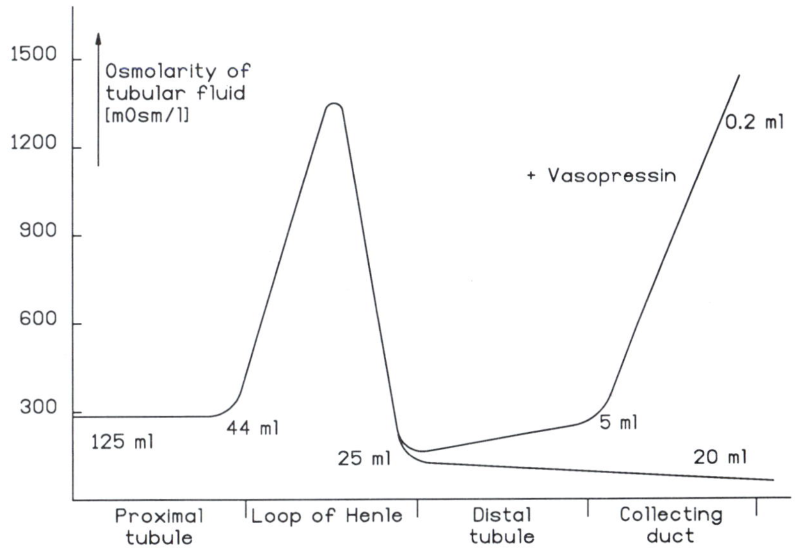

Figure 12.11.
Osmolarity of the tubular fluid as it passes through the nephron. Redrawn with permission from Vander AJ, *Renal Physiology*. New York, NY: McGraw-Hill Book Co; 1975.

Na^+ reabsorption. The mechanism of this intrarenal autoregulation, known as **glomerulotubular balance,** is not completely understood.

The reabsorption of the remaining 2% of the filtered Na^+ ions (about 33.4 g/day) is under the control of aldosterone and **atrial natriuretic peptide (ANP).** Aldosterone promotes Na^+ reabsorption, whereas ANP increases Na^+ excretion.

Aldosterone induces the synthesis of additional Na^+ channels in the cell membranes lining the distal tubules and the collecting ducts. This results in an increased rate of Na^+ reabsorption. The synthesis and release of aldosterone from the adrenal cortex is controlled by the renin-angiotensin system (discussed in detail in Section 13.10). **Renin,** a proteolytic glycoprotein produced by the granular cells of the **juxtaglomerular apparatus** of the nephron, converts the plasma protein **angiotensinogen** to **angiotensin I,** which is converted in the lung and other tissues to **angiotensin II.** Angiotensin II directly stimulates the adrenal cortex to secrete aldosterone. The precise mechanism of control of renin secretion is still controversial. Several factors seem to participate in controlling the release of renin:

1. A decreased pressure within the afferent renal arterioles (low ECF volume) stimulates renin release.
2. A low Na^+ flux through the early distal tubules (juxtaglomerular apparatus) promotes renin secretion.

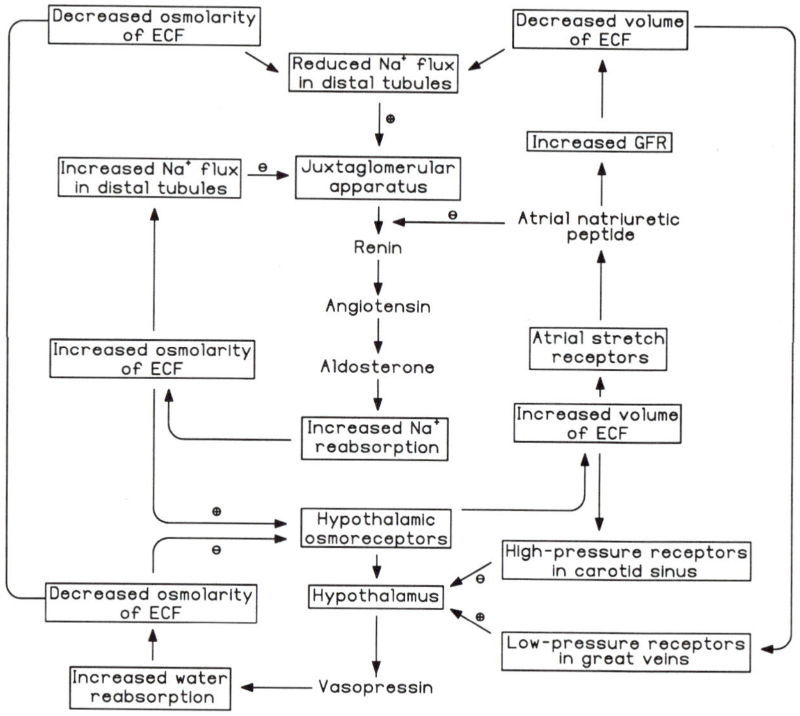

Figure 12.12. Control of volume and osmolarity of ECF.

3. Renal sympathetic stimulation and circulating catecholamines stimulate renin secretion.
4. Prostaglandins promote renin release.
5. ANP inhibits renin secretion.

Human atrial natriuretic peptide (ANP) is a 28-amino-acid peptide with an internal disulfide bond (see Table 13.1). It is released from the muscle cells of the atria in response to the stimulation of stretch receptors (increased ECF). It has been postulated that ANP produces **natriuresis** by binding to ANP receptors in the glomeruli, thereby causing an increase in the surface area available for filtration. It is also possible that ANP directly stimulates the secretion of Na^+ in the tubules.

It is obvious that both a change in plasma osmolarity and plasma volume affect the release of ANP and renin, thus controlling either directly (via ANP), or through the secretion of aldosterone, the rate of Na^+ reabsorption. In addition, osmoreceptors in the hypothalamus monitor plasma osmolarity and control the release of vasopressin, which stimulates water reabsorption (Figure 12.12). (For volume effects on vasopressin release, see Section 13.1.)

12.4 RENAL CONTROL OF ACID-BASE BALANCE

The kidneys contribute to the regulation of the acid-base balance mainly by raising or lowering the plasma bicarbonate concentration. This is a complex process and involves, in addition to the HCO^- ion, the metabolism of H^+, Na^+, K^+, HPO_4^{2-}, $H_2PO_4^-$, NH_4^+, and sometimes organic acids.

The average North American, whose diet contains a high proportion of animal protein, excretes an acid urine because of the production of the strong "fixed acid" H_2SO_4 derived from the degradation of cysteine and methionine. Vegetarians may excrete alkaline urine because a substantial proportion of their diet contains salts (mostly potassium) of organic acids (e.g., citrus fruits). As the organic acid component is metabolized, it leaves HCO_3^- in the body as an alkaline residue.

12.4.1 Mechanism of Reabsorption of Bicarbonate

Bicarbonate is freely filtered at the glomerulus. In contrast to the chloride ion, HCO_3^- can only be reabsorbed through the **secretion of H^+ ions** by the tubular cells. In the tubular lumen, the H^+ ion combines with the filtered HCO_3^- ion to form carbonic acid, which is quickly converted by **carbonic anhydrase,** located on the brush border of the proximal tubular cell membrane, to CO_2 and H_2O. CO_2 then leaves the tubular lumen by diffusion.

The secretion of H^+ ions is an active process requiring energy and can lower the pH in the tubular fluid to about 6.7 in the proximal tubule (determined in rats) and down to 4.5 in the distal tubule. The largest

H^+ concentration gradient generated by this process thus is about 800:1.

The transport mechanism for H^+ ions is believed to be localized in the lumenal membrane. The secretion of H^+ ions is electrically balanced by the diffusion of tubular Na^+ into the tubular cell, which, at this part of the nephron, is highly permeable for that ion. The process, however, is not an "Na-linked pump" because H^+ secretion can take place in the absence of Na^+. Net H^+ secretion takes place throughout the nephron until the limiting pH gradient is reached.

The H^+ ion is probably generated intracellularly from CO_2 by the action of carbonic anhydrase:

$$CO_2 + H_2O \rightleftharpoons H_2CO_3 \rightleftharpoons HCO_3^- + H^+ \text{ (secreted)}$$

The secretion of an H^+ ion into the tubular lumen is always accompanied by the creation of a bicarbonate ion inside the tubular cell. The overall process of bicarbonate reabsorption is illustrated in Figure 12.13.

Acetazolamide is a potent inhibitor of carbonic anhydrase. When administered intravenously, the drug depresses the reabsorption of bicarbonate. As a consequence, the excretion of bicarbonate and Na^+ ions increases, and the urine becomes alkaline. In addition, acetazolamide produces a marked increase in potassium excretion. The most logical explanation for this effect is that K^+ ions are exchanged for Na^+ ions as counter ions for the urinary bicarbonate (see Section 12.5).

Normally, all filtered bicarbonate is reabsorbed (90% in the proximal tubule, 10% in the loop of Henle and the distal tubule) and none is excreted until the plasma level reaches 26–28 m*M*, a level known as the **renal bicarbonate threshold.** Urinary bicarbonate appears in increasing amounts if this level is exceeded.

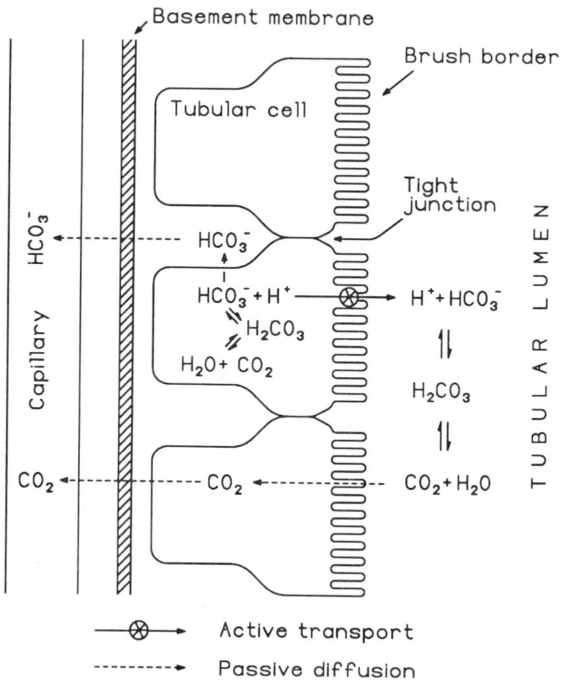

Figure 12.13.
Mechanism of bicarbonate reabsorption.

The renal bicarbonate threshold is variable. At least four different factors influence it:

1. An increase in **arterial P_{CO_2}** causes increased HCO_3^- reabsorption, and a decrease in P_{CO_2} reduces the rate of HCO_3^- reabsorption. This effect is probably based on the intracellular availability of H^+ ions for secretion.
2. **Plasma potassium concentration** is inversely related to the rate of bicarbonate reabsorption. This effect may be explained by the movement of K^+, Na^+, and H^+ ions across the peritubular membranes of the tubular cells. Experiments with muscle cells have shown that for every three K^+ ions that cross the cellular membrane in one direction, two Na^+ ions and one H^+ ion cross in the opposite direction. This overall stoichiometry of ion movement into and out of cells seems to hold for most cells, including tubular cells. At a low plasma concentration of K^+, for example, potassium moves out of the cell, and Na^+ and H^+ move in. The increased intracellular H^+ concentration then facilitates H^+ secretion in the tubular cells, leading to bicarbonate reabsorption. This raises plasma pH and creates the paradoxical situation of an intracellular acidosis with simultaneous extracellular alkalosis known as **hypokalemic metabolic alkalosis** or **paradoxical aciduria.**

Extracellular Fluid			Membrane		Intracellular Fluid
HCO_3^-	2 Na^+	→→		→→	2 Na^+
CL^-	3 K^+	←←		←←	3 K^+
	H^+	→→		→→	H^+
Extracellular alkalosis					Intracellular Acidosis

3. **Plasma chloride concentration** is also inversely related to the rate of bicarbonate reabsorption. Chloride and bicarbonate are the only anions of significance in plasma. To balance the relatively constant number of Na^+ ions, HCO_3^- has to be increased whenever there is a loss of Cl^-, and vice versa.
4. Increased concentrations of **aldosterone** promote K^+ excretion (see Section 12.5). The depletion of K^+ ions causes the intracellular H^+ concentration to rise, thus leading to increased bicarbonate reabsorption.

12.4.2 Urinary Buffer Systems

Most of the H^+ secretion (over 98%) is normally coupled to the reabsorption of filtered bicarbonate. At a glomerular filtration rate of 180 L/day (125 ml/min), a total of 180 × 24 mEq of bicarbonate is filtered and has to be reabsorbed, requiring an equal amount of H^+ to be secreted. Any additional H^+ ions secreted by the kidney will lower the urinary pH and have to be buffered. The presence of tubular buffers keeps the pH high, and H^+ ions can be secreted until all the buffers are titrated to the limiting pH of 4.5.

An important urinary buffer is HPO_4^{2-}. This anion (conjugate base) accepts H^+ ions to form $H_2PO_4^-$ with a pK value of 6.8. In plasma (pH = 7.4) the ratio of HPO_4^{2-} to $H_2PO_4^-$ is 4:1. The amount of H^+ ions that can be buffered by phosphate depends on the concentration of phosphate (controlled by parathyroid hormone) as well as the pH of the urine. At a urinary pH of 6.2, for example, the ratio of HPO_4^{2-} to $H_2PO_4^-$ would be 1:4, just the reverse of the ratio found in plasma. This means that three out of four HPO_4^{2-} ions in the original glomerular filtrate have accepted a proton and thus were converted to $H_2PO_4^-$.

Additional H^+ ions can be buffered by ammonia, which the kidney is capable of generating on demand (see below). NH_3 reacts with H^+ to form the ammonium ions NH_4^+. The pK of this reaction is 9.2. In the range of urinary pH values (4.5–8.0), the NH_4^+ ion is the predominant species.

The extent of titration of the individual buffer components can be predicted from the **isohydric principle,** which states that all buffers in a system are in equilibrium with the same H^+ ion concentration.

$$\text{pH} = 6.1 + \log \frac{[HCO_3^-]}{\alpha P_{CO_2}} = 6.8 + \log \frac{[HPO_4^{2-}]}{[H_2PO_4^-]} = 9.2 + \log \frac{[NH_3]}{[NH_4^+]}$$

In the proximal tubule, the predominant buffer is bicarbonate; phosphate ions become important buffers as the urine gets more concentrated and most bicarbonate has been reabsorbed.

12.4.3 Urinalysis

The extent of H^+ secretion, indicating the renal contribution to the acid-base balance, can be determined by urinalysis. The following parameters are measured:

1. Titratable acid (number of OH^- equivalents required to titrate urine to pH 7.4)
2. Ammonium ions
3. Bicarbonate ions

The sum of 1 and 2 minus 3 equals the net gain or loss of bicarbonate to the body (negative values equal loss, positive values gain).

Normally, about 75% of the excess H^+ ions are excreted as NH_4^+, the remainder as titratable acid. The normal rate of fixed acid (mainly phosphate and sulfate) excretion is about 60 mEq/day. (Remember, however, that the rate of hydrogen ion secretion is considerably higher because of the need to reabsorb bicarbonate.)

In disease states, such as diabetes mellitus, large variations in H^+ excretion are observed. This is due to the production and subsequent excretion of strong organic acids (ketone bodies), which have to be buffered in the urine. For example, the pK of β-hydroxybutyric acid is 4.4. Most of its H^+ ions have to be excreted as ammonium ions.

12.4.4 Renal Ammonia Production

Most of the ammonia produced in tubular cells is derived from glutamine. Glutamine is hydrolytically cleaved by **glutaminase I** to NH_3 and glutamate. The glutamate is subsequently oxidatively deaminated by **glutamate dehydrogenase** to yield another NH_3 molecule and α-ketoglutarate.

In another reaction, glutamine transaminase transfers the α nitrogen of glutamine to a variety of α-keto acids. The reaction product, **α-ketoglutaramate,** loses its amide nitrogen, yielding NH_3 and α-ketoglutarate. The enzyme complex of glutamine transaminase and the amidase is known as **glutaminase II.** Other transaminases covert α-ketoglutarate back to glutamate, using the α amino group from another amino acid. Most amino acids probably contribute to the ammonia production through transamination to glutamate, except glycine, which may be oxidatively deaminated by D-amino acid oxidase, and asparagine, which has its amide nitrogen removed by asparaginase (Figure 12.14).

Figure 12.14.
Generation of ammonia in the kidney.

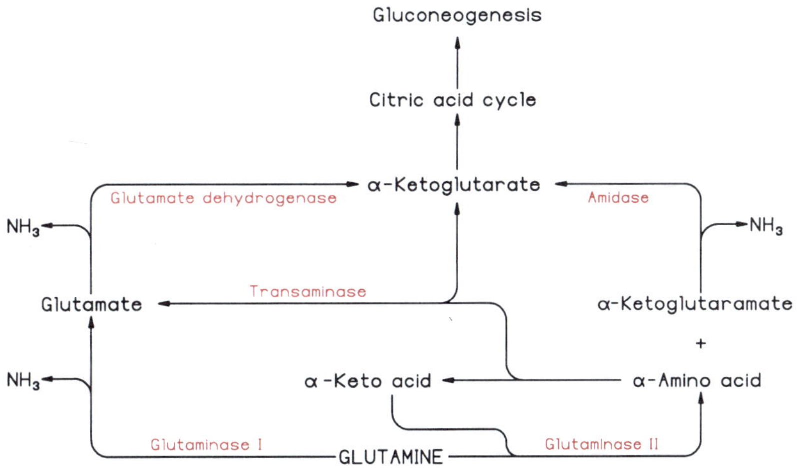

Normally, some 30–50 mEq of ammonia are excreted daily in the urine. This value may increase 10-fold in severe acidosis. The control of ammonia production is not completely understood. The following theory has received the most attention.

In acidosis, **renal gluconeogenesis** is **increased** significantly because of an increase in the concentration of phosphoenolpyruvate carboxykinase (PEPCK). This enzyme converts oxaloacetate to phosphoenolpyruvate, an important intermediate in gluconeogenesis. As a consequence, the concentrations of all renal intermediates of the Krebs cycle, including α-ketoglutarate, are decreased. This relieves the inhibition of glutamate dehydrogenase and thus increases the rate of conversion of glutamate to α-ketoglutarate and NH_3. The reduced concentration of glutamate, in turn, relieves the inhibition of glutaminase I and increases the conversion of glutamine to glutamate and NH_3.

Ammonia diffuses both into the urine and into peritubular capillary blood. When urine is acid, the NH_3 that enters the tubular fluid is trapped as NH_4^+ ions.

12.5 RENAL CONTROL OF POTASSIUM BALANCE

Most of the potassium of the body is contained within cells but is readily exchangeable. Only 2% is found in extracellular fluid. The normal diet contains 50–150 mmol of K^+/day. Most of this is excreted in the urine, the remainder (2 mmol/day) in the feces.

Potassium homeostasis is vitally important for normal nerve and muscle function. K^+ ions can be reabsorbed throughout the nephron (except the descending limb of the loop of Henle) and are secreted in the distal tubule and the collecting duct. Most of the filtered K^+ is reabsorbed in the proximal tubule. The reabsorption of potassium is an active process.

The secretion of K^+ ions in the distal tubule is thought to occur by active transport into the cell from the interstitial fluid (Na^+/K^+ exchange pump) followed by passive diffusion out of the cell, along an electrochemical gradient, into the tubular lumen. The lumenal membrane potential, which opposes K^+ diffusion, is low (because of the high membrane conductivity for Na^+, g_{Na}), and the chemical concentration gradient favors K^+ movement out of the cell. In contrast, the peribular membrane potential is high and approaches the equilibrium potential for K^+. The minimal leakage of K^+ that occurs through this membrane (owing to a small electrochemical gradient) is counteracted by the Na^+/K^+ exchange pump (Figure 12.15).

The rate of K^+ secretion in the distal tubules can be affected in three ways:

1. By a **change in the chemical gradient** (concentration gradient) for the diffusion of K^+ ions from cell to lumen. An increased gradient, which would accelerate the diffusion of K^+ ions from the cell into the lumen, could be the result of an increased activity of the Na^+/K^+ exchange pump or of an increased rate of flow of tubular fluid, which could keep the lumenal K^+ concentration low. The intracellular K^+ concentration is influenced by the acid-base status; it rises when intracellular pH rises. Intracellular pH rapidly changes in response to changes in P_{CO_2} but changes more slowly upon metabolic changes of the acid-base status.
2. By a **change in the permeability of the lumenal membrane to K^+.** The permeability for K^+ ions has been shown to rise when plasma K^+ concentration is increased and also when the plasma bicarbonate concentration is raised.
3. By a **change in the translumenal membrane potential,** E_l, which opposes K^+ diffusion out of the cell into the lumen. For example, an increase in the lumenal Na^+ concentration causes E_l to become less negative (because E_{Na}, the equilibrium potential for sodium, rises). This causes increased K^+ diffusion into the lumen. Similarly, a fall in the Cl^- concentration of tubular fluid (usually coinciding with an increased bicarbonate concentration) reduces E_l (due to a decrease in E_{Cl}). The membrane is impermeable to bicarbonate. A change in membrane permeability to K^+ ions (see item 2 above) will, of

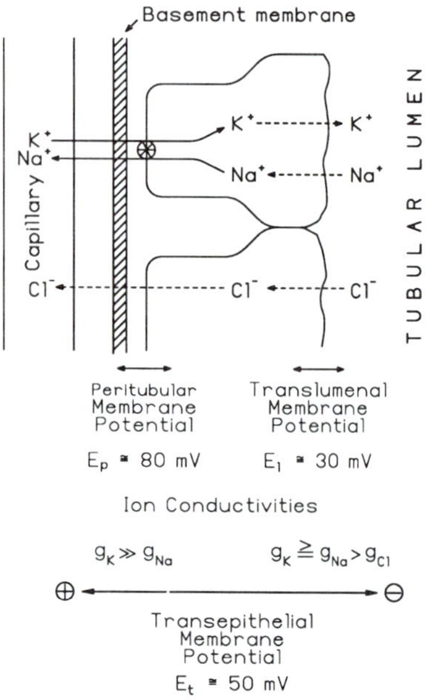

Figure 12.15.
Mechanism of K^+ secretion by the distal tubular cell. (For an explanation, see the text.)

course, also alter E_i. (For a more thorough discussion of the membrane potentials, consult physiology texts.)

An increase in the plasma K^+ concentration stimulates **aldosterone** secretion by the adrenal cortex. Aldosterone stimulates the secretion of K^+ ions. The exact mechanism of this control is not known. The regulation of K^+ balance by aldosterone is separate from the regulation of Na^+ homeostasis and volume control by the same hormone. However, other stimuli such as dehydration, which also cause aldosterone secretion, may lead to hypokalemia. The diuretic drug **spironolactone** specifically inhibits the effect of aldosterone on the distal tubule. Many diuretics that inhibit Na^+ reabsorption, and thus increase the flow of urine through the distal tubules, also cause the loss of K^+ ions. Spironolactone is a **potassium-sparing diuretic,** since it also inhibits K^+ secretion.

Gastrointestinal secretions contain two to five times the K^+ concentration of plasma. Diarrhea or vomiting thus cause a loss of potassium that may become severe because of the secretion of aldosterone (due to the loss of ECF volume) and the disturbance of the acid-base balance.

Most disturbances in acid-base balance eventually lead to loss of potassium.

In **respiratory acidosis,** K^+ excretion initially falls because of a decrease in K^+ concentration in the distal tubular cells (H^+/K^+ exchange). However, in chronic respiratory acidosis, potassium excretion eventually increases because of a rise in plasma bicarbonate (due to renal compensation) and a corresponding fall of chloride concentrations. (The fall of Cl^- in tubular fluid lowers E_l, and the rise of HCO_3^- may increase the permeability of the tubular membrane to K^+.)

In **metabolic acidosis,** the loss of Na^+ ions (as counter-ions to urinary organic acids) induces the secretion of aldosterone; this hormone then effects increased potassium secretion.

In **respiratory alkalosis,** the decrease in P_{CO_2} causes the intracellular pH, and consequently the K^+ concentration, to rise (H^+/K^+ exchange), which leads to increased K^+ excretion. However, the potassium loss is usually minimal.

Finally, in **metabolic alkalosis,** plasma HCO_3^- concentration rises and Cl^- concentration falls. Both events increase K^+ secretion. The resulting K^+ deficiency is called **hypochloremic, hypokalemic alkalosis.** When correcting this condition, it is important to give the patient both Cl^- and K^+ in order to remove the stimulus for continued potassium secretion.

12.6 SUMMARY OF COMPENSATORY MECHANISMS

12.6.1 Renal Compensation of Respiratory Acidosis

A reduction in alveolar ventilation causes the P_{CO_2} of blood to rise (**hypercapnea**). The increased P_{CO_2} stimulates tubular H^+ secretion, which leads to an increase in plasma bicarbonate above its normal value. After several days, the plasma pH returns to normal because the ratio of bicarbonate to carbonic acid again approaches the normal ratio of 20:1. (Remember that in this compensatory state, both the level of plasma bicarbonate and the P_{CO_2} are above normal.) The extent of compensation is nearly complete for P_{CO_2} values of up to 65 mmHg. At higher values, the maximal rate of bicarbonate reabsorption becomes limiting, and the degree of compensation is less.

12.6.2 Renal Compensation of Respiratory Alkalosis

Chronic **hypocapnea** (decreased P_{CO_2} caused by **hyperventilation**) is infrequent but may occur in some forms of hysteria, at high altitude, during salicylate therapy (aspirin), or in heart disease when the arterial oxygen concentration is sufficiently low to stimulate chemoreceptors that in turn drive ventilation. Renal compensation usually is complete.

12.6.3 Renal Compensation of Metabolic Acidosis

In metabolic acidosis, the plasma concentration of bicarbonate is reduced because of its reaction with the excess H^+ ions. The kidneys respond by increasing the rate of H^+ secretion, and thereby create new bicarbonate ions in the plasma. This occurs even in the absence of an increased P_{CO_2}, the usual stimulus for H^+ secretion. On the contrary, the P_{CO_2} in this condition is usually below normal because of partial respiratory compensation of the acidosis (hyperventilation). The renal compensation takes place nevertheless because fewer H^+ ions are required to reabsorb all filtered bicarbonate, the amount of which is diminished in this condition. Additional H^+ ions that are secreted acidify the urine and create new bicarbonate ions in the plasma, thereby raising plasma pH.

12.6.4 Renal Compensation of Metabolic Alkalosis

Metabolic alkalosis may result from chronic vomiting and loss of gastric secretions. This results in the net addition of alkali to the blood by the gastrointestinal system, and the bicarbonate concentration in plasma rises. The respiratory system responds by increasing the P_{CO_2} (hypoventilation) and partially compensates for the alkalosis. Because of the large load of HCO_3^-, the kidney excretes bicarbonate despite the increased H^+ secretion (caused by an increased P_{CO_2}). Virtually no titratable acid or NH_4^+ ions are formed, and the urine becomes alkaline.

Complicating factors sometimes interfere with the renal response to metabolic alkalosis. For instance, the loss of NaCl and water during vomiting causes a reduction in the ECF volume. This increases the rate of salt and water reabsorption and is accompanied by an increased reabsorption of bicarbonate. Correction of the state of dehydration by administration of NaCl and water will, therefore, aid the kidney in correcting the alkalosis.

12.6.5 Comparison of Renal and Respiratory Compensation

Respiratory responses to metabolic disturbances in acid-base balance are rapid but result in only partial compensation. Renal responses take considerably more time but usually achieve complete compensation. Respiratory disturbances are fully compensated within several days. The extent of renal compensation of metabolic disturbances depends on the primary event causing the disturbance. In uncontrolled diabetes mellitus, for example, the renal compensation is never complete.

It should be noted that the result of the renal response to respiratory disturbances of acid-base balance is different from the response to metabolic disturbances. The former is a true "compensation" in the sense that the plasma bicarbonate concentration is "driven away" from the normal level in order to return the pH to normal. The latter is more appropriately called a "correction" because both the pH and the plasma bicarbonate concentration are returned to normal levels.

12.6.6 Influence of H^+ Secretion on NaCl Reabsorption

The ratio of chloride to bicarbonate in the glomerular filtrate is about 5:1. Usually, bicarbonate reabsorption occurs at about the same rate as Na^+ reabsorption. The result is a 1:1 exchange of H^+ for Na^+.

When Na^+ reabsorption proceeds faster than H^+ secretion (as observed in alkalosis), more chloride is reabsorbed. By contrast, if H^+ secretion is faster than Na^+ reabsorption (acidosis), more Na^+ is exchanged for H^+, and more Cl^- is excreted. As a consequence, an inverse correlation exists between the excretion rates of chloride and bicarbonate ions.

When bicarbonate reabsorption is incomplete, some Na^+ is lost in the urine. However, the sodium loss is less than the loss of bicarbonate because, due to the alkalosis, more K^+ is secreted in exchange for Na^+ ions.

Carbonic anhydrase inhibitors are used as diuretics because they reduce bicarbonate reabsorption, thus causing increased excretion of both Na^+ and HCO_3^-. The increased osmolarity of the urine causes loss of water. This leads to loss of K^+ because the Na^+ ions in the tubular fluid are exchanged in part for K^+ ions.

Further References

Cogan MG. Renal effects of atrial natriuretic factor. *Ann Rev Physiol.* 1990; 52:699–708.

Perutz MF. Mechanisms regulating the reactions of human hemoglobin with oxygen and carbon monoxidase. *Ann Rev Physiol.* 1990; 52:1–25.

Rabon EC, Reuben MA. The mechanism and structure of the gastric H,K-ATPase. *Ann Rev Physiol.* 1990; 52:321–344.

Schafer JA. Transepithelial osmolality differences, hydraulic conductivities, and volume absorption in the proximal tubule. *Ann Rev Physiol.* 1990; 52:709–726.

Wolff SD, and Balaban RS. Regulation of the predominant renal medullary organic solutes in vivo. *Ann Rev Physiol.* 1990; 52:727–746.

Zeidel ML. Renal actions of atrial natriuretic peptide: Regulation of collecting duct sodium and water transport. *Ann Rev Physiol.* 1990; 52:747–759.

CHAPTER

13 Hormones

OBJECTIVES

After studying this chapter, the student should be able to

1. Describe the principle of a radioimmune assay for determining hormone concentrations in body fluid.
2. Interpret a Scatchard plot.
3. List the known hypothalamic hormones and describe their principal functions.
4. Draw a graph illustrating the change in the levels of circulating LH, FSH, estradiol, and a progesterone during the menstrual cycle and describe the functions of these hormones.
5. Describe the mechanisms of the direct and indirect actions of growth hormone.
6. Explain why growth hormone is diabetogenic.
7. Describe the subunit structure of pituitary glycoprotein hormones and name an extrapituitary hormone of analogous structure.
8. Outline the steps involved in the biosynthesis of thyroid hormones.
9. Name several substances (including hormones) that affect the uptake of iodide into the thyroid.
10. Describe the functions of parathyroid hormone and calcitonin in calcium and phosphate homeostasis.
11. Explain the significance of C-peptide determination in evaluating pancreatic B cell function.
12. Compare the principal actions of insulin and glucagon on various tissues.
13. Describe the probable pathogenesis of type I and type II diabetes.
14. Explain why injection of insulin may lead to acute hypokalemia.
15. Describe the molecular mechanism of glucagon action.
16. Outline the pathways for inactivation of catecholamines.
17. Describe the different intracellular events that result from the interaction of catecholamines with α-adrenergic and β-adrenergic receptors.
18. Name the first intermediate in the biosynthesis of steroid hormones and describe its synthesis from cholesterol.
19. Compare the endocrine consequences of congenital 21-hydroxylase and 11-hydroxylase deficiencies.
20. Outline the control of glucocorticoid and mineralocorticoid biosynthesis and secretion.
21. Describe the intracellular events that follow the interaction of a steroid hormone with its target cell.
22. Explain the term "permissive action" of glucocorticoids.
23. Identify the tissues upon which cortisol exerts a catabolic effect and those upon which the hormone produces anabolic effects.
24. Describe the origins and functions of the hormones secreted during pregnancy.
25. Explain the mechanism by which aspirin inhibits platelet aggregation.

The classical definition of a hormone is a chemical substance synthesized and released into the bloodstream by an **endocrine gland.** (In contrast, **exocrine glands** release their products into the nonvascular extracellular space, e.g., digestive enzymes of the pancreas released into the pancreatic duct.) The blood carries the hormone to a specific **target tissue** where it is received as a chemical messenger capable of triggering specific biochemical reactions. The target tissue is recognized by the hormone through the presence of unique **receptors** on the plasma membrane or in the cytoplasm of the cell. Some hormones (e.g., protein hormones, catecholamines), after binding to the membrane receptor, exert their effect via the synthesis of a **second messenger** (e.g., a cyclic nucleotide) within the target cell. The second messenger then activates or inactivates a protein kinase that, in turn, activates specific enzymes.

Other hormones (e.g., steroid and thyroid hormones) function by forming a complex with an intracellular receptor protein, which then initiates transcription of specific genes followed by protein synthesis.

It is now clear that hormones are much more ubiquitous than previously thought and that the above definition of hormone function represents only a special case of a more universal function. Hormones, such as insulin, for example, which were thought to be unique to the pancreas have been found in brain, testes, liver, and other tissues. In contrast to the endocrine effects of the classical hormones, the hormones produced in these secondary tissues have only **paracrine effects,** that is, they act only upon cells in the immediate neighborhood of the secreting cell. A hormone may act on the same cell in which it is synthesized; this is called an **autocrine effect.** This wider definition of a hormone includes cell growth factors, some of which have only paracrine functions.

Hormones are distributed throughout nature. Insulinlike substances, for example, have been discovered in many different forms of life, even in protozoa and bacteria. Perhaps hormones have been utilized throughout evolution as means of cell-cell communication, acting initially only as growth factors for certain types of cells; but with the evolution of larger multicellular organisms, the need for endocrine communication arose.

Four types of chemical compounds are represented in hormones:

1. Proteins or oligopeptides
2. Derivatives of amino acids

Table 13.1. Human Hormones

Name	Structure	Comments
Protein and Polypeptide Hormones		
Hypothalamic releasing hormones		
Thyrotropin-releasing hormone (TRH)	Tripeptide, pyroGlu-His-Pro(NH_2)	Causes release of TSH
Gonadotropin-releasing hormone (GnRH,LHRH)	Decapeptide, pyroGlu-His-Trp-Ser-Tyr-Gly-Leu-Arg-Pro-Gly(NH_2)	Causes release of LH and FSH
Somatostatin (GIH)	Tetradecapeptide, Ala-Gly-Cys-Lys-Asn-Phe-Phe-Trip-Lys-Thr-Phe-Thr-Ser-Cys (there is an intramolecular S—S bond between the Cys residues)	Inhibits secretion of GH
Prolactin-inhibiting hormone (PIH)	Dopamine (not a peptide)	Inhibits release of prolactin
Corticotropin-releasing hormone (CRH)	Polypeptide, 41 amino acids	Causes release of ACTH
Growth-hormone-releasing hormone (GRH)	Polypeptide, 44 amino acids	Causes release of GH
Prolactin-releasing hormone (PRH)	Polypeptide, sequence not known	Causes release of prolactin
Neurohypophyseal hormones (synthesized in the hypothalamus)		
Vasopressin (antidiuretic hormone, ADH)	Nonapeptide, Cys-Tyr-Phe-Gln-Asn-Cys-Pro-Arg-Gly(NH_2) (the two cysteines are linked by a disulfide bond)	Promotes water reabsorption in kidney, causes vasoconstriction
Oxytocin	Nonapeptide, Cys-Tyr-Ile-Gln-Asn-Cys-Pro-Leu-Gly(NH_2) (the two cysteines are linked by a disulfide bond)	Causes uterine contraction, milk ejection into ducts of the breast
Adenohypophyseal hormones (anterior pituitary)		
Follicle-stimulating hormone (FSH)	Glycoprotein, α and β subunits	Induces spermatogenesis in male testis, maturation of follicles in ovaries, synthesis of estrogens in theca cells
Luteinizing hormone (LH)	Glycoprotein, α and β subunits	Induces testosterone synthesis in male testis, ovulation in the female, estrogen and progesterone synthesis in the corpus luteum
Thyroid-stimulating hormone (TSH)	Glycoprotein, α and β subunits	Induces synthesis and release of thyroid hormones by the thyroid gland
Growth hormone (GH; somatotropin)	Protein, 191 amino acids	Increases growth in most tissues, lipolytic in adipose cells
Prolactin	Protein, 198 amino acids	Induces lactation
Adrenocorticotropic hormone (ACTH)	Oligopeptide, 39 amino acids	Induces glucocorticoid synthesis in adrenal cortex
β-Lipotropin (β-LPH)	Polypeptide, 93 amino acids	Function unknown
Melanocyte-stimulating hormone (β-MSH)	Oligopeptide, 18 amino acids	Stimulates pigmentation in amphibia, function in man unknown (two smaller forms also exist)
Calciotropic hormones		
Parathyroid hormone (PTH)	Protein, 115 amino acids	Increases plasma Ca^{2+}, lowers plasma phosphate
Calcitonin	Oligopeptide, 32 amino acids	Synthesized by the parafollicular cells of the thyroid, lowers plasma Ca^{2+}
Pancreatic hormones		
Insulin	Protein: A chain, 21 amino acids; B chain, 30 amino acids, 3 disulfide bridges	Increases cellular uptake of glucose in muscle and adipose cells; anabolic actions: glycogen synthesis in the liver, fat synthesis in adipose cells
Glucagon	Polypeptide, 29 amino acids	Catabolic hormone, increases blood glucose, acts via cAMP

Table 13.1. (*Continued*)

Name	Structure	Comments
Somatostatin	Tetradecapeptide	Inhibits secretion of pancreatic hormones
Pancreatic polypeptide	Polypeptide, 36 amino acids	Function unknown
Placental hormones		
Human chorionic gonadotropin (hCG)	Glycoprotein similar to pituitary glycoproteins	LH activity, maintains corpus luteum during early pregnancy
Human chorionic somatomammotropin (hCS), also called human placental lactogen (hPL)	Protein, 191 amino acids	GH activity, lactogenic
Relaxin	Protein similar to insulin, A and B chains	Secreted by corpus luteum of pregnancy, relaxes pelvic joints, dilates cervix
Gastrointestinal hormones		
Cholecystokinin (CCK)	Polypeptide, 4 forms: 58, 39, 33, and 8 amino acids	Synthesized in duodenum, causes contraction of gall bladder, secretion of pancreatic enzymes; all biological activity is contained in carboxy-terminal octapeptide; sulfation of Tyr-5 from carboxyl terminus is required for full biological activity
Gastrin	Polypeptide, 3 active forms: 14, 17, and 34 amino acids	Synthesized in antrum of stomach, stimulates HCl and pepsin secretion
Secretin	Polypeptide, 27 amino acids, sequence similar to that of glucagon	Synthesized in duodenum, stimulates pancreatic secretions (high bicarbonate, low enzymes)
Vasoactive intestinal peptide (VIP)	Polypeptide, 28 amino acids, sequence similar to glucagon	Present throughout GI tract, stimulates intestinal secretions, dilatation of peripheral vessels
Gastric inhibitory peptide (GIP)	Polypeptide, 43 amino acids, extensive sequence homology with glucagon	Synthesized in duodenum and jejunum, secretion stimulated by glucose and fat in duodenum, stimulates insulin secretion
Glucagon	Polypeptide, 29 amino acids	Secreted by mucosa of stomach and duodenum
Motilin	Oligopeptide, 22 amino acids	Found in endocrine cells of the mucosa of duodenum and jejunum, stimulates gastric pepsin secretion, increases peristalsis
Somatostatin	Tetradecapeptide (structure see above)	Found throughout GI tract
Gastric-releasing peptide (GRP; bombesinlike immunoreactivity)	Oligopeptide, 14 amino acids	Probably released from nerve endings throughout GI tract
Substance P	Oligopeptide, 11 amino acids	Probably released from nerve endings throughout GI tract
Vasoactive peptide hormones		
Angiotensin II	Octapeptide, Asp-Arg-Val-Tyr-Ile-His-Pro-Phe	Powerful vasoconstrictor, made from angiotensin I by converting enzyme, stimulates aldosterone secretion
Bradykinin	Nonapeptide, Arg-Pro-Pro-Gly-Phe-Ser-Pro-Phe-Arg	Generated from kininogen by kallikrein, powerful vasodilator (lysylkallikrein, a decapeptide, has similar actions)
Atrial natriuretic hormone (ANP)	Oligopeptide, several forms, (α-ANP), 28 amino acids: Ser-Leu-Arg-Arg-Ser-Ser-Cys-Phe-Gly-Gly-Arg-Met-Asp-Arg-Ile-Gly-Ala-Gln-Ser-Gly-Leu-Gly-Cys-Asn-Ser-Phe-Arg-Tyr (the two cysteines are linked by a disulfide bond)	Causes natriuresis and fall in blood pressure, released from cardiac atria in response to stretch receptor activation (increased blood volume)

Table 13.1. (*Continued*)

Name	Structure	Comments
Amino Acid–Derived Hormones		
Thyroid hormones		
Thyroxine (T4), tetraiodothyronine	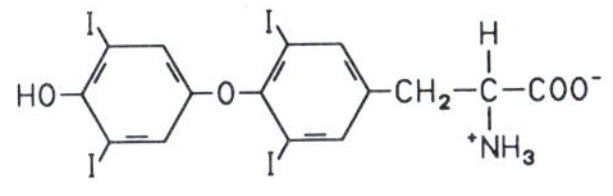	Stimulates metabolic rate in most cells
Triiodothyronine (T3)	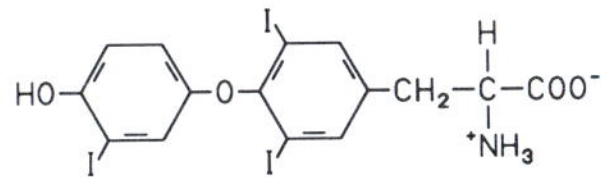	Same actions as T4, more active
Hormones of the adrenal medulla		
Epinephrine (adrenaline)	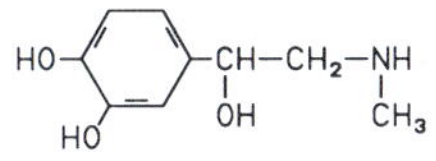	Synthesized in chromaffin cells, stimulation of metabolism, rise in blood glucose and FFA, vasoactive
Norepinephrine (noradrenaline)	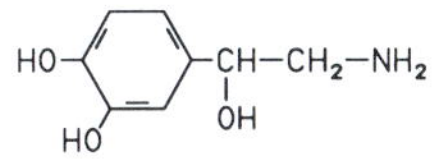	Similar actions, also released at nerve endings
Pineal hormone, Melatonin	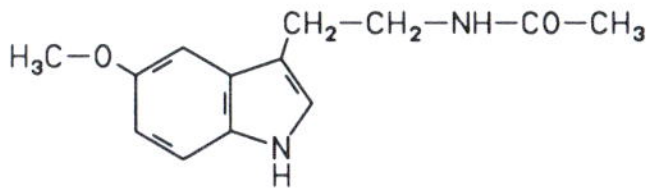	Synthesized by the pineal gland, causes pigment aggregation in amphibia (lightening of the skin), function in humans unknown
Steroid Hormones		
Adrenal cortex hormones		
Cortisol	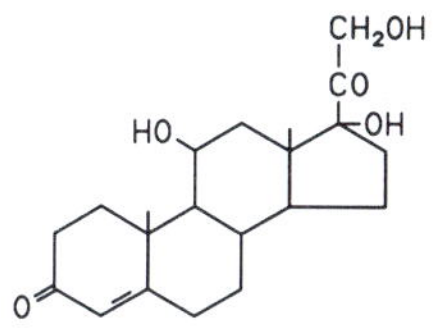	Synthesized in zona fasciculata and zona reticularis, causes increased carbohydrate storage, permissive action for other hormones, anti-inflammatory
Aldosterone	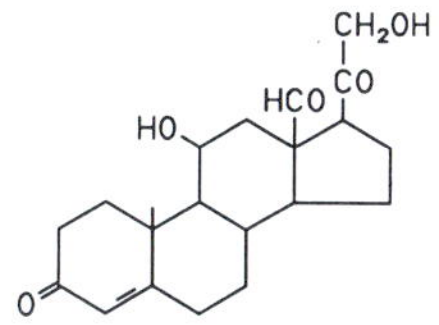	Synthesized in zona glomerulosa, stimulates reabsorption of Na^+ by renal tubules
Ovarian hormones		
Progesterone	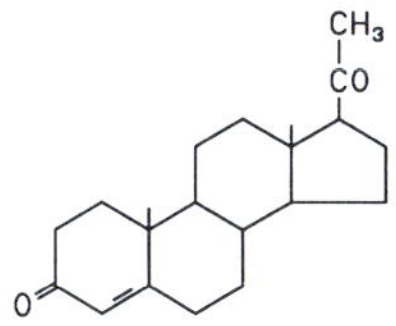	Synthesized in corpus luteum, maintains pregnancy; is also synthesized by the placenta
Estradiol	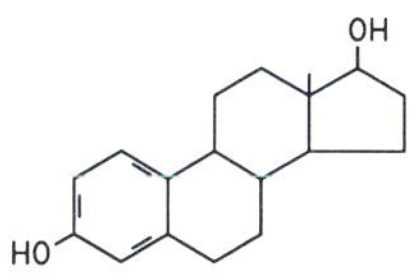	Synthesized in Graafian follicle and corpus luteum, principal female sex hormone
Testicular hormones		
Testosterone	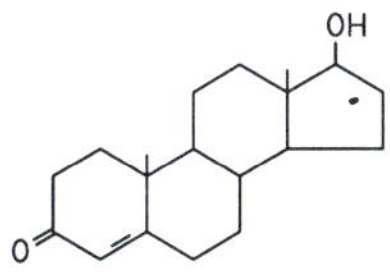	Synthesized in Leydig cells of testis, principal male sex hormone

Table 13.1. *(Continued)*

Name	Structure	Comments
Prostaglandins and Related Hormones		
Prostaglandins		
PGF$_2$	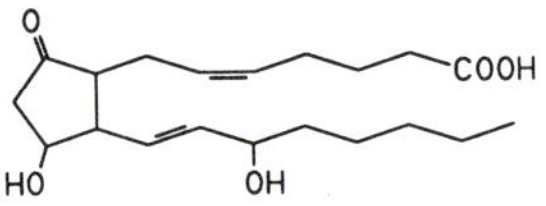	Relaxes smooth muscle, dilates bronchial tubes
PGF$_{2\alpha}$	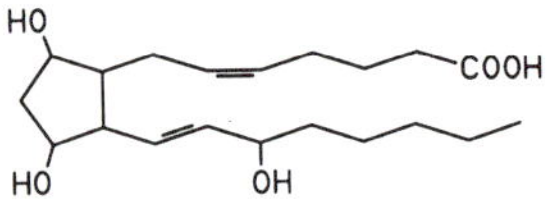	Stimulates uterine contraction, causes bronchial constriction, causes luteolysis
PGD$_2$	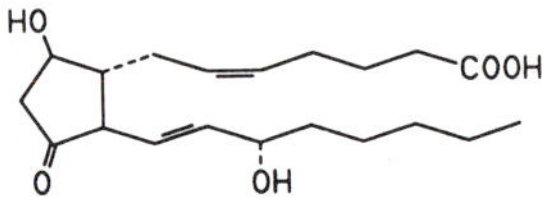	Actions similar to PGF$_{2\alpha}$
Thromboxanes		
Thromboxane A$_2$ (TXA$_2$)	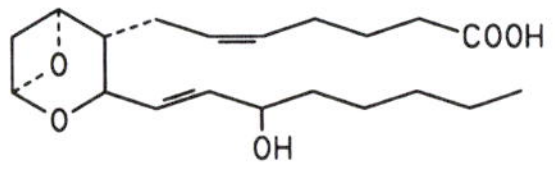	Released by platelets, promotes platelet aggregation, constricts smooth muscle
Icosanoids		
Prostacyclin (PGI$_2$)	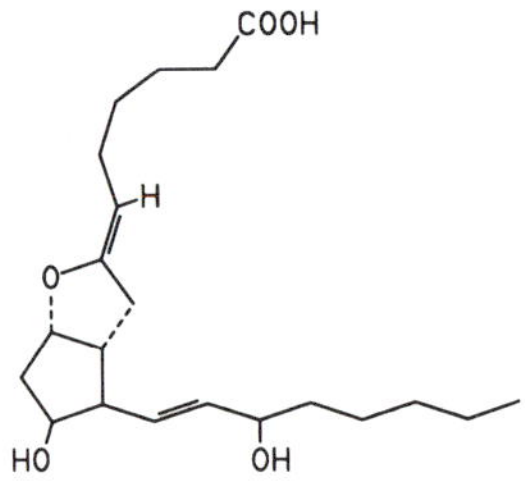	Released by vascular endothelium, inhibits platelet aggregation, relaxes smooth muscle
Endoperoxide PGG$_2$	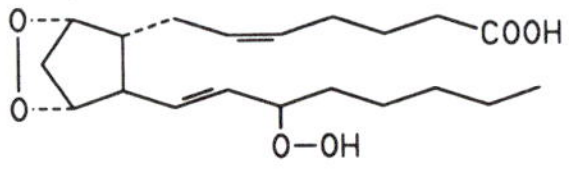	
5-Hydroperoxyeicosa-tetraenoic acid 5-HPETE	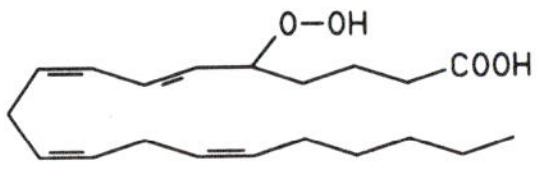	
12-Hydroperoxyeicosa-tetraenoic acid 12-HPETE	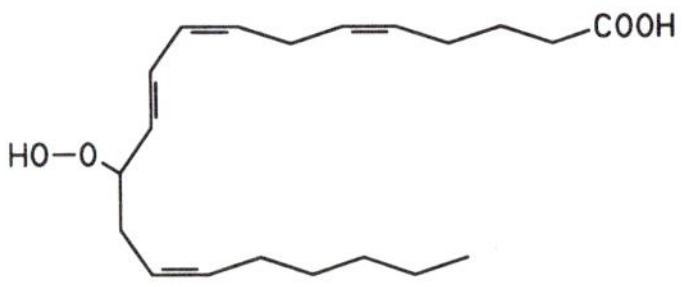	
Leukotrienes		
Leukotriene (LTC$_4$)	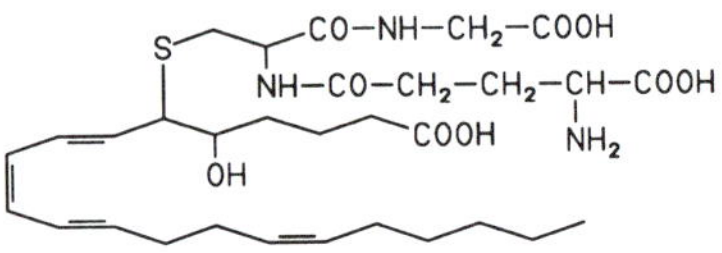	Slow-reacting substance of anaphylaxis, powerful vasoconstrictors, increase vascular permeability

3. Steroids
4. Prostaglandins (derivatives of fatty acids)

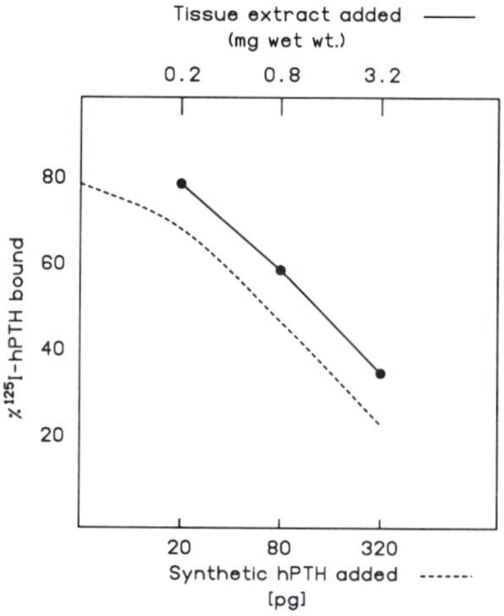

Figure 13.1.
Example of a radioimmune assay for human parathyroid hormone (hPTH). The dashed line represents a standard curve obtained with a known amount of pure hPTH.

Table 13.1 presents an overview of the known human hormones. The list does not include the many cellular growth hormones that have been discovered in recent years.

The concentration of a particular hormone in a body fluid may be determined (1) by chemical means (low sensitivity, low specificity), (2) by biological assay, in which the hormone's biological effect upon an animal or isolated target organ is measured (low specificity), or (3) by competitive binding assays such as the **radioimmune assay (RIA).** In this last, most sensitive and specific assay, a known amount of a radioactively labeled hormone is mixed with an unknown sample and with antibodies against this hormone. Since the unlabeled hormone competes with the labeled hormone for antibody binding sites, the amount of antibody-bound radioactivity is inversely proportional to the concentration of the unlabeled hormone in the sample (Figure 13.1).

The concentration of most hormones in blood is 10^{-11} to $10^{-9}M$. In order to respond to circulating hormones in a very specific way, the cellular receptors must have a high affinity for the hormone. Hormone-hormone receptor binding constants are determined experimentally from the **Scatchard plot.** The dissociation constant K_d of the complex is

$$K_d = \frac{[H][R]}{[HR]} \tag{13.1}$$

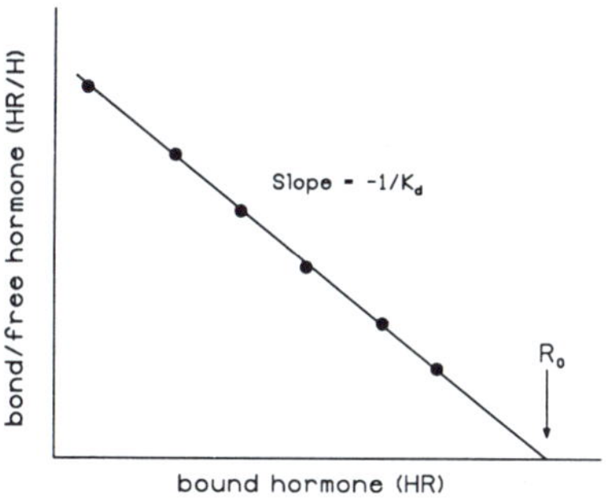

Figure 13.2.
Scatchard plot.

where [H] is the concentration of free hormone, [R] is the concentration of free receptor, and [HR] is the concentration of the hormone-receptor complex. At high concentrations of hormone, all receptor is in the form HR. Thus, [HR] equals the total concentration of receptor, R_0. Equation (13.1) then becomes

$$\frac{[HR]}{[H]} = \frac{R_0 - [HR]}{K_d} = \frac{[HR]}{K_d} + \frac{R_0}{K_d} \tag{13.2}$$

The last member of equation (13.2) is a constant. Thus, if bound/free hormone ([HR]/[H]) is plotted versus bound hormone ([HR]), the slope of the resulting line equals $-1/K_d$, and its intersection with the abscissa is the total receptor concentration, R_0 (Figure 13.2).

13.1 HYPOTHALAMIC HORMONES

The **hypothalamus** is an important region in the base of the brain serving as a control center for the integration of autonomic mechanisms concerned with homeostatic regulation. It is located above the **pituitary gland (hypophysis)** and is connected to it by the pituitary stalk. Specialized hypothalamic neurons produce, upon neural stimulation, a series of hormones, mostly oligopeptides, and release them into a network of capillaries that empty into a system of portal veins leading to the anterior lobe of the pituitary gland. Specific target cells in the adenohypophysis (anterior lobe of the pituitary) react to the presence of hypothalamic hormones by increasing or decreasing the secretion of a variety of polypeptide hormones into the general circulation (see Section 13.2). Seven different **hypothalamic releasing and inhibiting hormones** can be distinguished:

1. Thyrotropin-releasing hormone (TRH)
2. Gonadotropin-releasing hormone (GnRH)
3. Somatostatin or growth hormone–inhibiting hormone (GIH)
4. Prolactin-inhibiting hormone (PIH)
5. Corticotropin-releasing hormone (CRH)
6. Growth hormone–releasing hormone (GRH)
7. Prolactin-releasing hormone (PRH)

The chemical structures of the first four hormones are known. **TRH** is a tripeptide, **GnRH** is a decapeptide, **somatostatin** is a tetradecapeptide, and **PIH** is dopamine (see Table 13.1). Note that many of the oligopeptide hormones are blocked at both termini. They carry an amide group at their carboxyl end and a cyclized amino acid (pyroglutamate) at the amino terminus. Most of the hypothalamic hormones appear to affect only the secretion of one or more pituitary hormones. Somatostatin, dopamine, and TRH, however, are also found in other parts of the body. In the brain, they probably function as neurotransmitters. Somatostatin is also found in the pancreas, where it inhibits the secretion of insulin and glucagon, and in the gastrointestinal mucosa, where it interferes with the secretion of gastrin, secretin, and other gastrointestinal hormones.

In addition to releasing and inhibiting hormones, hypothalamic neurons also synthesize two hormones, **vasopressin** and **oxytocin,** which are released into the general circulation via the posterior lobe of the pituitary. Synthesis of these two nonapeptide hormones occurs within the neuron cell bodies in the hypothalamus in the form of larger precursor molecules (prohormones). These **prohormones** are then transported, packaged in neurosecretory vesicles, along the axon to terminals located in the posterior lobe of the pituitary. Before secretion, the peptide hormones are cleaved from their precursors. Upon neural stimulation, both hormones are released into the circulation, together with their remnant peptides, called **neurophysins.** The sequences of the two hormones differ in only two amino acids. After being released, their half-life is about 10 minutes. In addition to the neurophysin peptide, the vasopressin prohormone contains the sequence of a 39-amino-acid glycopeptide of unknown function.

Vasopressin, more appropriately called **antidiuretic hormone (ADH),** functions in the control of water balance. Its target cells are the collecting ducts of the kidney, where the hormone increases the cells' permeability to water and thus allows reabsorption of water from the urine. As is the case with most peptide hormones, the binding of vasopressin to cell membrane receptors activates **adenylate cyclase.** The increased levels of cyclic adenosinc monophosphate (cAMP) then probably activate a protein kinase that, through the phosphorylation of certain proteins, causes the insertion of **water channels** into the cell membrane. This insertion is very fast and probably involves the fusion of intracellular vesicles containing functional water channels with the plasma membrane. As soon as the vasopressin levels decrease, as occurs after the consumption of alcohol, which inhibits vasopressin release from the pituitary, the water channels are internalized and stored again within cytoplasmic vesicles located just under the cell membrane (Figure 13.3).

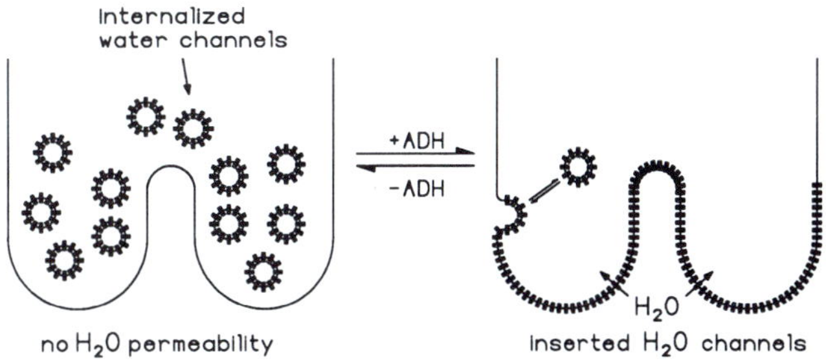

Figure 13.3.
The probable mechanism of insertion of water channels into the lumenal plasma membranes of distal tubular cells under the influence of vasopressin.

The release of vasopressin is controlled by **osmoreceptor cells** in the anterior hypothalamus that respond to small changes in the osmolarity of blood. In addition, **low-pressure receptors** in the **great veins** and **high-pressure receptors** in the **carotid sinuses,** registering changes in volume of extracellular fluid, affect vasopressin secretion. In large doses, vasopressin raises arterial blood pressure. In form of a nasal spray, vasopressin is used to stop esophageal bleeding, since arteriolar vasoconstriction reduces portal venous pressure. **Diabetes insipidus,** characterized by excessive loss of fluid, may be caused by hypothalamic lesions or by an inability of the kidney to respond to vasopressin (nephrogenic diabetes insipidus).

Oxytocin, the other hypothalamic hormone secreted by the posterior hypophysis, causes uterine contraction and milk ejection into the ducts of the breast. Its secretion is triggered by neural stimuli from the uterus (dilatation of the cervix) and the breast (suckling). The hormone is used clinically for the induction of labor.

13.2 PITUITARY HORMONES

In response to hypothalamic releasing hormones, the human anterior pituitary, or adenohypophysis, releases at least seven different hormones. Most of these hormones stimulate the production of additional hormones by other endocrine organs; for example, the adrenals, the gonads, the thyroid. A few pituitary hormones induce metabolic changes directly in their target cells. Pituitary hormones can be divided into three groups according to their chemical structure: glycoproteins, proteins, and oligopeptides.

13.2.1 Glycoproteins

This group includes the two **gonadotropins, follicle-stimulating hormone (FSH)** and **luteinizing hormone (LH),** as well as thyroid-stimulating hormone (TSH). All three proteins contain about 15% carbohydrate and consist of two subunits (α subunit, 22 kd; β subunit, 34 kd). The α subunits are identical in all three hormones. In contrast, the β subunits are distinct from each other and are responsible for the specificity of the hormones. Both gonadotropins are secreted in pulsatile fashion from the pituitary in response to pulses of GnRH from the hypothalamus. This episodic secretion is necessary in order to avoid down-regulation of GnRH receptors in the pituitary. Long-acting synthetic GnRH analogs are used clinically to suppress LH and FSH secretion. The binding of gonadotropins to their receptors activates adenylate cyclase.

Follicle-Stimulating Hormone

In the male, FSH stimulates testicular growth and promotes the synthesis of **androgen-binding protein** by the **Sertoli cells.** This protein binds testosterone and thereby keeps the intratesticular androgen concentration at the level required for **spermatogenesis.** Spermatogenesis requires both FSH and LH. FSH probably stimulates the Sertoli cells in their support of the maturation of spermatids into spermatozoa. It may also stimulate estrogen synthesis in the same cells. Sertoli cells synthesize the protein **inhibin** that, through feedback, inhibits the release of FSH. Secretion of LH is inhibited directly by testosterone.

In the female, FSH causes the **maturation of Graafian follicles** in the ovaries. Each follicle contains an immature oocyte suspended in a fluid that is surrounded by an epithelium of **granulosa and theca cells.** FSH, in cooperation with LH, stimulates the synthesis of **estrogen** by theca cells. Follicular granulosa cells also synthesize inhibin, thereby controlling their stimulation by FSH.

Luteinizing Hormone

LH stimulates testosterone synthesis in the **Leydig cells (interstitial cells)** of the testis. In the Leydig cells, it stimulates the synthesis of **inhibin,** a protein that selectively inhibits the secretion of FSH by the pituitary. In women, LH stimulates **estrogen** and **progesterone synthesis** in the ovary. A sharp peak in LH levels on day 15 of the menstrual cycle causes rupture of a Graafian follicle (ovulation), release of the ovum, and proliferation of the granulosa and theca cells of the empty follicle that forms the **corpus luteum,** an endocrine organ that produces both estrogen and progesterone. Progesterone reduces the rate of release of both FSH and LH by feedback inhibition at the level of the hypothalamus and pituitary. This prevents the maturation of additional follicles. If no fertilization takes place, the corpus luteum degenerates within 2 weeks, allowing plasma levels of FSH to rise again, thus initiating the maturation of another follicle (Figure 13.4). (The events that occur after fertilization are discussed in Section 13.8.)

The Hypothalamic-Pituitary-Gonadal Axis

The release of LH and FSH by the pituitary is controlled by the level of circulating steroid hormones (Figure 13.5). Negative feedback control occurs at both the hypothalamus (long feedback loop, through the release of GnRH) and the pituitary (short feedback loop). Postmeno-

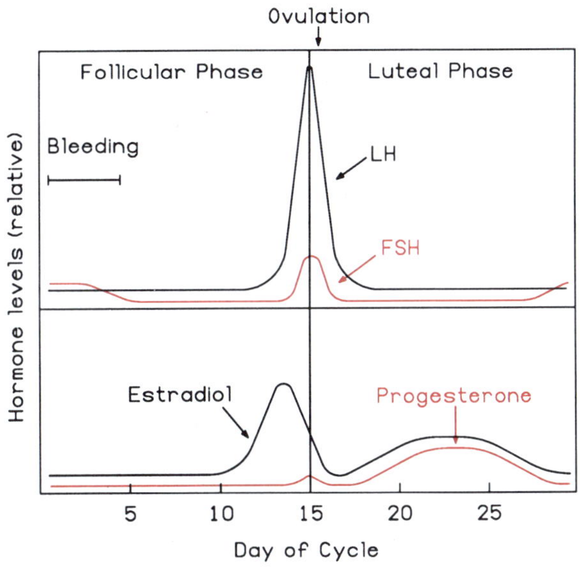

Figure 13.4. Hormone levels during the menstrual cycle. Absolute levels of individual hormones cannot be compared from this graph.

pausal women have high levels of circulating FSH and LH. Estrogen appears to exert a double control. At low plasma levels, a rapid rise of estrogen has a positive effect upon LH release, whereas at constant levels it exerts a negative control. It has been postulated that estradiol sensitizes the pituitary to GnRH by increasing the number of GnRH receptors.

The control of FSH and LH release may be interrupted by positive and negative stimuli from the central nervous system (CNS). Stress, for example, may interfere with gonadotropin release and cause amenorrhea. On the other hand, coitus may trigger spontaneous ovulation. Oral contraceptives, containing a mixture of estrogen and progesterone, prevent ovulation by inhibiting, via the hypothalamus, FSH and LH release.

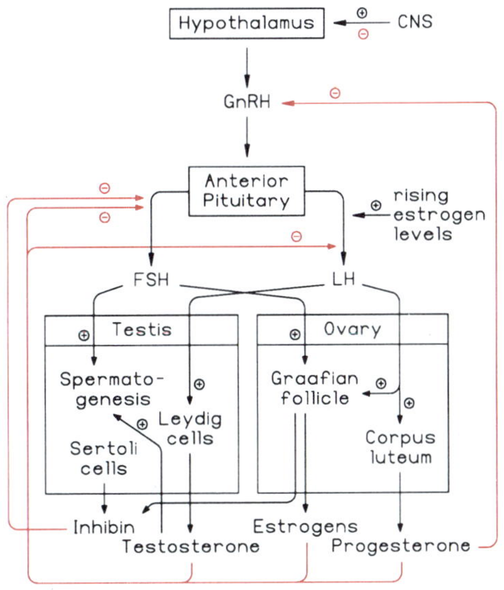

Figure 13.5. The hypothalamic-pituitary-gonadal axis for the regulation of FSH and LH release from the pituitary.

Thyroid-Stimulating Hormone

TSH is required for the synthesis of thyroid hormones in the thyroid gland. Its mechanism of action involves the activation of adenylate cyclase. The control of TSH release is discussed in Section 13.3.

13.2.2 Proteins

The pituitary synthesizes two protein hormones that do not contain carbohydrate side chains: growth hormone and prolactin.

Growth Hormone

Growth hormone (GH) or somatotropin, is a 191-amino-acid protein with two intrachain disulfide bonds. Its sequence shows considerable homology to that of **prolactin** and **human placental lactogen (hPL),** also called **chorionic somatomammotropin (hCS),** a placental hormone, suggesting a common ancestral gene for the three hormones.

The metabolic effects of GH are complex. They are still not completely understood because of the interplay of GH with several other hormones. In general, GH promotes growth. **Hypopituitarism** (underfunction of the pituitary) causes **dwarfism,** while **hyperpituitarism** (usually due to a pituitary tumor) leads to **gigantism** or **acromegaly.**

Direct and indirect actions of GH can be distinguished (Figure 13.6). The **direct actions** appear to be mediated through the activation of adenylate cyclase in the target cell membrane. This causes increased lipolysis and reduced glucose uptake in adipose tissue, leading to a rise in blood glucose and free fatty acids (FFA). The stimulation of glucagon secretion from the pancreatic A cells by GH enhances this **diabetogenic action** of the hormone. In muscle and liver cells, GH stimulates RNA and protein synthesis; in fibroblasts it increases the rate of collagen synthesis. These direct **anabolic effects** of GH are further enhanced by the action of glucocorticoids and androgens (see Sections 13.7 and 13.8).

The **indirect actions** of GH are mediated through the synthesis of several growth-promoting factors called **somatomedins.** These and other cell-specific growth factors, some of which appear to depend on GH, are summarized in Table 13.2

Somatomedins are released into the circulation by liver cells, as well as other cells, in response to the stimulation by GH. Somatomedins are

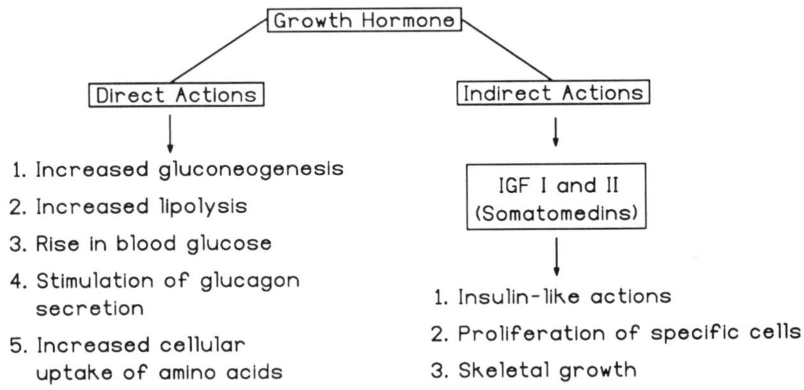

Figure 13.6.
Actions of growth hormone.

Table 13.2. Examples of Growth Factors Released into the Circulation by the Liver and Other Cells

Growth Factor	Action
IGV I	Stimulates sulfate uptake and DNA synthesis in cartilage, makes pancreas responsive to insulinogenic stimuli
IGF II	Stimulates sulfate uptake in chondrocytes
Nerve growth factor (NGF)	Promotes growth of sympathetic neurons
Epidermal growth factor (EGF)	Promotes growth of epidermal cells
Ovarian growth factor (OGF)	Promotes growth of ovarian cells
Fibroblast growth factor (FGF)	Promotes growth of fibroblasts

Note: Only IGF I and IGF II appear to be under the control of GH

now generally called **insulinlike growth factors (IGFs). IGF I** (somatomedin C) is under direct control of GH hormone. It makes the pancreas more responsive to stimuli that increase insulin secretion. An early effect of GH administration, therefore, is a reduction of serum glucose concentration. This effect is later supplanted by the direct diabetogenic effect of GH (via glucagon secretion). The synthesis of the other somatomedin, IGF II (formerly called somatomedin A), is less dependent upon GH. Both IGF I and IGF II are important in fetal growth and development. Their action is responsible for the increased level of **chondrogenesis** (cartilage formation) and **osteogenesis** observed after GH administration. The receptor for IGF I is a tyrosine-specific protein kinase located in the cell's plasma membrane. It shows extensive homology with the insulin receptor. In contrast, the receptor for IGF II is entirely different, with a large extracellular domain consisting of 15 repeat sequences of about 150 amino acids each. It does not exhibit protein kinase activity but has a domain that is homologous to fibronectin.

Normal plasma levels of GH range between 0 and 3 ng/ml. The half-life of GH is only about 30 minutes. GH appears to be secreted in bursts. Usually during the early phase of sleep there is a sharp increase in GH release. Stress also stimulates GH release, as does hypoglycemia, increased plasma levels of amino acids, and anesthesia. FFA, glucocorticoids, and increased plasma glucose levels inhibit GH release (Figure 13.7).

Prolactin

Human prolactin has 198 amino acids and contains three disulfide bonds. Normal plasma levels of prolactin are 0.5 ng/ml in males and 0.8 ng/ml in females. They are increased during exercise, sleep, and pregnancy. Section of the pituitary stalk results in overproduction of prolactin, suggesting that the effects of PIH (dopamine) normally exceed the effects of PRH. Similarly, drugs that block dopamine receptors (e.g., chlorpromazine) increase prolactin secretion.

Prolactin induces the secretion of milk and inhibits the action of gonadotropins upon the ovaries. Thus, **hyperprolactinemia,** which is often seen in **anorexia nervosa,** is accompanied by amenorrhea.

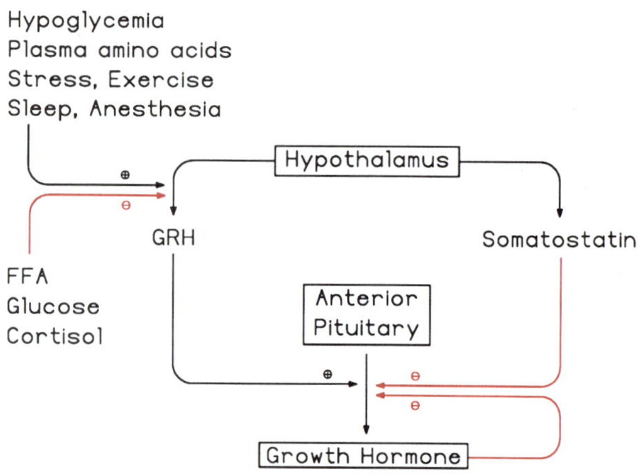

Figure 13.7.
Control of GH secretion.

13.2.3 Oligopeptides

The oligopeptide hormones, **adrenocorticotropic hormone (ACTH)** and **β-lipotropin (β-LPH),** are derived from a large precursor molecule called **pro-opiomelanocortin.** This 263-amino-acid prohormone contains several Lys-Arg, Lys-Lys, and Arg-Arg dipeptide sequences, sites that function in a large number of prohormones as cleavage sites for the production of individual oligopeptide hormones (Figure 13.8).

In humans, the cells of the anterior pituitary normally cleave only ACTH, β-LPH, and β-endorphin from pro-opiomelanocortin. In animals, whose intermediary lobe of the pituitary is developed (rudimentary in humans), several additional oligopeptide hormones are cleaved from the precursor: α-melanocyte–stimulating hormone (α-MSH), β-MSH, and corticotropinlike intermediate lobe peptide (CLIP). In the human pituitary, Met-enkephalin, whose sequence is present in β-LPH, is not produced. It is made, however, from the same or a similar prohormone polypeptide in the hypothalamus.

ACTH consists of 39 amino acids. Only the first 23 amino-terminal residues are required for biological activity. (A synthetic 23-amino-acid analog has full potency.) Secretion of ACTH is induced by stress. The hormone stimulates the synthesis of glucocorticoids in the adrenal cortex via activation of adenylate cyclase. In the absence of ACTH (hypophysectomy), synthesis of glucocorticoids is reduced to very low levels.

Pituitary tumors that secrete excess amounts of ACTH cause **Cushing's syndrome** (excessive production of cortisol; see Section 13.7). Hypersecretion of ACTH may also be due to nonpituitary tumors (e.g., oat cell carcinoma of the lung; **ectopic ACTH syndrome**).

The regulation of ACTH release from the pituitary is discussed in connection with the hormones of the adrenal cortex in Section 13.7.

β-LPH, a 93-amino-acid polypeptide, is a weak hormone that appears to cause lipolysis in adipose tissue. Its physiologic role is not clear.

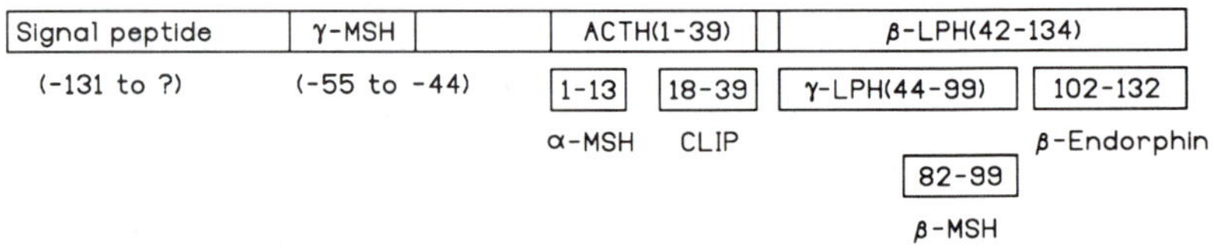

Figure 13.8.
Structure of pro-opiomelanocortin and the peptide hormones derived from it.

13.3 THYROID HORMONES

Thyroid hormones, although not essential for life, are required for normal growth and development. In general, they stimulate cellular metabolism. Under the influence of TSH, the thyroid gland secretes two thyroid hormones: tetraiodothyronine, or **thyroxin (T4),** and **triiodothyronine (T3).**

13.3.1 Biosynthesis and Secretion

The **thyroid gland** contains many follicles, called **acini,** each acinus consisting of a single-layered shell of cells that surround a central space filled with **colloid.** Colloid is largely composed of **thyroglobulin (TG),** a two-subunit glycoprotein (molecular weight [MW] = 660,000) that contains about 110 tyrosine residues. TG is synthesized by the follicular cells and, after iodination, is secreted into the colloid space (see Figure 13.9).

Thyroid cells take up iodide ions from the bloodstream by an Na^+-coupled transport system called the **iodide pump.** The ratio of thyroid-

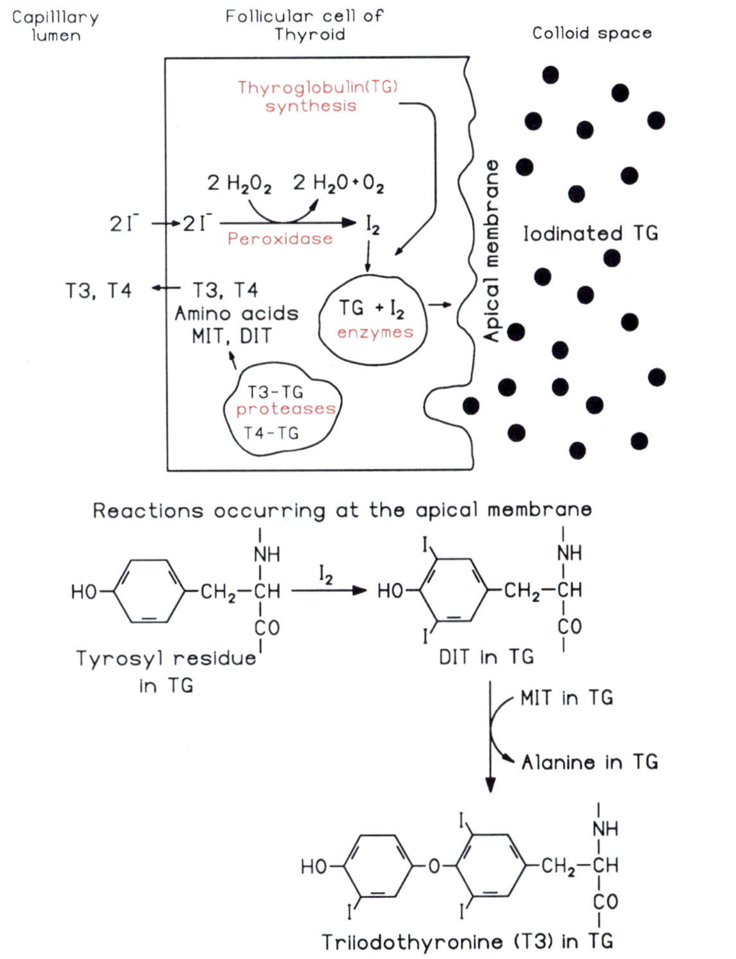

Figure 13.9.
Synthesis of thyroid hormones.

trapped iodide to free serum iodide, called the **T/S ratio,** normally is over 25. After uptake into the thyroid cell, the iodide is oxidized, most likely to a free iodine radical, which is immediately incorporated into tyrosine residues within TG. The reaction is catalyzed by **thyroid peroxidase,** located in the apical membrane of thyroid cells. The electron generated in the reaction is transferred to hydrogen peroxide. The same enzyme also catalyzes the transfer of an iodinated *p*-hydroxyphenyl ring from one tyrosine residue to the hydroxyl group of another tyrosine residue. The removal of the *p*-hydroxyphenyl group leaves an alanine residue in the peptide chain and creates an iodinated **thyronine** residue at the insertion site. Coupling of **3-monoiodotyrosine (MIT)** to **3,5-diiodotyrosine (DIT)** yields a triiodothyronine (T3) residue, whereas two DITs produce a thyroxine (T4) residue, both of which are still part of the TG molecule (Figure 13.9).

Secretion of thyroid hormones begins with the transport of an iodinated TG molecule via lipid vesicles into the cell where the protein is hydrolyzed inside lysosomes to amino acids and free T4 and T3 as well as MIT and DIT. MIT and DIT are deiodinated by the microsomal enzyme **iodotyrosine dehalogenase** and recycled, whereas T4 and T3 are released into the bloodstream.

13.3.2 Transport and Catabolism

The normal human adult secretes about 90 μg of T4 and 30 μg of T3 per day. To maintain this rate of production, the thyroid gland takes up 120 μg of iodide per day (minimum daily requirement).

In the bloodstream, both T4 and T3 are bound to three plasma proteins, **thyroxine-binding globulin (TBG), thyroxine-binding prealbumin (TBPA),** and **albumin** (in order of decreasing affinity). Only a small fraction (0.02% of T4 and 0.2% of T3) is free in solution. However, it is this fraction that is biologically active and interacts with target cells. Free thyroid hormones are in equilibrium with protein-bound hormones. Levels of TBG are increased in pregnancy, infectious hepatitis, and hypothyroidism; they are decreased by glucocorticoids, and hyperthyroidism.

The total plasma concentration of thyroid hormones is 6–12 μg/100 ml. Although only 1% of it is T3, because of its lower affinity for plasma proteins, it can diffuse out of the vascular compartment more easily than T4, and thus is more important at the cellular level than T4. It had been suggested that T3 is the only active thyroid hormone and that T4 represents a prohormone that is deiodinated to T3 at the cellular level. It now appears, however, that T4 is itself a hormone.

Determination of **protein-bound iodine (PBI)** may be used, instead of the more accurate radioimmune assay, to estimate the level of circulating thyroid hormones. However, PBI includes iodine-containing radioopaque dyes (bound to albumin) that might be present in the patient's plasma for months after injection.

The half-lives of T4 and T3 are 6 days and 1 day, respectively. Catabolism of thyroid hormones begins with their deiodination in many tissues. The released iodide is excreted in the urine. About 25% of the thyroid hormones are conjugated in the liver to glucuronate or sulfate and then excreted into the bile.

The catabolism of T4 by deiodination produces, in addition to T3, **reverse T3 (rT3)** in which the inner ring of thyronine is deiodinated. Reverse T3 is biologically inactive.

13.3.3 Actions of Thyroid Hormones

Thyroid hormones increase the basal metabolic rate (BMR). This **calorigenic action,** measured as an increase in the rate of oxygen consumption, is observed in most tissues that are metabolically active: for example, heart, liver, kidney, and gastric mucosa. The effect is not seen in adult brain, gonads, or spleen. Thyroid hormones diminish oxygen consumption by the anterior pituitary. Thyroid hormones also increase the overall oxygen consumption by raising the levels of 2,3-diphosphoglycerate (2,3-DPG) in erythrocytes, thus facilitating the dissociation of O_2 from hemoglobin in the tissues.

Catecholamines (epinephrine, norepinephrine) act synergistically with thyroid hormones. It appears that thyroid hormones function in a permissive way, increasing the effects of catecholamines on a variety of tissues. For example, adaptation to cold temperatures requires not only an increase in heat production but also a decrease in heat loss through the skin. This is accomplished by cutaneous vasoconstriction mediated by the action of catecholamines. Thyroid hormones appear to facilitate this action. (This was concluded from the fact that hypothyroid patients show a poor tolerance for cold.)

Metabolic actions of thyroid hormones include the acceleration of **glucose catabolism.** The effect of epinephrine upon **lipolysis** is also increased. In young people, **protein synthesis** is increased. On the other hand, excessive doses of thyroid hormones cause muscle wasting, resulting in creatinuria. **Cholesterol** biosynthesis is stimulated by thyroid hormones. However, the excretion of cholesterol into the bile is increased even more. Thus, an excess of thyroid hormones results in a reduction of plasma cholesterol levels.

In early life, thyroid hormones are essential for the development of the central nervous system. Hypothyroidism in the newborn causes irreversible damage and results in mental retardation. Appropriate treatment may avert this tragic outcome. Later in life, hypothyroidism only slows down mental processes. Thyroid hormones potentiate the effects of growth hormone on normal growth and development.

13.3.4 Mechanism of Action

Thyroid hormones bind to nuclear receptors and initiate the transcription of specific groups of genes. The gene for the thyroid receptor is the c-*erb*-A proto-oncogene.

In addition to this nuclear mode of action, there are probably others. For example, both epinephrine and thyroid hormones increase cardiovascular activity. Certain experiments suggest that heart cells contain two distinct adenylate cyclases, one responsive to norepinephrine and the other to thyroid hormones. This action of thyroid hormones is rather fast, whereas the other action involving de novo synthesis of proteins is considerably slower.

13.3.5 Control of Thyroid Hormone Secretion

TSH, released by the pituitary, stimulates the thyroid, via activation of adenylate cyclase, to increase iodide uptake, synthesis of thyroglobulin, iodination of TG, production of T4 and T3, and secretion of these hormones. In the presence of growth hormone, corticosteroids, and insulin, prolonged exposure to high levels of TSH causes hypertrophy of the thyroid **(goiter).** Release of TSH, in turn, is stimulated by TRH, a tripeptide (pyroGlu-His-Pro-amide) that is made in the hypothalamus from a 30 kd precursor containing five copies of the sequence ···Lys-Arg-Gln-His-Pro-Gly-Arg-Arg··· . TSH secretion is feedback inhibited by cirlating free T4 and T3. (Only T3 is directly inhibitory, whereas T4 must first be converted to T3). TRH release, in turn, is stimulated by exposure to cold (more pronounced in infants) and inhibited by stress and fasting. Free thyroid hormones also directly inhibit their own synthesis (uptake of iodide) and release by the thyroid gland (Figure 13.10).

13.3.6 Disorders of Thyroid Function

Hypothyroidism, or deficiency of thyroid hormone secretion, may have several causes:

1. A hypothalamic defect resulting in low TRH levels (very rare)
2. Pituitary failure resulting in insufficient TSH (rare)
3. A disease of the thyroid gland (most frequent), which may be due to several factors:
   a. A congenital defect in the biosynthesis of the hormone
   b. Iodine deficiency in the diet
   c. Presence of antithyroid drugs that interfere with thyroid hormone synthesis
   d. Thyroiditis (e.g., **Hashimoto's disease,** caused by the presence of antithyroid antibodies)

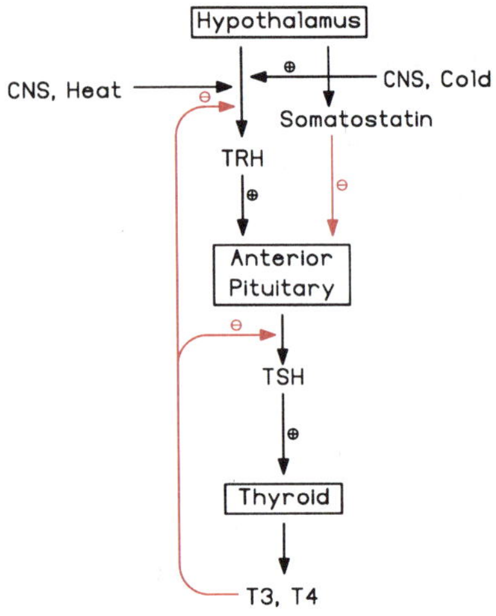

Figure 13.10.
Control of thyroid hormone secretion.

Hypothyroidism due to thyroid deficiency can be recognized by abnormally high levels of plasma TSH (lack of feedback inhibition). In children, hypothyroidism produces **cretinism.** The patients are dwarfed and mentally retarded. In adults, hypothyroidism is called **myxedema,** a disease characterized by thickened edematous skin, mental apathy, etc.

Dietary deficiency of iodine, found in many inland regions of the world, called **goiter** belts (e.g., Switzerland, the Great Lakes region), produces hypertrophy (goiter) because of the increased TSH secretion (lack of feedback inhibition). In this type of goiter, the T/S ratio may increase to 250. The thyroid may succeed in maintaining a marginal state of euthyroidism. If there is too little iodine, hypothyroidism develops.

Hyperthyroidism may be caused by TSH-secreting pituitary tumors. More common, however, is **Grave's disease (exophthalmic goiter),** in which the thyroid is enlarged and hyperplastic because of the presence of antibodies that bind to TSH receptors in the thyroid. These **thyroid-stimulating immunoglobulins (TSI)** include **long-acting**

thyroid stimulator (LATS), which stimulates the thyroid to secrete T4 and T3.

Uptake of radioactive ^{131}I into the thyroid is used in the evaluation of thyroid function. In hypothyroidism, iodine uptake is rather slow; in hyperthyroidism, it is very fast. Several monovalent ions (e.g., chlorate, periodate, nitrate) interfere with the transport of iodide into the follicular cell (competitive inhibition). Thiocyanate also blocks iodide transport but, in contrast to the other ions, is not accumulated in the thyroid. These drugs may be used clinically to treat **thyrotoxicosis,** an extreme form of hyperthyroidism. The T/S ratio falls to unity in the presence of these drugs.

Thiourea and its derivatives (thiocarbamides) inhibit the iodination of tyrosine residues in thyroglobulin and block their coupling to thyronine. The T/S ratio increases temporarily to near 250 and then falls again, since the trapped iodide is not retained in the thyroid.

A large dose of iodide also inhibits thyroid hormone secretion in the hyperthyroid but has little effect in the euthyroid. This is used clinically to differentiate between these cases.

13.4 PARATHYROID HORMONE AND CALCITONIN

Three hormones are responsible for calcium homeostasis: vitamin D_3 (considered a hormone because it can be synthesized by the human body), parathyroid hormone, and calcitonin. The biochemistry of vitamin D_3 is discussed in Section 8.3.

13.4.1 Parathyroid Hormone

Parathyroid hormone (PTH) is a polypeptide hormone synthesized by the **chief cells** of the **parathyroid glands** (closely associated with the thyroid). It is synthesized as a 113-amino-acid **pre-pro-PTH** molecule from which a 23-amino-acid N-terminal signal peptide is removed to yield **pro-PTH** (90 amino acids). Pro-PTH is further processed to PTH (84 amino acids) by removal of six more amino acids from the N-terminus. Circulating PTH has a half-life of only 20 minutes; it is removed from the circulation by the Kupffer cells of the liver and cleaved into two fragments. The smaller amino-terminal fragment (34 amino acids) retains full activity before it is further degraded.

PTH stimulates the mobilization of Ca^{2+} from bone (resorption), increases renal reabsorption of filtered Ca^{2+} and Mg^{2+}, and, as a consequence, raises plasma Ca^{2+} levels. The bone resorption requires the participation of $1,25(OH)_2D_3$ (vitamin D_3). The renal reabsorption of phosphate and bicarbonate is lowered by PTH. Thus, the phosphate that is released from bone together with Ca^{2+} is removed from the serum where it would complex the released Ca^{2+} ions.

The molecular mechanism of PTH action on tubular cells involves the activation of adenylate cyclase. Secretion of the hormone is inhibited by high plasma concentrations of Ca^{2+} and stimulated by low Ca^{2+} (Figure 13.11).

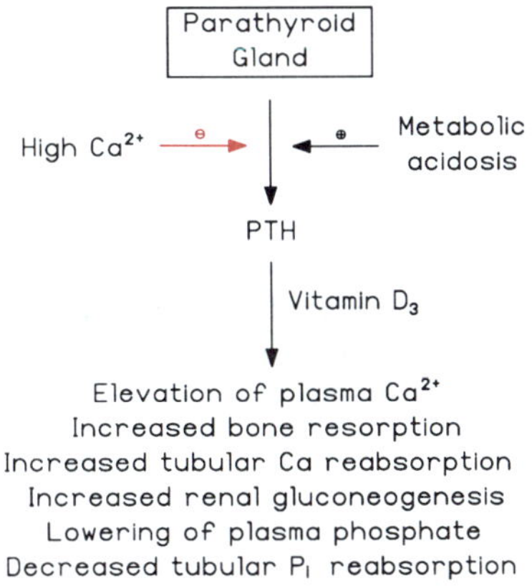

Figure 13.11.
Actions of parathyroid hormone.

Hyperparathyroidism, caused by tumors, can lead to excessive loss of Ca^{2+} into the urine because the amount of Ca^{2+} that is filtered exceeds the amount that can be reabsorbed. Secondary hyperparathyroidism may develop as the result of chronic renal disease or rickets (vitamin D deficiency) because the lowered plasma levels of Ca^{2+} cause hypertrophy of the parathyroid glands followed by oversecretion of PTH.

Hypoparathyroidism occurs most frequently as a result of thyroidectomy. It leads to hypocalcemia, which causes symptoms of neuromuscular hyperactivity.

13.4.2 Calcitonin

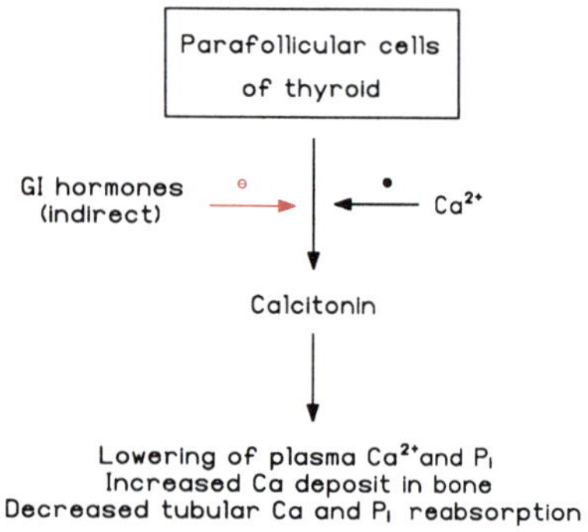

Figure 13.12.
Actions of calcitonin.

Calcitonin (CT) is synthesized by the **parafollicular cells** of the **thyroid** gland (also known as C cells) as a large precursor molecule that is cleaved into calcitonin and katacalcin, a peptide of unknown function found in the circulation together with CT. CT is a 32-amino-acid polypeptide with an intramolecular disulfide bridge and a blocked carboxyl terminus (amide group).

Calcitonin has a half-life of 10 minutes. It lowers plasma Ca^{2+} levels by promoting Ca^{2+} deposition in bone and reducing Ca^{2+} reabsorption in the kidney (Figure 13.12).

Secretion of calcitonin is regulated both directly and indirectly. **Direct regulation** involves plasma Ca^{2+} levels; high Ca^{2+} stimulates secretion of the hormone. **Indirect regulation** occurs via the gastrointestinal hormones gastrin, cholecystokinin, glucagon, and secretin, all of which stimulate calcitonin secretion. The indirect control mechanism possibly prevents a rise of plasma Ca^{2+} caused by the absorption of calcium-containing food.

The role of calcitonin in regulating Ca^{2+} homeostasis in human adults is probably minor. The hormone's main function may be to protect against hypercalcemia after a meal.

13.5 HORMONES SECRETED BY THE PANCREAS

The pancreas, a large exocrine organ that secretes digestive enzymes into the intestine, is also an endocrine gland secreting four different hormones into the bloodstream. The hormones are produced in the **islets of Langerhans,** small islands of cells dispersed throughout the pancreas (3% of total mass), that contain four distinct types of cells. The **A cells** (20%) manufacture **glucagon,** the **B cells** (70%) synthesize **insulin,** the **D cells** (about 5%) make **somatostatin,** and the **F cells** (2%) release a **pancreatic polypeptide** whose function is still unknown.

13.5.1 Insulin

Insulin consists of two polypeptide chains, the A and the B chain, connected by two disulfide bridges. (An additional disulfide bridge is located within the A chain.) The hormone is synthesized as a single polypeptide, **pre-proinsulin**, which is processed to **proinsulin** by removal of a 23-amino-acid signal sequence from the amino terminus. Proinsulin already possesses the three-dimensional folding of insulin. It is stored in secretory vesicles (granules) in the B cells in the form of a complex with Zn^{2+}. The connecting peptide **(C peptide)** is removed within the granules and is secreted together with the hormone. Plasma levels of C peptide may thus be used to determine B cell function in patients receiving injections of insulin, which is devoid of C peptide (Figure 13.13).

The half-life of circulating insulin is about 5 minutes. Most cells of the body contain **insulin receptors** on their membranes. However, since insulin is released into the portal vein, most insulin is taken up by the liver. The insulin receptor is a large tetrameric glycoprotein consisting of two different types of subunit. The larger α subunits, exhibiting insulin binding sites, are entirely extracellular. They are anchored via disulfide bonds to a pair of disulfide-linked β subunits that are transmembrane proteins. The intracellular domains of the β subunits are tyrosine-specific protein kinases capable of phosphorylating themselves as well as other cellular proteins. After combining with insulin, the receptor is internalized by endocytosis and degraded by lysosomal enzymes. It is still not known whether or not internalization of the insulin-receptor complex is required for hormone action. Prolonged exposure of cells to high concentrations of insulin decreases the number of membrane receptors **(down regulation),** whereas low insulin concentrations have the opposite effect. Glucocorticoids (see Section 13.7) decrease the affinity of the receptors for insulin.

Actions of Insulin

All actions of insulin appear to be triggered by its binding to the membrane receptors. The intracellular action of the insulin-receptor complex, however, is still not understood. If a "second messenger" is involved, it has not been discovered yet. There is some evidence that in hepatocytes cAMP levels fall, and in fact insulin causes the dephosphorylation of many proteins, thereby activating them (see Table 13.3).

Figure 13.13.
Synthesis of insulin.

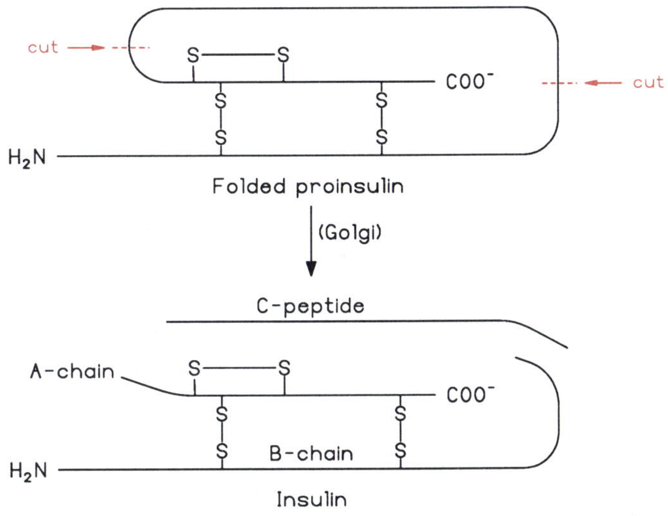

Table 13.3. Tissue-Specific Actions of Insulin and Glucagon

	Action	
Tissue	Increased by Insulin Decreased by Glucagon	Increased by Glucagon Decreased by Insulin
Liver	Protein synthesis	Ketogenesis
	Lipid synthesis	Gluconeogenesis
	Glycogen synthesis	Glycogenolysis
	Activity of	*Activity of*
	Glycogen synthase	Phosphorylase
	Phosphofructokinase*	Fructose 1,6-bisphosphatase*
	Pyruvate dehydrogenase	
	Enzyme levels of	
	Glucokinase	
	Phosphofructokinase	
	Pyruvate kinase	
	Pyruvate dehydrogenase	
	Citrate lyase	
Adipose tissue	Glucose transport	
	Fatty acid synthesis	
	Glycerol phosphate synthesis	
	Activity of	*Activity of*
	Lipoprotein lipase	Hormone-sensitive lipase
	Phosphofructokinase	
	Pyruvate kinase	
	Pyruvate dehydrogenase	
	Citrate lyase	
	Acetyl CoA carboxylase	
	Enzyme levels of	
	Citrate lyase	
	Acetyl CoA carboxylase	
	Fatty acid synthetase	
Muscle	Glucose transport	Protein degradation
	Amino acid transport	
	Keto acid transport	
	Glycogen synthesis	
	K^+ uptake	
	Enzyme levels	
	Similar to liver tissue	

Note: All enzymes are active in the dephosphorylated form and active in the phosphorylated form, with the exception of phosphorylase and hormone-sensitive lipase, both of which are active in the phosphorylated form.
* These two enzymatic activities are part of the same polypeptide.

In other cells, however, there is little correlation between cAMP levels and effects of insulin.

Insulin is an **anabolic hormone** that acts on most cells, directly or indirectly. Its primary target tissues, however, are the liver, skeletal muscle, and adipose tissue.

1. **Liver:** Insulin promotes anabolism and inhibits catabolism. More specifically, it stimulates glycogen and triglyceride synthesis as well as glycolysis. This is accompanied through activation of key enzymes by dephosphorylation (see Table

13.3) as well as through enzyme induction by gene activation. At the same time gluconeogenesis, glycogenolysis, and ketogenesis are inhibited.

2. **Muscle:** Insulin promotes protein and glycogen synthesis by increasing the transport of amino acids and glucose into muscle cells. During exercise, the transport of glucose into muscle cells becomes independent of insulin.
3. **Fat tissue:** Insulin promotes triglyceride synthesis. This is accomplished through induction of lipoprotein lipase, the enzyme bound to the endothelial cells of adipose tissue, which catalyzes the hydrolysis of triglycerides from circulating lipoproteins. Degradation of triglycerides is inhibited by phosphorylation of the intracellular lipase, the hormone-sensitive lipoprotein lipase. Insulin also increases glucose transport into adipocytes. The glucose is used as a source for α-glycerol phosphate.

Other actions of insulin include the stimulation of the Na^+/K^+ exchange pump in muscle and adipose tissue. This promotes the uptake of K^+ ions and thereby increases the membrane potential of these cells. Table 13.3 summarizes the actions of both insulin and glucagon on liver, muscle, and fat cells.

One distinguishes **slow and fast actions** of insulin. The induction of enzymes involving de novo protein synthesis is a relatively slow process, whereas the direct activation of already existing enzymes by dephosphorylation is fast. Only two enzymes are activated by phosphorylation: glycogen phosphorylase and hormone-sensitive lipase. Both are catabolic enzymes; their dephosphorylation in the presence of insulin inactivates them.

Control of Insulin Secretion

Secretion of insulin is stimulated by many different factors, the single most important one being glucose. Although the mechanism of this stimulation is still unclear, it appears that a metabolite of glucose is involved, since 2-deoxyglucose, which inhibits glucose metabolism, prevents the stimulation of insulin secretion by glucose. Insulin release requires calcium, which in B cells is released from mitochondria in response to glucose stimulation.

In addition to glucose and mannose, amino acids (especially leucine and arginine) and β-keto acids stimulate insulin release. The gastrointestinal hormones secretin, cholecystokinin (CCK), gastrin, glucagon, and gastric inhibitory peptide have also been shown to stimulate insulin secretion. This explains why ingested glucose is a more powerful stimulant for insulin release than injected glucose (Figure 13.14).

The release of insulin is biphasic. There is an initial short burst of insulin release **(early phase)** after which insulin temporarily slows down, only to rise again later to reach a steady level **(late phase).**

Epinephrine and norepinephrine inhibit insulin secretion by reducing intracellular cAMP in B cells (via α_2-adrenergic receptors, see Section 13.6). Drugs that raise cAMP levels in B cells (e.g., theophylline, inhibiting phosphodiesterase) stimulate insulin secretion. **Tolbutamide,**

Figure 13.14.
Regulation of insulin secretion.

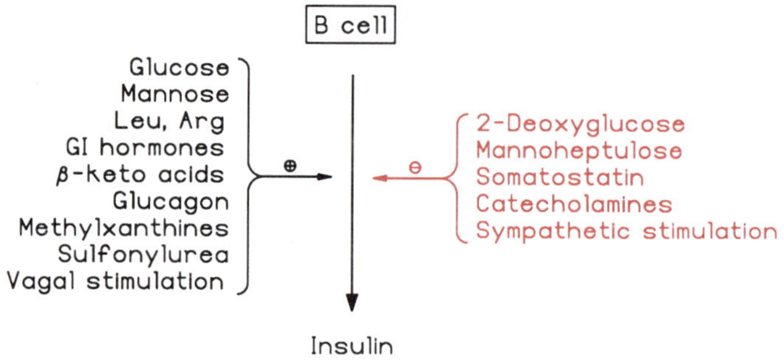

a sulfonylurea drug used in the treatment of type II diabetes, probably stimulates insulin secretion by this mechanism.

Diabetes Mellitus

There are two major forms of diabetes mellitus that differ in their severity. **Type I diabetes,** also called **insulin-dependent diabetes** or **juvenile diabetes,** most often appears at an early age. It is characterized by the virtually complete absence of circulating insulin. Patients require continued injections of insulin and careful monitoring of their blood glucose levels. This type of diabetes may be caused by the destruction of B cells, possibly as the result of a viral infection, through autoantibodies.

In contrast, **type II diabetes,** also called **non–insulin dependent diabetes** or **maturity-onset diabetes,** is usually mild and is more easily controlled, often without injection of insulin. There is a genetic predisposition to developing the type II form of diabetes. Over 80% of patients with type II diabetes are obese. They produce large amounts of insulin in response to their constant eating, but their insulin receptors are down-regulated and thus have reduced sensitivity.

Diabetes mellitus is characterized by **hyperglycemia,** caused either by a deficiency in insulin production or by a lack of response of cellular receptors to the action of insulin. Diabetes mellitus can be diagnosed by the **glucose tolerance test.** In this test, a defined amount of glucose ($40 g/m^2$ of body surface) is given orally after a 12-hour period of fasting. Blood glucose is determined before and at several times after ingestion. Normally, the fasting glucose level (80 mg/100 ml) rises within 0.5 hour to about 130 mg/100 ml and then, within about 2 hours, falls back to the fasting level. If blood glucose continues to rise and values stay above 140 mg/100 ml after 2 hours, this is diagnostic for diabetes mellitus.

It has been suggested that diabetes mellitus may develop from an **exhaustion of islet B cells** because of excessive ingestion of carbohydrates. As with other endocrine cells, B cells at first respond to continued stimulation with hypertrophy, but unlike other endocrine cells, B cells may soon become exhausted and eventually atrophy. This may be one of the mechanisms by which diabetes mellitus (type II) develops in obese persons, who require more insulin to regulate their glucose metabolism to the same extent as lean persons and, therefore, show continuously elevated insulin levels. In obese persons, insulin has a diminished effect on glucose levels because of the down regulation of insulin receptors. Hyperthyroidism may cause diabetes mellitus for the same reason (chronically elevated blood glucose levels).

Even though there is an excess of glucose in the blood of diabetics, many cells starve because they cannot take up glucose. As a result, the cells are forced to fill their energy needs from fat and protein stores. This raises the plasma level of free fatty acids, which are converted to acetyl CoA in the liver. Since the citric acid cycle is overtaxed, large amounts of acetyl CoA are converted to acetoacetate and other ketone bodies. This causes metabolic acidosis and leads to the loss of Na^+ and water in the urine.

Amino acids derived from protein degradation are converted to glucose in the liver. This glucose is added to the already high levels of glucose in the bloodstream. If the plasma glucose concentration exceeds the renal reabsorption maximum (200 mg/100 ml), glucose is lost in the urine, causing excessive loss of water (osmotic diuresis).

All of these symptoms can be treated with insulin. However, the amount of injected insulin must be precisely measured, since too large a dose might cause extreme hypoglycemia with accompanying coma. Also, it is necessary to keep in mind that insulin lowers the extracellular K^+ concentration. Patients with diabetic acidosis, who are already low on potassium stores, might develop acute **hypokalemia** upon injection of insulin.

During physical exercise, glucose can enter muscle cells even in the absence of insulin. Diabetics, therefore, must reduce their insulin dosage when engaging in physical exercise in order to avoid developing hypoglycemia. In general, a physically active life will keep the need for insulin down.

13.5.2 Glucagon

Glucagon is a 29-amino-acid polypeptide hormone. It is synthesized by the **A cells** of the islets in the form of a larger **proglucagon** from which a signal peptide is removed before secretion. Its half-life is about 10 minutes; it is inactivated mostly by the liver through removal of the N-terminal His residue.

In many respects glucagon has the opposite physiologic effects of insulin. As a catabolic, calorigenic hormone, it causes blood glucose levels to rise. This is accomplished by the activation of liver glycogen phosphorylase, increased liver gluconeogenesis, and fat breakdown in adipose tissue. Glucagon is thus **ketogenic** (Table 13.3).

Glucagon Secretion

The main stimulus for glucagon secretion is low blood sugar. The hypoglycemia increases the release of catecholamines that, in turn, stimulate the A cells via β-adrenergic receptors (see Section 13.6). Exercise, stress, or cortisol also stimulate glucagon secretion. Glucose inhibits glucagon secretion probably indirectly via the release of insulin or somatostatin, both of which inhibit glucagon release. Glucagon secretion is also increased after a protein meal. The principal amino acids that stimulate glucagon secretion are arginine, which releases both insulin and glucagon, and alanine, which primarily stimulates glucagon release.

Figure 13.15.
Regulation of a glucagon secretion.

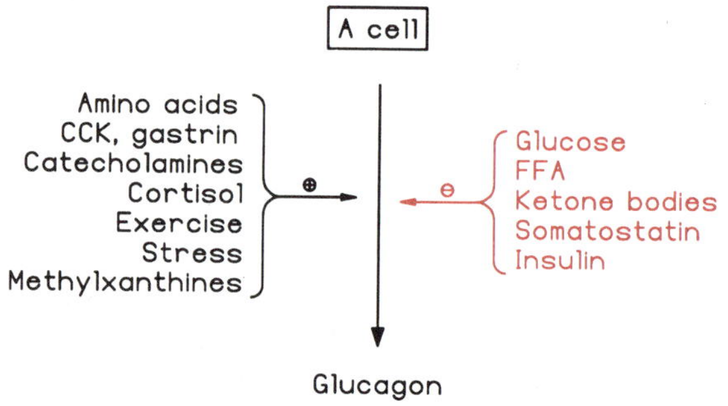

Leucine, a strong stimulator of insulin release, has no effect upon glucagon release. The released glucagon then promotes hepatic gluconeogenesis from the ingested amino acids (Figure 13.15).

Molecular Mechanism of Glucagon Action

Glucagon has the same target organs as insulin. After binding to specific membrane receptors, it exerts its effect on cellular metabolism via a second messenger. In liver and adipose cells, this second messenger is cAMP. The mechanism of activation of adenylate cyclase is as follows.

After binding to glucagon, the receptor protein interacts with another membrane protein, the **G protein.** G proteins consist of three subunits, an α chain, G_α, and a complex of a β and a γ subunit, $G_{\beta\gamma}$, which anchors the G protein in the cell membrane. There is an entire family of different G proteins all functioning in transmembrane **signal transduction.** They share the same β and γ subunits but contain different α subunits. It is the α subunits that are activated by the hormone receptor. In their activated form, they can stimulate as well as reduce the activity of different cellular proteins (see next section).

The G protein associated with the glucagon receptor is a stimulatory G protein, G_s. When the glucagon receptor is unoccupied, G_s exists as an inactive complex with guanosine diphosphate (GDP), bound to the $G_{s\alpha}$ subunit. Upon binding glucagon, the receptor causes a conformational change in $G_{s\alpha}$ allowing it to release GDP and bind guanosine triphosphate (GTP) instead. This, in turn, causes the $G_{s\alpha}$ subunit to dissociate from the $G_{\beta\gamma}$ complex and bind to adenylate cyclase, thereby activating it. The cAMP generated by adenylate cyclase then activates **protein kinase A,** an enzyme that phosphorylates, and thereby inactivates (see Table 13.3; exceptions: glycogen phosphorylase and hormone-sensitive lipase), key metabolic enzymes.

The $G_{s\alpha}$-GTP complex has a half-life of less than 1 minute. Its intrinsic **GTPase** activity hydrolyzes GTP to GDP, which causes $G_{s\alpha}$ to dissociate from adenylate cyclase and reform a complex with $G_{\beta\gamma}$. The cAMP generated by adenylate cyclase is also slowly hydrolyzed by **cAMP phosphodiesterase.** Methylxanthines (caffeine, theobromine, theophylline) inhibit this phosphodiesterase and, therefore, mimic the actions of hormones that use cAMP as second messenger (Figure 13.16).

Cholera toxin transfers ADP-ribose from NAD^+ to the $G_{s\alpha}$ subunit, thereby inhibiting the intrinsic GTPase. As a result, the $G_{s\alpha}$ remains activated indefinitely. This causes continuous high levels of cAMP in intestinal mucosal cells, resulting in a large efflux of Na^+ and water into the gut (diarrhea).

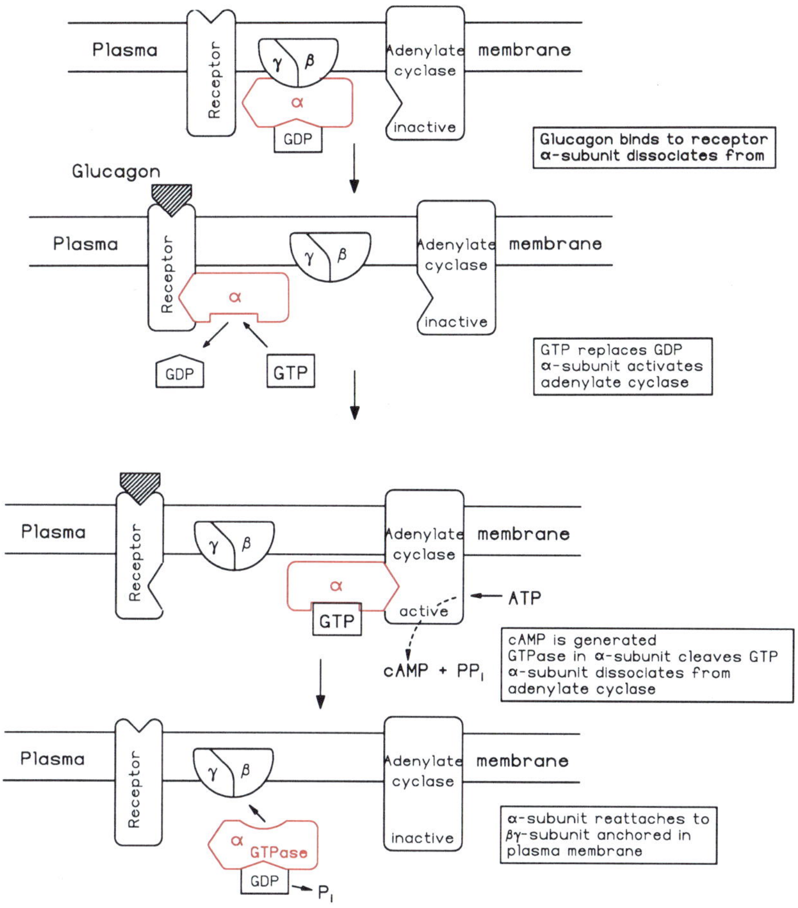

Figure 13.16.
Activation of adenylate cyclase.

13.5.3 Other Pancreatic Islet Hormones

The tetradecapeptide **somatostatin** is synthesized from a 28-amino-acid precursor molecule in the D cells of the islets of Langerhans. It is an inhibitory hormone that may function primarily in paracrine fashion; that is, it inhibits the release of hormones from the neighboring A and B cells. Its only endocrine effect may be to slow down movement of nutrients from the intestinal tract into the circulation.

Pancreatic polypeptide, a 36-amino-acid polypeptide, is secreted by the F cells in response to hypoglycemia, fasting, exercise, and ingestion of protein. Its physiologic function is not known.

13.6 HORMONES OF THE ADRENAL MEDULLA

The adrenal gland consists of two separate endocrine organs: the **adrenal medulla,** secreting **catecholamines,** and the **adrenal cortex,** secreting several types of **steroid hormones.** The discussion in this section will be limited to the medullary hormones.

The adrenal **medulla** may be considered a **specialized sympathetic ganglion** (same embryonic origin) which, in response to signals from

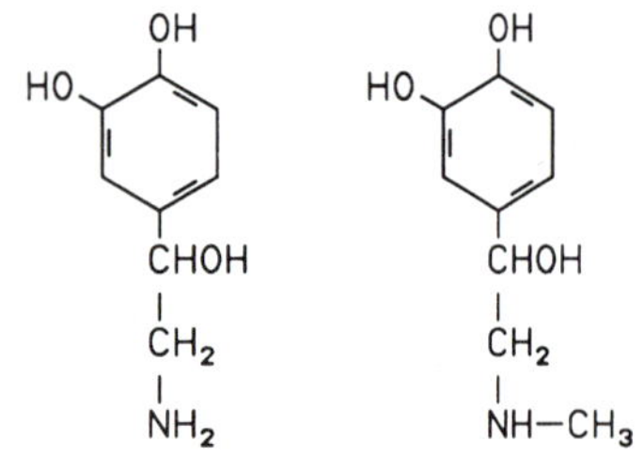

Norepinephrine Epinephrine

Figure 13.17.
Structures of epinephrine and norepinephrine.

cholinergic preganglionic nerve fibers, releases the two catecholamine hormones **epinephrine** and **norepinephrine** (Figure 13.17).

13.6.1 Biosynthesis and Metabolism

The two hormones epinephrine and norepinephrine are synthesized in two types of **chromaffin cells.** The biosynthetic pathway from tyrosine via dopa and dopamine to catecholamines is discussed in Section 4.11. Dopamine is synthesized in the cytoplasm of the cell and is then transported by an active process into intracellular vesicles where the enzyme **dopamine hydroxylase** converts it to norepinephrine. The enzyme is released into the bloodstream together with norepinephrine and can be used as an indicator for assessment of the functioning of catecholamine-producing tissues.

The conversion of norepinephrine to epinephrine occurs in the cytoplasm by the enzyme **phenylethanolamine-*N*-methyl transferase (PNMT).** This enzyme is induced by glucocorticoids. Epinephrine is then pumped back into vesicles (granules) for storage and later release.

About 80% of the catecholamines released by the human adrenal medulla is epinephrine, the remainder being norepinephrine. (Remember that sympathetic nerve terminals release only norepinephrine). Extraadrenal sites of chromaffin cells (usually associated with the small intestine or dorsal aorta) may predominantly release norepinephrine.

In the bloodstream, catecholamines appear to circulate predominantly in free form. Their half-life is only about 2 minutes. The **inactivation** of catecholamines by **monoamine oxidase (MAO)** and **catecholamine-*O*-methyl transferase (COMT)** to **(nor)metanephrine** and **vanillyl mandelate** is described in Section 4.11. Both degradation products are found in the urine in addition to small amounts of free catecholamines.

13.6.2 Actions of Catecholamines

Adrenal catecholamines are not essential for life but enable the body to cope with emergency situations. In the liver and in skeletal muscle, catecholamines cause breakdown of glycogen; in adipose tissue, they promote lipolysis. As a result, blood glucose and FFA rise. This stimulation of metabolism (**calorigenic action**) by catecholamines depends on the simultaneous presence of thyroid hormones and glucocorticoids.

In addition to these metabolic actions, both catecholamines stimulate cardiovascular activity. Excess levels of these hormones may cause serious arrhythmias. Epinephrine causes dilatation of blood vessels in muscle and liver and vasoconstriction in other tissues. Norepinephrine produces vasoconstriction in most tissues.

The actions of catecholamines are mediated by two types of cellular receptors: **α-adrenergic receptors** and **β-adrenergic receptors.** Each type of receptor has two subtypes with different modes of action.

The **α_1-adrenergic receptors,** when bound to catecholamine, activate a G protein, G_p, whose α subunit then activates a **phosphoinositide-specific phospholipase.** This enzyme, in turn, cleaves phosphorylated phosphatidyl inositol, a minor component of mammalian cell membranes, releasing two second messengers, **diacylglycerol** and **inositol 1,4,5-trisphosphate,** into the cytoplasm. Diacylglycerol activates **protein kinase C,** an enzyme with a different specificity than protein kinase A. Inositol trisphosphate, on the other hand, causes the release of Ca^{2+} from intracellular calcium-sequestering compartments (Figure 13.18).

The **α_2-adrenergic receptors** activate an **inhibitory G protein, $G_{i\alpha}$**. The α subunit ($G_{i\alpha}$) of this G protein inhibits adenylate cyclase and thus causes a lowering of intracellular cAMP levels. The **pertussis toxin,** made by the bacterium responsible for whooping cough, specifically ADP-ribosylates $G_{i\alpha}$, thereby inhibiting its GTPase activity, analogous to the cholera toxin that modifies $G_{s\alpha}$.

Both **β-adrenergic receptors,** β_1 and β_2, use the $G_{s\alpha}$ for signal transduction across the cell membrane. They raise intracellular cAMP.

Differential distribution of the various types of adrenergic receptors produces different results in different tissues. For example, adipose and skeletal muscle cells carry mostly β-adrenergic receptors. The increase in intracellular cAMP owing to the binding of catecholamines to these receptors causes activation of hormone-sensitive lipase in adipose cells and activation of glycogen phosphorylase (accompanied by inactivation of glycogen synthase) in muscle cells (see Section 2.11 and 3.7). In contrast, the effect of catecholamines upon liver cells is mediated by α_1 receptors. (This action by catecholamines is probably physiologically insignificant compared with their indirect effect via glucagon.) Pancreatic islet B cells predominantly express α_2 receptors, whereas A cells carry β receptors. As a consequence, catecholamines inhibit insulin secretion by reducing intracellular cAMP but stimulate glucagon secretion by increasing cAMP levels (Table 13.4).

Adrenal medullary hormones are released upon stimulation from the CNS (hypothalamus). The acetylcholine released from the nerve endings makes the chromaffin cells permeable to Ca^{2+} ions. This facilitates membrane-membrane fusion and thereby promotes exocytosis of catecholamines stored in intracellular vesicles.

When a person stands up from a recumbent position, plasma levels of norepinephrine rise from about 30 ng/100 ml to 600 ng/100 ml. Even the anticipation of physical exercise stimulates the secretion of epinephrine from the adrenal medulla.

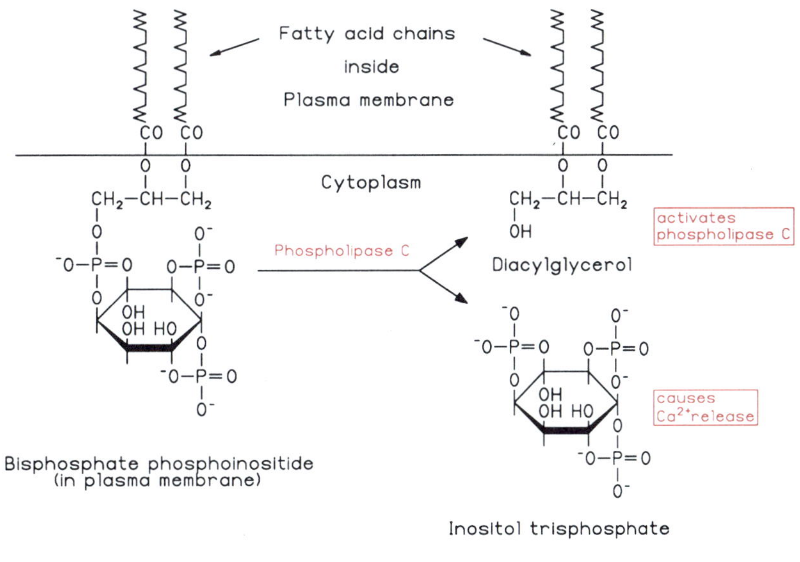

Figure 13.18.
The release of inositol trisphosphate and diacyglycerol from the plasma membrane in response to activation of α_1-adrenergic receptors.

Table 13.4. Tissue-Specific Actions of Circulating Catecholamines

Tissue	Action: Increased	Action: Decreased
Liver	*Activity of* Phosphorylase kinase (via α_2 receptors and Ca^{2+}) Glycogen phosphorylase *Rate of* Gluconeogenesis Proteolysis FFA oxidation	*Activity of* Glycogen synthase Pyruvate kinase
Adipose tissue	*Activity of* Hormone-sensitive lipase	*Rate of* Glucose transport
Muscle	*Rate of* Glucose transport Glycogenolysis (via β receptors and cAMP) Glycolysis to lactate	
Endocrine pancreas	Glucagon secretion (via β receptors)	Insulin secretion (via α_1 receptors)

Long-term stimulation increases the synthesis of tyrosine hydroxylase, dopamine hydroxylase, and PNMT. ACTH also appears to stimulate de novo synthesis of these key enzymes of catecholamine biosynthesis.

13.7 HORMONES OF THE ADRENAL CORTEX

Surrounding the adrenal medulla are three distinct zones of cells collectively called the **adrenal cortex.** Each zone produces predominantly one type of steroid hormone:

Zona glomerulosa (outermost layer): mineralocorticoids

Zona fasciculata (intermediate layer): glucocorticoids

Zona reticularis (inner layer): sex hormones

13.7.1 Nomenclature of Steroid Hormones

Steroid hormones are derived from cholesterol. In humans there are four different types of steroid hormones:

1. Corticosteroids (C_{21})
2. Progestins (C_{21})
3. Androgens (C_{19})
4. Estrogens (C_{18})

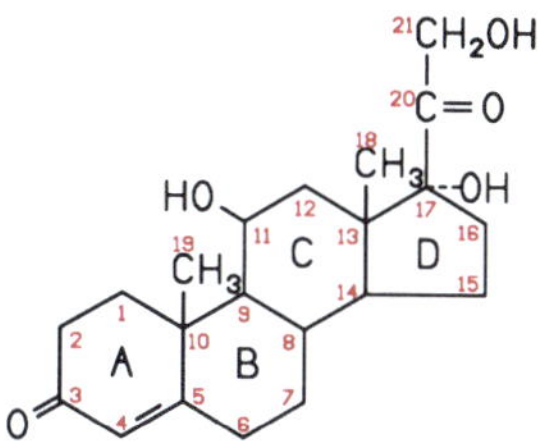

Figure 13.19.
Structure of cortisol, a C_{21} steroid.

The structure of cortisol, a C_{21} steroid, including the C-numbering convention, is shown in Figure 13.19.

Cortisol has a hydroxyl group attached to carbon 17 that lies below the plane of the steroid ring to which it is bound. Such a group is called an α-OH group and is drawn as a dotted line in Figure 13.19. Groups that lie above the plane of the ring are called β groups and are drawn as solid lines. Most adrenal steroids with 17-hydroxy groups are in the α-configuration (except testosterone and estradiol); 3-, 11-, and 21-hydroxy groups are in the β configuration.

13.7.2 Biosynthesis

The adrenal cortex secretes three main types of steroids:

1. Two **glucocorticoids,** so named because of their profound effects on carbohydrate metabolism:
   - **Cortisol:** total plasma concentration 5–25 μg/100 ml, free 1 μg/100 ml, daily output 8–24 mg.
   - **Corticosterone:** total plasma concentration 0.5 μg/100 ml, free 0.1 μg/100 ml, daily output 1.5–4 mg.

2. Two **mineralocorticoids** functioning in the homeostasis of Na^+, K^+, and water:
   - **Aldosterone:** total plasma concentration 5–10 ng/100 ml, free 5 ng/100 ml, daily output 0.04–0.2 mg.
   - **Deoxycorticosterone:** has only 3% of the activity of aldosterone, similar plasma concentrations.

3. One weak **androgen** whose function is not known:
   - **Dehydroepiandrosterone:** total plasma concentration 50 μg/100 ml.

The structures of these steroids are shown in Figure 13.20.

In addition, small amounts of several other steroids, mostly biosynthetic intermediates, are also secreted by the adrenal cortex. They include pregnenolone, progesterone, 18-hydroxycorticosterone, Δ4-androstene-3,17-dione, and testosterone. The majority of the sex hormones are produced by gonadal tissues (see Section 13.8).

Biosynthesis of steroid hormones begins with the uptake of cholesterol into the mitochondrion, followed by its conversion into **pregnenolone** by the **cholesterol side chain cleavage enzyme system.** This enzyme system includes three enzymes: a 20 α-hydroxylase, a 22 R-hydroxylase (both are mixed-function oxidases requiring the reduced form of nicotinamide-adenine dinucleotide phosphate [NADPH], O_2, a flavoprotein, a nonheme-iron protein, and cytochrome P_{450}), and a lyase (desmolase). The synthesis of pregnenolone is the rate-limiting step in steroid biosynthesis. The reaction sequence is shown in Figure 13.21.

The enzymes responsible for the conversion of pregnenolone into the other steroids are located either in the mitochondrion or in the endoplasmic reticulum (ER). The general pathway of steroid biosynthesis is shown in Figure 13.20.

The adrenal cortex contains a high concentration of ascorbic acid

Figure 13.20.
Biosynthesis of steroid hormones.

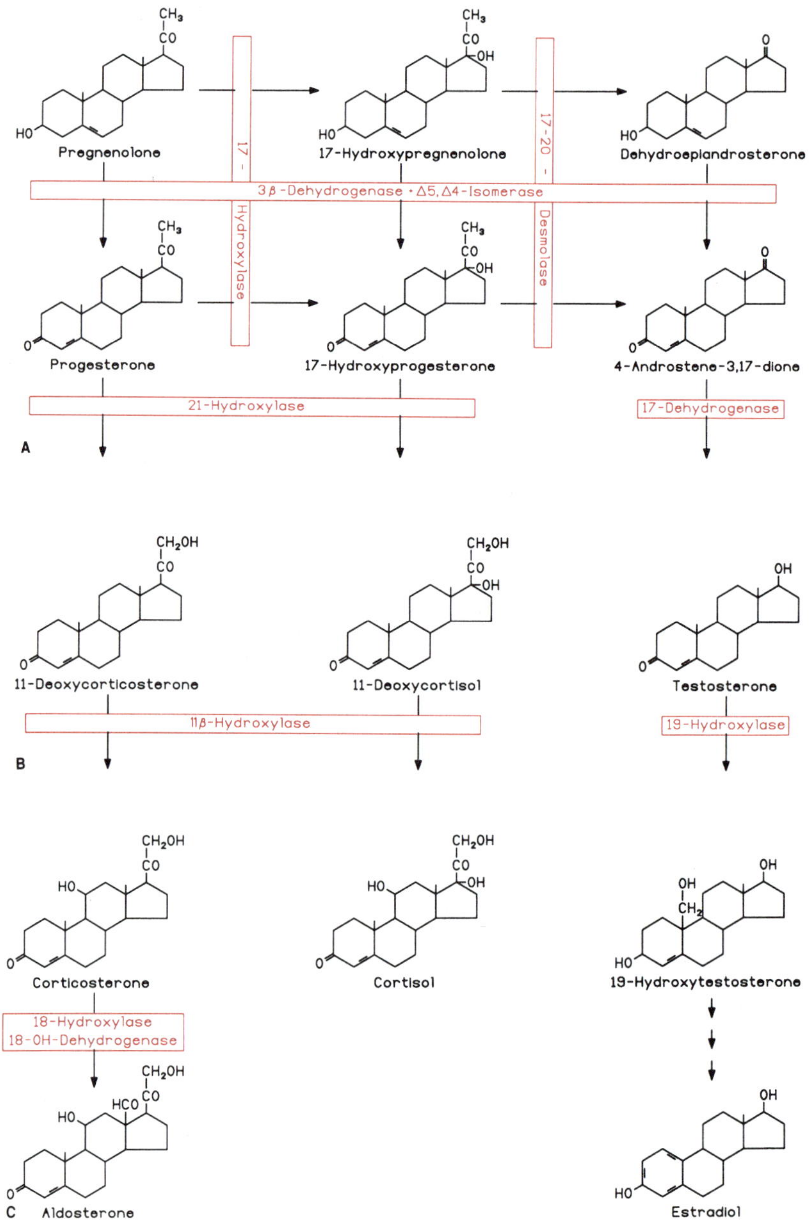

(vitamin C) that decreases upon induction of steroid synthesis by ACTH. Its function in steroid biosynthesis is not known.

13.7.3 Transport and Catabolism

Apparently, steroid hormones are not stored in the cortical cells; they are made on demand and immediately excreted into the bloodstream. There, corticosteroids are mostly bound to **corticosteroid-binding globulin (CBG)** and albumin. Other steroids do not bind effectively to CBG. Only late in pregnancy, when its levels are very high, will progesterone occupy about 25% of the binding sites on CBG. Albumin has a much higher binding capacity but has a lower affinity for steroid hormones. Bound steroids are in equilibrium with small amounts of free

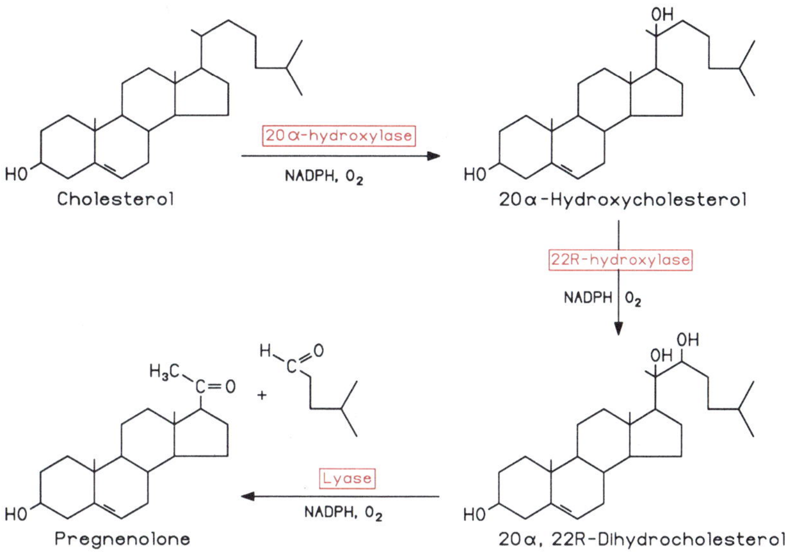

Figure 13.21.
Biosynthesis of pregnenolone, the common precursor of all steroid hormones.

hormones. (Dehydroepiandrosterone is the only steroid that is not secreted in free form but conjugated to sulfate.)

Catabolism of steroids occurs in the liver (glucocorticoids), kidney (aldosterone), and gastrointestinal tract. It involves reduction of the keto groups to hydroxy groups and saturation of double bonds. The hydroxy groups are subsequently conjugated to glucuronate or sulfate. The conjugates are excreted in the urine as well as in the bile (15%).

13.7.4 Actions of Glucocorticoids

Glucocorticoids, especially cortisol, are essential for life. **Cortisol** induces and controls the synthesis of important metabolic enzymes in a variety of tissues. The hormone is taken up into the cytoplasm of the target cell, where it binds to a **receptor protein.** The hormone-receptor complex then migrates to the nucleus and activates specific sets of genes. All actions of steroid hormones appear to involve transcription and de novo protein synthesis. This is why it takes at least half an hour for steroid hormones to produce visible results in the target cell.

The synthesis of new cellular enzymes has profound effects on the intermediary metabolism of the cell. In most tissues (muscle, connective tissue, lymphoid cells), cortisol has a **catabolic effect.** Degradation of protein is increased. In adipose tissue, fat degradation is stimulated as well. At the same time, glucose uptake by these cells is decreased. These actions raise plasma levels of glucose and amino acids and thus can be considered anti-insulin effects. In **liver** cells, however, cortisol exhibits an **anabolic,** insulinlike **effect.** Gluconeogenesis is increased, and the synthesized glucose is incorporated into glycogen.

In addition to these direct effects, glucocorticoids exhibit what is called a **permissive action.** The hormones glucagon and catecholamines require the presence of small amounts of glucocorticoids to be effective. Furthermore, the ability of the kidney to excrete water depends on the presence of glucocorticoids. (This action of glucocorticoids is different from that of aldosterone.) The number of circulating white blood cells is reduced by cortisol, mainly because of the removal of these cells from the circulation. Many of these actions are best ex-

Table 13.5. Actions of Glucocorticoids

Tissue	Actions
General	Change in enzyme pattern in cell due to induction of specific enzyme synthesis Permissive action on glucagon and catecholamines Reduced affinity of insulin receptors for insulin Considered a catabolic hormone (in contrast to testosterone) in all cells except the liver
Liver	Increased gluconeogenesis due to induction of pyruvate carboxylase, PEP carboxykinase, fructose 1,6-diphosphate, glucose 6-phosphatase Induction of glycogen synthesis
Adipose tissue	Inhibition of glucose uptake Induction of lipolysis Induction of proteolysis
Muscle and other tissue	Inhibition of glucose uptake Induction of proteolysis

plained by the ability of glucocorticoids to act as inducers of specific proteins and enzymes in a variety of cells, thus changing their overall enzyme pattern (Table 13.5).

In unphysiologic large doses, glucocorticoids inhibit inflammatory and allergic reactions.

13.7.5 Control of Glucocorticoid Secretion

Synthesis and secretion of glucocorticoids (and some androgens) in the **zona fasciculata** and **zona reticularis** are stimulated by **ACTH.** The molecular mechanism of this stimulation involves cAMP. The rate-limiting step in steroid biosynthesis is the cleavage of the side chain from cholesterol. There is some evidence that cAMP activates this step, probably through protein kinase A. ACTH is also required for the maintenance of the adrenal cortex. After hypophysectomy, both the zona fasciculata and the zona reticularis atrophy. (The zona glomerulosa remains intact because it is stimulated by angiotensin II.)

Release of ACTH from the pituitary is feedback-inhibited by circulating glucocorticoids both at the pituitary and the hypothalamus (via release of CRH). Stress, including low blood sugar, stimulates ACTH secretion via the hypothalamus (Figure 13.22). The release of cortisol is diurnal, with a very low level in late evening that is followed by a large peak in the early morning hours. During the afternoon there are several secondary but smaller peaks in response to eating and exercise.

ACTH secretion, and consequently glucocorticoid secretion, is not uniform throughout the day; it occurs in bursts that are most frequent around 6:00 A.M. and least frequent after midnight.

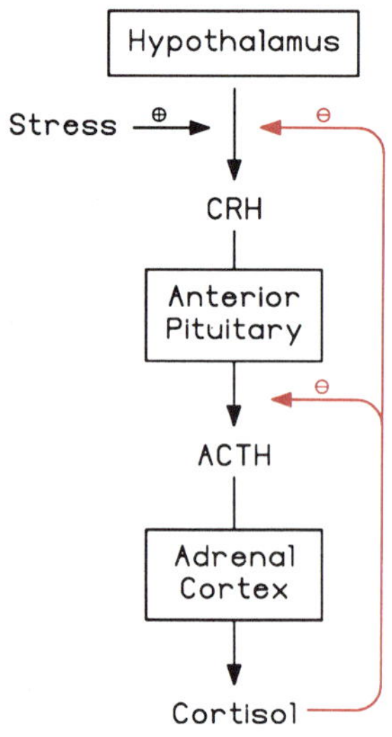

Figure 13.22.
Regulation of glucocorticoid secretion.

13.7.6 Adrenal Androgens

In addition to glucocorticoids, the zona fasciculata and zona reticularis synthesize a small amount of androgens. These include androstenedione, dehydroepiandrosterone (DHEA), and DHEA sulfate. Compared with testosterone they have little androgenic activity but can be converted peripherally into testosterone and dihydrotestosterone. In males, only about 5% of all androgens are derived from the adrenal cortex, the remainder being synthesized in the testis. In females, however, adrenal androgens account for a large portion of total androgen production, the rest coming from the ovary.

13.7.7 Mineralocorticoids

Aldosterone, the major human mineralocorticoid, is synthesized exclusively in the **zona glomerulosa** of the adrenal cortex. The hormone increases the rate of Na^+ reabsorption in the distal renal tubules and causes loss of K^+ ions. It achieves this by stimulating the synthesis of Na^+ channel proteins. The resulting increase in the number of Na^+ channels on the lumenal side of the tubular cell allows more Na^+ ions to enter the cell. The rate of Na^+ reabsorption appears to be determined solely by the number of Na^+ channels; the Na^+/K^+ exchange pump, located in the cell membrane facing the plasma, has a very large reserve capacity, thus allowing the Na^+ ions to pass into the bloodstream.

Aldosterone synthesis and secretion is mainly regulated by the **renin-angiotensin system.** In response to a drop in renal arterial pressure as well as to neural stimuli, the proteolytic hormone **renin** is released from the juxtaglomerular cells of the kidney. Renin cleaves **angiotensinogen,** a circulating α_2-globulin, producing the decapeptide **angiotensin I.** Angiotensin I, in turn, is converted to the octapeptide **angiotensin II** by **converting enzyme** located in the lung, liver, and kidney. A reduction in the volume of extracellular fluid, such as that caused by hemorrhage, will thus activate the renin-angiotensin system and stimulate aldosterone secretion. A rise in plasma K^+ by as little as 1 mEq/L will also stimulate aldosterone secretion. ACTH has a small but distinctive effect upon aldosterone secretion.

13.7.8 Disorders of Adrenal Function

Overproduction of glucocorticoids, also known as **Cushing's syndrome,** is often due to excessive secretion of ACTH (75% of cases), either by the pituitary or from ectopic sources (tumors). Excess secretion of glucocorticoids may also be caused by adrenal cortical tumors (adenomas or carcinomas). More frequently, the symptoms of Cushing's syndrome are the result of chronic glucocorticoid therapy; they include moon-faced appearance, obesity, protein depletion, osteoporosis, mental abnormalities, and hypertension and are often accompanied by diabetes mellitus.

Adrenocortical insufficiency is usually caused by atrophy of the adrenocortical cells because of tuberculosis and perhaps autoimmune disease. The defect, called **Addison's disease,** involves all three zones of the adrenal cortex. Symptoms include hypoglycemia, weight loss, weakness, pigmentation (because ACTH and β-MSH levels rise in the absence of cortisol), hypotension, and hyponatremia (because of a lack of aldosterone). Complete absence of adrenocorticoids is lethal.

Primary aldosteronism, characterized by excessive secretion of aldosterone, is usually due to an adrenocortical adenoma. The disease produces hypertension because of Na^+ and water retention. It is accompanied by severe K^+ depletion, followed by alkalosis and loss of Ca^{2+} ions.

The **adrenogenital syndrome** is a congenital adrenal hyperplasia caused by the absence or reduced activity of one of the hydroxylases that function in the pathway from progesterone and 17-hydroxyprogesterone to aldosterone and cortisol (see Figure 13.21). Because of the enzymatic block, the steroid intermediates before the block accumulate and are released into the bloodstream. The absence of cortisol causes excessive ACTH secretion (loss of feedback inhibition), which in turn stimulates cortical hyperplasia.

There are two main types of adrenogenital syndrome, one **salt-losing form** and one **hypertensive form,** both characterized by excessive androgen production:

1. **21-Hydroxylase deficiency** (autosomal recessive) allows very little production of aldosterone, cortisol, or corticosterone. The precursors 17-hydroxypregnenolone and 17-hydroxyprogesterone are shunted toward production of Δ-4-androstene-3,17-dione, which can be converted by many tissues to testosterone, the most active androgen. In young males, this causes precocious **pseudopuberty;** in young females it causes **pseudohermaphroditism;** in older females it causes **virilization.** The lack of aldosterone produces hyponatremia and **hypotension.**
2. **11β-Hydroxylase deficiency** produces similar symptoms except that 11-deoxycorticosterone with mineralocorticoid properties accumulates, producing hypernatremia and **hypertension.**

17α-Hydroxylase deficiency (also expressed in gonadal tissue) is characterized by absence of both male and female sex hormones. Children born with this defect have a female phenotype regardless of genetic sex.

Adrenal function can be monitored by **determination of 17-keto steroids** in a 24-hour sample of urine. The principle of this determination is to convert most urinary steroids to 17-keto steroids, which can be measured colorimetrically by reaction with *m*-dinitrobenzene (**Zimmermann reaction**). C_{21} steroids containing a 17-hydroxy group (mostly cortisol and derivatives) are converted into C_{19} 17-keto steroids by treatment with Na-borohydride (which reduces 20-keto groups to hydroxy groups) followed by oxidation with periodate. C_{19} androgens and C_{18} estrogens are themselves 17-keto steroids or are converted into them by periodate.

13.8 GONADAL AND PLACENTAL HORMONES

Sex hormones are largely produced by the cells of gonadal tissues. In the male, the androgen **testosterone** is synthesized by the **Leydig cells** of the **testis;** in the female, the most potent estrogen, **estradiol,** is produced by the **follicular cells** of the **ovaries.** Sex hormones are responsible for the development and maintenance of gonadal tissues as well as secondary sex characteristics. In addition, they function as important **anabolic hormones** during development.

The ovaries also secrete **progesterone,** the hormone that prepares the endometrium of the uterus for the implantation of the embryo and maintains the endometrial function during pregnancy.

During pregnancy, the **placenta** secretes two protein hormones, **human chorionic gonadotropin (hCG)** and **human chorionic somatomammotropin (RCS),** also called **human placental lactogen (hPL).** In addition, the adrenal cortex of the developing fetus contributes important steroid intermediates to the maternal circulation, where they are converted to estrogens.

13.8.1 Testosterone

In the **Leydig cells** of the testis, all enzymes involved in adrenal steroid biosynthesis are present, with the exception of 21-hydroxylase and 11-hydroxylase (see Figure 13.21). Steroid biosynthesis, therefore, is shunted toward androgen synthesis. Synthesis and secretion of testosterone is stimulated by LH (via cAMP). LH, in turn, is feedback-inhibited by circulating testosterone (see Figure 13.5).

In the plasma, testosterone is mostly (97%) bound to **gonadal steroid-binding globulin (GBG)** and albumin. The total plasma concentration of testosterone is 0.65 μg/100 ml in males and 0.03 μg/100 ml in females (largely produced by the adrenal cortex).

Most of the 17-keto steroids found in the urine originate from the adrenal cortex. Only about 30% are derived from testosterone (inactivated in the liver).

In many of its male-specific target tissues, testosterone is reduced to **dihydrotestosterone (DHT),** which binds, like all steroid hormones, to intracellular receptor proteins and thereby stimulates transcription and protein synthesis. Together with FSH, testosterone is required for spermatogenesis. The development of secondary sex characteristics is due to the action of testosterone as well as other androgens (17-keto steroids).

Testosterone possesses strong **anabolic effects** (increased protein synthesis and growth) on the human body. These actions are produced by testosterone itself, not DHT. The increased rate of growth in boys during puberty is largely due to this androgenic action.

13.8.2 Estradiol

The principal estrogen synthesized and secreted by the **theca interna** cells of the maturing ovarian follicle is estradiol. It is synthesized from testosterone via a series of hydroxylation reactions (see Figure 13.23). Note that all estrogens have an aromatic A ring.

Estradiol is also secreted by the **corpus luteum** and the **placenta** and to a minor degree by the adrenal cortex and the testis.

In plasma, estradiol is bound to albumin and GBG; only about 3% is free. It is in equilibrium with its oxidized form, **estrone,** which is further metabolized to **estriol** in the liver (Figure 13.23).

Plasma levels of estradiol in females vary from 3 ng/100 ml to 25 ng/

Figure 13.23.
Biosynthesis of estrogens.

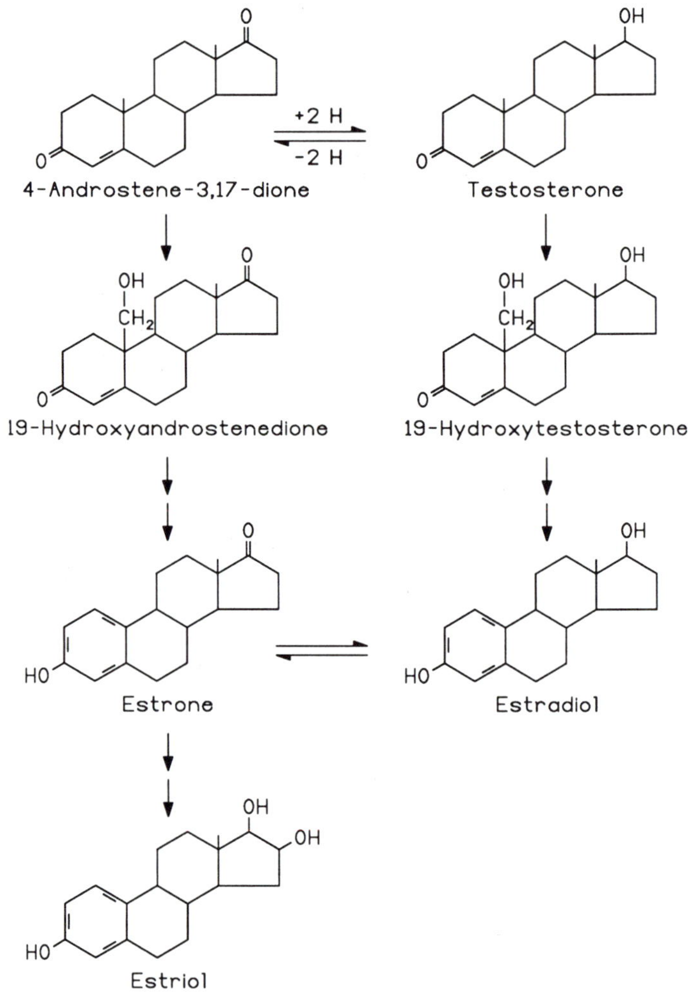

100 ml the day before ovulation (see Figure 13.4). In men, estrogen levels are below 2 ng/100 ml.

In addition to supporting the growth and maintenance of the female sex organs and secondary sex characteristics, estrogens stimulate the proliferation of the endometrial cells of the uterus during the follicular phase of the menstrual cycle (proliferative phase of the uterus). They also enhance the contractile activity of the myometrium and increase the activity of the secretory and ciliated cells of the fallopian tubes.

Like androgens, estrogens exhibit anabolic actions, although less pronounced. They effectively lower plasma cholesterol levels and thus may contribute to the lower incidence of atherosclerosis in premenopausal women.

The control of estrogen secretion by LH and FSH has been discussed before (see Figure 13.5).

13.8.3 Progesterone

Progesterone is an important intermediate in steroid biosynthesis and may be released in small quantities by all tissues that produce steroid hormones. The primary sources of progesterone, however, are the **corpus luteum** and the **placenta.**

Plasma levels of progesterone in women vary from 90 ng/100 ml in the follicular phase to about 1.8 μg/100 ml in midluteal phase. A small rise in plasma progesterone levels is observed at ovulation. It may be responsible for the synthesis of a proteolytic enzyme that degrades the collagen network of the follicular wall and thus causes rupture of the follicle.

The progesterone that is released in increasing quantities from the corpus luteum after ovulation initiates the **secretory phase** of the endometrium in which endometrial glands secrete carbohydrate-rich material thought to serve as nourishment for the fertilized egg. The increased levels of progesterone also cause the myometrium to decrease its contractions and cause ciliary activity in the fallopian tubes to subside.

If no fertilization takes place, the corpus luteum slowly decreases its secretion of progesterone (and estrogen) until, about 2 weeks after ovulation, the low levels of progesterone and estrogens cause the endometrium to become ischemic and to shed its cells (menstrual bleeding).

The regulation of progesterone secretion and its inhibitory effect on FSH and LH secretion have already been discussed (see Figure 13.5).

13.8.4 Hormones Secreted by the Fetoplacental Unit

About 4–6 days after fertilization of the egg, the developing blastocyst becomes implanted in the endometrium, and a **syncytiotrophoblast** develops around which the placenta is formed. Three or 4 days after implantation, the syncytiotrophoblast cells begin to secrete the placental hormone **human chorionic gonadotropin.** hCG is a glycoprotein very similar in structure to the pituitary glycoprotein hormones FSH, LH, and TSH, consisting of an α and a β subunit. The α subunit is almost identical to that of the pituitary glycoprotein hormones. hCG has LH-like activity. Because of its activity, the corpus luteum does not degenerate but increases its secretion of estrogen and progesterone, which support the continued growth of the endometrium and prevent the onset of menstruation.

hCG secretion reaches a peak around the sixth week of pregnancy and then declines rather sharply to a lower level that remains constant for the remainder of the pregnancy.

hCG is excreted in the urine and can be detected there as early as 2 weeks after fertilization. It forms the basis of a popular early pregnancy test.

The placenta also secretes **human chorionic somatomammotropin**, also called **human placental lactogen,** a protein hormone very similar in structure to growth hormone (161 out of 191 amino acids are identical). hCS appears to supplement pituitary GH in its growth-promoting actions in the pregnant woman. (Secretion of GH is not increased during pregnancy.) It also has some TSH activity and is lactogenic.

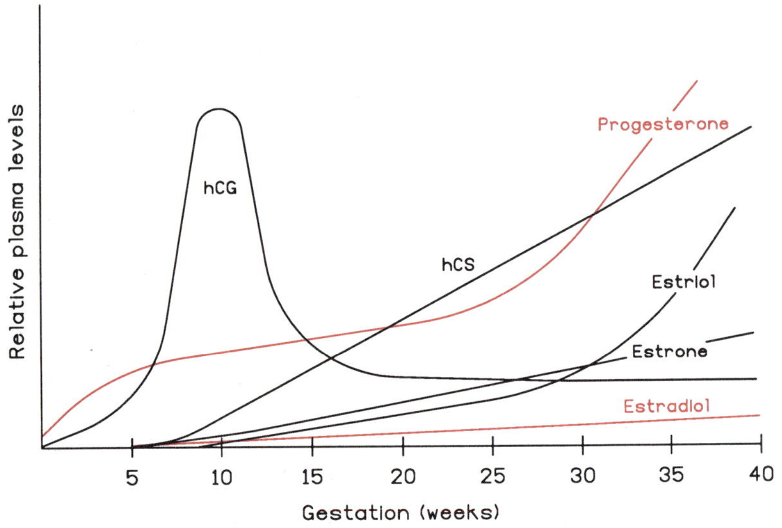

Figure 13.24.
Plasma levels of hormones during pregnancy.

Plasma concentrations of hCS increase almost linearly throughout pregnancy. For this reason, hCS can be used as a diagnostic indicator of placental function (Figure 13.24).

Relaxin is a polypeptide hormone secreted by the corpus luteum. It consists of two polypeptide chains linked together by two disulfide bonds. The molecule resembles insulin and has an additional disulfide bridge within its A chain. Relaxin relaxes the pelvic joints and dilates the cervix of the uterus.

The fetoplacental unit also participates in steroid hormone biosynthesis. In the placenta, cholesterol is converted to pregnenolone and progesterone, which can enter both the fetal and the maternal circulation. In the fetal adrenals, pregnenolone is converted via 17α-hydroxypregnenolone to dehydroepiandrosterone and 16α-hydroxy-DHEA. The two steroids pass back to the placenta in the form of sulfate conjugates. There they are deconjugated, converted to estradiol and estriol, and released into the maternal circulation (Figure 13.25).

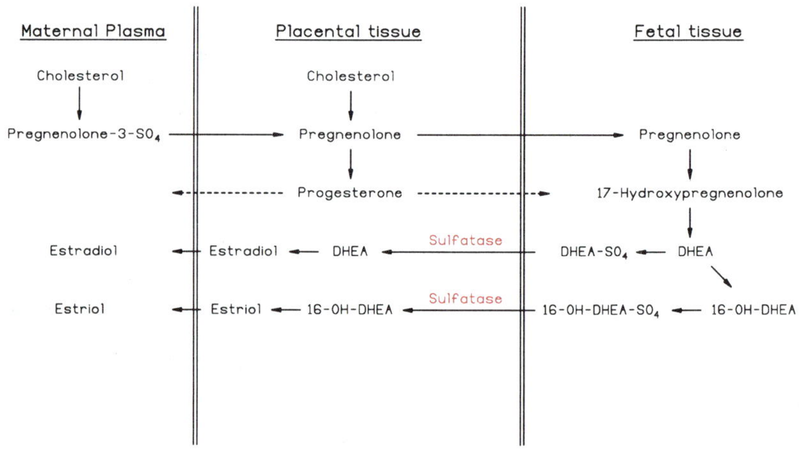

Figure 13.25.
Steroid biosynthesis in the fetoplacental unit.

13.9 PROSTAGLANDINS

Prostaglandins are a family of acidic lipids with profound physiologic and biochemical effects in many different tissues. Their actions occur at or very near the site of their synthesis. Prostaglandins are present in most tissues of the body. Especially high concentrations are found in human semen. The name "prostaglandin," derived from "prostate," actually is a misnomer because the source of prostaglandins in semen are the seminal vesicles, not the prostate.

13.9.1 Nomenclature

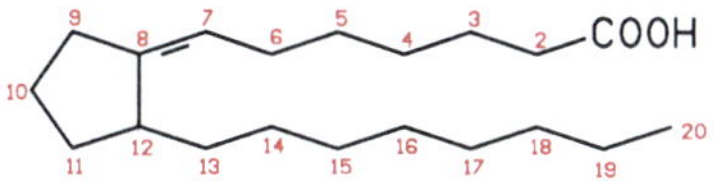

Figure 13.26.
Prostanoic acid.

Prostaglandins are derivatives of **prostanoic acid,** a hypothetical C_{20} acid (Figure 13.26).

There are several groups or series of primary prostaglandins, each group being identified by a capital letter. They differ in the nature of their substituents on the cyclopentane ring. Prostaglandins of the D, E, and F series are found in human tissues. Other series (A, B, C) may or may not be metabolites and are found in many other biological systems.

Individual members within each group are identified by subscript numerals indicating the number of double bonds in the two side chains. The Greek letters α and β, following the numbers in the F series, indicate the orientation of a hydroxyl group at carbon 9 (α, below the plane of the cyclopentane ring; β, above; see examples in Figure 13.27).

In addition to the primary prostaglandins (PG) there are several other derivatives of arachidonic acid, some of them exhibiting even more powerful bioregulatory actions. They include **5-hydroxy-eicosatetraenoic acids (HETEs), PG endoperoxides, thromboxanes, prostacyclins,** and **leukotrienes.** Collectively, all these compounds are called **icosanoids** because they contain 20 carbon atoms.

13.9.2 Biosynthesis

All members of the prostaglandin family are derived from **arachidonic acid,** a polyunsaturated C_{20} fatty acid. Their biosynthetic pathway is known as the arachidonic acid cascade and is illustrated in Figure 13.27.

Arachidonic acid is a component of cell membrane phospholipids and can be released by the action of **phospholipase A_2**. The release of arachidonic acid may be triggered by hormones, inflammatory or immunological stimuli, or perturbation of the cell membrane in general. The vasodilator oligopeptide **bradykinin** (see Section 13.10) appears to activate phospholipase A_2. Glucocorticoids inhibit phospholipase A_2.

The enzyme arachidonic acid cyclooxygenase converts arachidonic acid to the unstable **endoperoxide PGG_2**, which is immediately reduced by the same enzyme, acting as a peroxidase, to PGH_2, the common precursor of prostacyclin (PGI_2), prostaglandins (PGE_2, PGD_2, $PGF_{2\alpha}$), and thromboxane A_2 (TXA_2). **Arachidonic acid cyclooxygenase** is inhibited by **aspirin** and indomethacin, two widely used anti-inflammatory drugs.

The biosynthesis of leukotrienes and 5-hydroxy-eicosatetraenoic acids from arachidonic acid is catalyzed by **lipoxygenases** that are not inhibited by aspirin.

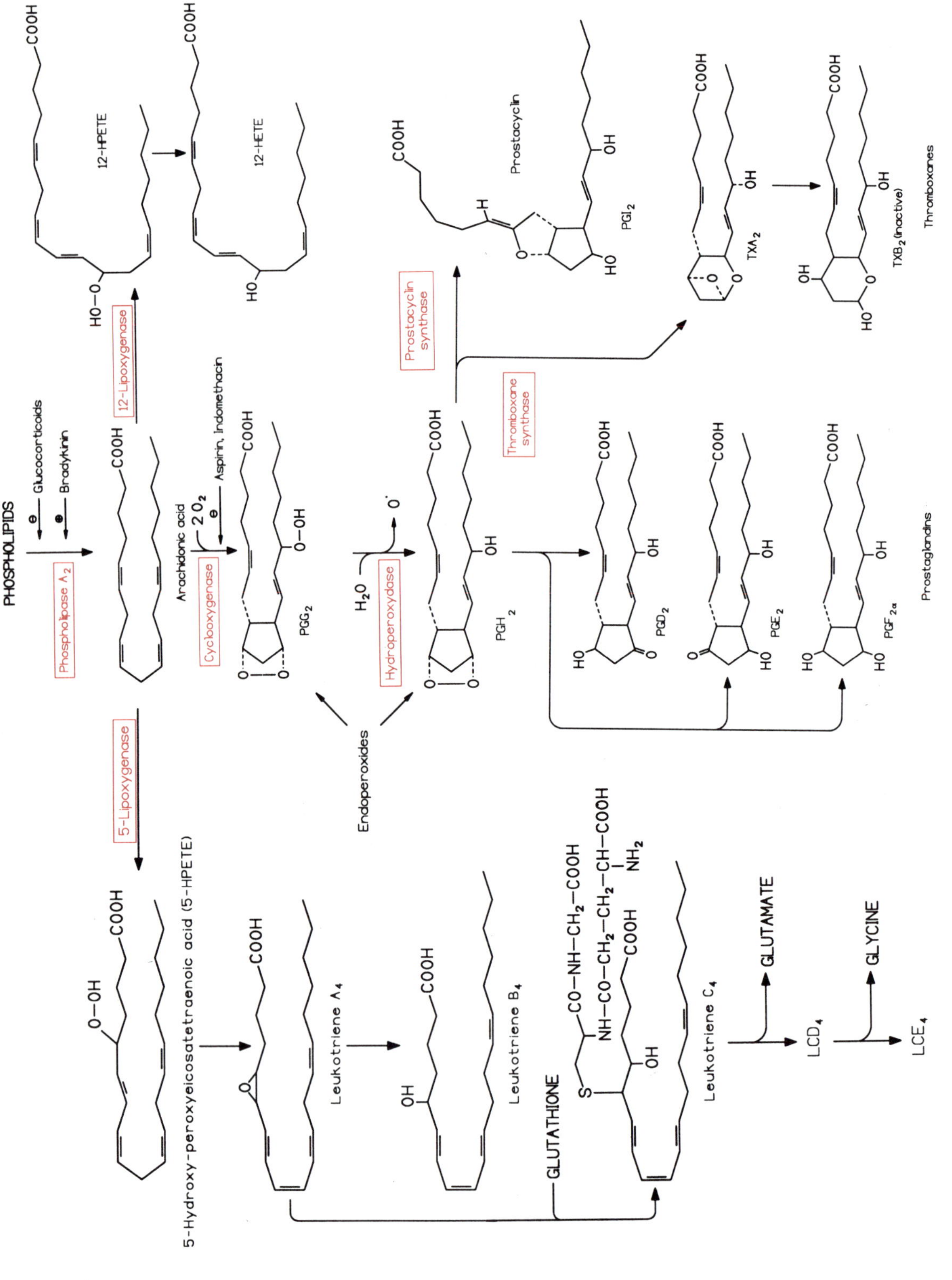

Figure 13.27.
The arachidonic acid cascade.

13.9.3 Actions

The actions of prostaglandins and related compounds as bioregulators in different tissues are complex and, in many cases, not yet completely understood. A complete description of what is known to date cannot be given here. Instead, only a few examples of specific actions are discussed.

Most eicosanoids affect the intracellular levels of the cyclic nucleotides cAMP or **HcGMP**. Often the same eicosanoid has different effects upon different cells because of the presence of different receptors. The half-life of eicosanoids usually is less than 1 minute. They are active at extremely low concentrations.

Prostaglandin E_2 (PGE_2) has been shown to relax bronchial smooth muscle. It also increases Na^+ excretion in the urine and thereby increases urine flow. **PGE_1**, on the other hand, inhibits adenylate cyclase in adipocytes and thereby prevents the effects of glucagon and insulin upon hormone-sensitive lipase. In some other cells, PGE_1 stimulates adenylate cyclase.

Prostaglandin $F_{2\alpha}$ ($PGF_{2\alpha}$) stimulates uterus smooth muscle contraction and causes luteolysis of a functioning corpus luteum. It also causes bronchodilation. PGD_2 has similar actions.

Thromboxane A_2 (TXA_2) is released by platelets. It promotes platelet aggregation and constricts smooth muscle. In contrast, **prostacyclin (PGI_2),** released by the vascular endothelium as well as by other tissues, inhibits platelet aggregation and relaxes smooth muscle. The opposing effects of these two eicosanoids may be important in limiting thrombus formation to the site of vascular injury. Thromboxane A_2 has a half-life of only 30 seconds and is converted to the biologically inert $TBXB_2$.

The **endoperoxides PGG_2** and **PGH_2** in some cases appear to be even more active than their metabolites. However, it is possible that their activity is actually due to their conversion to TXA_2. Endoperoxides and/or their derivatives participate in the **inflammatory process.** Biosynthesis of prostaglandins can be inhibited by glucocorticoids and by aspirin, both important anti-inflammatory drugs. One of the side effects of aspirin (bleeding) is the inhibition of platelet aggregation due to the blocked thromboxane A_2 synthesis. Aspirin may be used to prevent premature labor, since uterine contraction is mediated through $PGF_{2\alpha}$.

Leukotrienes are the active components present in the **slow-reacting substances of anaphylaxis (SRS).** (SRS causes slow contractions of smooth muscle.) The three leukotrienes, LTC_4, LTD_4, and LTE_4, initially cause arteriolar constriction, which is followed by increased vascular permeability. (LTC_4 is a very powerful constrictor of coronary arteries.) **HETEs** function as chemotactic agents to attract neutrophils and eosinophils to sites of inflammation.

13.10 VASOACTIVE PEPTIDE HORMONES

Kinins and **angiotensins** are small oligopeptides with powerful vasodilator and vasoconstrictor properties, respectively. They are generated from larger precursor proteins by the action of specific proteases.

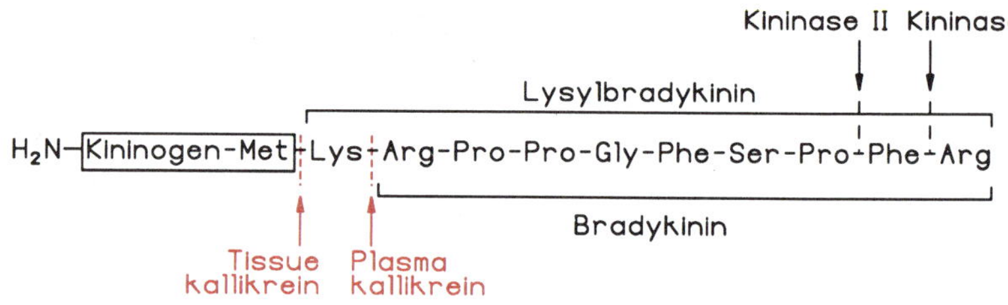

Figure 13.28.
Generation of kinins from kininogens.

13.10.1 Kinins

Plasma bradykinin is a nonapeptide that is cleaved from **high molecular weight (HMW) kininogen** by the proteolytic enzyme **kallikrein. Plasma kallikrein,** in turn, is formed from an inactive precursor, **prekallikrein.** The activation of prekallikrein involves the activated **clotting factor XIIa** (see Section 9.5).

Another kinin, the decapeptide **lysylbradykinin,** is formed from low-molecular-weight kininogens in tissues by the action of **tissue kallikrein.** Although lysylbradykinin is active as a vasodilator by itself, it is converted in the plasma to bradykinin by an aminopeptidase (Figure 13.28).

Kinins relax vascular smooth muscle and increase the permeability of capillaries; they produce constriction of visceral smooth muscle. Their half-lives are extremely short, and they are inactivated by two **kininases.** (Kininase II is identical to **converting enzyme.**)

13.10.2 Angiotensins

Angiotensin I is a decapeptide released from the plasma α_2-globulin **angiotensinogen** by the proteolytic enzyme **renin.**

Renin is secreted into the circulation by the juxtaglomerular cells of the kidney. It is a glycoprotein (MW = 40,000) derived from a larger precursor, **prorenin,** by the action of tissue kallikrein. The half-life of circulating renin is 80 minutes.

Angiotensin I is itself inactive. Its activation to **angiotensin II** occurs predominantly in the lung by **converting enzyme.** The octapeptide angiotensin II causes arteriolar constriction. This effect raises the blood pressure. The half-life of angiotensin II is only about 15 seconds, the approximate time of one complete circulation. Angiotensin II also stimulates aldosterone secretion by the adrenal cortex. A plasma amino-

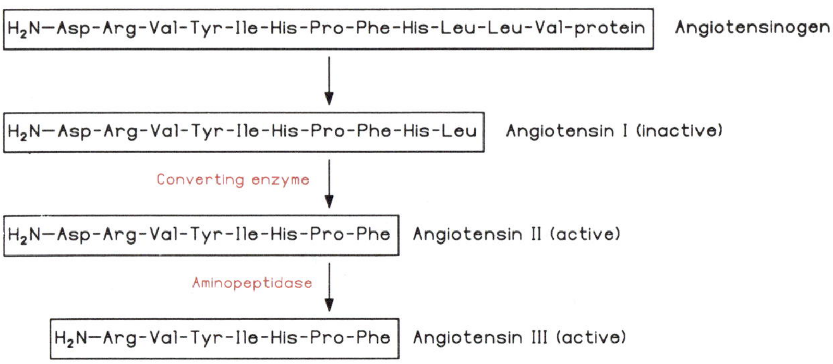

Figure 13.29.
Generation of angiotensins.

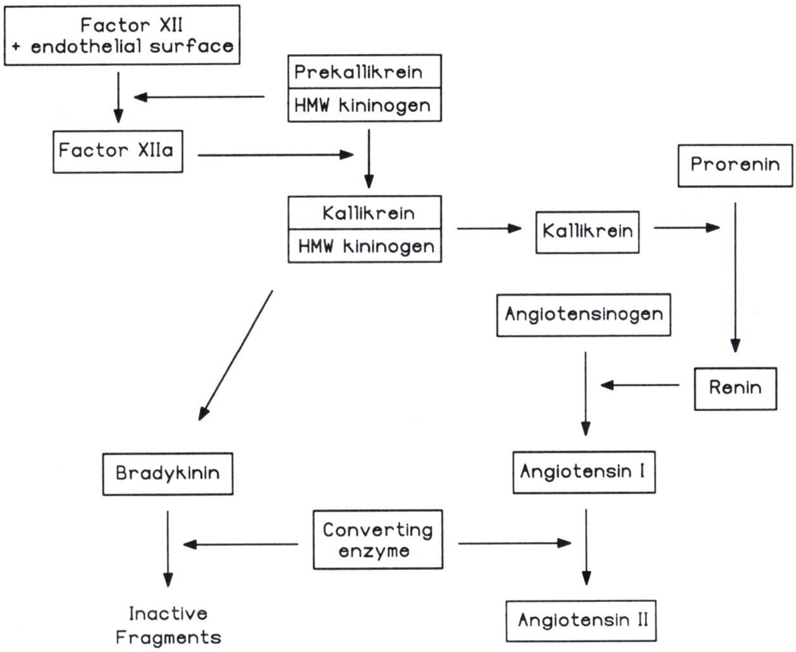

Figure 13.30.
Interplay of kinin and angiotensin systems.

peptidase removes an Asp residue from angiotensin II to produce **angiotensin III.** It is possible that angiotensin III is the more potent mediator of aldosterone secretion (Figure 13.29). The interplay of the kinin and angiotensin systems is shown in Figure 13.30.

13.10.3 Atrial Natriuretic Hormone

Another peptide hormone affecting blood pressure is **atrial natriuretic hormone** released from cardiac atria in response to the activation of stretch receptors by a large blood volume. There are several natriuretic peptides found in circulation ranging from 22 to 28 amino acids in length. It is not clear at this time which peptide is released into circulation. Natriuretic hormone causes natriuresis and a fall in blood pressure (as a result of reduced cardiac output).

13.11 GASTROINTESTINAL AND NEURAL PEPTIDE HORMONES

In recent years it has become evident that many peptide hormones found in the gut are also produced in the brain and vice versa. Examples are somatostatin, GnRH, and cholecystokinin. It appears that such peptides can serve both as neurotransmitters to nearby target cells and as hormones for more distant target cells. Since peptides are synthesized on ribosomes in the cell body of the neuron and, therefore, must be transported along the axon before they can be released at the nerve terminus, it is obvious that such a release cannot be repeated too many times over a short time interval. In contrast, neurotransmitters such as norepinephrine or acetylcholine are synthesized close to the nerve terminus and thus can be released in rapid succession over an extended period of time. The fast-action neurotransmitters usually alter the permeability of the target cell membrane to small cations. Neuropeptides appear to act differently. They seem to modify the target cell's response to other signals. Some neurons actually release both a fast-acting transmitter and a neuropeptide.

Table 13.1 lists several peptides found in the gastrointestinal system. **Cholecystokinin (CCK)** and **gastrin** share the same five carboxyl-terminal amino acids. **Glucagon, gastric inhibitory peptide (GIP), secretin,** and **vasoactive intestinal peptide (VIP)** have even more amino acid sequences in common. All of these peptides are also found in neural tissues. Their actions are summarized in Table 13.1.

Gastrin-releasing peptide (GRP), the mammalian form of **bombesin** (found in frog skins), and **substance P** probably serve in the gut only as neuropeptides released from nerve endings. **Motilin,** a 22-amino-acid peptide, found in endocrine cells of the upper small intestine causes concentration of intestinal smooth muscle when applied in vitro.

Other types of neuropeptides are the **endorphins** and **enkephalins.** (The word "endorphin" comes from "endogenous morphin.") Endorphins bind to **opiate receptors** located in regions of the brain that are implicated in the perception of pain. Containing 15 or 16 amino acids, they are found exclusively in cells of the hypothalamus and the anterior lobe of the pituitary.

Enkephalins contain only five amino acids. The two known enkephalins, **Met-enkephalin** and **Leu-enkephalin,** share the sequence Tyr-Gly-Gly-Phe and have Met or Leu, respectively, at their carboxyl end. Cells that contain enkephalins are found throughout the central nervous system. As with endorphins, they bind to opiate receptors and have analgesic effects.

The sequences of endorphins and enkephalins are usually contained within larger polypeptide chains from which they are released by proteolytic enzymes (compare pro-opiomelanocortin, Section 13.2.3).

For a more extensive discussion of liver functions, consult the following textbooks:

Greenspan, Forsham. *Basic and Clinical Endocrinology,* 2nd ed. Los Altos, Ca: Lange Medical Publications; 1990.

Mathews et al. (1990): Chapter 38

Stryer (1988): Chapter 12

Further References

Atkinson MA, Maclaren NK. What causes diabetes? *Sci Am.* 1990; 263:62–71.

Bansal S, Majerus PW. Phosphatidylinositol-derived precursors and signals. *Ann Rev Cell Biol.* 1990; 6:41–68.

Berger FG, Watson G. Androgen-regulated gene expression *Ann Rev Physiol.* 1989; 51:51–65.

Brent GA, Moore DD, Larsen PR. Thyroid hormone regulation of gene expression. *Ann Rev Physiol.* 1991; 53:17–35.

Collins S, Caron MG, Lefkowitz RJ. Regulation of adrenergic receptor responsiveness through modulation of receptor gene expression. *Ann Rev Physiol.* 1991; 53:497–508.

Homcy CJ, Vatner SF, Vatner DE. β-Adrenergic receptor regulation in the heart in pathophysiologic states: Abnormal adrenergic responsiveness in cardiac disease. *Ann Rev Physiol.* 1991; 53:137–159.

Kaziro Y, et al. Structure and function of signal-transducing GTP-binding proteins. *Ann Rev Biochem.* 1991; 60:349–400.

Majerus PW, Ross TS, Cunningham TW, Caldwell KK, Jefferson AB, Bausal VS. Recent insights in phosphatidyl inositol signaling. *Cell.* 1990; 63:459–465.

Rassmussen H. The cycling of calcium as an intracellular messenger. *Sci Am.* 1989; 261 (4):66–73.

Rosenzweig A, Seidman CE. Atrial natriuretic factor and related peptide hormones. *Ann Rev Biochem.* 1991; 60:229–256.

Ullrich A, Schlessinger J. Signal transduction by receptors with tyrosine kinase activity. *Cell.* 1990; 61:203–212.

Weinberger C, Bradley DJ. Gene regulation by receptors binding lipid-soluble substances. *Ann Rev Physiol.* 1990; 52:823–840.

Williamson JR, Monck, JR. Hormone effects on cellular Ca^{2+} fluxes. *Ann Rev Physiol.* 1989; 51:107–124.

INDEX

INDEX

INDEX

J

K

T